Progress in Mathematics

Volume 206

Victor P. Snaith

Algebraic K-Groups as Galois Modules

Springer Basel AG

Author:

Victor P. Snaith
Faculty of Mathematical Sciences
University of Southampton
Highfield
Southampton, SO17 1BJ
UK
e-mail: V.P.Snaith@maths.soton.ac.uk

2000 Mathematics Subject Classification 11R33 (primary); 11M06, 11R23, 11R34, 11R65, 11R70, 11S25, 11S31, 11S40, 12G05, 16G99, 16H05, 19A31, 19D99, 19F27, 20J06 (secondary)

A CIP catalogue record for this book is available from the Library of Congress, Washington D.C., USA

Deutsche Bibliothek Cataloging-in-Publication Data

Snaith, Victor P.:
Algebraic K-groups as Galois modules / Victor P. Snaith. – Basel ; Boston ; Berlin : Birkhäuser, 2002
(Progress in mathematics ; Vol. 206)
ISBN 978-3-0348-9473-9 ISBN 978-3-0348-8207-1 (eBook)
DOI 10.1007/978-3-0348-8207-1

Originally published by Birkhäuser Verlag in 2000
Printed on acid-free paper produced of chlorine-free pulp. TCF ∞

ISBN 978-3-0348-9473-9

9 8 7 6 5 4 3 2 1 www.birkhauser-science.com

Contents

Preface

This volume began as the last part of a one-term graduate course given at the Fields Institute for Research in the Mathematical Sciences in the Autumn of 1993. The course was one of four associated with the 1993-94 Fields Institute programme, which I helped to organise, entitled "Artin L-functions". Published as [132], the final chapter of the course introduced a manner in which to construct class-group-valued invariants from Galois actions on the algebraic K-groups, in dimensions two and three, of number rings. These invariants were inspired by the analogous Chinburg invariants of [34], which correspond to dimensions zero and one. The classical Chinburg invariants measure the Galois structure of classical objects such as units in rings of algebraic integers. However, at the "Galois Module Structure" workshop in February 1994, discussions about my invariant ($\Omega_1(L/K, 3)$ in the notation of Chapter 5) after my lecture revealed that a number of other higher-dimensional cohomological and motivic invariants of a similar nature were beginning to surface in the work of several authors. Encouraged by this trend and convinced that K-theory is the archetypical motivic cohomology theory, I gratefully took the opportunity of collaboration on computing and generalizing these K-theoretic invariants. These generalizations took several forms – local and global, for example – as I followed the prevalent trends in the "Galois Module Structure" part of number theory and arithmetic geometry. This book aims to give the story so far – mainly through the study of Galois invariants attached to algebraic K-groups and étale cohomology groups.

When contemplating a "story-so-far" treatment one is forced to ask whether the story has evolved far enough, or even too far, for such a treatment to be apt. I must confess that this question did cause me some concern. After all, one can easily continue ad nauseam constructing imitations of the classical invariants, ramifying them with intriguing conjectures and partial computations, but how to know whether all this activity will lead somewhere? Finally I salved my conscience by the inclusion of Chapters 6 and 7. For in those chapters, the constructions of the earlier chapters are shown to lead back, not only to a strategy for evaluating all our K-theory Galois module invariants in terms of special values of L-functions, but also to a strategy for attacking the fundamental Coates-Sinnott Conjecture and perhaps even the Brumer Conjecture.

Now I shall describe the contents in more detail.

Chapter 1 briefly reviews the properties of Artin L-functions and Artin root numbers. Then the Lichtenbaum conjecture is described, since it represents the first and most basic connection between number theory and Galois actions on K-groups. Finally, we describe without proof some numerical examples of the invariants which are constructed in Chapter 5.

Chapter 2 describes the background which we shall need from K-theory and involves two distinct aspects. Since we are concerned with Galois module invariants of higher-dimensional Quillen K-groups, a brief description of their construction and properties is given. However, our Galois structure invariants will lie in absolute or relative K_0-groups attached to the integral group-ring of the Galois group. Therefore a description is given of the constructions with perfect and nearly perfect complexes which give rise to "Euler characteristics" lying in these K-groups. The chapter concludes with the Hom-descriptions, due to Fröhlich, of these K_0's, which are very important because they allow us to define K_0-classes in terms of functions (such as special values of L-functions) on Galois representations.

The classical Chinburg invariants ([34], [77]) come in global and local forms, denoted in this book by $\Omega_0(L/K,3)$ and $\Omega_0(L/K,2)$, respectively. In Chapter 3 the higher K-theory version of $\Omega_0(L/K,2)$ is constructed. This uses local fundamental classes for higher-dimensional K-groups of local fields which originated in ([134], [135], [136], [137]) and generalize the local fundamental classes of class field theory [128]. The chapter concludes with a description of what happens when one attempts to apply the construction of the local fundamental classes to two-dimensional local fields.

The invariants, $\Omega_r(L/K,2)$, usually involve regulators, which can be a troublesome source of calculational errors. However, in characteristic $p > 0$ there are no regulators. In this case, assuming only tame ramification, we give a computation of $\Omega_1(L/K,2)$ in Chapter 4. This invariant is constructed using the K_2/K_3 local fundamental classes. On the other hand, the motivic complex $\Gamma(2,L)$ also gives a 2-extension lying in $\mathrm{Ext}^2_{\mathbf{Z}[G(L/K)]}(K_2(L),K_3(L))$ when L/K is a Galois extension of local fields. We conclude Chapter 4 with a comparison of these 2-extensions in characteristic $p > 0$.

Chapter 5 gives the construction of the invariants $\Omega_n(L/K,3)$ which originated in ([36]; see also [38]) and which generalize $\Omega_1(L/K,3)$ of ([132] Chapter VII). Since the examples featured in §1.3 are $\Omega_1(L/K,3)$'s when L is a quaternion extension of the rationals, we give another description of $\Omega_1(L/K,3)$ which is better suited to calculations and describe how the example calculations are made. In addition we describe, in readiness for the material of Chapter 7, how our invariants lift to classes in the relative K-group, $K_0(\mathbf{Z}[G(L/K)],\mathbf{Q})$, when L/K is a totally real Galois extension of number fields.

Chapter 6 describes the Iwasawa polynomial, giving a functorial version of the treatment in [64]. Under mild hypotheses, the Main Conjecture of Iwasawa theory, as proved in [166], leads to a p-adically unit-valued function, given by the ratio of the p-adic L-function to the Iwasawa polynomial. This Galois invariant function we call the Wiles Unit. The simplest manner in which to make such functions is via the determinants of units in the p-adic group-ring. The Wiles Unit Conjecture (Conjecture 6.3.4) is presented with some supporting evidence. In particular, this conjecture was formulated because it leads to a calculation of $\Omega_n(L/K,3)$ with the anticipated answer (see Conjecture 5.2.18). In the section on Hida modular forms with group-ring coefficients, we show how to construct the Galois representation which would be necessary to attack the Wiles Unit Conjecture via generalizations of the techniques of [166].

In Chapter 7 we turn our attention to annihilator ideals. It is shown how chains of annihilator relations can be derived from elements in $K_0(\mathbf{Z}[G],\mathbf{Q})$ and how the lifted invariants of Chapter 5 give annihilator relations which resemble those of the Coates-Sinnott and Brumer Conjectures. In fact, it is shown that the Wiles Unit conjecture implies the Coates-Sinnott conjecture. In addition, it is explained how these constructions might be applied to the (seemingly more difficult) Brumer Conjecture. In the real cyclotomic case we use [27] to circumvent the Wiles Unit Conjecture to give a proof of the original Coates-Sinnott conjecture, at odd primes, concerning annihilators of the étale cohomology groups which are expected to equal $K'_{2n}(\mathcal{O}_{L,S})\otimes\mathbf{Z}_p$. The chapter closes with a proof that the higher Stickelberger ideals and the annihilators of the even dimensional K-groups which are supposedly related by the Coates-Sinnott conjecture, actually have the same radical.

I am very grateful to a number of mathematicians for their interest in this material, for their helpful comments and for their explanations of their own relevant results. Here they are in alphabetical order: David Burns introduced me to relative K_0 and a series of significant invariants lying in them ([22], [23], [24], [25], [26], [140]) and explained [27] to me; Ted Chinburg spent many hours in collaborative discussions about these Galois module structure invariants, sharing many excellent insights such as the realization that the Wiles Unit Conjecture would evaluate $\Omega_n(L/K,3)$ and, in addition, taught me the results about the radical of the Stickelberger ideals which conclude Chapter 7; Thomas Geisser explained to me his work with Marc Levine [60]; Karl Gruenberg, Juergen Ritter and Al Weiss explained to me their series ([68], [121], [122]) dealing with lifted invariants and 2-extensions which are related to the Brumer Conjecture by Chapter 7; Lars Hesselholt explained to me his work with Ib Madsen ([73], [74]) and Michael Spiess explained to me how to use ([81], [92], [146]) to give an affirmative answer to Question 4.3.2 for *all* local fields. Also I had a great deal of help from my collaborators Manfred Kolster and Georges Pappas. Andrew Wiles helped considerably by explaining [166], by discussing approaches to the Wiles Unit Conjecture and by pointing out to me that the results of [167] may fail when p divides the order of the Galois group. Andrew Sydenham provided me with calculations of Stickelberger elements and long-sufferingly put up with many vague lunch-time conversations about Λ-adic modular forms. Finally, it was a pleasure to find that my homotopy-theorist's mathematical peregrinations have eventually brought me to a topic which I can discuss with my old friend, John Coates, whose work originally got me interested in algebraic K-theory 30 years ago!

I am also grateful to the sponsors and organizers of the Ferran Sunyer y Balaguera Prize competition, which provided the impetus behind this monograph. The opportunity to write much of this monograph was provided by a visit, sponsored by the Royal Society Joint Projects Scheme, to the Euler Institute in St. Petersburg in September 2000 and I am very grateful to Igor Zhukov for helpful collaboration and hospitality during that visit.

Victor Snaith,
University of Southampton,
December 2000.

Chapter 1

Galois Actions and L-values

The aim of this chapter is to introduce by example the study of the manner in which the Galois structure on various motivic objects from arithmetic influence the special values of associated L-functions. In these examples our motivic object will be K_2 of rings of algebraic integers in a Galois extension, whose Galois structure influences the value of the associated Artin L-function at $s = -1$. Our examples will be a family of totally real number fields given by quaternion extensions of the rationals.

We begin by recalling the Artin L-function.

1.1 Analytic prerequisites

1.1.1 *Artin L-functions*

Let us recall the construction and properties of the *Artin L-function.* For further details the reader is referred to the survey article of Martinet [101].

Let L/K be a Galois extension of number fields and let

1.1.2
$$\rho : G(L/K) \longrightarrow GL(V)$$

be a finite-dimensional complex Galois representation.

Let $\mathcal{O}_K$ denote the ring of algebraic integers of K. If $P \triangleleft \mathcal{O}_K$ is a (finite) prime of K let Q denote a choice of a prime of L which divides P. The associated *decomposition group* of P is defined to be

$$D_P = \{g \in G(L/K) \mid g(Q) = Q\} \cong G(L_Q/K_P),$$

where K_P denotes the local field given by the completion of K in the P-adic topology. Hence $D_P \leq G(L/K)$ is well defined up to conjugation in $G(L/K)$, since $G(L/K)$ acts transitively on the set of primes of L lying over P. The inertia group, $G_0(L_Q/K_P) \triangleleft D_P$, satisfies

$$D_P/G_0(L_Q/K_P) \cong G(\overline{L}_Q/\overline{K}_P)$$

which is the cyclic Galois group of the residue field extension, $\overline{L}_Q/\overline{K}_P$, generated by the Frobenius automorphism, F_P. Hence F_P is well defined up to multiplication by elements of $G_0(L_Q/K_P)$ and therefore there is a well-defined automorphism

$$F_P \in GL(V_P),$$

where $V_P = V^{G_0(L_Q/K_P)}$ denotes the $G_0(L_Q/K_P)$-fixed subspace of V.

The Artin L-function, $L_K(s,\rho)$, is the unique meromorphic function of a complex variable, s, which is given for $Re(s) > 1$ by the *Euler product*

$$L_K(s,\rho) = \prod_{\substack{P \lhd O_K \\ \text{prime}}} (\det(1 - N(P)^{-s}\rho(F_P \mid V_P)))^{-1}.$$

Here $N(P) = |O_K/P|$ is the *absolute norm* of P. Note that F_P is characterised (modulo $G_0(L_Q/K_P)$) by

$$F_P(z) \equiv z^{N(P)} \quad (\text{modulo } Q)$$

for all $z \in O_L$.

The *Dedekind zeta function* of K is the important case when $\rho = 1$, the one-dimensional trivial representation of $G(L/K)$, in 1.1.2 so that

$$\zeta_K(s) = L_K(s,1).$$

Proposition 1.1.3

The Artin L-function, $L_K(s,\rho)$, is a meromorphic function of the complex variable, s, satisfying the following properties:

(i) $L_K(s, \rho_1 \oplus \rho_2) = L_K(s,\rho_1)L_K(s,\rho_2)$.

(ii) *If $K \subset L \subset N$ is a chain of finite Galois extensions and $G(N/K) \longrightarrow G(L/K)$ is the canonical map then*

$$L_K(s, \mathrm{Inf}_{G(L/K)}^{G(N/K)}(\rho)) = L_K(s,\rho).$$

(iii) *If F is an intermediate field of L/K and $\rho : G(L/F) \longrightarrow GL(W)$ is a representation then*

$$L_K(s, Ind_{G(L/F)}^{G(L/K)}(\rho)) = L_F(s,\rho).$$

1.1.4 *The Artin Conductor*

Let $P \lhd O_K$ be a prime of K. The representation, ρ, of 1.1.2 yields a representation of the decomposition group

$$\mathrm{Res}_{G(L_Q/K_P)}^{G(L/K)}(\rho) : G(L_Q/K_P) \longrightarrow GL(V).$$

The *Artin conductor* of this local Galois representation is defined to be

$$n(\rho, P) = \sum_{i=0}^{\infty} [G_0(L_Q/K_P) : G_i(L_Q/K_P)]^{-1}(\dim(V) - \dim(V^{G_i(L_Q/K_P)})).$$

For an unramified prime $n(\rho, P) = 0$ so that we may define the *Artin conductor ideal* by the formula

$$f(\rho, L/K) = f(\rho) = \prod_{\substack{P \triangleleft \mathcal{O}_K \\ \text{prime}}} P^{n(\rho,P)} \triangleleft \mathcal{O}_K.$$

The Artin conductor satisfies the following properties, which are derived from corresponding properties for the function $n(\rho, P)$ (c.f. [128], Ch. VI, §2).

Proposition 1.1.5

The Artin conductor ideal, $f(\rho, L/K)$, satisfies the following properties:

(i) $f(\rho_1 \oplus \rho_2, L/K) = f(\rho_1, L/K) f(\rho_2, L/K)$.

(ii) *If $K \subset L \subset N$ is a chain of finite Galois extensions and $G(N/K) \longrightarrow G(L/K)$ is the canonical map then*

$$f(\mathrm{Inf}_{G(L/K)}^{G(N/K)}(\rho), L/K) = f(\rho, M/K).$$

(iii) *If F is an intermediate field of L/K and $\rho : G(L/F) \longrightarrow GL(W)$ is a representation then*

$$f(Ind_{G(L/F)}^{G(L/K)}(\rho), L/K) = N_{F/K}(f(\rho, L/F)) D(F/K)^{\dim(\rho)}$$

where $D(F/K)$ is the discriminant of the extension, F/K.

1.1.6 If d_K is the *absolute discriminant* of $K/\mathbf{Q}$ we define

$$A(\rho) = |d_K|^{\dim(V)} N_{K/\mathbf{Q}}(f(\rho)).$$

Now write

$$\gamma(s) = \pi^{-s/2}\Gamma(s/2)$$

where $\Gamma(s)$ is the classical *Gamma function.* Define

$$\gamma_\rho(s) = \prod_v \gamma_\rho^v(s)$$

where v ranges over the Archimedean places of K and $\gamma_\rho^v(s)$, the *local factor at infinity*, is defined in the following manner. If v is complex we set

$$\gamma_\rho^v(s) = [\gamma_\rho(s)\gamma_\rho(s+1)]^{\dim(V)}.$$

If v is real then choose a place, w of L, lying above v. The order of the decomposition group, D_v, is either 1 or 2 and the generator plays the role of the Frobenius automorphism. Write $V = V^+ \oplus V^-$ where $V^\pm$ is the (± 1)-eigenspace for the generator of D_v. Finally set

$$\gamma_\rho^v(s) = \gamma_\rho(s)^{\dim(V^+)}\gamma_\rho(s+1)^{\dim(V^-)}.$$

Definition 1.1.7 Define the *extended Artin L-function*, $\Lambda_K(s,\rho)$, by the formula

$$\Lambda_K(s,\rho) = A(\rho)^{s/2}\gamma_\rho(s)L_K(s,\rho).$$

Proposition 1.1.8

The extended Artin L-function, $\Lambda_K(s,\rho)$, is a meromorphic function of the complex variable, s, satisfying the following properties:

(i) $\Lambda_K(s,\rho_1 \oplus \rho_2) = \Lambda_K(s,\rho_1)\Lambda_K(s,\rho_2)$.

(ii) *If $K \subset L \subset N$ is a chain of finite Galois extensions and $G(N/K) \longrightarrow G(L/K)$ is the canonical map then*

$$\Lambda_K(s, \mathrm{Inf}_{G(L/K)}^{G(N/K)}(\rho)) = \Lambda_K(s,\rho).$$

(iii) *If F is an intermediate field of L/K and $\rho : G(L/F) \longrightarrow GL(W)$ is a representation then*

$$\Lambda_K(s, \mathrm{Ind}_{G(L/F)}^{G(L/K)}(\rho)) = \Lambda_F(s,\rho).$$

(iv) *If $\overline{\rho}$ denotes the complex conjugation of ρ then there is a functional equation of the form*

$$\Lambda_K(1-s,\rho) = W_K(\rho)\Lambda_K(s,\overline{\rho})$$

where $W_K(\rho)$ is a complex number of modulus one.

Definition 1.1.9 The complex number of absolute value one, $W_K(\rho)$, is called the *Artin root number* of ρ. The following properties are immediate from Proposition 1.1.8.

Proposition 1.1.10

The Artin root number, $W_K(\rho)$, is a complex number of norm one which satisfies the following properties:

(i) $W_K(\rho_1 \oplus \rho_2) = W_K(\rho_1)W_K(\rho_2)$.

(ii) *If $K \subset L \subset N$ is a chain of finite Galois extensions and $G(N/K) \longrightarrow G(L/K)$ is the canonical map then*

$$W_K(\mathrm{Inf}_{G(L/K)}^{G(N/K)}(\rho)) = W_K(\rho).$$

(iii) *If F is an intermediate field of L/K and $\rho : G(L/F) \longrightarrow GL(W)$ is a representation then*

$$W_K(\mathrm{Ind}_{G(L/F)}^{G(L/K)}(\rho)) = W_F(\rho).$$

1.2 The Lichtenbaum conjecture

1.2.1 The Lichtenbaum conjecture relates the value of the leading term in the Taylor series for the Dedekind zeta function at a negative integer to the algebraic K-theory of the ring of algebraic integers, by analogy with the analytic class number formula.

Let $\zeta_K(s)$ denote the Dedekind zeta function of the number field K, as in §1.1.1. If $r_1, 2r_2$ are the number of distinct real and complex embeddings of K then the analytic class number formula computes the residue at $s = 1$. It takes the form ([163] p. 37)

$$\mathrm{Res}_{s=1}\ \zeta_K(s) = \frac{2^{r_1+r_2}\pi^{r_2}|\,\mathrm{TorsK_0}(\mathcal{O}_K)|R_1(K)}{|\,\mathrm{TorsK_1}(\mathcal{O}_K)|\sqrt{d_K}}$$

where d_K is the discriminant of K, TorsH is the torsion subgroup of A and $R_1(K)$ is the Dirichlet regulator given by the covolume of the image of Dirichlet's logarithmic map

$$\rho_1 : \mathcal{O}_K^* \longrightarrow \mathbf{R}^{r_1+r_2-1}.$$

To transform this equation to its classical form one observes that $K_0(\mathcal{O}_K) \cong \mathcal{CL}(\mathcal{O}_K) \oplus \mathbf{Z}$ and $K_1(\mathcal{O}_K) \cong \mathcal{O}_K^*$.

Applying the functional equation of Proposition 1.1.8(iv), one finds that the leading term at $s = 0$ in the Taylor series for $\zeta_K(s)$, denoted by $\zeta_K(0)^*$, is given by

$$\zeta_K(0)^* = \lim_{s\to 0}\ s^{1-r_1-r_2}\zeta_K(s) = -\frac{|\,\mathrm{TorsK_0}(\mathcal{O}_K)|R_1(K)}{|\,\mathrm{TorsK_1}(\mathcal{O}_K)|}.$$

In higher dimensions the algebraic K-groups $K_j(\mathcal{O}_K)$ are finitely generated and there is a Borel regulator, $R_m(K) \in \mathbf{R}$, associated to $K_{2m-1}(\mathcal{O}_K)$ ([17], [18], [19], [116]) and $K_{2m}(\mathcal{O}_K)$ is finite for $m \geq 1$.

Conjecture 1.2.2 By analogy with the case $m = 1$ of §1.2.1 the Lichtenbaum conjecture states that for all integers $m \geq 2$

$$\zeta_K(1-m)^* = \pm 2^{\epsilon}\frac{|K_{2m-2}(\mathcal{O}_K)|R_m(K)}{|\,\mathrm{TorsK}_{2m-1}(\mathcal{O}_K)|}$$

for some integer $\epsilon \geq 0$.

When $m = 2$ and K is totally real the conjecture is true as a consequence the results of Wiles [166]. In this case the regulator is trivial, $K_3(\mathcal{O}_K)$ is a finite group and $\zeta_K(-1)$ is a non-zero rational number which satisfies

$$\zeta_K(-1) = \pm 2^{\epsilon}\frac{|K_2(\mathcal{O}_K)|}{|K_3(\mathcal{O}_K)|}.$$

In addition, there are similar conjectures about the leading terms of Artin L-functions which have been studied in special cases. These generalizations are discussed in [90] and [91] (see also §1.3.1) and are usually formulated prime by prime in terms of étale cohomology of $\mathcal{O}_K$. Let p be a prime. The abiding conjectures in algebraic K-theory predict that $K_{2m-e}(\mathcal{O}_K) \otimes \mathbf{Z}_p$ and $H^e_{\text{étale}}(\operatorname{Spec} \mathcal{O}_K; \mathbf{Z}_p(m))$ should essentially coincide for $e = 1, 2$. This conjecture has been proved when $p = 2$ [160] (see also [123]).

In this setting a typical generalization of the Lichtenbaum conjecture would be the following: fix an odd prime p and let L/K be a totally complex abelian Galois extension with K totally real. Let χ be a character of the Galois group $G(L/K)$ – that is, a one-dimensional complex representation – such that $\chi(c) = (-1)^m$, where c denotes complex conjugation. View χ as a p-adic character, as in [166], and let $d_\chi = [\mathbf{Q}_p[\chi] : \mathbf{Q}_p]$. Then the relation

$$L_K(1-m, \overline{\chi})^{d_\chi} \sim_p \frac{|H^2_{\text{étale}}(\operatorname{Spec} \mathcal{O}_L; \mathbf{Z}_p(m))^\chi|}{|H^1_{\text{étale}}(\operatorname{Spec} \mathcal{O}_L; \mathbf{Z}_p(m))^\chi|}$$

follows from the Main Conjecture of Iwasawa theory, proved in [166] (see [90] §4). Here $A \sim_p B$ means that the orders of the finite groups A and B are divisible by the same power of p and A^χ denotes the subgroup of A on which the Galois group acts via χ.

1.3 Examples of Galois structure invariants

1.3.1 In this section we shall discuss the promised examples in which Galois module structure of K-groups influences special values of L-functions. They should be viewed as part of a general connection between L-values and Galois motivic actions which is complementary to the Lichtenbaum conjecture formulae mentioned at the end of §1.2.2.

Although the examples are the result of some difficult calculations, the Galois structure/L-value connection which they manifest is a rather humble one. Two possible Galois structures will distinguish between the two possibilities for the image of the odd part of $L_{\mathbf{Q}}(-1, \nu)$ in $(\mathbf{Z}/4)^*$, the group of units modulo 4.

In defence of such modest examples, I should point out that distinguishing between two possibilities has some respectable precedents in number theory. If p is an odd prime and $\chi : \mathbf{F}_p^* \longrightarrow \{\pm 1\}$ is the non-trivial character then the associated quadratic Gauss sum is given by

$$\tau = \sum_{t \in \mathbf{F}_p^*} \chi(t) e^{2\pi i t/p}$$

and is easily seen to satisfy $\tau^2 = (-1)^{(p-1)/2} p$. Gauss was sufficiently intrigued by the problem of deciding between the two possible values of τ that, starting in May 1801, hardly a week passed without his thinking hard about the problem. He testifies to this in a letter of September 3, 1805 to his friend Olbers which triumphantly concludes with the words "Wie der Blitz einschlägt, hat sich das

Räthsel gelöst"[1]; in fact,

$$\tau = \begin{cases} \sqrt{p} & p \equiv 1 \text{ (modulo 4)}, \\ i\sqrt{p} & p \equiv 3 \text{ (modulo 4)}, \end{cases}$$

$\sqrt{p}$ being the positive square root of p.

Example 1.3.2 *([37] §2)*

For each prime $p \equiv 3$ (modulo 8) we shall now construct a Galois extension $N/\mathbf{Q}$ whose group, $G(N/\mathbf{Q})$, is isomorphic to the quaternion group of order eight

$$Q_8 = \{x, y \mid x^2 = y^2,\ xyx^{-1} = y^{-1}\}.$$

Set $E = \mathbf{Q}(\sqrt{2}, \sqrt{p})$. Since $p \equiv 3$ (modulo 8), -2 is a norm from $\mathbf{Q}(\sqrt{p})$ and we may find u, v so that $-2 = u^2 - v^2 p$. If β satisfies

$$\beta^2 = ((\sqrt{p})/(u + v\sqrt{p})) \cdot (\sqrt{2}(1 + \sqrt{2}))$$

then $N = E(\beta) = \mathbf{Q}(\sqrt{2}, \sqrt{p}, \beta)$ is the required extension. The Galois action is given by

$$x(\beta) = \beta/(\sqrt{2} + 1), \quad y(\beta) = \beta((u + v\sqrt{p})/\sqrt{2}).$$

Also $N/\mathbf{Q}$ ramifies only at the primes 2 and p.

Example 1.3.3 *([39] §2)*

Let $N = \mathbf{Q}(\sqrt{2}, \sqrt{3}, \beta)$ denote the quaternion field which was constructed in Example 1.3.2 when $p = 3$. Hence

$$\beta^2 = \frac{\sqrt{3}(2 + \sqrt{2})}{(1 + \sqrt{3})} \in \mathbf{Q}(\sqrt{2}, \sqrt{3}) = E.$$

Then $N/\mathbf{Q}$ is a quaternion extension with Galois action given by the formulae

$$\begin{array}{ll} x(\beta) = \frac{\beta}{1+\sqrt{2}}, & y(\beta) = \frac{\beta(1+\sqrt{3})}{\sqrt{2}}, \\ x(\sqrt{2}) = -\sqrt{2}, & y(\sqrt{2}) = \sqrt{2}, \\ x(\sqrt{3}) = \sqrt{3}, & y(\sqrt{3}) = -\sqrt{3}. \end{array}$$

Let q be any prime satisfying $q \equiv 7$ or 17 (modulo 24). Then $N(\sqrt{q})/\mathbf{Q}$ is a Galois extension with group isomorphic to $Q_8 \times \mathbf{Z}/2$, where Q_8 fixes $\sqrt{q}$ and acts on N as before while the remaining generator, τ, fixes N and satisfies $\tau(\sqrt{q}) = -\sqrt{q}$.

Set $N_q = N(\sqrt{q})^{x^2\tau}$, the fixed field of $x^2\tau$. Hence $N_q = \mathbf{Q}(\sqrt{2}, \sqrt{3}, \beta\sqrt{q})$, $N_q/\mathbf{Q}$ is ramified only at the primes $2, 3, q$ and $G(N_q/\mathbf{Q}) \cong Q_8$.

[1] As lightning strikes was the puzzle solved!

1.3.4 Now let us consider the Galois module structure invariant

$$\Omega_1(M/\mathbf{Q},3) \in \mathcal{CL}(\mathbf{Z}[G(M/\mathbf{Q})]) \cong (\mathbf{Z}/4)^*$$

when $M/\mathbf{Q}$ is one of the quaternion fields of Examples 1.3.2 and 1.3.3. This invariant was introduced in ([132] Chapter 7) and generalized in ([36], [38]). We shall return to the general case in Definition 5.2.11 (see also Conjecture 5.2.18).

Let S be a finite Galois-invariant set of primes of $\mathcal{O}_M$ which contains all primes above those which ramify in $M/\mathbf{Q}$. There exists a canonical 2-extension of $\mathbf{Z}[G(M/\mathbf{Q})]$-modules of the form

$$K_3^{\text{ind}}(\mathcal{O}_{M,S}) \longrightarrow A \longrightarrow B \longrightarrow K_2(\mathcal{O}_{M,S})'$$

where $\mathcal{O}_{M,S}$ denotes the ring of S-integers of M and

$$K_2(\mathcal{O}_{M,S})' = \operatorname{Ker}(K_2(\mathcal{O}_{M,S}) \longrightarrow \oplus_{v,a\ real\ place}\, K_2(M_v)).$$

Here $K_3^{\text{ind}}(\mathcal{O}_{M,S}) \cong K_3^{\text{ind}}(M) = K_3(M)/(K_1(M) \cdot K_1(M) \cdot K_1(M))$ denotes the indecomposable K_3-group of M. This canonical 2-extension may be represented by a sequence in which A and B are finite, cohomologically trivial $\mathbf{Z}[G(M/\mathbf{Q})]$-modules. In this case A and B each have a two-term projective $\mathbf{Z}[G(M/\mathbf{Q})]$-module resolution whose Euler characteristics define classes $[A]$ and $[B]$ in the class-group. The Galois module structure invariant we require is defined by

$$[A] - [B] = \Omega_1(M/\mathbf{Q},3) \in \mathcal{CL}(\mathbf{Z}[G(M/\mathbf{Q})]) \cong (\mathbf{Z}/4)^*$$

which is independent of the choice of S.

Being isomorphic to the quaternion group of order eight, $G(M/\mathbf{Q})$ has a unique two-dimensional irreducible complex representation, $\nu(M)$. This is a symplectic representation so that its Artin root number (see Definition 1.1.9) satisfies

$$W_{\mathbf{Q}}(\nu(M)) \in \{\pm 1\} \cong \mathcal{CL}(\mathbf{Z}[Q_8]).$$

In all the quaternion examples, $M/\mathbf{Q}$, of Examples 1.3.2 and 1.3.3 the Artin root number, $W_{\mathbf{Q}}(\nu(M))$, is trivial ([37] §3; [39] §7).

The central conjecture of ([36] p. 1436; [38] p. 321) reduces in this special case to the following:

Conjecture 1.3.5 In §1.3.4 when $M/\mathbf{Q}$ is one of the quaternion fields of Examples 1.3.2 and 1.3.3

$$\Omega_1(M/\mathbf{Q},3) = W_{\mathbf{Q}}(\nu(M)) \in (\mathbf{Z}/4)^* \cong \mathcal{CL}(\mathbf{Z}[G(M/\mathbf{Q})]).$$

The evidence for this conjecture comes mainly from the fact that similar conjectures were posed for the analogous Galois structure invariants $\Omega_0(M/\mathbf{Q},2)$ and $\Omega_0(M/\mathbf{Q},3)$ in [34] which have been proved for many quaternion fields ([35] [77]).

The value of $\Omega_1(M/\mathbf{Q},3)$ has been calculated for all the quaternion fields of Examples 1.3.2 and 1.3.3.

Theorem 1.3.6 *([37] Theorem 1.1; see also §5.4.5)*
Let S denote any set of rational primes containing $\{2, p\}$ with $p \equiv 3$ (modulo 8) and let $N/\mathbf{Q}$ be the associated quaternion extension of Example 1.3.2. Then

$$\Omega_1(N/\mathbf{Q}, 3) = -\tilde{L}_{\mathbf{Q},S}(-1, \nu(N))(-1)^{(1/4)\log_2(z)} \in \mathcal{CL}(\mathbf{Z}[G(N/\mathbf{Q})]) \cong (\mathbf{Z}/4)^*$$

where $z = \tilde{L}_{\mathbf{Q},S}(-1, \mathrm{Ind}_{\langle C\rangle}^{G(N/\mathbf{Q})}(1))$. Here $L_{\mathbf{Q},S}(s,\rho)$ denotes the Artin L-function of a representation of $G(N/\mathbf{Q})$ with the Euler factors at 2 and p removed and $C = G(N/\mathbf{Q}(\sqrt{2}, \sqrt{p}))$, the centre of $G(N/\mathbf{Q})$. Also, when the L-value is a non-zero rational number, $\tilde{L}_{\mathbf{Q},S}(s,\rho)$ denotes its odd-primary part.
The element $(-1)^{(1/4)\log_2(z)}$ is equal to $\chi_+(z)$ where $\chi_+ : (\mathbf{Z}/8)^* \longrightarrow (\mathbf{Z}/4)^*$ is the surjective homomorphism with kernel equal to $\{\pm 1\}$.

1.3.7 When $p = 3, 11, 19$ computer calculations ([37] pp. 27/28; see also [39] §8) yield the following table of values for the L-functions which appear in Theorem 1.3.6. In all these examples we choose $S = \{2, p\}$ and observe that $L_{\mathbf{Q}}(-1, \nu(N)) = L_{\mathbf{Q},S}(-1, \nu(N))$.

p	$L_{\mathbf{Q}}(-1, \mathrm{Ind}_{\langle C\rangle}^{Q_8}(1))$	$L_{\mathbf{Q},S}(-1, \mathrm{Ind}_{\langle C\rangle}^{Q_8}(1))$	$L_{\mathbf{Q},S}(-1, \nu(N))$
3	1	$(1-2)(1-3^2)$	$2^2 \cdot 71$
11	$3^{-1} \cdot 7 \cdot 23$	$3^{-1} \cdot 7 \cdot 23 \cdot (1-2) \cdot (1-11^2)$	$2^2 \cdot 3 \cdot 11 \cdot 109$
19	$3^{-1} \cdot 19 \cdot 41$	$3^{-1} \cdot 19 \cdot 41 \cdot (1-2) \cdot (1-19^2)$	$2^2 \cdot 3^2 \cdot 1993$

Corollary 1.3.8 *([36] Corollary 1.2)*
Conjecture 1.3.5 is true for $N/\mathbf{Q}$ in Example 1.3.2 when $p = 3, 11$ or 19. In fact, $\Omega_1(N/\mathbf{Q}, 3)$ and $W_{\mathbf{Q}}(\nu(N))$ are both trivial in these cases.

Proof. As mentioned in §1.3.4, $W_{\mathbf{Q}}(\nu(N)) = 1$ in these examples. On the other hand, calculating modulo 4, using the formula of Theorem 1.3.5 when $p = 3$ we obtain

$$\Omega_1(N/\mathbf{Q}, 3) \equiv -71\chi_+(1) \equiv 1 \in (\mathbf{Z}/4)^*.$$

Similarly, when $p = 11$

$$\Omega_1(N/\mathbf{Q}, 3) \equiv -3 \cdot 11 \cdot 109\chi_+(3^{-1} \cdot 7 \cdot 23 \cdot 3 \cdot 5) \equiv -\chi_+(5) \equiv 1 \in (\mathbf{Z}/4)^*$$

and for $p = 19$

$$\Omega_1(N/\mathbf{Q}, 3) \equiv -3^2 \cdot 1993\chi_+(3^{-1} \cdot 19 \cdot 41 \cdot 9 \cdot 5) \equiv -\chi_+(5) \equiv 1 \in (\mathbf{Z}/4)^*$$

as required. □

1.3.9 Now consider the quaternion fields, $N_q/\mathbf{Q}$, of Example 1.3.3. From ([39] Proposition 2.4) we have an isomorphism of the form

$$K_2(\mathcal{O}_{N_q})' \otimes \mathbf{Z}_2 \cong \mathbf{Z}/2 \times \mathbf{Z}/2.$$

From the K-theory localisation sequence it was shown in ([39] Corollary 2.5) that

$$K_2(\mathcal{O}_{N_q}[1/q])' \otimes \mathbf{Z}_2 \cong \mathbf{Z}/2^{5+e} \oplus \mathbf{Z}/2^{5+e}$$

where the prime $q \equiv 7, 17$ (modulo 24) of Example 1.3.3 satisfies $q^2 - 1 = 2^{4+e}(2s+1)$ for some integers, e and s.

In addition, $K_2(\mathcal{O}_{N_q}[1/q])' \otimes \mathbf{Z}_2$ is a cohomologically trivial $\mathbf{Z}[Q_8]$-module and there are only two possibilities (see [39] §2.6). If $x, y \in G(N_q/\mathbf{Q})$ are the two generators whose Galois action is described in Examples 1.3.2 and 1.3.3 then either

$$K_2(\mathcal{O}_{N_q}[1/q])' \otimes \mathbf{Z}_2 \cong \mathrm{Ind}_{\langle y\rangle}^{G(N_q/\mathbf{Q})}(\mathbf{Z}/2^{5+e})$$

where $\mathbf{Z}/2^{5+e}$ is the (unique) cohomologically trivial $\mathbf{Z}[\langle y\rangle]$-module or the Q_8-action on $\mathbf{Z}/2^{5+e} \oplus \mathbf{Z}/2^{5+e}$ is given (up to isomorphism) by left multiplication by matrices of the form

$$x = \begin{pmatrix} 0 & q^2 \\ 1 & 0 \end{pmatrix}, \quad y = \begin{pmatrix} q + d2^{4+e} & 2^{4+e} \\ 2^{4+e} & q^3 + d2^{4+e} \end{pmatrix}$$

for $d = 0$ or 1.

Being cohomologically trivial and finite, $K_2(\mathcal{O}_{N_q}[1/q])'$ defines a class in the class-group.

Proposition 1.3.10 *([39] Proposition 2.9)*
In the notation of §1.3.9 the class

$$K_2(\mathcal{O}_{N_q}[1/q])' \otimes \mathbf{Z}_2 \in \mathcal{CL}(\mathbf{Z}[G(N_q/\mathbf{Q})]) \cong (\mathbf{Z}/4)^*$$

is trivial if and only if the module is induced from $\mathbf{Z}[\langle y\rangle]$.

Theorem 1.3.11 *([39] Theorem 1.1)*
Let S denote the set of rational primes $\{2, 3, q\}$ with $q \equiv 17$ (modulo 96) and let $N_q/\mathbf{Q}$ be the quaternion extension of Example 1.3.3. Then

$$\Omega_1(N_q/\mathbf{Q}, 3) + [K_2(\mathcal{O}_{N_q}[1/q])' \otimes \mathbf{Z}_2] = -\tilde{L}_{\mathbf{Q},S}(-1, \nu(N_q))(-1)^{(1/4)\log_2(z)}$$

in $\mathcal{CL}(\mathbf{Z}[G(N_q/\mathbf{Q})]) \cong (\mathbf{Z}/4)^*$, where $z = \tilde{L}_{\mathbf{Q},S}(-1, \mathrm{Ind}_{\langle x^2\rangle}^{G(N_q/\mathbf{Q})}(1))$. Here $L_{\mathbf{Q},\Sigma}(s, \rho)$ denotes the Artin L-function of a representation of $G(N_q/\mathbf{Q})$ with the Euler factors at 2, 3 and q removed and $\langle x^2\rangle = G(N_q/\mathbf{Q}(\sqrt{2}, \sqrt{3}))$, the centre of $G(N_q/\mathbf{Q})$. Also, when the L-value is a non-zero rational number, $\tilde{L}_{\mathbf{Q},S}(s, \rho)$ denotes its odd-primary part. The element $(-1)^{(1/4)\log_2(z)}$ is equal to $\chi_+(z)$ where $\chi_+ : (\mathbf{Z}/8)^* \longrightarrow (\mathbf{Z}/4)^*$ is the surjective homomorphism with kernel equal to $\{\pm 1\}$.

Remark 1.3.12 Note that, in the statement of Theorem 1.3.11 we have been forced to restrict from $q \equiv 7, 17$ (modulo 24) as in Example 1.3.3 to $q \equiv 17$ (modulo 96). This is done for technical reasons involved in the very computational proof of ([39] Proposition 8.4) – I believe that the formula is valid for all $q \equiv 7, 17$ (modulo 24).

1.3.13 *Galois actions and L-values*

Let $N_q/\mathbf{Q}$ denote the quaternion field of Example 1.3.3 when $q \equiv 17$ (modulo 96). Let $N = \mathbf{Q}(\sqrt{2}, \sqrt{3}, \beta)$ denote the quaternion field which was constructed in Example 1.3.2 when $p = 3$.

Since $W_{\mathbf{Q}}(\nu(N)) = W_{\mathbf{Q}}(\nu(N_q)) = 1$, Conjecture 1.3.5 implies that

$$\Omega_1(N_q/\mathbf{Q}, 3) = \Omega_1(N/\mathbf{Q}, 3) \in (\mathbf{Z}/4)^*.$$

Since N_q and N have the same biquadratic subfield there is a common factor, $(-1)^{(1/4)\log_2(z)}$, in the formulae of Theorems 1.3.6 and 1.3.11 provided that we use $S = \{2, 3, q\}$ in Theorem 1.3.6. In this case the conjecture would predict that, in $(\mathbf{Z}/4)^*$,

$$\tilde{L}_{\mathbf{Q},S}(-1, \nu(N)) = \begin{cases} \tilde{L}_{\mathbf{Q},S}(-1, \nu(N_q)) & \text{if } K_2(\mathcal{O}_{N_q}[1/q])' \otimes \mathbf{Z}_2 \\ & \qquad \cong \mathrm{Ind}_{\langle y \rangle}^{G(N_q/\mathbf{Q})}(\mathbf{Z}/2^{5+e}), \\ -\tilde{L}_{\mathbf{Q},S}(-1, \nu(N_q)) & \text{otherwise.} \end{cases}$$

Since $K_2(\mathcal{O}_{N_q})' \otimes \mathbf{Z}_2$ is elementary abelian of rank two, $K_2(\mathcal{O}_{N_q}) \otimes \mathbf{Z}_2$ has order 2^{10}. By [89], the Birch-Tate conjecture is true for such fields (cf. [37] Proposition 2.3)

$$\zeta_{N_q}(-1) = \frac{|K_2(\mathcal{O}_{N_q})|}{|K_3^{\mathrm{ind}}(\mathcal{O}_{N_q})|}$$

and so $\zeta_{N_q}(-1) = 2^{10}u/2^4 = 2^6 u$ where $u \in \mathbf{Q}^*$ is a non-zero rational 2-unit. Also

$$\zeta_{N_q}(-1) = L_{\mathbf{Q}}(-1, \mathrm{Ind}_{\{1\}}^{Q_8}(1)) = L_{\mathbf{Q}}(-1, \nu(N_q))^2 \cdot \zeta_{\mathbf{Q}(\sqrt{2},\sqrt{3})}(-1).$$

By ([37] §5.6), $\zeta_{\mathbf{Q}(\sqrt{2},\sqrt{3})}(-1) = 1$ so that

$$\zeta_{N_q}(-1) = L_{\mathbf{Q}}(-1, \mathrm{Ind}_{\{1\}}^{Q_8}(1)) = L_{\mathbf{Q}}(-1, \nu(N_q))^2$$

and, since $W_{\mathbf{Q}}(\nu(N_q)) = 1$,

$$L_{\mathbf{Q}}(-1, \nu(N_q)) = 2^3\alpha$$

where $0 < \alpha \in \mathbf{Q}^*$ is a 2-unit. Similarly, by ([37] §5.6),

$$2^4 \cdot 71^2 = \zeta_N(-1) = L_{\mathbf{Q}}(-1, \nu(N))^2$$

so that

$$2^2 \cdot 71 = L_{\mathbf{Q}}(-1, \nu(N)).$$

Now let us calculate $L_{\mathbf{Q},S}(-1, \nu(N))$ and $L_{\mathbf{Q},S}(-1, \nu(N_q))$. At the prime, q, the decomposition subgroup of $G(N/\mathbf{Q})$ is $\langle y \rangle$ and the inertia subgroup is trivial.

Hence, by ([37] §5.6), if Frob_p denotes the Frobenius at p and I_p denotes the inertia group

$$\begin{aligned} &L_{\mathbf{Q},S}(-1,\nu(N)) \\ &= L_{\mathbf{Q}}(-1,\nu(N))\textstyle\prod_{p=2,3,q}\det(1-p(\mathrm{Frob}_p\,|\nu(N)^{I_p})) \\ &= 2^2\cdot 71\cdot(1)\cdot(1)\cdot\det\begin{pmatrix} 1+qi & 0 \\ 0 & 1-qi \end{pmatrix} \\ &= 2^2\cdot 71\cdot(1+q^2) \\ &= 2^3\cdot 71\cdot\tfrac{(1+q^2)}{2}, \end{aligned}$$

where $\frac{(1+q^2)}{2}$ is of the form $8t+1$ since $q \equiv 17$ (modulo 96).

Similarly, in $G(N_q/\mathbf{Q})$, the decomposition group for q is $\langle y\rangle$ and the inertia group is $\langle y^2\rangle$. Combining this with the calculations of ([37] §5.6) one sees that

$$L_{\mathbf{Q},S}(-1,\nu(N_q)) = L_{\mathbf{Q}}(-1,\nu(N_q)) = 2^3\alpha.$$

Hence, in $(\mathbf{Z}/4)^*$, Conjecture 1.3.5 is true for $N_q/\mathbf{Q}$ with $q \equiv 17$ (modulo 96) if and only if

$$\alpha \equiv \begin{cases} 3 & \text{if } K_2(\mathcal{O}_{N_q}[1/q])'\otimes\mathbf{Z}_2 \text{ is induced from } \langle y\rangle, \\ 1 & \text{otherwise.} \end{cases}$$

In other words these examples illustrate the manner in which the Galois module structure of the 2-primary part of $K_2(\mathcal{O}_{N_q}[1/q])'$ is expected to determine $L_{\mathbf{Q}}(-1,\nu(N_q))$ (modulo 32).

For the sake of comparison with alternative approaches, I believe that the modular forms technique of [50] gives only congruences modulo 16 for these examples.

Chapter 2

K-groups and Class-groups

In this chapter we shall recall the salient facts about low-dimensional K-groups and localisation sequence for the integral group-ring, $\mathbf{Z}[G]$, of a finite group G. This material is recounted just as well, if not better, in many other places. Nonetheless, repeating it here will serve both to remind the reader and also to establish our conventions, particularly with regard to the Hom-descriptions of Theorem 2.5.25, Theorem 2.5.32 and §2.5.34. For further details the reader is referred to ([3], [49], [62], [84], [109], [117], [124], [133], [150], [155]).

Later we shall need low-dimensional K-groups of group-rings because these are the groups which contain the Galois module structure invariants we are going to study. On the other hand we shall need the higher-dimensional K-groups of arithmetic-geometric objects as the principal source of motivic Galois modules which possess interesting Galois module structure invariants.

2.1 Low-dimensional algebraic K-theory

Definition 2.1.1 Let $\mathcal{C}$ be a small abelian category. For each object, $A \in \mathrm{obj}(\mathcal{C})$ let $[A]$ denote the isomorphism class of A. Then $K_0(\mathcal{C})$ is defined to be the abelian group generated by the set

$$\{[A] \mid A \in \mathrm{obj}(\mathcal{C})\}$$

subject to the relations that $[A] = [A'] + [A'']$ whenever there exists a short exact sequence in $\mathcal{C}$ of the form

$$0 \longrightarrow A' \longrightarrow A \longrightarrow A'' \longrightarrow 0.$$

In the group $K_0(\mathcal{C})$ it is easy to see that $[0]$ is the identity element, that $[A] = [B]$ if $A \cong B$ and that $[A \oplus B] = [A] + [B]$.

Example 2.1.2 (i) Now let R be a ring. If $\mathcal{P}(R)$ denotes the category of finitely generated projective (left) R-modules then $K_0(\mathcal{P}(R))$ is traditionally abbreviated to $K_0(R)$.

Let G be a finite group. We shall be particularly interested in the case when $R = \mathbf{Z}[G]$, the integral group-ring. In this case, if P is a finitely generated, projective $\mathbf{Z}[G]$-module then $P \otimes \mathbf{Q}$ is a finitely generated, free $\mathbf{Q}[G]$-module and the *rank* of P is the integer defined by the relation

$$\mathrm{rank}(P) \cdot |G| = \dim_{\mathbf{Q}}(P \otimes \mathbf{Q}).$$

Hence the rank gives a surjection of the form

$$rk : K_0(\mathbf{Z}[G]) \longrightarrow \mathbf{Z}.$$

The reduced K_0 of $\mathbf{Z}[G]$, $\tilde{K}_0(\mathbf{Z}[G])$, is defined to be the kernel of rk so that

$$K_0(\mathbf{Z}[G]) \cong \tilde{K}_0(\mathbf{Z}[G]) \oplus \mathbf{Z}.$$

Each finitely generated, projective $\mathbf{Z}[G]$-module P has the form $P \cong \mathbf{Z}[G]^{rk(P)-1} \oplus J$ for some ideal $J \lhd \mathbf{Z}[G]$ [3]. Therefore every element of $\tilde{K}_0(\mathbf{Z}[G])$ may be written in the form $[J] - [\mathbf{Z}[G]]$ for some $J \lhd \mathbf{Z}[G]$.

This situation is analogous to the case when $R = \mathcal{O}_K$, the ring of algebraic integers in a number field. Being a Dedekind domain, every finitely generated projective $\mathcal{O}_K$-module, P, satisfies $P \cong \mathcal{O}_K^{rk(P)-1} \oplus J$ for some $J \lhd \mathcal{O}_K$ [59]. Sending P to J defines an isomorphism between $\tilde{K}_0(\mathcal{O}_K)$ and $\mathcal{CL}(\mathcal{O}_K)$, the ideal class group given by ideals modulo principal ideals with group multiplication induced by taking the product of ideals [109].

Prompted by the analogy with the case of rings of algebraic integers we shall follow the Galois module structure convention and refer to $\tilde{K}_0(\mathbf{Z}[G])$ as the *class-group* of $\mathbf{Z}[G]$, denoted by $\mathcal{CL}(\mathbf{Z}[G])$. In addition, if P is a finitely generated projective $\mathbf{Z}$[G]-module, we shall write $[P] \in \mathcal{CL}(\mathbf{Z}[G])$ as an abbreviation for $[P] - rk(P) \cdot [\mathbf{Z}[G]] \in \tilde{K}_0(\mathbf{Z}[G])$.

(ii) Let $\mathcal{C}$ denote the full subcategory of the category of finitely generated (left) $\mathbf{Z}[G]$-modules whose objects are the cohomologically trivial modules. That is, for $M \in \mathrm{obj}(\mathcal{C})$ the Tate cohomology groups, $\hat{H}^i(U; M)$ ([132] p. 3), vanish for all integers i and all subgroups, $U \subseteq G$.

Any such module, M, has a projective resolution of the form [49]

$$0 \longrightarrow P_1 \longrightarrow P_0 \longrightarrow M \longrightarrow 0$$

and sending $[M]$ to the *Euler characteristic*, $[P_0] - [P_1]$, induces an isomorphism of the form

$$K_0(\mathcal{C}) \xrightarrow{\cong} K_0(\mathbf{Z}[G]).$$

Let $\mathcal{C}_{fin}$ denote the full subcategory of $\mathcal{C}$ whose objects are the finite cohomologically trivial modules. For such an M the ranks of P_0 and P_1 are equal. In this case the map, $([M] \mapsto [P_0] - [P_1])$, induces a homomorphism of the form

$$K_0(\mathcal{C}_{fin}) \longrightarrow \tilde{K}_0(\mathbf{Z}[G]) = \mathcal{CL}(\mathbf{Z}[G]).$$

Since a finite module is canonically the sum of its Sylow p-subgroups, there is an isomorphism of the form

$$K_0(\mathcal{C}_{fin}) \cong \oplus_{p \text{ prime}} K_0(\mathcal{C}_{fin,p})$$

where $\mathcal{C}_{fin,p}$ is the full subcategory of $\mathcal{C}$ whose objects are the finite cohomologically trivial modules of order a power of a prime, p.

(iii) Let $\mathcal{C}_{lf}$ denote the full subcategory of the category of finitely generated (left) $\mathbf{Z}[G]$-modules whose objects are the *locally free* modules. Hence a finitely generated $\mathbf{Z}[G]$-module, M, belongs to $\mathrm{obj}(\mathcal{C}_{lf})$ if and only if $R \otimes_{\mathbf{Z}} M$ is a free $R[G]$-module for $R = \mathbf{Q}$, the rationals and for $R = \mathbf{Z}_p$, the p-adic integers for each prime p.

It is a result of Swan [49] that P is a finitely generated projective $\mathbf{Z}[G]$-module if and only if it is locally free. This result does not hold if $\mathbf{Z}$ is replaced by an arbitrary Dedekind domain. However, in the case of $\mathbf{Z}[G]$ we obtain an isomorphism of the form

$$K_0(\mathbf{Z}[G]) \xrightarrow{\cong} K_0(\mathcal{C}_{lf}).$$

(iv) If $\mathcal{C}$ is the category of finitely generated R-modules then $K_0(\mathcal{C})$ is usually denoted by $G_0(R)$ (or $K_0'(R)$ in [117]). The forgetful map

$$c : K_0(R) \longrightarrow G_0(R)$$

is called the *Cartan homomorphism*.

(v) Suppose that R is a Dedekind domain with fraction field F. Let A be an R-algebra. Then an A-module M is called locally free if $F \otimes_R M$ is a free $F \otimes_R A$-module and, for each prime ideal $P \lhd R$, $R_P \otimes_R M$ is a free $R_P \otimes_R A$-module. Here R_P denotes the completion of R in the P-adic topology (cf. §2.5.1). Denote by $\mathcal{C}_{loc.fr.}$ the category of finitely generated, locally free A-modules with associated K-group

$$K_0^{loc.fr.}(A) = K_0(\mathcal{C}_{loc.fr.}).$$

When $R = \mathcal{O}_K$, the ring of algebraic integers of a number field, K, and G is a finite group then an $\mathcal{O}_K[G]$-module is locally free if and only if it is projective [149]. Hence there is an isomorphism of the form

$$K_0^{loc.fr.}(\mathcal{O}_K[G]) \cong K_0(\mathcal{O}_K[G]).$$

This fact will be important in Theorem 2.5.25. The associated class-group is defined by

$$\mathcal{CL}(\mathcal{O}_K[G]) = K_0(\mathcal{O}_K[G])/\langle[\mathcal{O}_K[G]]\rangle,$$

the quotient of $K_0(\mathcal{O}_K[G])$ by the infinite cyclic subgroup generated by the class of $\mathcal{O}_K[G]$. When $\mathcal{O}_K = \mathbf{Z}$ this class-group is canonically isomorphic to that of Example 2.1.2(i)–(ii).

Definition 2.1.3 Let R be a ring with an identity element and let $GL_n(R)$ denote the general linear group of all invertible $n \times n$ matrices with entries in R. Let $GL(R) = \bigcup_n GL_n(R)$ denote the infinite general linear group, given by the direct limit of the sequence of inclusions

$$GL_1(R) \subset GL_2(R) \subset \ldots GL_{n-1}(R) \subset GL_n(R) \subset \ldots$$

where $GL_k(R)$ is injected into $GL_{k+1}(R)$ by the map

$$X \mapsto \begin{pmatrix} X & 0 \\ 0 & 1 \end{pmatrix}.$$

An *elementary matrix* is an element $e^{\lambda}_{a,b} \in GL(R)$, for distinct integers $a, b \geq 1$, satisfying $Y_{i,i} = 1$ for all i, $Y_{a,b} = \lambda \in R$ and $Y_{i,j} = 0$ for all other pairs $(i,j) \neq (a,b)$. The subgroup generated by the elementary matrices is denoted by $E(R)$. In fact, $E(R) \lhd GL(R)$ and the Whitehead Lemma ([109] p. 25) states that $E(R) = [GL(R), GL(R)]$, the commutator subgroup of $GL(R)$.

Define

$$K_1(R) = GL(R)/E(R) = GL(R)^{ab},$$

the abelianisation of $GL(R)$. This group was introduced in the work of J.H.C. Whitehead on simple homotopy type and for that reason is sometimes known as the Whitehead group of R.

Example 2.1.4 (i) When R is commutative the determinant homomorphism yields a splitting of the form

$$K_1(R) \cong SK_1(R) \oplus R^*$$

where R^* denotes the group of units of R and $SK_1(R) = \mathrm{Ker}(\det : K_1(R) \longrightarrow R^*)$.

In general $SK_1(R)$ is very difficult to compute. However when R is a field it is easy to show that SK_1 is trivial. Much more difficult is the vanishing, proved in [4], of $SK_1(\mathcal{O}_K)$ for the ring of integers in a number field.

(ii) Let G be a finite abelian group. Then $\mathbf{Q}[G]$ is isomorphic to a product of fields $\mathbf{Q}[G] \cong \prod_i F_i$ and, by the previous example,

$$K_1(\mathbf{Q}[G]) \cong \prod_i F_i^*,$$

since K_1 commutes with finite products.

Definition 2.1.5 Let $f : R \longrightarrow S$ be a homomorphism of rings. Following ([150] p. 214) let F_f denote the category whose objects are triples (A, g, B) with $A, B \in \mathrm{obj}(\mathcal{P}(R))$ and g an isomorphism of (left) S-modules of the form

$$g : S \otimes_R A \xrightarrow{\cong} S \otimes_R B.$$

A morphism in F_f from (A, g, B) to (A', g', B') consists of a pair of (left) R-module homomorphisms, $u_A : A \longrightarrow A'$ and $u_B : B \longrightarrow B'$ such that

$$g' \cdot (1 \otimes u_A) = (1 \otimes u_B) \cdot g : S \otimes_R A \longrightarrow S \otimes_R B'.$$

A sequence

$$0 \longrightarrow (A', g', B') \longrightarrow (A, g, B) \longrightarrow (A'', g'', B'') \longrightarrow 0$$

is called exact in F_f if both

$$0 \longrightarrow A' \longrightarrow A \longrightarrow A'' \longrightarrow 0$$

and

$$0 \longrightarrow B' \longrightarrow B \longrightarrow B'' \longrightarrow 0$$

are exact in the category $\mathcal{P}(R)$.

The *relative K-group* $K_0(R, f)$ is defined to be the abelian group with generators

$$\{[A, g, B] \mid (A, g, B) \in \mathrm{obj}(F_f)\}$$

and the following relations:

(i) $[A, g, B] = [A', g', B'] + [A'', g'', B'']$ if there exists an exact sequence

$$0 \longrightarrow (A', g', B') \longrightarrow (A, g, B) \longrightarrow (A'', g'', B'') \longrightarrow 0$$

and

(ii) $[A, gh, C] = [A, h, B] + [B, g, C]$. In fact ([150] Lemma 15.6) every element of $K_0(R, f)$ has the form $[A, g, B]$.

There exists a homomorphism

$$\partial : K_1(S) \longrightarrow K_0(R, f)$$

defined by $\partial[X] = [R^n, (- \cdot X), R^n]$ where $[X]$ is represented by a matrix, $X \in GL_n(S)$. There is also a homomorphism

$$\pi : K_0(R, f) \longrightarrow K_0(R)$$

given by $\pi[A, g, B] = [A] - [B]$.

Theorem 2.1.6 *([150] Theorem 15.5)*

Let $f : R \longrightarrow S$ be a homomorphism of rings with identity. Then there is an exact sequence of the form

$$K_1(R) \xrightarrow{f_*} K_1(S) \xrightarrow{\partial} K_0(R, f) \xrightarrow{\pi} K_0(R) \xrightarrow{f_*} K_0(S).$$

Example 2.1.7 Let G be a finite group. Let R be an integral domain and let $f : \mathbf{Z}[G] \longrightarrow R[G]$ be the corresponding inclusion of group-rings. Then the exact sequence of Theorem 2.1.6 takes the form

$$\ldots \longrightarrow K_1(\mathbf{Z}[G]) \xrightarrow{f_*} K_1(R[G]) \xrightarrow{\partial} K_0(\mathbf{Z}[G], R) \xrightarrow{\pi} K_0(\mathbf{Z}[G]) \xrightarrow{f_*} K_0(R[G]).$$

Here, because we are hoping to do arithmetic with these sequences, we have adopted the (arithmetical) notation of [23] and written $K_0(\mathbf{Z}[G], R)$ for the relative K-group denoted by $K_0(\mathbf{Z}[G], f)$ in Definition 2.1.5.

Now consider the situation when $R = \mathbf{Q}$, the rational field, which is the case of particular interest to us. Incidentally, in this case the relative K-theory exact sequence is just the low-dimensional end of a localisation sequence (cf. [117] §5 Theorem 5, [62] p. 233; see also §2.4.6). The categorical construction of the localisation sequence in [117] shows that there is an isomorphism of the relative K-groups of the form

$$K_0(\mathbf{Z}[G]; \mathbf{Q}) \cong \oplus_{p \text{ prime}} \; K_0(\mathbf{Z}_p[G]; \mathbf{Q}_p)$$

sending $[A, \phi, B]$ to $\{[A \otimes \mathbf{Z}_p, \phi, B \otimes \mathbf{Z}_p] \mid p \text{ prime }\}$. Here $K_0(\mathbf{Z}_p[G]; \mathbf{Q}_p)$ is defined in an analogous manner to $K_0(\mathbf{Z}[G]; \mathbf{Q})$, replacing $\mathbf{Z}$ by $\mathbf{Z}_p$ and $\mathbf{Q}$ by $\mathbf{Q}_p$.

Example 2.1.8 *Constructing elements*

We shall close this section by giving two constructions of elements in the K-groups of Definitions 2.1.1 and 2.1.5. Let G be a finite group. There is one other important construction of elements in $K_0(\mathbf{Z}[G])$, namely the Euler characteristic of a *nearly perfect complex*. That construction is rather involved and will be given in Definition 2.3.13.

(i) Suppose that X and Y are finitely generated $\mathbf{Z}[G]$-modules and that

$$X \longrightarrow A \longrightarrow B \longrightarrow Y$$

is a 2-extension of $\mathbf{Z}[G]$-modules representing an element of $\mathrm{Ext}^2_{\mathbf{Z}[G]}(Y, X)$ inducing, via cup-product, isomorphisms between the Tate cohomology groups $\hat{H}^i(G; Y)$ and $\hat{H}^{i+2}(G; X)$ for all i. Under these circumstances a representative 2-extension may be chosen for which A and B are finitely generated, cohomologically trivial $\mathbf{Z}[G]$-modules. The Euler characteristic of such a 2-extension is defined to be the element

$$[A] - [B] \in \mathcal{CL}(\mathbf{Z}[G])$$

as in Example 2.1.2(ii). This element of the class-group depends only on the class of the 2-extension in the set $Aut_G(Y) \backslash \mathrm{Ext}^2_{\mathbf{Z}[G]}(Y, X) / Aut_G(X)$ (see Theorem 2.2.5).

(ii) Suppose that we have a bounded, perfect cochain complex of $\mathbf{Z}[G]$-modules. That is, a bounded cochain complex of finitely generated, projective $\mathbf{Z}[G]$-modules of the form

$$P^* : 0 \longrightarrow P^n \xrightarrow{d_n} P^{n+1} \xrightarrow{d_{n+1}} \ldots \xrightarrow{d_{m-1}} P^m \longrightarrow 0 \, ,$$

together with a given isomorphism

$$\psi : \oplus_j \; H^{2j+1}(P^*) \otimes \mathbf{Q} \xrightarrow{\cong} \oplus_j \; H^{2j}(P^*) \otimes \mathbf{Q}.$$

Associated to this data is a well-defined element – called a Reidemeister-Whitehead torsion – of $K_0(\mathbf{Z}[G], \mathbf{Q})$, $[P^{\mathrm{od}}, \phi, P^{\mathrm{ev}}]$ in the following manner. Set $P^{\mathrm{od}} = \oplus_j P^{2j+1}$ and $P^{\mathrm{ev}} = \oplus_j P^{2j}$.

Choose $\mathbf{Q}[G]$-module splittings for the right-hand epimorphisms in each of the short exact sequences:

$$0 \longrightarrow \mathrm{Ker}(d_j) \otimes \mathbf{Q} \longrightarrow P^j \otimes \mathbf{Q} \longrightarrow d_j(P^j) \otimes \mathbf{Q} \longrightarrow 0,$$
$$0 \longrightarrow d_{j-1}(P^{j-1}) \otimes \mathbf{Q} \longrightarrow \mathrm{Ker}(d_j) \otimes \mathbf{Q} \longrightarrow H^j(P^*) \otimes \mathbf{Q} \longrightarrow 0.$$

Using these chosen splittings the rational isomorphism, ϕ, is given by the following composition

$$\begin{aligned}
&\oplus_j \; P^{2j+1} \otimes \mathbf{Q} \\
&\cong (\oplus_j \; \mathrm{Ker}(d_{2j+1}) \otimes \mathbf{Q}) \oplus (\oplus_j \; d_{2j+1}(P^{2j+1}) \otimes \mathbf{Q}) \\
&\cong (\oplus_j \; d_{2j}(P^{2j}) \otimes \mathbf{Q}) \oplus (\oplus_j \; H^{2j+1}(P^*) \otimes \mathbf{Q}) \oplus (\oplus_j \; d_{2j+1}(P^{2j+1}) \otimes \mathbf{Q}) \\
&\overset{\psi}{\cong} (\oplus_j \; d_{2j}(P^{2j}) \otimes \mathbf{Q}) \oplus (\oplus_j \; H^{2j}(P^*) \otimes \mathbf{Q}) \oplus (\oplus_j \; d_{2j+1}(P^{2j+1}) \otimes \mathbf{Q}) \\
&\cong (\oplus_j \; d_{2j}(P^{2j}) \otimes \mathbf{Q}) \oplus (\oplus_j \; \mathrm{Ker}(d_{2j}) \otimes \mathbf{Q}) \\
&\cong \oplus_j \; P^{2j} \otimes \mathbf{Q}.
\end{aligned}$$

The Euler characteristic of the perfect complex is defined to be $\chi(P^*) = \sum_j (-1)^{j+1}[P^j] \in K_0(\mathbf{Z}[G])$ so that $\pi_*[P^{\mathrm{od}}, \phi, P^{\mathrm{ev}}] = \chi(P^*)$ in the exact sequence of Theorem 2.1.6.

The class, $[P^{\mathrm{od}}, \phi, P^{\mathrm{ev}}]$ does not depend on the choice of rational splittings but only on the rational cohomology isomorphism and the quasi-isomorphism class of the chain complex, P^* (see Proposition 2.5.35).

2.2 Perfect complexes

In this section we shall briefly review the role of perfect chain complexes of $\mathbf{Z}[G]$-modules in the construction of elements in the class-group, $\mathcal{CL}(\mathbf{Z}[G])$, of the group-ring of a finite group G. We have already met a special case in Example 2.1.8(i). The contents of this section will justify the claims made in that example.

Definition 2.2.1 Let G be a finite group. A *perfect complex* of $\mathbf{Z}[G]$-modules is a cochain complex of $\mathbf{Z}[G]$-module homomorphisms of the form

$$P^* : \; \ldots \longrightarrow P^n \xrightarrow{d_n} P^{n+1} \xrightarrow{d_{n+1}} \ldots \xrightarrow{d_{m-1}} P^m \longrightarrow \ldots$$

in which each P^i is a finitely generated, projective $\mathbf{Z}[G]$-module.

A *cohomologically trivial complex* is a similar cochain complex in which each P^i is cohomologically trivial (cf. Example 2.1.2(ii)). Since projective modules are cohomologically trivial a perfect complex is also a cohomologically trivial complex.

A cochain complex is *bounded* if P^i is zero for all but finitely many values of i.

The *Euler characteristic* of a bounded cohomologically trivial complex, P^*, is the element of the class-group defined by

$$\chi(P^*) = \sum_i (-1)^i [P_i] \in \mathcal{CL}(\mathbf{Z}[G])$$

where $[P_i]$ is interpreted as in Example 2.1.2(i)/(ii).

A *quasi-isomorphism* of complexes, $\{f_n : P^n \longrightarrow Q^n \mid n \in \mathbf{Z}\}$, is a cochain map (that is, $d_n \cdot f_n = f_{n+1} \cdot d_n : P^n \longrightarrow Q^{n+1}$ for all n) whose induced map on cohomology

$$f_* : H^n(P^*) \longrightarrow H^n(Q^*) = \frac{\mathrm{Ker}(d_n : Q^n \longrightarrow Q^{n+1})}{Im(d_{n-1} : Q^{n-1} \longrightarrow Q^n)}$$

is an isomorphism for all n.

Lemma 2.2.2

Let $f : P^* \longrightarrow Q^*$ be a quasi-isomorphism of $\mathbf{Z}[G]$-module complexes. Then the mapping cone complex given by

$$C(f) : \ldots \longrightarrow Q_{n-1} \oplus P_n \xrightarrow{\tilde{d}_n} Q_n \oplus P_{n+1} \longrightarrow \ldots$$

is exact. Here

$$\tilde{d}_n(a, b) = (d_{n-1}(a) + (-1)^n f_n(b), d_n(b))$$

for all $a \in Q_{n-1}, b \in P_n$. The converse is also true.

Proof. The short exact sequence of complexes

$$0 \longrightarrow Q^{*-1} \longrightarrow C(f)^* \longrightarrow P^* \longrightarrow 0$$

yields a long exact cohomology sequence of the form

$$\ldots \longrightarrow H^n(P^*) \xrightarrow{f_*} H^n(Q^*) \longrightarrow H^{n+1}(C(f)) \longrightarrow H^{n+1}(P^*) \xrightarrow{f_*} \ldots .$$

Therefore $H^*(C(f)) = 0$ and $(C(f), \tilde{d})$ is an exact complex. □

Corollary 2.2.3

Let $f : P^* \longrightarrow Q^*$ be a $\mathbf{Z}[G]$-module quasi-isomorphism of bounded, cohomologically trivial complexes. Then

$$\chi(P^*) = \chi(Q^*) \in \mathcal{CL}(\mathbf{Z}[G]).$$

Proof. By Lemma 2.2.2, $0 = \chi(C(f)) = \sum_n (-1)^n([Q^{n-1}] + [P^n]) \in \mathcal{CL}(\mathbf{Z}[G])$. □

Proposition 2.2.4

Let (C^*, d) be a bounded, cohomologically trivial complex of $\mathbf{Z}[G]$-modules such that $H^i(C^*)$ is a finitely generated $\mathbf{Z}[G]$-module for each integer, i.

(i) *There exists a quasi-isomorphism, $f : P^* \longrightarrow C^*$, in which P^* is a bounded perfect complex of $\mathbf{Z}[G]$-modules.*

(ii) *Suppose that $g : Q^* \longrightarrow D^*$ is a second quasi-isomorphism, as in (i), from a bounded perfect complex to a bounded cohomologically trivial complex with finitely generated cohomology groups. Suppose that P^k is a free $\mathbf{Z}[G]$-module unless Q^j, C^j, D^j are trivial for all $j \leq k$. Then for any $\mathbf{Z}[G]$-module cochain map, $j : C^* \longrightarrow D^*$, there exists a cochain map*

$$\{h_k : P^k \longrightarrow Q^k \mid k \in \mathbf{Z}\}$$

and a cochain homotopy

$$\{s_k : P^k \longrightarrow D^{k-1} \mid k \in \mathbf{Z}\}$$

such that

$$g_k h_k - j_k f_k = ds_k + s_{k+1} d : P^k \longrightarrow D^k$$

for all k.

Proof. For part (i), suppose that $C^k = 0$ for $k > m$ and that C^m is non-trivial. Choose a finitely generated projective $\mathbf{Z}[G]$-module, P^m, together with a surjection

$$\tilde{f} : P^m \longrightarrow H^m(C^*) = C^m/d(C^{m-1}).$$

Since $\pi : C^m \longrightarrow C^m/d(C^{m-1})$ is surjective we may lift $\tilde{f}$ to $f_m : P^m \longrightarrow C^m$ such that $\pi \cdot f_m = \tilde{f}$.

Now, by induction, suppose that we have constructed $\mathbf{Z}[G]$-module homomorphisms, $f_j : P^j \longrightarrow C^j$ for $j \geq r$ where

$$P^r \xrightarrow{d} P^{r+1} \xrightarrow{d} \dots \xrightarrow{d} P^m \longrightarrow 0$$

is a cochain complex of finitely generated, projective $\mathbf{Z}[G]$-modules such that, for all $j \geq r$,

$$f_j(\mathrm{Ker}(d : P^j \longrightarrow P^{j+1})) \subseteq \mathrm{Ker}(d : C^j \longrightarrow C^{j+1})$$
$$f_{j+1}(d(P^j)) \subseteq d(C^j)$$

and

$$H^j(P^*) = \frac{\mathrm{Ker}(d : P^j \longrightarrow P^{j+1})}{d(P^{j-1})} \xrightarrow{f_*} \frac{\mathrm{Ker}(d : C^j \longrightarrow C^{j+1})}{d(C^{j-1})} = H^j(C^*)$$

is an isomorphism for all $j \geq r+1$. The induction starts with $r = m$.

For the induction step, enlarge P^r by adding a direct summand with trivial differential and amending f_r so that

$$\tilde{f}_r : \mathrm{Ker}(d : P^r \longrightarrow P^{r+1}) \xrightarrow{f_r} \mathrm{Ker}(d : C^r \longrightarrow C^{r+1}) \longrightarrow H^r(C^*)$$

is surjective. Choose a finitely generated, projective $\mathbf{Z}[G]$-module, P^{r-1} together with a surjection, $\tilde{d} : P^{r-1} \longrightarrow \mathrm{Ker}(\tilde{f}_r)$. Since $d : C^{r-1} \longrightarrow d(C^{r-1})$ is surjective and $f_r(\tilde{d}(P^{r-1})) \subseteq d(C^{r-1})$ we may lift $f_r\tilde{d}$ to $f_{r-1} : P^{r-1} \longrightarrow C^{r-1}$ so that $df_{r-1} = f_r\tilde{d}$. Set $d : P^{r-1} \longrightarrow P^r$ equal to $\tilde{d}$ followed by the inclusion of $\mathrm{Ker}(\tilde{f}_r)$ into P^r. This completes the induction step.

Using the induction step we may construct a complex of finitely generated, projective $\mathbf{Z}[G]$-modules together with a quasi-isomorphism to C^*. However, if $C^i = 0$ for $i < n$ we may truncate the complex P^* to look like

$$0 \longrightarrow \mathrm{Ker}(d) \longrightarrow P^n \longrightarrow P^{n+1} \longrightarrow \ldots$$

which still maps to C^* by a quasi-isomorphism. The exactness of the mapping cone complex, proved in Lemma 2.2.2,

$$0 \longrightarrow \mathrm{Ker}(d) \longrightarrow P^n \longrightarrow C^n \oplus P^{n+1} \longrightarrow \ldots$$

shows that $\mathrm{Ker}(d)$ is a finitely generated, torsion-free, cohomologically trivial $\mathbf{Z}[G]$-module, which is therefore projective ([49] II). This establishes part (i).

For part (ii) suppose that P^k, Q^k, C^k, D^k are trivial for $k > m$ and that P^m is non-trivial. Then the homomorphisms

$$P^m/d(P^{m-1}) \xrightarrow{f_*} C^m/d(C^{m-1}) \xrightarrow{j_*} D^m/d(D^{m-1}) \xleftarrow{g_*} Q^m/d(Q^{m-1})$$

are all isomorphisms. Since $\pi_Q : Q^m \longrightarrow Q^m/d(Q^{m-1})$ is surjective, we may lift

$$P^m \xrightarrow{\pi_P} P^m/d(P^{m-1}) \xrightarrow{g_*^{-1}j_*f_*} Q^m/d(Q^{m-1})$$

to $h_m : P^m \longrightarrow Q^m$ such that $\pi_Q h_m = g_*^{-1}j_*f_*\pi_P$.

Therefore $(g_m h_m - j_m f_m)(P^m) \subseteq d(D^{m-1})$ and, since $d : D^{m-1} \longrightarrow d(D^{m-1})$ is surjective, we may lift this to a homomorphism, $s_m : P^m \longrightarrow D^{m-1}$, such that $g_m h_m - j_m f_m = ds_m : P^m \longrightarrow D^m$.

Now we proceed to construct the h_k, s_k's by induction, starting with $k = m$. Suppose for all $k \geq n+1$ we have constructed h_k and s_k as required. For any $z \in P^n$, $h_{n+1}(d(z)) \in Q^{n+1}$ satisfies $d(h_{n+1}(d(z))) = h_{n+2}(d(d(z))) = 0$ so that $h_{n+1}(d(z)) \in \mathrm{Ker}(d : Q^{n+1} \longrightarrow Q^{n+2})$. However, the cohomology class $[h_{n+1}(d(z))] \in H^{n+1}(Q^*)$ is zero because

$$\begin{aligned} g_*[h_{n+1}(d(z))] &= [g_{n+1}(h_{n+1}(d(z)))] \\ &= [j_{n+1}(f_{n+1}(d(z))) + d(s_{n+1}(d(z))) + s_{n+2}(d(d(z)))] \\ &= [j_{n+1}(f_{n+1}(d(z)))] \\ &= [d(j_n(f_n(z)))] \\ &= 0. \end{aligned}$$

Hence $h_{n+1}(d(P^n)) \subseteq d(Q^n)$ and, since π_Q is surjective, we may lift $h_{n+1}d$ to obtain a homomorphism, $h'_n : P^n \longrightarrow Q^n$ such that $dh'_n = h_{n+1}d : P^n \longrightarrow Q^{n+1}$.

Next we observe that

$$\begin{aligned} d(g_n h'_n - j_n f_n - s_{n+1}d) &= g_{n+1}h_{n+1}d - j_{n+1}f_{n+1}d - ds_{n+1}d \\ &= ds_{n+1}d + s_{n+2}dd - ds_{n+1}d \\ &= 0 \end{aligned}$$

and so we obtain a homomorphism

$$g_n h'_n - j_n f_n - s_{n+1}d : P^n \longrightarrow \mathrm{Ker}(d : D^n \longrightarrow D^{n+1}).$$

For each $z \in P^n$ there exists a cocycle, $w \in \mathrm{Ker}(d : Q^n \longrightarrow Q^{n+1})$, such that

$$[(g_n h'_n - j_n f_n - s_{n+1}d)(z)] = g_*[w] \in H^n(D^*).$$

If P^n is free, let $z_1, \ldots, z_t$ be a free basis. Note that, in the case when P^n is projective but not free, we may choose $h_n = 0 = s_n$. For each z_i choose $w_i \in \mathrm{Ker}(d : Q^n \longrightarrow Q^{n+1})$ satisfying

$$[(g_n h'_n - j_n f_n - s_{n+1}d)(z_i)] = g_*[w_i].$$

Set $h_n(z_i) = h'_n(z_i) - w_i \in Q_n$ and extend linearly to a homomorphism, h_n. Then $d(h_n(z_i)) = d(h'_n(z_i)) - d(w_i) = d(h'_n(z_i)) = h_{n+1}(d(z_i))$ for all i and so $dh_n = h_{n+1}d$, as required. Also

$$0 = [(g_n h'_n - j_n f_n - s_{n+1}d)(z_i)] - g_*[w_i] = [(g_n h_n - j_n f_n - s_{n+1}d)(z_i)] \in H^n(D^*)$$

so that we have a homomorphism

$$g_n h_n - j_n f_n - s_{n+1}d : P^n \longrightarrow d(D^{n-1})$$

which we may lift to a homomorphism, $s_n : P^n \longrightarrow D^{n-1}$ such that $g_n h_n - j_n f_n - s_{n+1}d = ds_n$, as required. □

Theorem 2.2.5

Let (C^*, d) be a bounded, cohomologically trivial complex of $\mathbf{Z}[G]$-modules such that $H^i(C^*)$ is a finitely generated $\mathbf{Z}[G]$-module for each integer, i. Suppose, for $j = 1, 2$, that

$$f(j) : P_j^* \longrightarrow C^*$$

are quasi-isomorphisms in which P_j^* is a bounded perfect complex of $\mathbf{Z}[G]$-modules. Then

$$\chi(P_1^*) = \chi(P_2^*) \in \mathcal{CL}(\mathbf{Z}[G]).$$

Proof. An elementary complex is a cochain complex of the form

$$0 \longrightarrow R \xrightarrow{1} R \longrightarrow 0.$$

By adding elementary perfect complexes to P_1^* we may construct a bounded, perfect complex P^* in which the modules P^n are free for all $n \geq N_0$ where N_0 is chosen so that $P_2^j = 0 = C^j$ for all $j \leq N_0$. Since an elementary perfect complex has trivial Euler characteristic we have $\chi(P^*) = \chi(P_1^*)$. In addition, by extending $f(1)$ by zero on the elementary perfect complexes which have been added we obtain a quasi-isomorphism of the form $f : P^* \longrightarrow C^*$. Now, setting $Q^* = P_2^*$, $D^* = C^*$, $j = 1$ and $g = f(2)$ in Proposition 2.2.4, we obtain a cochain homomorphism

$$h : P^* \longrightarrow P_2^*$$

with $f(2)h$ chain homotopic to f. Hence $f(2)_* h_* = f_*$ so that h is a quasi-isomorphism and

$$\chi(P_1^*) = \chi(P^*) = \chi(P_2^*) \in \mathcal{CL}(\mathbf{Z}[G])$$

by Corollary 2.2.3. □

Definition 2.2.6 Let (C^*, d) be a bounded, cohomologically trivial complex of $\mathbf{Z}[G]$-modules such that $H^i(C^*)$ is a finitely generated $\mathbf{Z}[G]$-module for each integer, i. Then the Euler characteristic, $\chi(C^*)$, of (C^*, d) is defined by

$$\chi(C^*) = \chi(P^*) \in \mathcal{CL}(\mathbf{Z}[G])$$

where $f : P^* \longrightarrow C^*$, is a quasi-isomorphism in which P^* is a bounded perfect complex of $\mathbf{Z}[G]$-modules. This definition is well defined by virtue of Theorem 2.2.5.

Example 2.2.7 *(Cf. Example 2.1.8(i))*

Let X and Y be finitely generated $\mathbf{Z}[G]$-modules and let $\alpha \in \mathrm{Ext}^n_{\mathbf{Z}[G]}(Y, X)$ be an n-extension such that the cup-product on Tate cohomology

$$(\alpha \cup -) : \hat{H}^i(G; Y) \longrightarrow \hat{H}^{i+n}(G; X)$$

is an isomorphism for all i. Suppose that α is represented by an exact sequence of the form

$$0 \longrightarrow X \longrightarrow C^0 \longrightarrow C^1 \longrightarrow \ldots \longrightarrow C^{n-1} \longrightarrow Y \longrightarrow 0.$$

Choose a projective resolution of Y of the form

$$\ldots \longrightarrow P^{-2} \longrightarrow P^{-1} \longrightarrow P^0 \longrightarrow \ldots P^{n-2} \longrightarrow P^{n-1} \longrightarrow Y \longrightarrow 0$$

and a cochain map from the resolution to the original n-extension which is the identity on Y. This cochain map induces a map from the n-extension

$$\mathrm{Ker}(d : P^0 \longrightarrow P^1) \longrightarrow P^0 \longrightarrow \ldots P^{n-1} \longrightarrow Y$$

to the original representative of α. Furthermore we may arrange that the induced map, $\tilde{f} : \mathrm{Ker}(d : P^0 \longrightarrow P^1) \longrightarrow X$, is surjective. This shows that the n-extension

$$\frac{\mathrm{Ker}(d : P^0 \longrightarrow P^1)}{\mathrm{Ker}(\tilde{f})} \longrightarrow \frac{P^0}{\mathrm{Ker}(\tilde{f})} \longrightarrow \ldots P^{n-1} \longrightarrow Y$$

also represents α.

The cup-product with α is given by splitting the n-extension up into n short exact sequences and composing the resulting n cohomology coboundary homomorphisms. From this one easily sees that $\tilde{f}_*$ is an isomorphism on $\hat{H}^i(H; -)$ for all i and all $H \subseteq G$. Therefore $\mathrm{Ker}(\tilde{f}) \subseteq P^0$ is a finitely generated, torsion-free, cohomologically trivial $\mathbf{Z}[G]$-module and hence projective [49]. Therefore

$$0 \longrightarrow \frac{P^0}{\mathrm{Ker}(\tilde{f})} \longrightarrow P^1 \longrightarrow \ldots P^{n-1} \longrightarrow 0$$

is a bounded, cohomologically trivial complex with finitely generated cohomology groups – only two are non-zero and they are isomorphic to X and Y.

The Euler characteristic associated to the n-extension, α, is equal to

$$[\frac{P^0}{\mathrm{Ker}(\tilde{f})}] - [P^1] + \ldots + (-1)^{n-1}[P^{n-1}] = [\mathrm{Ker}(\tilde{f})] \in \mathcal{CL}(\mathbf{Z}[G]).$$

From Theorem 2.2.5 it is easy to see that this Euler characteristic depends only on the class of α in $Aut_G(Y) \backslash \mathrm{Ext}^n_{\mathbf{Z}[G]}(Y, X)/Aut_G(X)$.

Proposition 2.2.8

Suppose that

$$0 \longrightarrow C_1^* \xrightarrow{f} C_2^* \xrightarrow{g} C_3^* \longrightarrow 0$$

is a short exact sequence of bounded, cohomologically trivial complexes of $\mathbf{Z}[G]$*-modules then*

$$\chi(C_1^*) + \chi(C_3^*) = \chi(C_2^*) \in \mathcal{CL}(\mathbf{Z}[G]).$$

Proof. The mapping cone complex of f (see Lemma 2.2.2) is given by

$$C(f)^{n-1} = C_2^{n-1} \oplus C_1^n \xrightarrow{d} C_2^n \oplus C_1^{n+1} = C(f)^n$$

where $d(a, b) = (d(a) + (-1)^n f(b), d(b))$ for all $a \in C_2^{n-1}, b \in C_1^n$. The homomorphisms, $\{h_{n-1} : C(f)^{n-1} \longrightarrow C_3^{n-1} \mid n \in \mathbf{Z}\}$, given by $h_{n-1}(a, b) = g(a)$ form a quasi-isomorphism of the form $h : C(f)^* \longrightarrow C_3^*$. This is by verifying that the identity maps on $H^*(C_1^*)$ and $H^*(C_2*)$ together with $\pm h_*$ send the long exact cohomology sequence for the original short exact sequence isomorphically to that of

$$0 \longrightarrow C_2^{*-1} \longrightarrow C_f^{*-1} \longrightarrow C_1^* \longrightarrow 0.$$

Since both $C(f)^*$ and C_3^* are bounded, cohomologically trivial $\mathbf{Z}[G]$-module complexes, Theorem 2.2.5 implies that $\chi(C(f)^*) = \chi(C_3^*)$. Applying Proposition 2.2.4 to $f : C_1^* \longrightarrow C_2^*$ we obtain quasi-isomorphisms from bounded, perfect complexes P_i^* $(i = 1, 2)$

$$\alpha : P_1^* \longrightarrow C_1^* \qquad \text{and} \qquad \beta :: P_2^* \longrightarrow C_2^*$$

together with cochain homomorphism

$$j : P_1^* \longrightarrow P_2^*$$

such that $f\alpha$ is chain homotopic to βj. Let $s : P_1^* \longrightarrow C_2^{*-1}$ be the cochain homotopy so that $\beta_k j_k - f_k \alpha_k = ds_k + s_{k+1}d$ for all k. The homomorphisms $\{i_{n-1} : C(j)^{n-1} \longrightarrow C(f)^{n-1}\}$ given by

$$i_{n-1}(u, v) = (\beta_{n-1}(u) + (-1)^n s_n(v), \alpha_n(v))$$

form a cochain map since

$$\begin{aligned}
d(i_{n-1}(u, v)) &= d(\beta_{n-1}(u) + (-1)^n s_n(v), \alpha_n(v)) \\
&= (d(\beta_{n-1}(u)) + (-1)^n d(s_n(v)) + (-1)^n f_n(\alpha_n(v)), d(\alpha_n(v))) \\
&= (\beta_n(d(u)) + (-1)^n \beta_n(j_n(v)) + (-1)^{n+1} s_{n+1}(d(v)), \alpha_{n+1}(d(v))) \\
&= i_n(d(u) + (-1)^n j_n(v), d(v)) \\
&= i_n(d(u, v)).
\end{aligned}$$

The induced map from the long exact cohomology sequences of $C(j)^*$ to that of $C(f)^*$ shows that they are quasi-isomorphic. Since $C(j)^*$ is a bounded, perfect complex we have, by Definition 2.2.6,

$$\begin{aligned}
\chi(C_3^*) &= \chi(C(f)^*) \\
&= \chi(C(j)^*) \\
&= \textstyle\sum_n (-1)^{n-1}([P_2^{n-1}] + [P_1^n]) \\
&= \chi(P_2^*) - \chi(P_1^*) \\
&= \chi(C_2^*) - \chi(C_1^*),
\end{aligned}$$

as required. □

2.3 Nearly perfect complexes

As usual, let G be a finite group. In this section we shall examine a generalization of the Euler characteristic of a bounded, cohomologically trivial complex of $\mathbf{Z}[G]$-modules, as promised in Example 2.1.8. If A is an abelian group, let A_{div} denote the maximal divisible subgroup of A and set $A_{\text{codiv}} = A/A_{\text{div}}$, the codivisible quotient of A.

Definition 2.3.1 A *nearly perfect complex* of $\mathbf{Z}[G]$-modules is a triple

$$(C^*, \{L_i\}_i, \{\tau_i\}_i)$$

of the following kind.

(a) C^* is a bounded, complex of cohomologically trivial $\mathbf{Z}[G]$-modules as in Example 2.1.2(ii).

$$C^* : \ldots \to C^m \to C^{m+1} \to \ldots \to C^n \to \ldots \tag{1}$$

If some of the C^i are non-zero, we let m (resp. n) be the smallest (resp. largest) integer for which $C^i \neq 0$. If all the $C^i = 0$, let $n = -1$ and $m = 0$. Define $\text{length}(C^*) = n - m + 1$.

(b) For each integer i, L_i is a torsion-free finitely generated $\mathbf{Z}[G]$-module. Let $H^i(C^*)$ be the i-th cohomology group of C^*. We require τ_i to be a $\mathbf{Z}[G]$-module isomorphism

$$\tau_i : \mathrm{Hom}_{\mathbf{Z}}(L^i, \mathbf{Q}/\mathbf{Z}) \longrightarrow H^i(C^*)_{\mathrm{div}}.$$

(c) For all i, the group $H^i(C^*)_{\mathrm{codiv}}$ is a finitely generated abelian group.

Definition 2.3.2 Two nearly perfect complexes

$$(C^*, \{L_i\}_i, \{\tau_i\}_i) \qquad \text{and} \qquad (C'^*, \{L'_i\}_i, \{\tau'_i\}_i)$$

are quasi-isomorphic if the following is true: (a) There is an isomorphism between C^* and C'^* in the derived category of the homotopy category of $\mathbf{Z}[G]$-modules (that is, a cochain map with an inverse cochain map up to chain homotopy). (b) There is a $\mathbf{Z}[G]$-module isomorphism $L'_i \to L_i$ for each i. (c) There is a commutative diagram of isomorphisms

$$\begin{array}{ccc}
\mathrm{Hom}_{\mathbf{Z}}(L^i, \mathbf{Q}/\mathbf{Z}) & \xrightarrow{\tau_i} & H^i(C^*)_{\mathrm{div}} \\
\downarrow & & \downarrow \\
\mathrm{Hom}_{\mathbf{Z}}(L'^i, \mathbf{Q}/\mathbf{Z}) & \xrightarrow{\tau'_i} & H^i(C'^*)_{\mathrm{div}}
\end{array}$$

in which the left vertical isomorphism is induced by (b) and the right one by (a).

Theorem 2.3.3 *Suppose* $(C^*, \{L_i\}_i, \{\tau_i\}_i)$ *is a nearly perfect complex.*

(a) *One can define an Euler characteristic* $\chi(C^*, \{L_i\}_i, \{\tau_i\}_i) \in K_0(\mathbf{Z}[G])$ *which depends only on the quasi-isomorphism class of* $(C^*, \{L_i\}_i, \{\tau_i\}_i)$.

(b) *The image of* $\chi(C^*, \{L_i\}_i, \{\tau_i\}_i)$ *in* $G_0(\mathbf{Z}[G])$ *(cf. Example 2.1.2(iv)) is equal to*

$$\sum_i (-1)^i \cdot ([H^i(C^*)_{\mathrm{codiv}}] - [\mathrm{Hom}_{\mathbf{Z}}(L^i, \mathbf{Z})]).$$

(c) *Suppose that the cohomology groups* $H^i(C^*)$ *are finitely generated* $\mathbf{Z}[G]$*-modules, so that the* L_i *and* τ_i *are trivial. In this case*

$$\chi(C^*, \{L_i\}_i, \{\tau_i\}_i) = \chi(C^*),$$

the Euler characteristic of Definition 2.2.6

Remark 2.3.4 Suppose R is a Dedekind ring and that the fraction field F of R is a number field. One can then replace $\mathbf{Z}$ by R and $\mathbf{Q}/\mathbf{Z}$ by F/R in Definitions 2.3.1 and 2.3.2 and in Theorem 2.3.3. This leads to Euler characteristics in $K_0(R[G])$ for nearly perfect complexes over $R[G]$.

2.3.5 The proof of Theorem 2.3.3 will occupy §§2.3.5–2.3.20.

We shall define $\chi(C^*)$ by induction on length(C^*).

Let $\delta_i : C^i \to C^{i+1}$ denote the i-th boundary map in C^*, and let $Z^i = \ker(\delta_i)$ be the i-th cycle group. Thus $Z^n = C^n$, since C^* has no terms above degree n. The 2-extension

$$Z^{n-1} \longrightarrow C^{n-1} \longrightarrow C^n \longrightarrow H^n(C^*) \qquad (2)$$

defines a class $\alpha = \alpha_n \in \mathrm{Ext}^2_{\mathbf{Z}[G]}(H^n(C^*), Z^{n-1})$. We now use α to construct a class in $\mathrm{Ext}^1_{\mathbf{Z}[G]}(M^n, Z^{n-1})$ for a suitable module, M^n.

We have a commutative diagram with short exact rows and columns of the following form:

$$\begin{array}{ccccc}
\mathrm{Hom}_{\mathbf{Z}}(L^n, \mathbf{Z}) & \longrightarrow & M^n & \longrightarrow & N^n \\
\downarrow & & \downarrow & & \downarrow \\
\mathrm{Hom}_{\mathbf{Z}}(L^n, \mathbf{Q}) & \rightarrow & \mathrm{Hom}_{\mathbf{Z}}(L^n, \mathbf{Q}) \oplus F^n & \longrightarrow & F^n \\
\downarrow & & \downarrow & & \downarrow \\
\mathrm{Hom}_{\mathbf{Z}}(L^n, \mathbf{Q}/\mathbf{Z}) & \longrightarrow & H^n(C^*) & \longrightarrow & H^n(C^*)_{\mathrm{codiv}}
\end{array}$$

To explain this diagram, note first that $\mathrm{Ext}^1_{\mathbf{Z}}(L^n, \mathbf{Z}) = 0$ since in Definition 2.3.1(b) L^n is a free finitely generated $\mathbf{Z}$-module. Hence the left column is exact. In the bottom row we have identified $\mathrm{Hom}_{\mathbf{Z}}(L^n, \mathbf{Q}/\mathbf{Z})$ with $H^n(C^*)_{\mathrm{div}}$ using the isomorphism τ_n of Definition 2.3.1(b). The module F^n is a finitely generated projective $\mathbf{Z}[G]$-module chosen to surject onto $H^n(C^*)_{\mathrm{codiv}}$, with $F^n = \{0\}$ if $H^n(C^*)_{\mathrm{codiv}} = 0$. We then lift this surjection to obtain a map to $H^n(C^*)$ to yield the middle column of diagram (3); the homomorphism $\mathrm{Hom}_{\mathbf{Z}}(L^n, \mathbf{Q}) \to H^n(C^*)$ in this column is the one compatible with the left column and the bottom row. The modules M^n and N^n are defined to be the kernels of the vertical homomorphisms, which implies that the columns of (3) are exact, by the long exact cohomology sequence associated to the short exact sequence of column complexes.

Applying the functor $\mathrm{Hom}_{\mathbf{Z}[G]}(-, Z^{n-1})$ to the middle column of (3) gives a long exact sequence of the form

$$\begin{aligned} \ldots \longrightarrow \mathrm{Ext}^1_{\mathbf{Z}[G]}(\mathrm{Hom}(L^n, \mathbf{Q}) \oplus F^n, Z^{n-1}) \longrightarrow \mathrm{Ext}^1_{\mathbf{Z}[G]}(M^n, Z^{n-1}) \\ \xrightarrow{\phi_1} \mathrm{Ext}^2_{\mathbf{Z}[G]}(H^n(C^*), Z^{n-1}) \longrightarrow \mathrm{Ext}^2_{\mathbf{Z}[G]}(\mathrm{Hom}(L^n, \mathbf{Q}) \oplus F^n, Z^{n-1}) \ldots \end{aligned} \tag{4}$$

Lemma 2.3.6

The module M^n in (3) is a finitely generated torsion-free $\mathbf{Z}[G]$-module and the boundary map

$$\mathrm{Ext}^1_{\mathbf{Z}[G]}(M^n, Z^{n-1}) \xrightarrow{\phi_1} \mathrm{Ext}^2_{\mathbf{Z}[G]}(H^n(C^*), Z^{n-1}) \tag{5}$$

in (4) is an isomorphism.

Proof. Since N^n and $\mathrm{Hom}_{\mathbf{Z}}(L^n, \mathbf{Z})$ are finitely generated abelian groups, diagram (3) shows M^n is also and since $\mathrm{Hom}_{\mathbf{Z}}(L^n, \mathbf{Q}) \oplus F^n$ is torsion-free, so is M^n.

To analyse (4) we use the spectral sequence

$$H^p(G; \mathrm{Ext}^q_{\mathbf{Z}}(D_1, D_2)) => \mathrm{Ext}^{p+q}_{\mathbf{Z}[G]}(D_1, D_2) \tag{6}$$

for $\mathbf{Z}[G]$-modules D_1 and D_2. The projective dimension of $\mathbf{Z}$ is one ([83] p. 171 and 191). Therefore

$$\mathrm{Ext}^q_{\mathbf{Z}}(D_1, D_2) = 0 \text{ if } q \geq 2. \tag{7}$$

If D_1 is finitely generated and torsion free, then $\mathrm{Ext}^1_{\mathbf{Z}}(D_1, D_2) = 0$. If D_1 is a finitely generated projective $\mathbf{Z}[G]$-module, then $\mathrm{Hom}_{\mathbf{Z}}(D_1, D_2)$ is a summand of an induced $\mathbf{Z}[G]$-module, so $\mathrm{Hom}_{\mathbf{Z}}(D_1, D_2)$ is cohomologically trivial. Any $\mathbf{Q}$-vector space is also a cohomologically trivial $\mathbf{Z}[G]$-module.

One finds from (6), (7) and the above remarks that if $q > 1$ then

$$\begin{aligned} &\mathrm{Ext}^q_{\mathbf{Z}[G]}(\mathrm{Hom}(L^n, \mathbf{Q}) \oplus F^n, Z^{n-1}) \\ &\quad = \mathrm{Ext}^q_{\mathbf{Z}[G]}(\mathrm{Hom}(L^n, \mathbf{Q}), Z^{n-1}) \oplus \mathrm{Ext}^q_{\mathbf{Z}[G]}(F^n, Z^{n-1}) = 0 \end{aligned} \tag{8}$$

and

$$\mathrm{Ext}^1_{\mathbf{Z}[G]}(M^n, Z^{n-1}) = H^1(G; \mathrm{Hom}_{\mathbf{Z}}(M^n, Z^{n-1})). \tag{9}$$

The homomorphism

$$\mathrm{Ext}^1_{\mathbf{Z}[G]}(\mathrm{Hom}(L^n, \mathbf{Q}) \oplus F^n, Z^{n-1}) \longrightarrow \mathrm{Ext}^1_{\mathbf{Z}[G]}(M^n, Z^{n-1}) \tag{10}$$

in the long exact sequence (4) is trivial, since its domain is a $\mathbf{Q}$-vector space and, by (9), its range is a group annihilated by the order of G. Thus (4), (9) and (10) establish the isomorphism stated in Lemma 2.3.6. □

Corollary 2.3.7

There is a unique class, $\beta \in \mathrm{Ext}^1_{\mathbf{Z}[G]}(M^n, Z^{n-1})$, which maps to the 2-extension $\alpha \in \mathrm{Ext}^2_{\mathbf{Z}[G]}(H^n(C^*), Z^{n-1})$ of the sequence (2) under the boundary isomorphism ϕ_1 of Lemma 2.3.6.

Definition 2.3.8 Let

$$0 \longrightarrow Z^{n-1} \longrightarrow D^{n-1} \longrightarrow M^n \longrightarrow 0 \tag{11}$$

be an exact sequence representing the extension class β of Corollary 2.3.7. Let $D^i = C^i$ if $i < n-1$, and let $D^i = 0$ for $i > n-1$. Let

$$D^* : \ldots \longrightarrow D^m \longrightarrow D^{m+1} \longrightarrow \ldots \longrightarrow D^{n-1} \longrightarrow 0 \longrightarrow \ldots$$

be the complex such that the boundary $\lambda_i : D^i \longrightarrow D^{i+1}$ is as follows. If $i < n-2$ then λ_i is the boundary map $\delta_i : C^i \longrightarrow C^{i+1}$ of C^*. If $i \geq n-1$, then λ_i is the zero homomorphism. Finally, if $i = n-2$ then $\lambda_{n-2} : D^{n-2} = C^{n-2} \longrightarrow D^{n-1}$ is the composition of the boundary map $\delta_{n-2} : C^{n-2} \longrightarrow Z^{n-1}$ with the inclusion $Z^{n-1} \longrightarrow D^{n-1}$ in sequence (11).

Proposition 2.3.9

The $\mathbf{Z}[G]$-module, D^{n-1}, is cohomologically trivial.

Proof. By Corollary 2.3.7, the cup product of β with the extension class of the middle column

$$0 \longrightarrow M^n \longrightarrow \mathrm{Hom}_{\mathbf{Z}}(L^n, \mathbf{Q}) \oplus F^n \longrightarrow H^n(C^*) \longrightarrow 0 \tag{12}$$

of (3) is ± 1 times the class, α, of sequence (2). Therefore splicing together (11) and (12) gives an exact sequence

$$0 \longrightarrow Z^{n-1} \longrightarrow D^{n-1} \longrightarrow \mathrm{Hom}_{\mathbf{Z}}(L^n, \mathbf{Q}) \oplus F^n \longrightarrow H^n(C^*) \longrightarrow 0 \tag{13}$$

which has extension class $\pm\alpha$. Since C^{n-1} and C^n in (2) were assumed to be cohomologically trivial $\mathbf{Z}[G]$-modules, cup product with $\pm\alpha$ induces isomorphisms in Tate cohomology

$$\hat{H}^j(\Gamma; H^n(C^*)) \longrightarrow \hat{H}^{j+2}(\Gamma; Z^{n-1}) \tag{14}$$

for all integers j and all subgroups $\Gamma \subseteq G$. Since $\mathrm{Hom}_{\mathbf{Z}}(L^n, \mathbf{Q}) \oplus F^n$ in sequence (13) is a cohomologically trivial $\mathbf{Z}[G]$-module, this implies D^{n-1} in (13) must also be cohomologically trivial. □

Corollary 2.3.10

Suppose $n = m$ in equation (1), so that $C^i = 0$ if $i \neq n$ and $C^n \neq 0$. Then $D^i = 0$ unless $i = n-1$. The module $D^{n-1} = M^n$ is a finitely generated projective $\mathbf{Z}[G]$-module, and

$$\begin{aligned} &\mathrm{rank}_{\mathbf{Z}[G]}(D^{n-1}) \\ &= \mathrm{rank}_{\mathbf{Z}[G]}(F^n) + |G|^{-1} \cdot (\mathrm{rank}_{\mathbf{Z}}(L^n) - \mathrm{rank}_{\mathbf{Z}}(H^n(C^*)_{\mathrm{codiv}})) \end{aligned} \tag{15}$$

where $C^n = H^n(C^*)$.

Proof. If $m = n$ then $Z^{n-1} = 0$. Hence sequence (11) shows $D^{n-1} = M^n$. By Lemma 2.3.6, M^n is a finitely generated torsion-free $\mathbf{Z}[G]$-module, and D^{n-1} is cohomologically trivial by Proposition 2.3.9. Hence $D^{n-1} = M^n$ must be a finitely generated, projective $\mathbf{Z}[G]$-module. The equality (15) follows from $D^{n-1} = M^n$, the top row and right column of (3), $\mathrm{rank}_{\mathbf{Z}}(\mathrm{Hom}_{\mathbf{Z}}(L^n, \mathbf{Z})) = \mathrm{rank}_{\mathbf{Z}}(L^n)$ and the fact that finitely generated projective $\mathbf{Z}[G]$-modules are locally free. □

Lemma 2.3.11

If $i < n-1$ then $H^i(D^*) = H^i(C^*)$ and if $i > n-1$ then $H^i(D^*) = 0$. Finally, there is an exact sequence of the form

$$0 \longrightarrow H^{n-1}(C^*) \longrightarrow H^{n-1}(D^*) \longrightarrow M^n \longrightarrow 0. \tag{16}$$

Proof. In view of (11), the exact sequence (16) is

$$0 \longrightarrow Z^{n-1}/B^{n-1} \longrightarrow D^{n-1}/B^{n-1} \longrightarrow D^{n-1}/Z^{n-1} \longrightarrow 0 \tag{17}$$

where $B^{n-1} \subset Z^{n-1}$ denotes the group of $n-1$ boundaries of both C^* and D^*. The rest of the result is clear. □

Corollary 2.3.12

Suppose $(C^*, \{L_i\}_i, \{\tau_i\}_i)$ is a nearly perfect complex, as in Definition 2.3.1. Let $L'_i = L_i$ if $i \le n-1$, and let $L'_i = 0$ if $i > n-1$. Define $\tau'_i = \tau_i :$ $\mathrm{Hom}_{\mathbf{Z}}(L_i, \mathbf{Q}/\mathbf{Z}) \longrightarrow H^i(C^*)_{\mathrm{div}} = H^i(D^*)_{\mathrm{div}}$ if $i \le n-1$, and let $\tau'_i = 0$ if $i > n-1$. Then $(D^*, \{L'_i\}_i, \{\tau'_i\})$ is a nearly perfect complex. If $n > m$, then

$$D^* : \ldots \longrightarrow 0 \longrightarrow D^m \longrightarrow \ldots D^{n-1} \longrightarrow 0 \longrightarrow \ldots \tag{18}$$

has $\mathrm{length}(D^*) \le n - m < \mathrm{length}(C^*) = n - m + 1$.

Proof. Since M^n is finitely generated by Lemma 2.3.6, sequence (16) gives an isomorphism

$$H^{n-1}(C^*)_{\mathrm{div}} \xrightarrow{\cong} H^{n-1}(D^*)_{\mathrm{div}}.$$

The Corollary is now clear from Lemma 2.3.11. □

Definition 2.3.13 If $\mathrm{length}(C^\bullet) = n - m + 1 = 0$, then C^* and D^* are both the zero complex and we let

$$\chi(C^*, \{L_i\}_i, \{\tau_i\}_i) = 0. \tag{19}$$

Suppose now that $\mathrm{length}(C^\bullet) = n - m + 1 = 1$. By Corollary 2.3.10, D^{n-1} has a class, $[D^{n-1}]$ in $K_0(\mathbf{Z}[G])$, so we can let

$$\chi(C^*, \{L_i\}_i, \{\tau_i\}_i) = (-1)^{n-1} \cdot [D^{n-1}] + (-1)^n \cdot [F^n]. \tag{20}$$

We now assume by induction that $n_0 \ge 0$ is an integer such that $\chi(C^*, \{L_i\}_i, \{\tau_i\}_i)$ has been defined whenever $\mathrm{length}(C^\bullet) = n - m + 1 \le n_0$. If $\mathrm{length}(C^*) = n_0 + 1$ then, using Corollary 2.3.12, we can define

$$\chi(C^*, \{L_i\}_i, \{\tau_i\}_i) = \chi(D^*, \{L'_i\}_i, \{\tau'_i\}_i) + (-1)^n \cdot [F_n]. \tag{21}$$

Proposition 2.3.14

The nearly perfect complex $(D^*, \{L_i'\}_i, \{\tau_i'\}_i)$ *depends on* $(C^*, \{L_i\}_i, \{\tau_i\}_i)$, *the projective module* F^n *and the homomorphism* $\lambda : F^n \to H^n(C^*)$ *induced by the middle column of diagram (3).*

The class $\chi(D^*, \{L_i'\}_i, \{\tau_i'\}_i) + (-1)^n \cdot [F_n]$ *does not depend on the choice of* F^n *or* $\lambda : F^n \to H^n(C^*)$. *Therefore in Definition 2.3.13,* $\chi(C^*, \{L_i\}_i, \{\tau_i\}_i)$ *depends only on the nearly perfect complex* $(C^*, \{L_i\}_i, \{\tau_i\}_i)$ *and not on further choices.*

2.3.15 Before beginning the proof of Proposition 2.3.14 we observe that the choice of F^n and $\lambda : F^n \to H^n(C^\bullet)$ determines the diagram (3). This fixes the long exact sequence (4) and so, from Corollary 2.3.7 and Definition 2.3.8, we see that $(D^*, \{L_i'\}_i, \{\tau_i'\}_i)$ is also determined by these choices together with the original nearly perfect complex $(C^*, \{L_i\}_i, \{\tau_i\}_i)$.

If C^* has length 0, then $\chi(C^*, \{L_i\}_i, \{\tau_i\}_i) = 0$ so Proposition 2.3.14 holds. We now assume C^* has strictly positive length and that, by induction, Proposition 2.3.14 is true for all nearly perfect complexes of length less than that of C^*.

In the course of the proof, we will show the following, by induction:

Hypothesis 2.3.16 Suppose $(C_1^*, \{L_{1,i}\}_i, \{\tau_{1,i}\}_i)$ and $(C_2^*, \{L_{2,i}\}_i, \{\tau_{2,i}\}_i)$ are nearly perfect complexes having the following properties.

(a) The length of each of these complexes is less than that of C^*.

(b) There is a morphism of complexes $C_1^* \longrightarrow C_2^*$ which is a term by term isomorphism.

(c) There are isomorphisms $L_{1,i} \longrightarrow L_{2,i}$ which are compatible with the cohomology isomorphisms $H^i(C_1^*) \longrightarrow H^i(C_2^*)$ induced by the morphism in (b), the $\tau_{1,i}$'s and the $\tau_{2,i}$'s. Then

$$\chi(C_2^*, \{L_{2,i}\}_i, \{\tau_{2,i}\}_i) = \chi(C_1^*, \{L_{1,i}\}_i, \{\tau_{1,i}\}_i) \tag{22}$$

where the two sides of this equality are well defined because, by assumption, Proposition 2.3.14 holds for complexes of length less than that of C^*.

This induction hypothesis is clearly true if C^* has length 1, since then C_1^* and C_2^* have length 0.

Lemma 2.3.17

Let $\tilde{\lambda} : F^n \longrightarrow H^n(C^*)_{\text{codiv}}$ *be the homomorphism induced by* $\lambda : F^n \to H^n(C^*)$, *so that* $\tilde{\lambda}$ *appears in the right column of diagram (3). Suppose* $\lambda' : F^n \to H^n(C^*)$ *is another homomorphism such that* $\tilde{\lambda} = \tilde{\lambda}'$. *Then assuming Hypothesis 2.3.16,* λ *and* λ' *lead to the same value for* $\chi(C^*, \{L_i\}_i, \{\tau_i\}_i)$. *Therefore* $\chi(C^\bullet, \{L_i\}_i, \{\tau_i\}_i)$ *depends only on the (surjective) homomorphism* $\tilde{\lambda} : F^n \longrightarrow H^n(C^*)_{\text{codiv}}$.

Proof. From diagram (3), we see that there is a homomorphism $\mu : F^n \longrightarrow \operatorname{Hom}_{\mathbf{Z}}(L^n, \mathbf{Q}/\mathbf{Z})$ such that $\lambda = \lambda' + i \cdot \mu$, where $i : \operatorname{Hom}_{\mathbf{Z}}(L^n, \mathbf{Q}/\mathbf{Z}) \longrightarrow H^n(C^*)$ is the inclusion in the bottom row of (3). Since F^n is projective and the homomorphism $\pi : \operatorname{Hom}_{\mathbf{Z}}(L^n, \mathbf{Q}) \longrightarrow \operatorname{Hom}_{\mathbf{Z}}(L^n, \mathbf{Q}/\mathbf{Z})$ in the left column of (3) is surjective, we can lift μ to a homomorphism $\mu' : F^n \rightarrow \operatorname{Hom}_{\mathbf{Z}}(L^n, \mathbf{Q})$ such that

$$\lambda = \lambda' + i \cdot \pi \cdot \mu'. \tag{23}$$

Let T be the automorphism of $\operatorname{Hom}_{\mathbf{Z}}(L^n, \mathbf{Q}) \oplus F^n$ defined by

$$T(a \oplus b) = (a + \mu'(b)) \oplus b. \tag{24}$$

Define $(3)'$ to be the diagram (3) when λ' is used instead of λ. Then we see that there is a unique isomorphism $\tilde{T}$ from diagram (3) to diagram $(3)'$ which fixes the left and right columns and which is the automorphism T on the module $\operatorname{Hom}_{\mathbf{Z}}(L^n, \mathbf{Q}) \oplus F^n$ in the middle of the diagram. This $\tilde{T}$ induces an automorphism of $H^n(C^*)$ and an isomorphism $M^n \longrightarrow M'^n$ from M^n to the corresponding module M'^n in diagram $(3)'$. Following through the construction in Definition 2.3.8, we see that there is a commutative diagram

$$\begin{array}{ccccccccc} 0 & \longrightarrow & Z^{n-1} & \longrightarrow & D^{n-1} & \longrightarrow & M^n & \longrightarrow & 0 \\ & & \downarrow & & \downarrow & & \downarrow & & \\ 0 & \longrightarrow & Z^{n-1} & \longrightarrow & D'^{n-1} & \longrightarrow & M'^n & \longrightarrow & 0 \end{array} \tag{25}$$

in which the left vertical homomorphism is the identity map, the right vertical homomorphism is the above isomorphism $M^n \rightarrow M'^n$, and the middle homomorphism is an isomorphism.

Let $(D^*, \{L_{1,i}\}_i, \{\tau_{1,i}\}_i)$ and $(D'^*, \{L_{2,i}\}_i, \{\tau_{2,i}\}_i)$ be the nearly perfect complex structures resulting from Definition 2.3.8 when one uses λ and λ', respectively. Via (25) we get a morphism of complexes $D^* \longrightarrow D'^*$ which is the identity on all terms except D^{n-1}, and which is the middle vertical isomorphism of (25) on D^{n-1}. This morphism induces a commutative diagram

$$\begin{array}{ccccccccc} 0 & \longrightarrow & H^{n-1}(C^*) & \longrightarrow & H^{n-1}(D^*) & \longrightarrow & M^n & \longrightarrow & 0 \\ & & \downarrow & & \downarrow & & \downarrow & & \\ 0 & \longrightarrow & H^{n-1}(C^*) & \longrightarrow & H^{n-1}(D'^*) & \longrightarrow & M'^n & \longrightarrow & 0 \end{array} \tag{26}$$

in which the left and right isomorphisms are those induced by $\tilde{T}$. By construction $\tilde{T}$ induces the identity isomorphism of

$$H^{n-1}(C^*)_{\text{div}} = H^{n-1}(D^*)_{\text{div}} = H^{n-1}(D'^*)_{\text{div}}.$$

Hence the identity isomorphism $L_{1,i} \longrightarrow L_{2,i}$ for each i is compatible with $D^* \longrightarrow D'^*$, $\tau_{1,i}$ and $\tau_{2,i}$. It follows now from Hypothesis 2.3.16 that

$$\begin{aligned} &\chi(D^*, \{L_{2,i}\}_i, \{\tau_{2,i}\}_i) + (-1)^n \cdot [F^n] \\ &= \chi(D'^*, \{L_{1,i}\}_i, \{\tau_{1,i}\}_i) + (-1)^n \cdot [F^n]. \end{aligned} \tag{27}$$

Therefore we get the same value for $\chi(C^*, \{L_i\}_i, \{\tau_i\}_i)$ whether we use λ or λ', which completes the proof of Lemma 2.3.17. □

2.3.18 (Proof of Proposition 2.3.14 and Hypothesis 2.3.16)

Let F_1^n be another choice of a projective $\mathbf{Z}[G]$-module and a surjective homomorphism $\lambda_1 : F_1^n \to H^n(C^*)$. Since both of the induced homomorphisms $\tilde{\lambda} : F^n \to H^n(C^*)_{\text{codiv}}$ and $\tilde{\lambda}_1 : F_1^n \to H^n(C^*)_{\text{codiv}}$ are surjective, we can form a pull back square

$$\begin{array}{ccc} F^n \oplus F_1^n & \longrightarrow & F_1^n \\ \downarrow & & \downarrow \tilde{\lambda}_1 \\ F^n & \xrightarrow{\tilde{\lambda}} & H^n(C^*)_{\text{codiv}}. \end{array} \tag{28}$$

It follows from Lemma 2.3.17 that in order to show $\tilde{\lambda}$ and $\tilde{\lambda}_1$ yield the same value for $\chi(C^*, \{L_i\}_i, \{\tau_i\}_i)$ we may reduce to the case in which $\tilde{\lambda}_1$ is the composition of a surjection $h : F_1^n \longrightarrow F^n$ and $\tilde{\lambda}$. Since the value of $\chi(C^*, \{L_i\}_i, \{\tau_i\}_i)$ does not depend on how $\tilde{\lambda}$ is lifted to λ, we can further assume that $\lambda_1 : F_1^n \longrightarrow H^n(C^*)$ is the composition of h and $\lambda : F^n \longrightarrow H^n(C^*)$.

Define $(3)_1$ to be the diagram (3) which results from using λ_1 instead of λ, and let M_1^n and N_1^n be the modules appearing in $(3)_1$ at the positions corresponding to M^n and N^n. We have a commutative exact diagram

$$\begin{array}{ccccccccc} & & 0 & \longrightarrow & 0 & \longrightarrow & 0 & & \\ & & \downarrow & & \downarrow & & \downarrow & & \\ 0 & \longrightarrow & F_2 & \longrightarrow & N_1^n & \longrightarrow & N^n & \longrightarrow & 0 \\ & & \downarrow & & \downarrow & & \downarrow & & \\ 0 & \longrightarrow & F_2 & \longrightarrow & F_1^n & \longrightarrow & F^n & \longrightarrow & 0 \\ & & \downarrow & & \downarrow \tilde{\lambda}_1 & & \downarrow \tilde{\lambda} & & \\ 0 & \longrightarrow & 0 & \longrightarrow & H^n(C^*)_{\text{div}} & \longrightarrow & H^n(C^*)_{\text{div}} & \longrightarrow & 0 \\ & & \downarrow & & \downarrow & & \downarrow & & \\ & & 0 & \longrightarrow & 0 & \longrightarrow & 0 & & \end{array} \tag{29}$$

in which the right columns are the right columns of diagrams $(3)_1$ and (3). The middle row of (29) splits because F^n is projective. Hence the top row of (29) also splits and F_2 is projective because F^n and F_1^n are. Thus we have compatible isomorphisms $N_1^n = N^n \oplus F_2$ and $M_1^n = M^n \oplus F_2$. Using the fact that F_2 is projective, we see that the construction of Definition 2.3.8 leads to an isomorphism $D_1^{n-1} \cong D^{n-1} \oplus F_2$, where D_1^* is the complex which results from using λ_1 instead of λ. The homomorphism $D_1^{n-2} = D^{n-2} = C^{n-2} \longrightarrow D_1^{n-1}$ is the composition of $D^{n-2} \longrightarrow D^{n-1}$ and the inclusion $D^{n-1} \longrightarrow D_1^{n-1} = D^{n-1} \oplus F_2$ which is the identity map onto the first summand of D^{n-1}. If $F_2 = 0$, then $\tilde{\lambda} = \tilde{\lambda}_1$ and there is nothing to prove because of Lemma 2.3.17. So we assume in what follows that F_2 is non-trivial.

We see from this description that $H^{n-1}(D_1^*) = H^{n-1}(D^*) \oplus F_2$ and that the nearly perfect complex structure $(D_1^*, \{L_i'\}_i, \{\tau_i'\}_i)$ of D_1^* is the one which results from $(D^*, \{L_i'\}_i, \{\tau_i'\}_i)$ by simply adding the projective module F_2 to D^{n-1}.

In view of Definition 2.3.13 and (29), the expressions for $\chi(C^*, \{L_i\}_i, \{\tau_i\}_i)$ which result from using $\tilde{\lambda}_1$ and $\tilde{\lambda}$, respectively, are

$$\chi(D_1^*, \{L_i'\}_i, \{\tau_i'\}_i) + (-1)^n \cdot ([F^n] + [F_2]) \tag{30}$$

and

$$\chi(D^*, \{L_i'\}_i, \{\tau_i'\}_i) + (-1)^n \cdot ([F^n]). \tag{31}$$

We will now show that (30) and (31) are equal.

Choose a free module F_3 and a G-morphism $\lambda_3 : F_3 \longrightarrow H^{n-1}(D^*)$ which induces a surjection $F_3 \longrightarrow H^{n-1}(D^*)_{\mathrm{codiv}}$. We further require $F_3 = \{0\}$ if $H^{n-1}(D^*)_{\mathrm{codiv}} = 0$. We compute $(D_1^*, \{L_i'\}_i, \{\tau_i'\}_i)$ by choosing the projective module $F_3 \oplus F_2$ together with the homomorphism

$$\lambda_4 : F_3 \oplus F_2 \longrightarrow H^{n-1}(D_1^*) = H^{n-1}(D^*) \oplus F_2$$

which is the identity map on the second summand and λ_3 on the first.

From the diagram of the form (3) which results and from Definitions 2.3.8 and 2.3.13, we see that there is a single nearly perfect complex $(D_2^*, \{L_i''\}_i, \{\tau_i''\}_i)$ of length less than length(C^*) such that

$$\chi(D^*, \{L_i'\}_i, \{\tau_i'\}_i) = \chi(D_2^*, \{L_i''\}_i, \{\tau_i''\}_i) + (-1)^{n-1} \cdot [F_3'] \tag{32}$$

and

$$\chi(D'^*, \{L_i'\}_i, \{\tau_i'\}_i) = \chi(D_2^*, \{L_i''\}_i, \{\tau_i''\}_i) + (-1)^{n-1} \cdot [F_3' \bigoplus F_2] \tag{33}$$

where $F_3' = D^{n-1}$ is projective if length$(C^*) = 1$, and $F_3' = F_3$ if length$(C^*) > 1$. Here we may use Hypothesis 2.3.16 to assert that $\chi(D_2^*, \{L_i''\}_i, \{\tau_i''\}_i)$ has the same value on the right sides of (32) and (33). It follows easily from (32) and (33) that (30) and (31) are equal.

The proof of Proposition 2.3.14 will now be complete once we show that our equation (22) of our inductive Hypothesis 2.3.16 holds for nearly perfect complexes $(C_1^*, \{L_{1,i}\}_i, \{\tau_{1,i}\}_i)$ and $(C_2^*, \{L_{2,i}\}_i, \{\tau_{2,i}\}_i)$ of the same length as C^*. This may be proved by using the inductive definition of $\chi(C_j^*, \{L_{j,i}\}_i, \{\tau_{j,i}\}_i)$ to reduce to the case of complexes of length less than that of C^*; we will leave the details to the reader. □

The proof of part(a) of Theorem 2.3.3 is now completed by the following result.

Proposition 2.3.19

Suppose that $(C^*, \{L_i\}_i, \{\tau_i\}_i)$ and $(C'^*, \{L_i'\}_i, \{\tau_i'\}_i)$ are quasi-isomorphic in the sense of Definition 2.3.2. Then $\chi(C^*, \{L_i\}_i, \{\tau_i\}_i) = \chi(C'^*, \{L_i'\}_i, \{\tau_i'\}_i)$.

Proof. By ([71] §1.3), we can reduce to the case in which there is a morphism of complexes $C^* \longrightarrow C'^*$ which gives rise to the quasi-isomorphism in the derived category which is referred to in Definition 2.3.2(a).

A shift by 1 to either the left or the right of the terms of $(C^*, \{L_i\}_i, \{\tau_i\}_i)$ multiplies $\chi(C^*, \{L_i\}_i, \{\tau_i\}_i)$ by -1. Thus we can reduce to the case in which at least one of C^* or C'^* is not the zero complex, $C^i = C'^i = 0$ if $i < 0$, and $C'^0 \neq 0$ or $C^0 \neq 0$. Let $n' \geq 0$ (resp. $n \geq 0$) be the largest non-negative integer for which $C'^n \neq 0$ (resp. $C^n \neq 0$). Suppose $n' > n$. We can add a non-zero finitely generated free module F to both C^n and $C^{n+1} = 0$ and use the identity map $F \longrightarrow F$ to construct a perfect complex $(C_2^*, \{L_i\}_i, \{\tau_i\}_i)$ from $(C^*, \{L_i\}_i, \{\tau_i\}_i)$ which has length one greater than $(C^*, \{L_i\}_i, \{\tau_i\}_i)$. Since $H^{n+1}(C_2^*) = 0$, we can use the trivial free module surjecting onto $H^{n+1}(C_2^*)$ to compute $\chi(C_2^*, \{L_i\}_i, \{\tau_i\}_i)$ via the counterpart of diagram (3). This leads to $\chi(C_2^*, \{L_i\}_i, \{\tau_i\}_i) = \chi(C^*, \{L_i\}_i, \{\tau_i\}_i)$. In this way we can increase the length of C^* to be able to assume $n' = n$; the case in which $n > n'$ initially is treated similarly.

We will prove Proposition 2.3.19 by induction on $n = n'$. Suppose $n = n' = 0$. Then C'^0 and C^0 are the only non-zero terms of C'^* and C^*, respectively. The morphism $C^* \longrightarrow C'^*$ must induce an isomorphism $C^0 = H^0(C^*) \longrightarrow H^0(C'^*) = C'^0$. Hence the case $n = n' = 0$ is treated by Hypothesis 2.3.16, which was proved in the course of the proof of Proposition 2.3.14.

We now suppose $n = n' > 0$ and that Proposition 2.3.19 is true for all pairs of complexes which have trivial terms in negative dimension and highest terms in dimension less than n. As in the definition of $\chi(C^*, \{L_i\}_i, \{\tau_i\}_i)$, we choose a free module F^n together with a morphism $F^n \longrightarrow H^n(C^*)$ inducing a surjection $F^n \longrightarrow H^n(C^*)_{\text{codiv}}$. We now use the isomorphisms $H^n(C^*) \longrightarrow H^n(C'^*)$ and $L^n \longrightarrow L'^n$ in Definition 2.3.2 to take the diagram (3) used to define $\chi(C^*, \{L_i\}_i, \{\tau_i\}_i)$ to the diagram (3)$'$ used to define $\chi(C'^*, \{L'_i\}_i, \{\tau'_i\}_i)$. In particular, we have an induced isomorphism $M^n \longrightarrow M'^{n-1}$. Via Lemma 2.3.6, Corollary 2.3.7 and Definition 2.3.8, this leads to a diagram

$$\begin{array}{ccccccccc} 0 & \longrightarrow & Z^{n-1} & \longrightarrow & D^{n-1} & \longrightarrow & M^n & \longrightarrow & 0 \\ & & \downarrow & & \downarrow & & \downarrow & & \\ 0 & \longrightarrow & Z'^{n-1} & \longrightarrow & D'^{n-1} & \longrightarrow & M'^n & \longrightarrow & 0 \end{array} \tag{34}$$

in which the right vertical arrow is an isomorphism. The morphism of complexes $C^* \longrightarrow C'^*$ together with (34) now gives a morphism of complexes $D^* \longrightarrow D'^*$ which induces cohomology isomorphisms $H^i(D^*) \cong H^i(D'^*)$ if $i \neq n-1$. When $i = n-1$, one has a commutative diagram

$$\begin{array}{ccccccccc} 0 & \longrightarrow & H^{n-1}(C^*) & \longrightarrow & H^{n-1}(D^*) & \longrightarrow & M^n & \longrightarrow & 0 \\ & & \downarrow & & \downarrow & & \downarrow & & \\ 0 & \longrightarrow & H^{n-1}(C'^*) & \longrightarrow & H^{n-1}(D'^*) & \longrightarrow & M'^n & \longrightarrow & 0 \end{array} \tag{35}$$

in which the left and right vertical homomorphisms are isomorphisms, so the middle vertical homomorphism is as well. This proves $D^* \longrightarrow D'^*$ is a quasi-isomorphism. The diagram (35) also shows that the nearly perfect complex structures on D^* and D'^* which are constructed in Corollary 2.3.12 are quasi-isomorphic in the sense of Definition 2.3.2. Since D^* and D'^* have highest non-zero terms in degree less than $n = n'$, the proof of Proposition 2.3.19 now follows by induction. □

2.3.20 (Completion of the proof of Theorem 2.3.3) *As noted before, Proposition 2.3.19 shows part (a) of Theorem 2.3.3. We now show parts (b) and (c) of Theorem 2.3.3 by induction on* $\text{length}(C^*) = n - m + 1$.

If $\text{length}(C^*) < 0$ *then* C^* *is the zero complex and* $\chi(C^*, \{L_i\}_i, \{\tau_i\}_i) = 0$ *by (19).*

From diagram (3) we have

$$[M^n] = [\text{Hom}_{\mathbf{Z}}(L^n, \mathbf{Z})] + [F^n] - [H^n(C^*)_{\text{codiv}}] \in G_0(\mathbf{Z}[G]). \tag{36}$$

Suppose $\text{length}(C^*) = 1$. By Corollary 2.3.10, $C^i = 0$ if $i \neq n$, and $H^n(C^*) = C^n \neq 0$. Furthermore, $D^{n-1} = M^n$ is a projective $\mathbf{Z}[G]$-module, and

$$\chi(C^*, \{L_i\}_i, \{\tau_i\}_i) = (-1)^{n-1} \cdot [M^n] + (-1)^n \cdot [F^n] \tag{37}$$

by (20). Combining (36) and (37) shows $\chi(C^*, \{L_i\}_i, \{\tau_i\}_i)$ has image

$$(-1)^{n-1} \cdot [M^n] + (-1)^n \cdot [F^n] = (-1)^n \cdot ([H^n(C^*)_{\text{codiv}}] - [\text{Hom}_{\mathbf{Z}}(L^n, \mathbf{Z})])$$

in $G_0(\mathbf{Z}[G])$ as asserted in Theorem 2.3.3(b). Suppose $H^n(C^*) = C^n$ is finitely generated as a $\mathbf{Z}[G]$-module. Then $L^n = 0$, and diagram (3) shows $[F^n] - [M^n] = [H^n(C^*)] = [C^n]$ in $K_0(\mathbf{Z}[G])$ when we identify $K_0(\mathbf{Z}[G])$ with the Grothendieck group of all finitely generated cohomologically trivial $\mathbf{Z}[G]$-modules, as in Example 2.1.2(ii). Thus (37) gives $\chi(C^*, \{L_i\}_i, \{\tau_i\}_i) = (-1)^n \cdot [C^n]$ in this case, which by Schanuel's Lemma [49] is equivalent to the assertion of Theorem 2.3.3(c).

Finally, suppose $\text{length}(C^*) > 1$, and let $(D^*, \{L_i'\}_i, \{\tau_i'\}_i)$ be a nearly perfect complex of the kind constructed in Corollary 2.3.12. By Definition 2.3.13,

$$\chi(C^*, \{L_i\}_i, \{\tau_i\}_i) = \chi(D^*, \{L_i'\}_i, \{\tau_i'\}_i) + (-1)^n \cdot [F_n]. \tag{38}$$

By induction, the image of $\chi(D^*, \{L_i'\}_i, \{\tau_i'\}_i)$ in $G_0(\mathbf{Z}[G])$ is

$$\sum_i (-1)^i \cdot ([H^i(D^*)_{\text{codiv}}] - [\text{Hom}_{\mathbf{Z}}(L'^i, \mathbf{Z})]).$$

In view of the construction of $(D^*, \{L_i'\}_i, \{\tau_i'\}_i)$, this equals

$$\begin{aligned} &(-1)^{n-1} \cdot ([H^{n-1}(D^*)_{\text{codiv}}] - [\text{Hom}_{\mathbf{Z}}(L^{n-1}, \mathbf{Z})]) \\ &\quad + \textstyle\sum_{i<n-1} (-1)^i \cdot ([H^i(C^*)_{\text{codiv}}] - [\text{Hom}_{\mathbf{Z}}(L^i, \mathbf{Z})]). \end{aligned} \tag{39}$$

By the exact sequence (16), we have

$$[H^{n-1}(D^*)_{\text{codiv}}] = [H^{n-1}(C^*)_{\text{codiv}}] - [M^n] \tag{40}$$

since M^n is finitely generated. Hence, on combining (38), (39) and (40), we see that $\chi(C^*, \{L_i\}_i, \{\tau_i\}_i)$ has image in $G_0(\mathbf{Z}[G])$ equal to

$$\sum_i (-1)^i \cdot ([H^i(C^*)_{\text{codiv}}] - [\text{Hom}_{\mathbf{Z}}(L^i, \mathbf{Z})])$$

as asserted in Theorem 2.3.3(b).

Suppose now that all the $H^i(C^*)$ are finitely generated. Then the groups $H^i(D^*)$ are as well by Lemmas 2.3.6 and 2.3.11. From Definition 2.3.8, the middle column of (3), and the fact that $L^n = 0$, we conclude that there is an exact sequence

$$0 \longrightarrow Z^{n-1} \longrightarrow D^{n-1} \longrightarrow F^n \to H^n(C^*) \longrightarrow 0 \tag{41}$$

which has the same extension class in $\text{Ext}^2_{\mathbf{Z}[G]}(H^n(C^*), Z^{n-1})$ as the exact sequence

$$0 \longrightarrow Z^{n-1} \longrightarrow C^{n-1} \longrightarrow C^n \longrightarrow H^n(C^*) \longrightarrow 0.$$

Let D'^* be the complex obtained from D^* by putting F^n in the n-th position and using the morphism $D^{n-1} \to F^n$ of (41). It follows that D'^* is isomorphic in the derived category to C^*.

Using Proposition 2.2.4 (see also [70] Lemma III.12.3) and ([107] p. 263), we now construct perfect complexes Q^* and P^* together with quasi-isomorphisms $Q^* \longrightarrow D^*$ and $P^* \to C^*$. We can furthermore assume that $Q^i = 0$ if $i > n-1$, since $D^i = 0$ for such i. Define Q'^* to be the complex obtained from Q^* by adjoining F^n in degree n and by letting the morphism $Q^{n-1} \longrightarrow F^n$ be the one induced by the isomorphism $Q^{n-1}/Z^{n-1}(Q^*) = H^{n-1}(Q^*) \longrightarrow H^{n-1}(D^*)$ followed by the map $H^{n-1}(D^*) = D^{n-1}/Z^{n-1} \longrightarrow F^n$ induced by the boundary map $D^{n-1} \longrightarrow F^n$ of D'^*. Then Q'^* is isomorphic to D'^* in the derived category and hence also to C^* and to P^*. It follows that

$$\sum_i (-1)^i [Q'^i] = \sum_i (-1)^i [P^i] \tag{42}$$

in $K_0(\mathbf{Z}[G])$. However, since D^* has smaller length than C^*, we know by induction on the statement of Theorem 2.3.3(c) that

$$\chi(D^*, \{L'_i\}_i, \{\tau'_i\}_i) = \sum_i (-1)^i [Q^i] = \sum_i (-1)^i [Q'^i] - (-1)^n [F^n]. \tag{43}$$

Now combining (38) with (42) and (43) shows

$$\chi(C^*, \{L_i\}_i, \{\tau_i\}_i) = \sum_i (-1)^i [P^i]$$

and this is the assertion of Theorem 2.3.3(c). □

Question 2.3.21 Suppose that

$$0 \to C(1)^* \to C(2)^* \to C(3)^* \to 0$$

is a short exact sequence of chain complexes of $\mathbf{Z}[G]$-modules where $(C(j)^*, \{L(j)_i\}_i, \{\tau(j)_i\}_i)$ are nearly perfect complexes, as in Definition 2.3.1, for $j = 1, 2, 3$. Under what conditions do we have additivity of Euler characteristics

$$\begin{aligned}&\chi(C(2)^*, \{L(2)_i\}_i, \{\tau(2)_i\}_i)\\&= \chi(C(1)^*, \{L(1)_i\}_i, \{\tau(1)_i\}_i) + \chi(C(3)^*, \{L(3)_i\}_i, \{\tau(3)_i\}_i) \in \mathcal{CL}(\mathbf{Z}[G])\end{aligned}$$

in Definition 2.3.13?

Certainly this type of additivity holds when each $C(j)^*$ is concentrated in dimension zero and $C(3)^0$ is finitely generated. This verification is an exercise in the use of the Snake Lemma.

Incidentally, one can make one-term nearly perfect complexes in the following manner. Choose a cyclic group, $\langle a \rangle \subseteq G$ of order n and let $\mathbf{Q}/\mathbf{Z}\tilde{\oplus}\mathbf{Z}$ be the direct sum, $\mathbf{Q}/\mathbf{Z} \oplus \mathbf{Z}$, on which the generator, a, acts by $a(\alpha + \mathbf{Z}, m) = (\alpha + m/n + \mathbf{Z}, m)$. Now set $C^0 = \mathrm{Ind}_{\langle a \rangle}^G(\mathbf{Q}/\mathbf{Z}\tilde{\oplus}\mathbf{Z})$.

2.4 Higher-dimensional algebraic K-theory

In this section we shall briefly recall from [117] the definition of higher-dimensional algebraic K-groups of an exact category.

2.4.1 *Categories and their classifying spaces*

Let $\mathcal{C}$ be a small category and denote by $N\mathcal{C}$ the simplicial set given by the nerve of $\mathcal{C}$. The n-simplices are the composable n-tuples of morphisms in $\mathcal{C}$

$$X_0 \longrightarrow X_1 \longrightarrow \ldots \longrightarrow X_n.$$

The classifying space of $\mathcal{C}$ is defined to be the geometrical realisation of $N_{\bullet}\mathcal{C}$ ([61], [111], [127])

$$B\mathcal{C} = |N\mathcal{C}|.$$

There is a canonical homotopy equivalence of the form $B(\mathcal{C} \times \mathcal{C}') \longrightarrow B\mathcal{C} \times B\mathcal{C}'$ [111].

A functor, $F : \mathcal{C} \longrightarrow \mathcal{D}$, induces a continuous map, $BF : B\mathcal{C} \longrightarrow B\mathcal{D}$ between the classifying spaces. Let $\mathcal{I}$ be the ordered set with two elements so that $B\mathcal{I}$ is the closed unit interval, I. Then a natural transformation, $\theta : F \longrightarrow G$, gives a functor from $\mathcal{C} \times \mathcal{I}$ to $\mathcal{D}$, which yields a homotopy between BF and BG.

Now suppose that $F : \mathcal{C} \longrightarrow \mathcal{D}$ is a functor and let $C_1, C_2, \ldots \in Obj(\mathcal{C})$, $D_1, D_2, \ldots \in Obj(\mathcal{D})$ denote objects. Fixing D_1, denote by $D_1 \backslash F$ the category of pairs, (C_1, f_1), with $f_1 \in \mathrm{Hom}_{\mathcal{D}}(D_1, F(C_1))$. Also let $F^{-1}(D_1) \subseteq \mathcal{C}$ denote the subcategory of all objects which map under F to D_1 and all morphisms which map to the identity of D_1. The category $F^{-1}(D_1)$ is called the *fibre* of F over D_1. We say that F makes $\mathcal{C}$ into a *prefibred category over* $\mathcal{D}$ if and only if for each object, D_1, the functor

$$\alpha : F^{-1}(D_1) \longrightarrow D_1\backslash F,$$

given by $\alpha(C_2) = (C_2, 1))$, has a right adjoint, β. That is, (cf. [70]) there are natural isomorphisms of the form

$$\mathrm{Hom}_{D_1\backslash F}(\alpha(C_1), (C_2, g)) \cong \mathrm{Hom}_{F^{-1}(D_1)}(C_1, \beta(C_2, g))$$

for all C_1, C_2 and g.

Write $\beta(C_2, g) = g^*(C_2) \in Obj(F^{-1}(D_1))$. Then, for each morphism, $g : D_1 \longrightarrow D_2$, we obtain a functor

$$g^* : F^{-1}(D_2) \longrightarrow F^{-1}(D_1)$$

called *base-change* by g. The prefibred category, $\mathcal{C}$ over $\mathcal{D}$, is called a *fibred category* if and only if for each pair of morphisms, $D_1 \xrightarrow{g} D_2 \xrightarrow{h} D_3$, in $\mathcal{D}$ the canonical natural transformation of functors $(h^*g^* \mapsto (gh)^*)$ is an isomorphism. There are dual notions of a precofibred category, a cofibred category and cobase-change by g ([117] p. 85).

The notions of base-change and cobase-change are important in the description of the homotopy-fibre of the map of classifying spaces, $BF : B\mathcal{C} \longrightarrow B\mathcal{D}$. Recall that, if $N = \pi^{-1}(b) \longrightarrow E \xrightarrow{\pi} B$, is a fibration [145] then there exists a long exact sequence of homotopy groups of the form

$$\ldots \longrightarrow \pi_{i+1}(B, b) \longrightarrow \pi_i(N, e) \longrightarrow \pi_i(E, e) \longrightarrow \pi_i(B, b) \longrightarrow \ldots$$

for $e \in N$. In addition, lifting a path in the base, B, yields a homotopy equivalence between the fibres above the two end-points. When trying to realise a fibration in terms of classifying spaces and maps induced by functors the role of path-lifting is played by base-change and cobase change along morphisms. In fact, more precisely one has the following result, which is a special case of ([117] Theorem B p. 89).

Theorem 2.4.2

Let $F : \mathcal{C} \longrightarrow \mathcal{D}$ be prefibred (respectively, precofibred) and suppose that every base-change (respectively, cobase-change) functor

$$g^* : F^{-1}(D_2) \longrightarrow F^{-1}(D_1)$$

is a homotopy equivalence. Then, for any $D \in Obj(\mathcal{D})$,

$$BF^{-1}(D) \longrightarrow B\mathcal{C} \xrightarrow{BF} B\mathcal{D}$$

is a fibration. In particular, there is a long exact sequence of homotopy groups of the form

$$\ldots \longrightarrow \pi_{i+1}(B\mathcal{D}, D) \longrightarrow \pi_i(BF^{-1}(D), C) \longrightarrow \pi_i(B\mathcal{C}, C) \longrightarrow \pi_i(B\mathcal{D}, D) \longrightarrow \ldots$$

where $F(C) = D$.

Recall that the functor, g^*, is a homotopy equivalence if and only if the induced maps on all homotopy groups of the associated classifying spaces are isomorphisms.

2.4.3 *Exact categories and their K-groups*

Consider the following example of an additive category with exact sequences.

Let $\mathcal{M}$ be an additive full subcategory of an abelian category, $\mathcal{C}$. Suppose that $\mathcal{M}$ is closed under extensions, which means that if $C \in Obj(\mathcal{C})$ has a subobject, $M' \in Obj(\mathcal{M})$, such that $M/M' \in Obj(\mathcal{M}) \subseteq Obj(\mathcal{C})$ then $M \in Obj(\mathcal{M})$.

Let $\mathcal{E}$ denote the class of short exact sequences in $\mathcal{C}$

$$0 \longrightarrow M_1 \stackrel{a}{\longrightarrow} M_2 \stackrel{b}{\longrightarrow} M_3 \longrightarrow 0$$

for which $M_1, M_2, M_3 \in Obj(\mathcal{M})$. In this case $\mathcal{E}$ satisfies:

(i) Any sequence in $\mathcal{M}$ which is isomorphic to a short exact sequence in $\mathcal{E}$ is also in $\mathcal{E}$. For any $M_1, M_2 \in Obj(\mathcal{M})$ the split sequence

$$0 \longrightarrow M_1 \longrightarrow M_1 \oplus M_2 \longrightarrow M_2 \longrightarrow 0$$

is in $\mathcal{E}$. For any short exact sequence in $\mathcal{E}$, as above, the morphism a (respectively, b) is a kernel (respectively, cokernel) for b (respectively, a) in $\mathcal{M}$.

(ii) Call a morphism in $\mathcal{M}$ an admissible monomorphism (respectively, admissible epimorphism) if it appears as an a (respectively, b) in a short exact sequence in $\mathcal{E}$. The class of admissible epimorphisms is closed under composition and under base-change (i.e. pull-back) by arbitrary morphisms of $\mathcal{M}$. Dually, the class of admissible monomorphisms is closed under composition and under cobase-change by arbitrary morphisms of $\mathcal{M}$.

(iii) Let $b : M_2 \longrightarrow M_3$ be a morphism in $\mathcal{M}$ which has a kernel (in $\mathcal{M}$). If there exists another morphism, $M \longrightarrow M_2$ in $\mathcal{M}$, such that the composition $M \longrightarrow M_2 \longrightarrow M_3$ is an admissible epimorphism then so is b. Dually for admissible monomorphisms, a.

An *exact category* is an additive category, $\mathcal{M}$, equipped with a family of short exact sequences, $\mathcal{E}$ as above, which satisfy properties (i)–(iii). Members of the family $\mathcal{E}$ are called the short exact sequences of $\mathcal{M}$. An *exact functor* is a functor between exact categories which sends exact sequences to exact sequences.

If $\mathcal{M}$ is an exact category, the associated category $Q\mathcal{M}$ has the same objects but an element of $\mathrm{Hom}_{Q\mathcal{M}}(M_1, M_2)$ is given by the equivalence class of a diagram in $\mathcal{M}$ of the form $M_1 \stackrel{b}{\longleftarrow} M \stackrel{a}{\longrightarrow} M_2$ in which b is an admissible epimorphism and a is an admissible monomorphism. Two such diagrams are equivalent if there is an isomorphism from one to the other which is the identity on M_1 and M_2. If $M_2 \stackrel{b'}{\longleftarrow} M' \stackrel{a'}{\longrightarrow} M_3$ is an element of $\mathrm{Hom}_{Q\mathcal{M}}(M_2, M_3)$ then the composition is the element of $\mathrm{Hom}_{Q\mathcal{M}}(M_1, M_3)$ given by the diagram

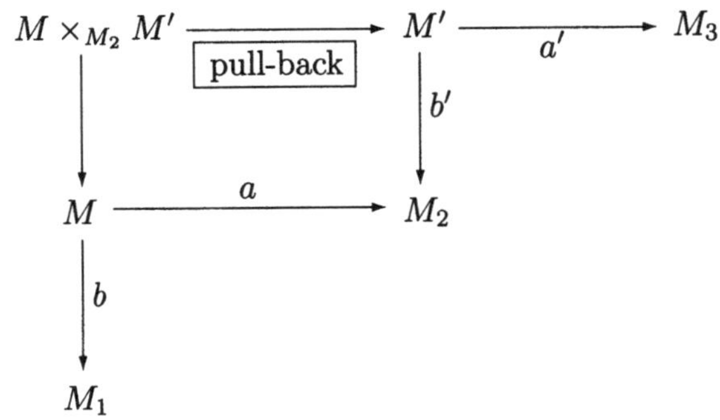

in which the left-hand vertical is an admissible epimorphism and the upper horizontal is an admissible monomorphism, because the square is a pull-back which base-changes the admissible epimorphism, b', and cobase-changes the admissible monomorphism, a.

Definition 2.4.4 Let $\mathcal{M}$ be an exact category, as in §2.4.3. Then, for $i \geq 0$, the i-th algebraic K-group of $\mathcal{M}$ is defined to be the homotopy group ([117] p. 95)

$$K_i(\mathcal{M}) = \pi_{i+1}(BQ\mathcal{M}, \underline{0})$$

where $\underline{0}$ is the zero object of $\mathcal{M}$.

When $i = 0$ there is an isomorphism ([117] Theorem 1 p. 94) between this homotopy-theoretic $K_0(\mathcal{M})$ and the group of Definition 2.1.1. In particular, if R is a ring then $K_i(R) = \pi_{i+1}(BQ\mathcal{P}(R), \underline{0})$ agrees with Example 2.1.2 when $i = 0$. It also agrees with Definition 2.1.3 when $i = 1$.

2.4.5 *Localisation*

Let R be a ring and let $S \subseteq R$ be a multiplicative set of central elements which are not zero-divisors in R. Let $\mathcal{H}$ denote the exact category of finitely generated R-modules, M, such that the localisation, M_S is zero and the projective dimension of M satisfies $\mathrm{projdim}_R(M) \leq 1$. The following result is an application of Theorem 2.4.2 to the functor, $F : \mathcal{P}(R) \longrightarrow \mathcal{P}(R_S)$ given by $F(M) = R_S \otimes_R M = M_S$. Here R_S denotes the localisation of R given by inverting the elements of S.

Theorem 2.4.6 *([62] p. 233)*

Let R be a ring and $S \subseteq R$ a multiplicative set of central non-zero-divisors. Then, in the notation of §2.4.5, there is a long exact sequence of the form

$$\ldots \longrightarrow K_{i+1}(R_S) \longrightarrow K_i(\mathcal{H}) \longrightarrow K_i(R) \xrightarrow{F_*} K_i(R_S) \longrightarrow \ldots.$$

Corollary 2.4.7

Let R be a Dedekind domain with fraction field F. Then there is an exact sequence of the form

$$\ldots \longrightarrow K_{i+1}(F) \xrightarrow{\partial} \oplus_{P \lhd R,\ \mathrm{prime}} K_i(R/P) \longrightarrow K_i(R) \longrightarrow K_i(F) \xrightarrow{\partial} \ldots.$$

Proof. In Theorem 2.4.6 take $S = R - \{0\}$ so that $F = R_S$. The category $\mathcal{H}$ splits as the product of the subcategories given by P-primary torsion modules, $\mathcal{H}_P$, for each prime ideal, $P \lhd R$. On the other hand, another application of Theorem 2.4.2 using the dévissage technique of ([117] §5) shows that $K_*(\mathcal{H}_P) \cong K_*(R/P)$. □

The following result is mentioned in [117] and was proved in [133].

Proposition 2.4.8

In Corollary 2.4.7, let $a, b \in F^* = F - \{0\} = K_1(F)$ and let $\{a, b\} \in K_2(F)$ denote the associated Steinberg symbol [109]. Let $P \lhd R$ be a prime ideal. Then the P-components of the boundary homomorphisms satisfy

(i) $\partial(a)_P = v_P(a) \in K_0(R/P) \cong \mathbf{Z}$

and

(ii) $\partial(\{a, b\}) = \overline{(-1)^{v_P(a)v_P(b)} \frac{a^{v_P(b)}}{b^{v_P(a)}}} \in (R/P)^* \cong K_1(R/P)$.

Here v_P denotes the P-adic valuation, $\overline{x}$ denotes the image of $x \in R$ in R/P and $\partial(\{a, b\})$ is called the P-adic tame symbol of a and b.

2.4.9 Suppose that R is a Dedekind domain with fraction field F and that S is a set of prime ideals of R. Then the localisation, R_S, is defined to be the subring of F given by

$$R_S = \{\frac{a}{b} \in F \mid bR = P_1^{\epsilon_1} \ldots P_n^{\epsilon_n},\ P_i \text{ prime}, P_i \in S\}.$$

For example, if S is equal to the set of all primes except one, P, then $R_S = R \cdot (R - P)^{-1}$ is a local ring with maximal ideal given by $P \cdot R_S$.

When $R = \mathcal{O}_K$, the ring of algebraic integers in a global field, K, then R_S is called the ring of S-integers of K and denoted by $\mathcal{O}_{K,S}$.

In this case there are localisation exact sequences of K-groups analogous to that of Corollary 2.4.7 whose boundary map is given by the formulae of Proposition 2.4.8

$$\ldots \longrightarrow K_{i+1}(F) \xrightarrow{\partial} \oplus_{P \lhd R,\ \text{prime}\ ,\ P \notin S} K_i(R/P) \longrightarrow K_i(R_S) \longrightarrow K_i(F) \xrightarrow{\partial} \ldots$$

and

$$\ldots \longrightarrow K_{i+1}(R_S) \xrightarrow{\partial} \oplus_{P \lhd R,\ \text{prime}\ ,\ P \in S} K_i(R/P) \longrightarrow K_i(R) \longrightarrow K_i(R_S) \xrightarrow{\partial} \ldots .$$

2.4.10 *K-theory of fields*

We close this section with a few remarks about the higher algebraic K-theory of fields.

Let L be a field. The Milnor K-theory of L was introduced in [110] and is defined to be the quotient of the graded ring given by the tensor algebra on L^* modulo the two-sided ideal generated by elements of the form $z \otimes (1 - z)$ for $z \in L^* - \{1\} = L - \{0, 1\}$. It is a graded ring and the n-dimensional part is denoted by $K_n^M(L)$. By Matsumoto's Theorem [109] $K_2^M(L) = K_2(L)$. For a local field, L, $K_n^M(L)$ is uniquely divisible for $n \geq 3$ [15].

Since $K_1^M(L) = K_1(L) = L^*$ and since Quillen K-theory, $K_*(L)$, is a graded ring there is a map of graded algebras, $K_*^M(L) \to K_*(L)$, which is the identity in dimension one. The cokernel of this map is called indecomposable K-theory, $K_*^{\text{ind}}(L)$. The norm residue gives a map of graded algebras

$$K_*(L) \otimes \mathbf{Z}/2 \longrightarrow H^*(L; \mathbf{Z}/2)$$

to the modulo 2 Galois cohomology of L. If L has characteristic zero this map is shown in [160] to be an isomorphism; this was formerly called the Milnor Conjecture.

There are other relations with Galois cohomology. In ([104], [105]) the norm residue symbol is shown to give an isomorphism

$$K_2(L) \otimes \mathbf{Z}/n \xrightarrow{\cong} H^2(L; \mu_n(2))$$

if $1/n \in L$. In ([96], [106]) it is shown that there is an isomorphism of the form

$$K_3^{\text{ind}}(L) \otimes \mathbf{Z}_p \xrightarrow{\cong} H^1(L; \mathbf{Z}_p(2))$$

when p is a prime different from the characteristic of L.

The Lichtenbaum-Quillen conjecture predicts similar results in all dimensions and for local fields these have been proved – in characteristic $p > 0$ in [60] and in characteristic zero in ([73], [74]).

2.5 Describing the class-group by representations

2.5.1 *Adèles and idèles*

Let K be an algebraic number field; that is, an extension of the rationals, $K/\mathbf{Q}$, of finite degree. Denote by $\mathcal{O}_K$ the ring of algebraic integers of K [94]. By a place of K we shall mean an embedding of K as a dense subfield into a complete discrete valuation field, $\hat{K}$. For example, $\hat{K}$ might be isomorphic to $\mathbf{R}$ or $\mathbf{C}$, the real or complex numbers, respectively. Embeddings into such fields are called infinite or Archimedean places of K. The finite places of K are comprised of all the non-Archimedean ones and they all come about by choosing a prime ideal, $P \triangleleft \mathcal{O}_K$, and completing K in the P-adic topology. This is the topology in which $\{x + P^n; n \geq 0\}$ is a base of neighbourhoods of the element, x, in $\mathcal{O}_K$. We denote this field by K_P.

Example 2.5.2 If p is a rational prime then the p-adic completion of $\mathbf{Z}$, the integers, is given by the inverse limit $\mathbf{Z}_p = \lim_{\substack{\leftarrow \\ n}} \mathbf{Z}/(p^n)$ and is called the ring of p-adic integers. Its field of fractions is given by $\mathbf{Q}_p = \mathbf{Z}_p[1/p]$, the field of p-adic rationals.

2.5.3 The completed fields introduced in §2.5.1 are called local fields ([54], [79]). In a Dedekind domain, such as $\mathcal{O}_K$, ideals have a unique primary decomposition. Let p be a rational prime. If $P \lhd \mathcal{O}_K$ is a prime such that the ideal and $p\mathcal{O}_K = P^e Q$ for some ideal $Q \lhd \mathcal{O}_K$ and some integer $e \geq 1$ then $K_P/\mathbf{Q}_p$ and $\mathcal{O}_K/\mathbf{Z}/p$ are finite field extensions and we say that P divides p or that P lies over p. If L/K is a number field extension then to say that one Archimedean prime (or place), $L \subset \hat{L}$, divides another, $K \subset \hat{K}$, means that the two inclusions of K into $\hat{L}$ are equal. If P is a finite prime then the algebraic integers of K_P, $\mathcal{O}_{K_P}$, are given by the P-adic completion of $\mathcal{O}_K$. By convention, if $\hat{K}$ is Archimedean we set $\mathcal{O}_{\hat{K}}$ equal to $\hat{K}$.

The adèle ring of a number field K is defined to be the ring given by the restricted product

2.5.4 $$J(K) = \prod\nolimits'_{P \text{ prime}} K_P.$$

In 2.5.4 $\prod'$ signifies that we take those elements of the topological ring $\prod_P K_P$ for which almost all entries lie in $\mathcal{O}_{K_P}$. The group of idèles is the group of units in $J(K)$,

2.5.5 $$J^*(K) =$$
$$\{(x_P) \in J(K) \mid x_P \neq 0 \quad \text{and almost everywhere} \quad x_P \in \mathcal{O}^*_{K_P}\}$$

where $\mathcal{O}^*_{K_P}$ denotes the multiplicative group of units in $\mathcal{O}_{K_P}$. The unit idèles is the subgroup

2.5.6 $$U(\mathcal{O}_K) = \prod\nolimits_{P \text{ prime}} \mathcal{O}^*_{K_P}.$$

Now let G be a finite group. We may extend the adèles and idèles to the group-rings, $\mathcal{O}_K[G]$ and $K[G]$. Define

2.5.7 $$\begin{cases} J(K[G]) & = \prod'_{P \text{ prime}} K_P[G], \\ J^*(K[G]) & = \{(\alpha_P) \in J(K[G]) \mid \alpha_P \in \mathcal{O}_{K_P}[G]^* \text{ for} \\ & \quad \text{almost all } P \text{ and } \alpha_P \in K_P[G]^* \text{ otherwise}\}, \\ U(\mathcal{O}_K[G]) & = \prod_{P \text{ prime}} \mathcal{O}_{K_P}[G]^*. \end{cases}$$

Now suppose that E/K is a finite Galois extension with Galois group, $G(E/K)$. In this case $G(E/K)$ acts on the set of primes of E and hence acts upon the groups $J^*(E), U(\mathcal{O}_E), J^*(E[G])$ and $U(\mathcal{O}_E[G])$. If Q is a prime of E which divides the prime, P, of K then $G(E_Q/K_P)$ is isomorphic to a subgroup of $G(E/K)$ which is called a decomposition group for P and depends only on P, up to conjugation in $G(E/K)$.

If E is large enough to contain all $|G|$-th roots of unity we call E a splitting field for G. In this case ([131] Theorem 4.1.9(i)) there is an isomorphism between representation rings of the form

2.5.8 $$R_E(G) = K_0(E[G]) \cong K_0(\mathbf{C}[G]) = R(G)$$

and so $G(E/K)$ acts upon $R(G)$ by the entry-by-entry action on a representation

2.5.9 $$T : G \longrightarrow GL_n(E).$$

Therefore we may consider the group of $G(E/K)$-equivariant maps

2.5.10 $$\mathrm{Hom}_{G(E/K)}(R(G), J^*(E)) =$$
$$\{f : R(G) \longrightarrow J^*(E) \mid \begin{array}{l} f(g(z)) = g(f(z)) \\ \text{for all } g \in G(E/K), z \in R(G)\}. \end{array}$$

More generally, if L/K is a Galois extension which contains E/K then $G(L/K)$ acts on $R(G)$ and on $J^*(E)$, since E/K is Galois and $G(E/K)$ is a quotient of $G(L/K)$. Therefore we may pass to the absolute Galois group of K, Ω_K, which is the topological group defined by

2.5.11 $$\Omega_K = \mathrm{inv\,lim}_{\substack{K \subset L \subset K^c \\ L/K \text{ Galois}}} G(L/K)$$

where K^c/K is a chosen algebraic closure of K. In this case we have

2.5.12 $$\mathrm{Hom}_{\Omega_K}(R(G), J^*(E)) \cong \mathrm{Hom}_{G(E/K)}(R(G), J^*(E)).$$

2.5.13 *$\mathcal{O}_K[G]$-modules and determinants*

A more extensive reference for the material of this section is ([49] II p. 334 et seq.).

Let M be a projective $\mathcal{O}_K[G]$-module of rank one. By [149] an $\mathcal{O}_K[G]$-module is projective if and only if it is locally free (cf. §2.1.2(v)). This means that $M \otimes_{\mathcal{O}_K} \mathcal{O}_{K_P}$ is a free $\mathcal{O}_{K_P}[G]$-module on one generator, x_P, for each prime, P, of K and that $M \otimes_{\mathcal{O}_K} K$ is a free $K[G]$-module on one generator, x_0. Since $K[G]$ and $\mathcal{O}_{K_P}[G]$ are subrings of $K_P[G]$ this means that there is a unit, $\lambda_P \in K_P[G]^*$ which is defined by

2.5.14 $$x_0 \cdot \lambda_P = x_P \in M \otimes_{\mathcal{O}_K} K_P.$$

In fact, λ_P will almost always lie in $\mathcal{O}_{K_P}[G]^*$ so that we obtain an idèle

2.5.15 $$(\lambda_P) \in J^*(K[G]).$$

Now suppose that T is a representation, as in 2.5.9. We may apply T to each λ_P to obtain

2.5.16 $$T(\lambda_P) \in M_n(K_P \otimes_K E)$$

where $M_n(A)$ denotes the $n \times n$ matrices with entries in A. There is a ring isomorphism of the form

2.5.17 $$K_P \otimes_K E \cong \prod_{\substack{Q|P \\ Q \text{ prime of } E}} E_Q.$$

Therefore we obtain an element

2.5.18 $$\det(T(\lambda_P)) \in \prod_{\substack{Q|P \\ Q \text{ prime of } E}} E_Q^*.$$

Since $(\lambda_P) \in J^*(K[G])$ we obtain an Ω_K-equivariant map, given by 2.5.18 at the primes Q of E which divide P,

2.5.19 $$\mathrm{Det}((\lambda_P)) \in \mathrm{Hom}_{\Omega_K}(R(G), J^*(E)).$$

Now let us consider the dependence of 2.5.19 upon the choices of x_0 and x_P in 2.5.14. If we replace x_P by another generator, x'_P, these choices will be related by an equation

2.5.20 $$x'_P = u_P x_P \text{ for some } u_P \in \mathcal{O}_{K_P}[G]^*$$

so that we obtain a unit idèle

2.5.21 $$u = (u_P) \in U(\mathcal{O}_K[G])$$

and 2.5.19 will be altered by multiplication by

2.5.22 $$\mathrm{Det}(u) \in \mathrm{Det}(U(\mathcal{O}_K[G])) \subset \mathrm{Hom}_{\Omega_K}(R(G), J^*(E)).$$

Also there is a diagonal embedding of E^* into $J^*(E)$ which induces an inclusion

2.5.23 $$\mathrm{Hom}_{\Omega_K}(R(G), E^*) \subset \mathrm{Hom}_{\Omega_K}(R(G), J^*(E)).$$

By a similar argument, changing x_0 to x'_0 will change $\mathrm{Det}((\lambda_P))$ by a function which lies in the subgroup of 2.5.23. Therefore we have associated to each locally free $\mathcal{O}_K[G]$-module of rank one, M, a well-defined element

2.5.24 $$\mathrm{Det}[M] \in \frac{\mathrm{Hom}_{\Omega_K}(R(G), J^*(E))}{\mathrm{Hom}_{\Omega_K}(R(G), E^*) \cdot \mathrm{Det}(U(\mathcal{O}_K[G]))}.$$

The construction of $\mathrm{Det}[M]$ extends to locally free modules of any finite rank. If M is a locally-free $\mathcal{O}_K[G]$-module of rank m one chooses bases and makes a similar construction in which λ_P is now an invertible matrix lying in $GL_m(K_P[G])$.

This construction yields the following "Hom-description" of the class-group, originally due to Fröhlich.

Theorem 2.5.25 *([49] II p. 334; [57])*

Let $\mathcal{CL}(\mathcal{O}_K[G])$ be the class-group of Example 2.1.2(v). With the notation introduced above there is an isomorphism

$$\mathrm{Det} : \mathcal{CL}(\mathcal{O}_K[G]) \xrightarrow{\cong} \frac{\mathrm{Hom}_{\Omega_K}(R(G), J^*(E))}{\mathrm{Hom}_{\Omega_K}(R(G), E^*) \cdot \mathrm{Det}(U(\mathcal{O}_K[G]))}$$

which sends a locally free module, M, to $\mathrm{Det}[M]$ *of 2.5.24.*

2.5.26 *The kernel group, $D(\mathcal{O}_K[G])$*

An $\mathcal{O}_K$-order, Λ in $K[G]$, is a subring containing $\mathcal{O}_K$, which is a finitely generated, projective $\mathcal{O}_K$-module such that $K \otimes_{\mathcal{O}_K} \Lambda = K[G]$. Suppose that Λ is a maximal $\mathcal{O}_K$-order of $K[G]$ then we may define the kernel group, $D(\mathcal{O}_K[G])$, by

2.5.27 $$D(\mathcal{O}_K[G]) = \ker(\mathcal{CL}(\mathcal{O}_K[G]) \longrightarrow \mathcal{CL}(\Lambda))$$

where $\mathcal{CL}(\Lambda)$ is the class-group of Λ, defined by a Grothendieck group construction analogous to those of Example 2.1.2. The group defined in this manner is independent of the choice of Λ. There is a Hom-description of this group also.

To describe this we need to introduce a subgroup

2.5.28 $$\mathrm{Hom}^{+}_{\Omega_K}(R(G), \mathcal{O}_E^*) \subset \mathrm{Hom}_{\Omega_K}(R(G), \mathcal{O}_E^*).$$

An irreducible representation, $T : G \longrightarrow GL_n(E)$, as in 2.5.19 gives rise to a complex representation, $T : G \longrightarrow GL_n(\mathbf{C})$, by choosing an embedding of E into $\mathbf{C}$. Let $\mathbf{H}$ denote the skewfield given by the quaternions. A (left) $\mathbf{H}$-vector space, V, of dimension m over $\mathbf{H}$, may be considered as a $2m$-dimensional complex vector space. If $GL_m(\mathbf{H})$ is the group of invertible $\mathbf{H}$-linear maps from V to itself then we obtain a map

2.5.29 $$c : GL_m(\mathbf{H}) \longrightarrow GL_{2m}(\mathbf{C}).$$

The representation, T, is called quaternionic or symplectic if $T : G \longrightarrow GL_{2m}(\mathbf{C})$ factors through the map, c, of 2.5.29. On the other hand, if T is symplectic then complex conjugation fixes $T \in R(G)$ so that, if $f \in \mathrm{Hom}_{\Omega_K}(R(G), E^*)$ and if K is a subfield of the real numbers then $f(T)$ lies in $\mathbf{R}$ for every Archimedean prime of E lying over $K \subset \mathbf{R}$. Therefore it makes sense to define $\mathrm{Hom}^{+}_{\Omega_K}(R(G), E^*)$ to be

2.5.30 $\{f \in \mathrm{Hom}_{\Omega_K}(R(G), E^*) \mid f(T) \text{ is positive}, T \text{ symplectic}\}$
where, in 2.5.30, "positive" means that $f(T)$ is positive under every Archimedean place of E which lies over a real place of K.

Similarly we may define

2.5.31 $$\mathrm{Hom}^{+}_{\Omega_K}(R(G), \mathcal{O}_E^*) = \{f \in \mathrm{Hom}^{+}_{\Omega_K}(R(G), E^*) \mid \mathrm{im}(f) \subset \mathcal{O}_E^*\}.$$

Theorem 2.5.32 *([49] II p. 334 et seq.; [57])*
The isomorphism of 2.5.25 induces an isomorphism

$$\mathrm{Det} : D(\mathcal{O}_K[G]) \xrightarrow{\cong} \frac{\mathrm{Hom}_{\Omega_K}(R(G), U(\mathcal{O}_E))}{\mathrm{Hom}^{+}_{\Omega_K}(R(G), \mathcal{O}_E^*) \cdot \mathrm{Det}(U(\mathcal{O}_K[G]))}.$$

Remark 2.5.33 For all primes, Q, which do not divide the order of G, $\mathcal{O}_{E_Q}[G]$ is a maximal $\mathcal{O}_{E_Q}$-order in $E_Q[G]$. Hence one may rewrite Theorem 2.5.32 as an isomorphism of the following form.

$$\mathrm{Det} : D(\mathcal{O}_K[G]) \xrightarrow{\cong} \frac{\mathrm{Hom}_{\Omega_K}(R(G), \prod_{Q||G|)} \mathcal{O}_{E_Q}^*)}{\mathrm{Hom}^{+}_{\Omega_K}(R(G), \mathcal{O}_E^*) \cdot \mathrm{Det}(\prod_{P||G|} \mathcal{O}_{K_P}[G]^*)}.$$

2.5.34 *The Hom-description of* $K_0(\mathbf{Z}[G]; \mathbf{Q})$

In Example 2.1.7 we introduced the relative K-group, $K_0(\mathbf{Z}[G]; \mathbf{Q})$. Comparing the exact sequences of Theorem 2.1.6 and Theorem 2.4.6 it is clear that there is a canonical isomorphism of the form

$$K_0(\mathbf{Z}[G]; \mathbf{Q}) \cong K_0(\mathcal{H})$$

where $\mathcal{H}$ is the category of finite, cohomologically trivial $\mathbf{Z}[G]$-modules. This follows from §2.4.5 together with the result of R.G. Swan ([49] II) that a finite $\mathbf{Z}[G]$-module is cohomologically trivial if and only if it has a projective resolution of length one. The decomposition of a finite abelian group into the sum of its p-Sylow subgroups yields a functorial decomposition

$$K_0(\mathbf{Z}[G]; \mathbf{Q}) \cong \oplus_{p \text{ prime}} \ K_0(\mathbf{Z}_p[G]; \mathbf{Q}_p).$$

In the terminology and notation of [155] we have an isomorphism of the form

$$K_0T(\mathbf{Z}[G]) \cong K_0(\mathcal{H})$$

since $\mathcal{H}$ is the category of finitely presented torsion $\mathbf{Z}[G]$-modules. Therefore from ([155] Theorem 3.3 p. 10) there is a Hom-description of the form

$$K_0(\mathbf{Z}_p[G]; \mathbf{Q}_p) \cong \frac{\mathrm{Hom}_{\Omega_{\mathbf{Q}_p}}(R(G), (\overline{\mathbf{Q}}_p)^*)}{\mathrm{Det}(\mathbf{Z}_p[G]^*)}.$$

Here $\Omega_{\mathbf{Q}_p}$ denotes the absolute Galois group and $\overline{\mathbf{Q}}_p$ an algebraic closure of the p-adic rationals, $\mathbf{Q}_p$. Taking the sum over all primes, p, we obtain a Hom-description of the form

$$K_0(\mathbf{Z}[G]; \mathbf{Q}) \cong \frac{\mathrm{Hom}_{\Omega_{\mathbf{Q}}}(R(G), J^*(\overline{\mathbf{Q}}))}{\mathrm{Det}(U(\mathbf{Z}[G]))}$$

where $\overline{\mathbf{Q}}$ is an algebraic closure of the rationals.

In terms of these Hom-descriptions, the homomorphism

$$K_0(\mathbf{Z}[G]; \mathbf{Q}) \longrightarrow K_0(\mathbf{Z}[G]) \longrightarrow \mathcal{CL}(\mathbf{Z}[G]),$$

coming from the homomorphism in the long exact sequence of Theorem 2.4.6 composed with the quotient map from K_0 to the class-group, sends the class of the function, f,

$$[f] \in \frac{\mathrm{Hom}_{\Omega_{\mathbf{Q}}}(R(G), J^*(\overline{\mathbf{Q}}))}{\mathrm{Det}(U(\mathbf{Z}[G]))}$$

to the class of the same function

$$[f] \in \frac{\mathrm{Hom}_{\Omega_{\mathbf{Q}}}(R(G), J^*(\overline{\mathbf{Q}}))}{\mathrm{Hom}_{\Omega_K}(R(G), \mathbf{Q}^*) \cdot \mathrm{Det}(U(\mathbf{Z}[G]))}.$$

In §2.1.8(ii) we showed how to make elements of $K_0(\mathbf{Z}[G]; \mathbf{Q})$ of the form $[P^{\mathrm{od}}, \phi, P^{\mathrm{ev}}]$ from a perfect complex of $\mathbf{Z}[G]$-modules of the form

$$P^* : 0 \longrightarrow P^n \xrightarrow{d_n} P^{n+1} \xrightarrow{d_{n+1}} \cdots \xrightarrow{d_{m-1}} P^m \longrightarrow 0$$

together with a given isomorphism

$$\psi : \oplus_j \ H^{2j+1}(P^*) \otimes \mathbf{Q} \xrightarrow{\cong} \oplus_j \ H^{2j}(P^*) \otimes \mathbf{Q}.$$

The Hom-description representative for $[P^{\mathrm{od}}, \phi, P^{\mathrm{ev}}]$ is constructed by choosing bases, as in the Hom-description underlying Theorem 2.5.25. For each prime p the module $P_i \otimes \mathbf{Z}_p$ is a free $\mathbf{Z}_p[G]$-module. Using a choice of $\mathbf{Z}_p[G]$-basis, the isomorphism

$$\phi \otimes 1 : \oplus_j P_{2j+1} \otimes_{\mathbf{Z}} \mathbf{Q}_p \cong \oplus_j P_{2j} \otimes_{\mathbf{Z}} \mathbf{Q}_p$$

gives rise to a matrix

$$\lambda_p \in GL_m(\mathbf{Q}_p[G])$$

where m is the $\mathbf{Z}_p[G]$-rank of $\oplus_j P_{2j+1}$. Applying the Det-construction of 2.5.19 yields the Hom-representative of $[P^{\mathrm{od}}, \phi, P^{\mathrm{ev}}]$

$$[P^{\mathrm{od}}, \phi, P^{\mathrm{ev}}] = \mathrm{Det}((\lambda_p)) \in K_0(\mathbf{Z}[G]; \mathbf{Q}) \cong \frac{\mathrm{Hom}_{\Omega_{\mathbf{Q}}}(R(G), J^*(\overline{\mathbf{Q}}))}{\mathrm{Det}(U(\mathbf{Z}[G]))}.$$

Proposition 2.5.35 *(See [22] §2)*
The element, $[P^{\mathrm{od}}, \phi, P^{\mathrm{ev}}]$ of §2.1.8(ii) and §2.5.34, does not depend on the choice of rational splittings but only on the rational cohomology isomorphism, ψ, and the quasi-isomorphism class of the chain complex, P^*

Proof. For notational convenience we shall switch to chain complexes. Let G be a finite group and suppose that

$$0 \longrightarrow F_k \xrightarrow{d_k} F_{k-1} \xrightarrow{d_{k-1}} \ldots \xrightarrow{d_2} F_1 \xrightarrow{d_1} F_0 \longrightarrow 0$$

is a perfect complex of $\mathbf{Z}[G]$-modules, as in Definition 2.2.1. Suppose that we are given a $\mathbf{Q}[G]$-module isomorphism,

$$A : \oplus_j \; H_{2j}(F_*) \otimes \mathbf{Q} \xrightarrow{\cong} \oplus_j \; H_{2j+1}(F_*) \otimes \mathbf{Q}.$$

As usual, let $Z_t = \mathrm{Ker}(d_t : F_t \longrightarrow F_{t-1})$ and $B_t = d_{t+1}(F_{t+1}) \subseteq F_t$ denote the $\mathbf{Z}[G]$-modules of t-dimensional cycles and boundaries, respectively. We have short exact sequences of the form

$$0 \longrightarrow B_i \xrightarrow{\lambda_i} Z_i \xrightarrow{\pi_i} H_i(F_*) \longrightarrow 0$$

and

$$0 \longrightarrow Z_i \xrightarrow{\phi_i} F_i \xrightarrow{d_i} B_{i-1} \longrightarrow 0$$

with rational splittings

$$0 \longleftarrow B_i \otimes \mathbf{Q} \xleftarrow{u_i} Z_i \otimes \mathbf{Q} \xleftarrow{\rho_i} H_i(F_*) \otimes \mathbf{Q} \longleftarrow 0$$

and

$$0 \longleftarrow Z_i \otimes \mathbf{Q} \xleftarrow{\psi_i} F_i \otimes \mathbf{Q} \xleftarrow{\eta_{i-1}} B_{i-1} \otimes \mathbf{Q} \longleftarrow 0.$$

Note that $u_i = 1 - \rho_i \pi_i$ and $\psi_i = 1 - \eta_{i-1} d_i$.

The rational isomorphism, $X : \oplus_j F_{2j} \otimes \mathbf{Q} \cong \oplus_j F_{2j+1} \otimes \mathbf{Q}$ is defined as the composition

$$\begin{aligned}
\oplus_j F_{2j} \otimes \mathbf{Q} \quad & \stackrel{\sum_j d_{2j}+\psi_{2j}}{\longrightarrow} \quad \oplus_j (Z_{2j} \oplus B_{2j-1}) \otimes \mathbf{Q} \\
& \stackrel{\sum_j u_{2j}+\pi_{2j}\oplus 1}{\longrightarrow} \quad \oplus_j (B_{2j} \oplus H_{2j} \oplus B_{2j-1}) \otimes \mathbf{Q} \\
& \stackrel{1\oplus A\oplus 1}{\longrightarrow} \quad \oplus_j (B_{2j} \oplus H_{2j-1} \oplus B_{2j-1}) \otimes \mathbf{Q} \\
& \stackrel{\sum_j 1\oplus(\rho_{2j-1}+\lambda_{2j-1})}{\longrightarrow} \quad \oplus_j (B_{2j} \oplus Z_{2j-1}) \otimes \mathbf{Q} \\
& \stackrel{\sum_j (\eta_{2j-2}+\phi_{2j-1})}{\longrightarrow} \quad \oplus_j (B_{2j} \oplus Z_{2j-1}) \otimes \mathbf{Q}.
\end{aligned}$$

Under the homomorphism, X, an element $w_{2j} \in F_{2j}$ is sent to

$$(d_{2j}(w_{2j}), \psi_{2j}(w_{2j})) \in (B_{2j-1} \oplus Z_{2j}) \otimes \mathbf{Q}$$

then to

$$(d_{2j}(w_{2j}), \pi_{2j}\psi_{2j}(w_{2j}), u_{2j}\psi_{2j}(w_{2j})) \in (B_{2j-1} \oplus H_{2j} \oplus B_{2j}) \otimes \mathbf{Q}$$

and then to

$$(d_{2j}(w_{2j}), A\pi_{2j}\psi_{2j}(w_{2j}), u_{2j}\psi_{2j}(w_{2j})) \in (B_{2j-1} \oplus (\oplus_s H_{2s-1}) \oplus B_{2j}) \otimes \mathbf{Q}$$

and so on.

Consider the maps, X and X', which are constructed using different splittings, η_{2k-1} and η'_{2k-1}, respectively, all other splittings being the same. We must examine the matrix, $\tilde{X}$, representing $X^{-1} \cdot X' : \oplus_j F_{2j} \otimes \mathbf{Q} \cong \oplus_j F_{2j} \otimes \mathbf{Q}$ with respect to a $\mathbf{Q}[G]$-basis. It will suffice to show for each representation, χ, that $\det(\chi(\tilde{X})) = 1$.

Now the change in η_{2k-1} affects only the part of the composition for X, X' written out above and then only through the change in ψ_{2k}. In other words, the only part of the composition to change is $F_{2k} \otimes \mathbf{Q} \longrightarrow (B_{2k-1} \oplus Z_{2k}) \otimes \mathbf{Q}$ and the difference in this part is the map, $w_{2k} \mapsto (0, (\eta_{2k-1} - \eta'_{2k-1})d_{2k}(w_{2k}))$. Now the inverses, both X^{-1} and $(X')^{-1}$, each map $(0, (\eta_{2k-1} - \eta'_{2k-1})d_{2k}(w_{2k}))$ to $(\eta_{2k-1} - \eta'_{2k-1})d_{2k}(w_{2k}) \in Z_{2k} \otimes \mathbf{Q} \subseteq F_{2k} \otimes \mathbf{Q}$. This difference is trivial on $2k$-cycles and maps every w_{2k} to a $2k$-cycle. This implies that the matrix, $\tilde{X}$, will be upper triangular with ones on the diagonal and so $\det(\chi(\tilde{X})) = 1$.

A similar discussion applies to changes in η_{2k} or ρ_j.

Let $f : F'_* \longrightarrow F_*$ be a quasi-isomorphism of perfect $\mathbf{Z}[G]$-module complexes. Then, as in Lemma 2.2.2, the mapping cone complex given by

$$C(f)_* : \ldots \longrightarrow F_n \oplus F'_{n-1} \stackrel{\tilde{d}_n}{\longrightarrow} F_{n-1} \oplus F'_{n-2} \longrightarrow \ldots$$

is exact. Here

$$\tilde{d}_n(a, b) = (d_n(a) + (-1)^n f_n(b), d_{n-1}(b))$$

for all $a \in F_n, b \in F'_{n-1}$. Since $C(f)_*$ is an exact sequence of finitely generated, projective $\mathbf{Z}[G]$-modules, we may choose a splitting of this complex which is defined over the integers. From the Hom-description such a splitting defines the trivial element of $K_0(\mathbf{Z}[G], \mathbf{Q})$. On the other hand, after tensoring with the rationals, another splitting is given by the sum of the rational isomorphism, X for F_*, and the inverse of the rational isomorphism corresponding to F'_*. The previous discussion shows that the function, $\chi \mapsto \det(\chi(X)) \cdot \det(\chi(X'))^{-1}$, also represents the trivial element of $K_0(\mathbf{Z}[G], \mathbf{Q})$. On the other hand it also represents the difference

$$[\oplus_j F_{2j}, X, \oplus_j F_{2j+1}] - [\oplus_j F'_{2j}, X', \oplus_j F'_{2j+1}] \in K_0(\mathbf{Z}[G], \mathbf{Q}),$$

as required. □

Example 2.5.36 *The quaternion group of order eight*

When $G = Q_8$, the quaternion group of order eight, all the complex representations have rational characters so that the Hom-description of §2.5.34 simplifies to become

$$K_0(\mathbf{Z}_p[Q_8]; \mathbf{Q}_p) \cong \frac{\mathrm{Hom}(R(Q_8), \mathbf{Q}_p^*)}{\mathrm{Det}(\mathbf{Z}_p[Q_8]^*)}.$$

Since determinantal functions on $R(Q_8)$ take values in $\mathbf{Z}_p^*$ there is a short exact sequence of the form

$$0 \longrightarrow \frac{\mathrm{Hom}(R(Q_8), (\mathbf{Z}_p)^*)}{\mathrm{Det}(\mathbf{Z}_p[Q_8]^*)} \longrightarrow \frac{\mathrm{Hom}(R(Q_8), (\mathbf{Q}_p)^*)}{\mathrm{Det}(\mathbf{Z}_p[Q_8]^*)} \longrightarrow \mathrm{Hom}(R(Q_8), \mathbf{Z}) \longrightarrow 0.$$

From this sequence we see that there is an isomorphism of the form

$$\mathrm{TorsK}_0(\mathbf{Z}_p[Q_8]; \mathbf{Q}_p) \cong \frac{\mathrm{Hom}(R(Q_8), \mathbf{Z}_p^*)}{\mathrm{Det}(\mathbf{Z}_p[Q_8]^*)}.$$

When p is an odd prime then every function in $\mathrm{Hom}(R(Q_8), \mathbf{Z}_p^*)$ is a determinantal function because in this case $\mathbf{Z}_p[Q_8]$ is a maximal order in $\mathbf{Q}_p[Q_8]$. Hence, if p is odd, then $\mathrm{TorsK}_0(\mathbf{Z}_p[Q_8]; \mathbf{Q}_p) = 0$.

Now let us examine the case when $p = 2$. Let $\mathrm{Hom}^+(R(Q_8), \{\pm 1\})$ denote the homomorphisms which are trivial on the unique two-dimensional irreducible complex representation, ν. The class-group of $\mathbf{Z}[Q_8]$,

$$\mathcal{CL}(\mathbf{Z}[Q_8]) = \tilde{K}_0(\mathbf{Z}[Q_8])$$

is cyclic of order two. Since $\mathcal{CL}(\mathbf{Z}[Q_8]) = D(\mathbf{Z}[Q_8])$, by Theorem 2.5.32, this group has a Hom-description of the form

$$\mathcal{CL}(\mathbf{Z}[Q_8]) \cong \frac{\mathrm{Hom}(R(Q_8), \mathbf{Z}_2^*)}{\mathrm{Det}(\mathbf{Z}_2[Q_8]^*)\,\mathrm{Hom}^+(R(Q_8), \{\pm 1\})}.$$

As explained in §2.5.34, the natural forgetful map

$$\mathrm{TorsK}_0(\mathbf{Z}_2[Q_8]; \mathbf{Q}_2) \cong \mathrm{TorsK}_0(\mathbf{Z}[Q_8]; \mathbf{Q}) \longrightarrow \tilde{K}_0(\mathbf{Z}[Q_8])$$

may be identified with the natural surjection

$$\frac{\mathrm{Hom}(R(Q_8), \mathbf{Z}_2^*)}{\mathrm{Det}(\mathbf{Z}_2[Q_8]^*)} \longrightarrow \frac{\mathrm{Hom}(R(Q_8), \mathbf{Z}_2^*)}{\mathrm{Det}(\mathbf{Z}_2[Q_8]^*)\,\mathrm{Hom}^+(R(Q_8), \{\pm 1\})}.$$

We shall need the following result in order to detect classes in $\mathrm{TorsK}_0(\mathbf{Z}[Q_8]; \mathbf{Q})$.

Proposition 2.5.37

The natural maps yield an isomorphism of the form

$$\mathrm{TorsK}_0(\mathbf{Z}[Q_8]; \mathbf{Q}) \xrightarrow{\cong} \mathcal{CL}(\mathbf{Z}[Q_8]) \oplus \mathrm{TorsK}_0(\mathbf{Z}[Q_8^{ab}]; \mathbf{Q}) \cong \mathbf{Z}/2 \times \mathbf{Z}/2.$$

Proof. Let $1, \chi_1, \chi_2, \chi_1\chi_2$ denote the four one-dimensional representations of Q_8. To be precise, if

$$Q_8 = \{x, y \mid x^2 = y^2, y^4 = 1, xyx = y\}$$

set $\chi_1(x) = -1 = \chi_2(y)$ and $\chi_1(y) = 1 = \chi_2(x)$.

Suppose that $f : R(Q_8) \longrightarrow \mathbf{Z}_2^*$ is a homomorphism. By inflation f induces a homomorphism on $R(Q_8^{ab})$. Since

$$\mathcal{CL}(\mathbf{Z}[Q_8^{ab}]) \cong \frac{\mathrm{Hom}(R(Q_8^{ab}), \mathbf{Z}_2^*)}{\mathrm{Det}(\mathbf{Z}_2[Q_8^{ab}]^*)\,\mathrm{Hom}(R(Q_8^{ab}), \{\pm 1\})}$$

is trivial there exists a unit, $v \in \mathbf{Z}_2[Q_8^{ab}]^*$ such that $\mathrm{Det}(v)(\chi)f(\chi) = \pm 1$ for all one-dimensional representations of Q_8^{ab}. However, there exists a unit, $u \in \mathbf{Z}_2[Q_8]^*$, which maps to v under the canonical quotient homomorphism. Hence, on $R(Q_8)$, $\mathrm{Det}(u)f$ maps each of $1, \chi_1, \chi_2, \chi_1\chi_2$ into $\{\pm 1\}$. This argument shows that both in

$$\frac{\mathrm{Hom}(R(Q_8^{ab}), \mathbf{Z}_2^*)}{\mathrm{Det}(\mathbf{Z}_2[Q_8^{ab}]^*)} \quad \text{and in} \quad \frac{\mathrm{Hom}(R(Q_8), \mathbf{Z}_2^*)}{\mathrm{Det}(\mathbf{Z}_2[Q_8]^*)}$$

elements may be represented by homomorphisms which take $1, \chi_1, \chi_2, \chi_1\chi_2$ into $\{\pm 1\}$. Replacing f by $h = f \cdot \mathrm{Det}(f(1)g)$ yields an equivalent homomorphism which satisfies

$$h(1) = 1,\ h(\chi_1) = f(1 + \chi_1)\chi_1(g),$$
$$h(\chi_2) = f(1 + \chi_2)\chi_2(g),\ h(\chi_1\chi_2) = f(1 + \chi_1\chi_2)\chi_1(g)\chi_2(g).$$

Choosing g suitably we may assume that h is trivial on all one-dimensional representations except perhaps χ_1.

If $a + bx + cy + dxy \in \mathbf{Z}_2[Q_8^{ab}]^*$ then

$$\mathrm{Det}(a + bx + cy + dxy)(1) = a + b + c + d,$$
$$\mathrm{Det}(a + bx + cy + dxy)(\chi_1) = a - b + c - d,$$
$$\mathrm{Det}(a + bx + cy + dxy)(\chi_2) = a + b - c - d,$$
$$\mathrm{Det}(a + bx + cy + dxy)(\chi_1\chi_2) = a - b - c + d.$$

If $1 = a+b+c+d = a+b-c-d = a-b-c+d$ then $c = -d = -b$. Hence $\pm 1 = 1+a+b-4b$ and $1+a+b = 1$ which implies that $b = 0 = c = d$ and $a = 1$. This shows that

$$\frac{\mathrm{Hom}(R(Q_8^{ab}), \mathbf{Z}_2^*)}{\mathrm{Det}(\mathbf{Z}_2[Q_8^{ab}]^*)} \cong \mathbf{Z}/2$$

is generated by the homomorphism which sends χ_1 to minus one and the other one-dimensional representations to one.

Now consider the classes of homomorphisms, $f : R(Q_8) \longrightarrow \mathbf{Z}_2^*$, in

$$\frac{\mathrm{Hom}(R(Q_8), \mathbf{Z}_2^*)}{\mathrm{Det}(\mathbf{Z}_2[Q_8]^*)}$$

which are trivial on all one-dimensional representations. These functions represent all the elements of

$$\mathrm{Ker}(\mathrm{TorsK}_0(\mathbf{Z}[Q_8]; \mathbf{Q}) \longrightarrow \mathrm{TorsK}_0(\mathbf{Z}[Q_8^{ab}]; \mathbf{Q}) \cong \mathbf{Z}/2).$$

The only determinantal functions which are trivial on all the one-dimensional representations are of the form $\mathrm{Det}(1 + (x^2 - 1)(a + bx + cy + dxy))$. However

$$\begin{aligned}
&\mathrm{Det}(1 + (x^2 - 1)(a + bx + cy + dxy))(\nu) \\
&= \det \begin{pmatrix} 1 - 2(a + bi) & -2(c + di) \\ -2(-c + di) & 1 - 2(a - bi) \end{pmatrix} \\
&= 1 - 4a + 4(a^2 + b^2 + c^2 + d^2) \in 1 + 4\mathbf{Z}_2.
\end{aligned}$$

Therefore multiplying by determinantal functions of this kind can only change $f(\nu)$ modulo four. Since the function which is minus one on ν and trivial on all the one-dimensional representations is a generator for the class-group of $\mathbf{Z}[Q_8]$ the previous argument shows that $\mathrm{TorsK}_0(\mathbf{Z}[Q_8]; \mathbf{Q})$ maps onto $\mathcal{CL}(\mathbf{Z}[Q_8]) \oplus \mathrm{TorsK}_0(\mathbf{Z}[Q_8^{ab}]; \mathbf{Q})$ and that the order of $\mathrm{TorsK}_0(\mathbf{Z}[Q_8]; \mathbf{Q})$ is at most four, which completes the proof. □

Corollary 2.5.38

$$\mathrm{TorsK}_0(\mathbf{Z}[\mathbf{Z}/2]; \mathbf{Q}) = 0.$$

Proof. The discussion in the proof of Proposition 2.5.37 shows that

$$\mathrm{TorsK}_0(\mathbf{Z}[\mathbf{Z}/2 \times \mathbf{Z}/2]; \mathbf{Q}) \longrightarrow \mathrm{TorsK}_0(\mathbf{Z}[\mathbf{Z}/2]; \mathbf{Q})$$

is onto. However, a Hom-description generator of $\mathrm{TorsK}_0(\mathbf{Z}[\mathbf{Z}/2 \times \mathbf{Z}/2]; \mathbf{Q})$ may be chosen to be trivial on the two one-dimensional representations inflated from $\mathbf{Z}/2$. Therefore this surjection is also the trivial map. □

Corollary 2.5.39 *In the notation used in the proof of Proposition 2.5.37 an isomorphism of the form*

$$\lambda : \frac{\mathrm{Hom}(R(Q_8^{ab}), \mathbf{Z}_2^*)}{\mathrm{Det}(\mathbf{Z}_2[Q_8^{ab}]^*)} \stackrel{\cong}{\longrightarrow} (\mathbf{Z}/4)^* = \{\pm 1\}$$

is given by

$$\lambda[f] = f(1 + \chi_1 + \chi_2 + \chi_1\chi_2) \qquad \text{(modulo } 4).$$

Proof. Let $a + bx + cy + dxy \in \mathbf{Z}_2[Q_8^{ab}]^*$. Then, as in the proof of Proposition 2.5.37, we have

$$\begin{aligned}
&\mathrm{Det}(a + bx + cy + dxy)(1 + \chi_1 + \chi_2 + \chi_1\chi_2) \\
&= (a+b+c+d)(a-b+c-d)(a+b-c-d)(a-b-c+d) \\
&= ((a+b)^2 - (c+d)^2)((a-b)^2 - (c-d)^2) \\
&= (a^2 + b^2 - c^2 - d^2 + (2ab - 2cd))(a^2 + b^2 - c^2 - d^2 - (2ab - 2cd)) \\
&\equiv (a^2 + b^2 - c^2 - d^2)^2 \qquad \text{(modulo } 4) \\
&= 1 \qquad \text{(modulo } 4)
\end{aligned}$$

since $a^2 + b^2 - c^2 - d^2$ is a unit in the 2-adic integers.

Therefore the homomorphism, λ, is well defined. It remains to observe, from the proof of Proposition 2.5.37, that the domain of λ is generated by the homomorphism, h, which sends χ_1 to minus one and the other one-dimensional representations to one. Thus $\lambda[h] = -1$, which completes the proof. □

2.5.40 *A generator for* $\mathrm{TorsK}_0(\mathbf{Z}[Q_8^{ab}]; \mathbf{Q})$

If Q_8^{ab} is generated by x and y then $e = (1 - x + y + xy)/2 \in \mathbf{Q}[Q_8^{ab}]$ satisfies $e^2 = 1$ since evaluating all the four one-dimensional representations on e^2 gives

$$\begin{aligned}
&1(e^2) = (2/2)^2 = 1, \chi_1(e^2) = ((1+1+1-1)/2)^2 = 1, \\
&\chi_2(e^2) = ((1-1-1-1)/2)^2 = 1, \chi_1\chi_2(e^2) = ((1+1-1+1)/2)^2 = 1.
\end{aligned}$$

Therefore the class $[\mathbf{Z}[Q_8^{ab}], e, \mathbf{Z}[Q_8^{ab}]] \in \mathrm{TorsK}_0(\mathbf{Z}[Q_8^{ab}]; \mathbf{Q})$ is either trivial or of order two since $e^2 = 1$. The Hom-description representative of this element is

$$\mathrm{Det}(e) : R(Q_8^{ab}) \longrightarrow \{\pm 1\} \subseteq \mathbf{Z}_2^* \subseteq \mathbf{Q}_2^*$$

and this is the function

$$\mathrm{Det}(e)(1) = 1, \ \mathrm{Det}(e)(\chi_1) = 1, \ \mathrm{Det}(e)(\chi_2) = -1, \ \mathrm{Det}(e)(\chi_1\chi_2) = 1$$

which, by Proposition 2.5.37 (or Corollary 2.5.39), is a function representing a generator. □

For completeness, we close this section with some facts about Hom-description representatives in $\mathcal{CL}(\mathbf{Z}[Q_8]) = D(\mathbf{Z}[Q_8]) \cong (\mathbf{Z}/4)^*$ and an algorithm for computing the class of a projective $\mathbf{Z}[Q_8]$-module.

Lemma 2.5.41 *([155] p. 88)*
If $g \in \mathrm{Hom}(R(Q_8), \mathbf{Z}_2^*)$ then the class

$$[g] \in D(\mathbf{Z}[Q_8]) = \mathcal{CL}(\mathbf{Z}[Q_8]) \cong (\mathbf{Z}/4)^*$$

corresponding to g in the Hom-description isomorphism

$$\mathcal{CL}(\mathbf{Z}[Q_8^{ab}]) \cong \frac{\mathrm{Hom}(R(Q_8^{ab}), \mathbf{Z}_2^*)}{\mathrm{Det}(\mathbf{Z}_2[Q_8^{ab}]^*)\, \mathrm{Hom}(R(Q_8^{ab}), \{\pm 1\})}$$

is given by the formula

$$[g] \equiv g(\nu)(-1)^{(1/4)\log_2(g(1+\chi_1+\chi_2+\chi_1\chi_2))} \text{ (modulo } 4).$$

2.5.42 *The class of a projective $\mathbf{Z}[Q_8]$-module*
It is useful to be able to determine the element of a finitely generated, projective $\mathbf{Z}[Q_8]$-module, M, in the class-group. For this there is an invariant, η, ([131] §5.2.9) which gives an isomorphism

$$\eta : \mathcal{CL}(\mathbf{Z}[Q_8]) \longrightarrow (\mathbf{Z}/4)^*$$

in terms of projective modules.

Define M_+ and M_- by the formulae

$$M_+ = \{m \in M \mid x^2(m) = m\}$$

and

$$M_- = \{m \in M \mid x^2(m) = -m\}.$$

Note that M_+ is naturally a module over the ring $\mathbf{Z}[Q_8]/(x^2-1) \cong \mathbf{Z}[\mathbf{Z}/2 \times \mathbf{Z}/2]$ while M_- is a module over $\mathbf{Z}[Q_8]/(x^2+1) \cong \mathbf{H}_{\mathbf{Z}}$, the ring of integral quaternions,

$$\mathbf{H}_{\mathbf{Z}} = \mathbf{Z}[i,j,k]/(i^2 = j^2 = k^2 = -1, ij = k, jk = i, ki = j).$$

However, over each of these quotient rings a finitely generated, projective module is free. Also, if M is a finitely generated, projective $\mathbf{Z}[Q_8]$-module then the quotient map

$$M/M_- \stackrel{\cong}{\longrightarrow} M_+$$

is an isomorphism of $\mathbf{Z}[\mathbf{Z}/2 \times \mathbf{Z}/2]$-modules. Therefore we may take a free $\mathbf{H}_{\mathbf{Z}}$-basis

$$\underline{a} = \{a_1, \ldots, a_k\} \in M_-$$

and we may choose

$$\underline{\alpha} = \{\alpha_1, \ldots, \alpha_k\} \in M$$

such that

$$\{\alpha_1 + M_-, \ldots, \alpha_k + M_-\}$$

is a free $\mathbf{Z}[\mathbf{Z}/2 \times \mathbf{Z}/2]$-basis for $M/M_- \cong M_+$. Now form

$$(1 - x^2)\underline{\alpha} = \{(1 - x^2)\alpha_1, \ldots, (1 - x^2)\alpha_k\} \subset M_-.$$

There is a $k \times k$ matrix, X, with entries in $\mathbf{H_Z}$, such that

$$\begin{pmatrix} (1-x^2)\alpha_1 \\ \vdots \\ (1-x^2)\alpha_k \end{pmatrix} = X \begin{pmatrix} a_1 \\ \vdots \\ a_k \end{pmatrix} \in M_-^k.$$

Let

$$c : M_k(\mathbf{H_Z}) \longrightarrow M_{2k}(\mathbf{Z}[i])$$

denote the complexification map, where $\mathbf{Z}[i]$ denotes the Gaussian integers. Now define $\eta([M])$ by the formula

$$\eta([M]) = \det(c(X)) \pmod 4 \in (\mathbf{Z}/4)^* \cong \mathbf{Z}/2.$$

Chapter 3

Higher K-theory of Local Fields

In this chapter we shall examine invariants of the Galois module structure on the higher algebraic K-groups of local fields. These will be constructed by the method of Example 2.1.8(i) as the Euler characteristic of suitable 2-extensions, called the *local fundamental classes* lying in $\mathrm{Ext}^2_{\mathbf{Z}[G(L/K)]}(K_{2r}(L), K_{2r+1}(L))$ for $r \geq 1$, where L/K is a Galois extension with group $G(L/K)$. Actually, when L is a p-adic local field $K_{2r+1}(L)$ is not a finitely generated $\mathbf{Z}[G(L/K)]$-module and so one applies the construction of Example 2.1.8(i) to the canonical corresponding element in $\mathrm{Ext}^2_{\mathbf{Z}[G(L/K)]}(K_{2r}(L), K_{2r+1}(L)/A)$ where A is a cohomologically trivial $\mathbf{Z}[G(L/K)]$-submodule chosen so that $K_{2r+1}(L)/A$ is finitely generated.

We shall begin with the construction of the local fundamental classes, which generalize the classical fundamental classes of local class field theory ([128] p. 202; see also [131] Chapter VII, [132] p. 9), which correspond to the case when $r = 0$.

3.1 Local fundamental classes and K-groups

3.1.1 Suppose that G is a finite group and that

$$\underline{\underline{E}}: \quad A \longrightarrow B \longrightarrow C \longrightarrow D$$

is a 2-extension of $\mathbf{Z}[G]$-modules in which B and C are cohomologically trivial. Such a sequence defines $[\underline{\underline{E}}] \in \mathrm{Ext}^2_{\mathbf{Z}[G]}(D, A)$ which induces cup-product isomorphisms in Tate cohomology

$$([\underline{\underline{E}}] \bigcup -) : \hat{H}^i(J; D) \xrightarrow{\cong} \hat{H}^{i+2}(J; A)$$

for all integers, i, and all subgroups, $J \subseteq G$.

There are two natural operations associated to a subgroup, $J \subseteq G$. The first – passage to subgroups – is merely to consider the modules as $\mathbf{Z}[J]$-modules. The second – passage to quotient groups – is more complicated and applies to the case of a normal subgroup, $J \lhd G$. In this case, let A^J, A_J denote the J-invariants and

J-coinvariants of A, respectively ([132] p. 3). We have a commutative diagram of $\mathbf{Z}[G/J]$-modules in which the rows and columns are exact and N_J denotes the norm, $N_J(x) = \sum_{g \in J} gx$,

$$\begin{array}{ccccccc} & & B_J & \longrightarrow & C_J & \longrightarrow & D_J \longrightarrow 0 \\ & & \cong \downarrow N_J & & \cong \downarrow N_J & & \\ 0 \longrightarrow & A^J & \longrightarrow B^J & \longrightarrow & C^J & & \end{array}$$

resulting in the associated J-invariant/coinvariant 2-extension

$$A^J \longrightarrow B^J \longrightarrow C^J \longrightarrow D_J$$

in which the $\mathbf{Z}[G/J]$-modules, B^J and C^J, are cohomologically trivial.

These operations are relevant to the naturality properties of the higher K-theory local fundamental classes (see Theorem 3.1.20).

3.1.2 Now we shall begin the construction of the local fundamental classes by considering the totally ramified case. Actually, in [134] and [135] we showed how to construct the canonical, natural 2-extension (depending up to isomorphism upon a choice of a root of unity, ξ_t) of Theorem 3.1.17 for finite Galois extensions of p-adic local fields. Furthermore, for K_{2r}, K_{2r+1} with $r \geq 2$ in this case we needed to assume that the Lichtenbaum-Quillen Conjecture ([135] p. 326) was true for the mod p algebraic K-theory of p-adic local fields. When [135] was written this was known for 2-adic fields by [160]. For p-adic fields when p is odd the conjecture was proved recently [74]. When L/K is a Galois extension of local fields of characteristic p the Lichtenbaum-Quillen Conjecture is now known to be true by [60], which shows that the K-theory of L has no p-torsion, combined with the results of [148]. These advances made it possible to construct the local fundamental classes associated to the higher K-groups of local fields without any assumptions (for further details, see [134], [135], [136] and [137]).

If A is an abelian group, we shall denote by $\mathrm{Tors}(A)$ the torsion subgroup of A. Also we set ${}_nA = \{a \in A \mid na = 0\}$, ${}_{p^\infty}A = \cup_{m=1}^{\infty} {}_{p^m}A$, $A/n = A/nA = A \otimes \mathbf{Z}/n$ and $K_i(X)$ will denote the i-dimensional Quillen K-group of the ring or scheme, X.

Let L/K be a totally ramified Galois extension of local fields (p-adic or of characteristic p; our construction is not very interesting for Archimdean local fields) with Galois group, $G(L/K)$. Let K_0/K denote the maximal unramified extension, obtained by adjoining all roots of unity of order prime to p, and set $L_0 = K_0L$. Let $F \in G(K_0/K) \cong \hat{\mathbf{Z}} \cong G(L_0/L)$ denote the topological generator given by the Frobenius automorphism. Then we have an isomorphism of the form

$$G(L_0/K) \cong G(L/K) \times G(K_0/K) \cong G(L/K) \times \hat{\mathbf{Z}}.$$

For each integer, s, let $K(s) \subset K_0$ denote the fixed field of $s\hat{\mathbf{Z}}$ and set $L(s) = K(s)L$. Hence $G(L(s)/L) \cong \mathbf{Z}/s$.

For $r \geq 2$, let

$$\kappa(s) = \mathrm{Ker}(N_{L(s)/L} : K_{2r-1}(L(s)) \longrightarrow K_{2r-1}(L)).$$

The map, $i_{L(st)/L(s)} : K_{2r-1}(L(s)) \longrightarrow K_{2r-1}(L(st))$, is injective. This is seen on the p-primary torsion subgroup by means of ([135] Proposition 2.2(i)) for p-adic fields. In characteristic p there is no p-primary torsion by [60]. By [148], the map on the prime-to-p torsion may be identified with the injective map on K_{2r-1} of the residue fields. For the torsion-free part it suffices to show that $K^{top}_{2r-1}(L(s)) \otimes \mathbf{Q}_p \longrightarrow K^{top}_{2r-1}(L(st)) \otimes \mathbf{Q}_p$ is injective and this follows from [162].

Therefore we may define

$$\mathcal{U} = \bigcup_s \kappa(s) \subseteq \bigcup_s K_{2r-1}(L(s)) = K_{2r-1}(L_0).$$

Similarly,

$$\mathrm{Ker}(1 - F : K_{2r-1}(L(s)) \longrightarrow K_{2r-1}(L(s)))$$

is isomorphic to $K_{2r-1}(L(s)^{G(L(s)/L)}) \cong K_{2r-1}(L)$. This follows by a transfer argument on the uniquely divisible part, from ([135] Proposition 2.2(ii)) on the p-adic part in the p-adic case (in characteristic p there is no p-adic part by [60]) and from the exactness of the sequence [148]

$$0 \longrightarrow K_{2r-1}(\mathcal{O}_L/\pi_L) \longrightarrow K_{2r-1}(\mathcal{O}_{L(s)}/\pi_{L(s)}) \stackrel{1-F}{\longrightarrow} K_{2r-1}(\mathcal{O}_{L(s)}/\pi_{L(s)})$$

on the prime-to-p torsion.

For $r \geq 2$, there exist exact sequences of the form ([135] §2)

$$\begin{array}{l} 0 \longrightarrow K_{2r-1}(L) \longrightarrow K_{2r-1}(L(s)) \stackrel{1-F}{\longrightarrow} \kappa(s) \\ \longrightarrow \mathrm{Ker}(K_{2r-2}(L) \longrightarrow K_{2r-2}(L(s))^{G(L(s)/L)}) \longrightarrow 0 \end{array}$$

whose direct limit yields the following result (cf. [135] Corollary 2.6).

Theorem 3.1.3

Let L/K be a totally ramified Galois extension of non-Archimedean local fields with Galois group, $G(L/K)$, as in §3.1.2. Then there is a natural 2-extension of $\mathbf{Z}[G(L/K)]$-modules of the form

$$K_{2r-1}(L) \longrightarrow K_{2r-1}(L_0) \stackrel{1-F}{\longrightarrow} \mathcal{U} \longrightarrow {}_{p^\infty}K_{2r-2}(L).$$

Furthermore, ${}_{p^\infty}K_{2r-2}(L)$ is trivial if L has characteristic $p > 0$.

3.1.4 As in §3.1.2, let L/K be a totally ramified Galois extension of local fields of residue characteristic p, with Galois group, $G(L/K)$.

If L is a local field, let $\overline{L}$ denote its residue field. Hence $\overline{L_0}$ is equal to the algebraic closure of $\mathbf{F}_p$ and $K_{2r-1}(\overline{L_0}) \cong \mathbf{Q}/\mathbf{Z}(r)[1/p]$ as a Galois module. Here $\mathbf{Q}/\mathbf{Z}(r)$ denotes the r-th Tate twist of the roots of unity module. Set

$$\mathcal{O}_{L_0} = \bigcup_n \mathcal{O}_{L(n)} \subset L_0$$

then the localisation sequence shows that the natural map, $K_{2r-1}(\mathcal{O}_{L_0}) \longrightarrow K_{2r-1}(L_0)$, is an isomorphism for all $r \geq 2$. By [148], the prime-to-p torsion in $K_{2r-1}(\mathcal{O}_{L_0})$ is isomorphic to $\mathbf{Q}/\mathbf{Z}(r)[1/p]$ via the homomorphism induced by the canonical quotient map, $\mathcal{O}_{L_0} \longrightarrow \overline{L_0}$. We have a chain of inclusions of $\mathbf{Z}[G(L/K)]$-modules

$$\mathbf{Q}/\mathbf{Z}(r)[1/p] \subseteq \mathcal{U} \subseteq K_{2r-1}(L_0)$$

inducing cohomology isomorphisms for all i

$$\hat{H}^i(G(L/K); \mathbf{Q}/\mathbf{Z}(2)[1/p]) \cong \hat{H}^i(G(L/K);\mathcal{U}) \cong \hat{H}^i(G(L/K); K_{2r-1}(L_0)).$$

In fact, since $G(L/K)$ acts trivially on $\mathbf{Q}/\mathbf{Z}(r)[1/p]$ these groups are given by

$$\hat{H}^i(G(L/K); K_{2r-1}(L_0)) \cong \begin{cases} \mathbf{Z}/[G(L/K) : G_1(L/K)] & \text{if } i \text{ is odd,} \\ 0 & \text{if } i \text{ is even.} \end{cases}$$

Here $G_i(L/K)$ denotes the i-th ramification group of L/K so that, in §3.1.2, $G(L/K) = G_0(L/K)$ and $G_1(L/K)$ is the first wild ramification group.

3.1.5 *Galois groups in the general case*

Let L/K be a Galois extension of local fields of residue characteristic p and Galois group, $G(L/K)$. We shall relax the total ramification condition of §3.1.2. Set $W = L \cap L_0$ so that W/K is the maximal unramified subextension of L/K. Hence L/W is totally ramified and the sequence

$$0 \longrightarrow K_{2r-1}(L) \longrightarrow K_{2r-1}(L_0) \stackrel{1-F}{\longrightarrow} \mathcal{U} \longrightarrow {}_{p^\infty}K_{2r-2}(L) \longrightarrow 0$$

of Theorem 3.1.3 is a 2-extension of $\mathbf{Z}[G(L/W)]$-modules.

We wish to construct a similar 2-extension of $\mathbf{Z}[G(L/K)]$-modules, following the method of ([134] §3) which was in turn an imitation of the fundamental class of local class field theory, as described in ([131] §7.1; [132] §1.2).

Notice that $G(L/W) = G_0(L/K)$ and $G_1(L/W) = G_1(L/K)$, by ([128] p. 62), so that $t = [G(L/W) : G_1(L/W)] = [G_0(L/K) : G_1(L/K)]$. If $G(L/E) = G_1(L/K)$ then

$$G(L/K)/G(L/E) \cong G(E/K) = \{a, g \mid a^t = 1, g^d = a^c, \mathrm{gag}^{-1} = a^v\}$$

where $\langle a\rangle = G(E/W)$, $d = [W : K]$ and $v = |\mathcal{O}_K/\pi_K| = |\overline{K}|$. Here $\mathcal{O}_K$ and π_K denote the integers and a prime element of K, respectively.

Now let us consider the Galois groups which will be involved in the construction. We have a map, given by the restriction of g to W and denoted by $(g|W)$, inducing a homomorphism

$$h_1 : G(K_0/K) \times G(L/K) \cong \hat{\mathbf{Z}} \times G(L/K) \longrightarrow G(W/K)$$

defined by $h_1(F^i, z) = (F^i|W) \cdot (z|W)^{-1}$. Here, as in §3.1.2, F denotes the Frobenius automorphism. To a pair, $(F^i, z) \in \mathrm{Ker}(h_1)$, we may associate the Galois automorphism of $L_0 = LK_0$ which is equal to F^i on K_0 and to z on L. This induces an isomorphism

$$\mathrm{Ker}(h_1) \xrightarrow{\cong} G(L_0/K).$$

If $d = [W : K]$ define $F_0 \in G(L_0/K)$ by $(F_0|K_0) = F^d$ and $(F_0|L) = 1$, the identity map.

As explained in ([131] pp. 303–304; [132] pp. 9–10), there is an isomorphism of K-algebras

$$\lambda : L \otimes_K K_0 \longrightarrow \oplus_{i=1}^{d} L_0$$

given by the formula $\lambda(\alpha \otimes \beta) = (F^{d-1}(\beta)\alpha, F^{d-2}(\beta)\alpha, \ldots, \beta\alpha)$ and fitting into a commutative diagram of the form

$$\begin{array}{ccc} L \otimes_K K_{nr} & \xrightarrow{(1 \otimes F)} & L \otimes_K K_{nr} \\ \downarrow \lambda & & \lambda \downarrow \\ \oplus_{i=1}^{d} L_0 & \xrightarrow{\hat{F}} & \oplus_{i=1}^{d} L_0 \end{array}$$

where $\hat{F}(x_1, \ldots, x_d) = (F_0(x_d), x_1, \ldots, x_{d-1})$. Since K-theory is additive there is an induced isomorphism

$$\lambda : K_s(L \otimes_K K_0) \longrightarrow \oplus_{i=1}^{d} K_s(L_0)$$

for each $s \geq 0$. This isomorphism induces a $G(L/K)$-action on the right-hand group, induced by the Galois action on L. Since the Frobenius acts on the other factor, K_0, in the tensor product we may define a $\mathbf{Z}[G(L/K)]$-homomorphism

$$\hat{F} : \oplus_{i=1}^{d} K_s(L_0) \longrightarrow \oplus_{i=1}^{d} K_s(L_0)$$

by the same formula as before.

The explicit formula for the group action on these modules is given by the following result, proved in ([134] Lemma 3.2).

Lemma 3.1.6

Suppose that $g \in G(L/K)$ satisfies $(g|W) = (F^j|W)$ for some $0 \leq j \leq d-1$. Then the action of g on $(u_1, u_2, \ldots, u_d) \in \oplus_{i=1}^d K_{2r-1}(L_0)$ is given by

$$g(u_1, u_2, \ldots, u_d) = \hat{F}^{-j}((F^j, g)(u_1), (F^j, g)(u_2), \ldots, (F^j, g)(u_d)).$$

Definition 3.1.7 Let $\mathcal{U} \subset K_{2r-1}(L_0)$ be as in §3.1.2. Define a subgroup, $\mathcal{W} \subset \oplus_{i=1}^d K_{2r-1}(L_0)$, by

$$\mathcal{W} = \{(u_1, u_2, \ldots, u_d) | \sum_{i=1}^{d} u_i \in \mathcal{U}\}.$$

Then $\mathcal{W} \subset \oplus_{i=1}^d K_{2r-1}(L_0)$ is a $\mathbf{Z}[G(L/K)]$-submodule and the image of $1 - \hat{F} : \oplus_{i=1}^d K_{2r-1}(L_0) \longrightarrow \oplus_{i=1}^d K_{2r-1}(L_0)$ lies in $\mathcal{W}$ (cf. [134] Lemma 3.4).

Proposition 3.1.8 *(cf. [134] Proposition 3.5)*

In the notation of §3.1.5, for $r \geq 2$ there is a natural 2-extension of $\mathbf{Z}[G(L/K)]$-modules of the form

$$0 \longrightarrow K_{2r-1}(L) \xrightarrow{\Delta} \oplus_{i=1}^d K_{2r-1}(L_0) \xrightarrow{1-\hat{F}} \mathcal{W} \xrightarrow{\pi} {}_{p^\infty}K_{2r-2}(L) \longrightarrow 0$$

where Δ denotes the diagonal map and π is the composition of the map of Theorem 3.1.3 with the addition of coordinates, $\mathcal{W} \longrightarrow \mathcal{U}$.

Theorem 3.1.9

With the notation of §3.1.5 the cohomology groups of the $\mathbf{Z}[G(L/K)]$-module, $\oplus_{i=1}^d K_{2r-1}(L_0)$, are given by

$$H^i(G(L/K); \oplus_{i=1}^d K_{2r-1}(L_0)) \cong \begin{cases} \mathbf{Z}/t & \text{if } i \text{ is odd,} \\ 0 & \text{if } 0 < i \text{ is even,} \\ K_{2r-1}(W_0) & \text{if } i = 0. \end{cases}$$

Here $t = [G_0(L/K) : G_1(L/K)]$.

Proof. The proof follows that of ([134] Theorem 3.6) with $q = |\mathcal{O}_K/\pi_K|^2 = v^2$ replaced by $q = v^r$ since the Frobenius acts on $(\mathbf{Q}/\mathbf{Z})(r)[1/p]$ by multiplication by v^r. □

3.1.10 Suppose that L/K is a totally ramified Galois extension of local fields of residue characteristic p with group, $G(L/K)$, and set $t = [G(L/K) : G_1(L/K)]$. Choose a primitive t-th root of unity, ξ_t. If X is any $\mathbf{Z}[G(L/K)]$-module containing ${}_t(\mathbf{Q}/\mathbf{Z})(r)[1/p]$, depending on the choice of ξ_t, we shall define a new module $X\tilde{\oplus}\mathbf{Z}$, and an extension of the form

$$0 \longrightarrow X \longrightarrow X\tilde{\oplus}\mathbf{Z} \longrightarrow \mathbf{Z} \longrightarrow 0.$$

For example, this construction may be applied with X equal to any module in the chain

$$\mathrm{Tors}(K_{2r-1}(L_0))[1/p] \cong (\mathbf{Q}/\mathbf{Z})(r)[1/p] \subset \mathcal{U} \subset K_{2r-1}(L_0).$$

Note that the action of $G(L/K)$ on $(\mathbf{Q}/\mathbf{Z})(r)[1/p]$ is trivial, since L/K is totally ramified, while the Frobenius, F of §3.1.2, acts like "multiplication" by v^r where $v = |\mathcal{O}_L/\pi_L|$.

As an abelian group $X\tilde{\oplus}\mathbf{Z}$ is just the direct sum, $X \oplus \mathbf{Z}$. As a $\mathbf{Z}[G(L/K)]$-module the embedding of X is given by sending $x \in X$ to $(x, 0)$.

If $G_1(L/K) = G(L/E)$, as in §3.1.5, the action on $(0,1) \in X\tilde{\oplus}\mathbf{Z}$ will factor through $G(L/K)/G(L/E) \cong G(E/K)$. Furthermore, if $G(E/K) \cong \mathbf{Z}/t$ with generator, g, then $g(0,1) = (\xi_t, 1)$. This action is well defined since $g^i(0,1) = (i\xi_t, 1)$. Here we have written the group operation in the X-coordinate in additive notation, following [135] (later, when X is the twisted roots of unity, we shall correct this eccentricity and revert to the multiplicative notation).

Similarly we may define a $\mathbf{Z}[G(L/K)]$-module, $X\tilde{\oplus}\mathbf{Z}[1/p]$, and an extension of the form

$$0 \longrightarrow X \longrightarrow X\tilde{\oplus}\mathbf{Z}[1/p] \longrightarrow \mathbf{Z}[1/p] \longrightarrow 0$$

by setting

$$X\tilde{\oplus}\mathbf{Z}[1/p] = \lim_{\rightarrow}(X\tilde{\oplus}\mathbf{Z} \stackrel{(1,v^r)}{\longrightarrow} X\tilde{\oplus}\mathbf{Z} \stackrel{(1,v^r)}{\longrightarrow} X\tilde{\oplus}\mathbf{Z} \stackrel{(1,v^r)}{\longrightarrow} \ldots),$$

the direct limit of iterations of the map sending (z, m) to (z, mv^r) where $v = |\mathcal{O}_L/\pi_L| = \overline{L}$. This is a $\mathbf{Z}[G(L/K)]$-module, since $v^r \equiv 1$ (*modulo* t).

Proposition 3.1.11 *([134] Proposition 3.8)*

In the notation of §3.1.10, let X be any of the $\mathbf{Z}[G(L/K)]$-modules in the chain

$$\mathrm{Tors}(K_{2r-1}(L_0))[1/p] \cong (\mathbf{Q}/\mathbf{Z})(r)[1/p] \subset \mathcal{U} \subset K_{2r-1}(L_0).$$

Then $X\tilde{\oplus}\mathbf{Z}[1/p]$ is a cohomologically trivial $\mathbf{Z}[G(L/K)]$-module.

3.1.12 In the situation of §3.1.5 we shall define a $\mathbf{Z}[G(L/K)]$-module

$$\tilde{\oplus}_{i=1}^{d} K_{2r-1}(L_0).$$

As an abelian group this module is simply the direct sum

$$(\oplus_{i=1}^{d} K_{2r-1}(L_0)) \oplus (\oplus_{i=1}^{d} \mathbf{Z}[1/p]).$$

The summand, $\oplus_{i=1}^{d} K_{2r-1}(L_0)$, has the $\mathbf{Z}[G(L/K)]$-module structure described in §3.1.5 and Lemma 3.1.6. Thus $g \in G(L/K)$ lifts to $(F, F) \in G(L_0/K)$ and acts like

$$g(a_1, \ldots, a_d) = (F(a_2), F(a_3), \ldots, F(a_d), F_0^{-1}(F(a_1))).$$

Any element, $h \in G(L/W)$, acts component-by-component as $h(a_1, \ldots, a_d) = (h(a_1), \ldots, h(a_d))$.

Consider the $\mathbf{Z}[G(L/K)]$-submodule,

$$\oplus_{i=1}^{d}(\mathbf{Q}/\mathbf{Z})(r)[1/p] \quad \text{contained in } \oplus_{i=1}^{d} K_{2r-1}(L_0).$$

Since L/W is totally ramified, $G(L/W)$ acts trivially on this submodule. Hence $G(L/K)$ acts through the quotient, $G(W/K) = \langle g \rangle$, with

$$g(a_1, \ldots, a_d) = (F(a_2), \ldots, F(a_d), F_0^{-1}F(a_1)) = (qa_2, qa_3, \ldots, qa_d, q^{1-d}a_1)$$

where $q = v^r$ and $v = |\mathcal{O}_K/\pi_K|$. An arbitrary element of the $\mathbf{Z}[G(L/K)]$-module

$$\mathrm{Inf}_{G(W/K)}^{G(L/K)} \mathrm{Ind}_{\{1\}}^{G(W/K)}((\mathbf{Q}/\mathbf{Z})(r)[1/p])$$

has the form $z = 1 \otimes b_d + g \otimes b_{d-1} + \ldots + g^{d-1} \otimes b_1$ and $g(z) = 1 \otimes b_1 + g \otimes b_d + \ldots + g^{d-1} \otimes b_2$. Hence, if we write $z = (b_1, \ldots, b_d)$, then $g(z) = (b_2, b_3, \ldots, b_d, b_1)$.

Define an isomorphism

$$\phi : \oplus_{i=1}^{d}(\mathbf{Q}/\mathbf{Z})(r)[1/p] \xrightarrow{\cong} \mathrm{Inf}_{G(W/K)}^{G(L/K)} \mathrm{Ind}_{\{1\}}^{G(W/K)}((\mathbf{Q}/\mathbf{Z})(r)[1/p])$$

by $\phi(a_1, \ldots, a_d) = (q^{1-d}a_1, q^{2-d}a_2, \ldots, q^{-1}a_{d-1}, a_d)$. This is an isomorphism of $\mathbf{Z}[G(L/K)]$-modules, since

$$\begin{aligned} \phi(g(a_1, \ldots, a_d)) &= \phi(qa_2, qa_3, \ldots, qa_d, q^{1-d}a_1) \\ &= (q^{2-d}a_2, q^{3-d}a_3, \ldots, a_d, q^{1-d}a_1) \end{aligned}$$

and

$$\begin{aligned} g(\phi(a_1, \ldots, a_d)) &= g(q^{1-d}a_1, q^{2-d}a_2, \ldots, q^{-1}a_{d-1}, a_d) \\ &= (q^{2-d}a_2, q^{3-d}a_3, \ldots, a_d, q^{1-d}a_1). \end{aligned}$$

We have an isomorphism of $\mathbf{Z}[G(L/K)]$-modules

$$\mathrm{Inf}_{G(W/K)}^{G(L/K)} \mathrm{Ind}_{\{1\}}^{G(W/K)}((\mathbf{Q}/\mathbf{Z})(r)[1/p]) \cong \mathrm{Ind}_{G(L/W)}^{G(L/K)}((\mathbf{Q}/\mathbf{Z})(r)[1/p])$$

where $G(L/W)$ acts trivially on $(\mathbf{Q}/\mathbf{Z})(r)[1/p]$. Hence we may form the following push-out diagram, which defines the module $\tilde{\oplus}_{i=1}^{d} K_{2r-1}(L_0)$.

$$\begin{array}{ccc} \oplus_{i=1}^{d}(\mathbf{Q}/\mathbf{Z})(r)[1/p] \cong \mathrm{Ind}_{G(L/W)}^{G(L/K)}((\mathbf{Q}/\mathbf{Z})(r)[1/p]) & \longrightarrow & \oplus_{i=1}^{d} K_{2r-1}(L_0) \\ \downarrow & & \downarrow \\ \mathrm{Ind}_{G(L/W)}^{G(L/K)}((\mathbf{Q}/\mathbf{Z})(r)[1/p] \tilde{\oplus} \mathbf{Z}[1/p]) & \longrightarrow & \tilde{\oplus}_{i=1}^{d} K_{2r-1}(L_0) \end{array}$$

Proposition 3.1.13

The $\mathbf{Z}[G(L/K)]$-module, $\tilde{\oplus}_{i=1}^{d} K_{2r-1}(L_0)$, of §3.1.12 is cohomologically trivial.

Proof. Since the upper horizontal homomorphism induces isomorphisms in cohomology so does the lower. Therefore the result follows from Proposition 3.1.11. □

3.1.14 Now consider the $\mathbf{Z}[G(L/K)]$-submodule, $\mathcal{W} \subset \oplus_{i=1}^{d} K_{2r-1}(L_0)$, of Definition 3.1.7. Since $(\mathbf{Q}/\mathbf{Z})(r)[1/p] \subset \mathcal{U}$ we have $\oplus_{i=1}^{d}(\mathbf{Q}/\mathbf{Z})(r)[1/p] \subset \mathcal{W}$. Also $G(L/W)$ acts component-by-component on $\mathcal{W}$ so that $\mathcal{W} \cong (\oplus_{i=1}^{d-1} K_{2r-1}(L_0)) \oplus \mathcal{U}$ as a $\mathbf{Z}[G(L/W)]$-module. Therefore, by the cohomology isomorphisms of §3.1.4, the inclusion induces a cohomology isomorphism of the following form

$$H^t(G(L/W); \mathcal{W}) \xrightarrow{\cong} H^t(G(L/W); \oplus_{i=1}^{d} K_{2r-1}(L_0)).$$

Comparison of cohomology spectral sequences implies that the natural map

$$\hat{H}^t(G(L/K); \mathcal{W}) \longrightarrow \hat{H}^t(G(L/K); \oplus_{i=1}^{d} K_{2r-1}(L_0))$$

is also an isomorphism.

Define $\tilde{\mathcal{W}}$ by the pushout diagram

$$\begin{array}{ccc} \oplus_{i=1}^{d}(\mathbf{Q}/\mathbf{Z})(r)[1/p] \cong \mathrm{Ind}_{G(L/W)}^{G(L/K)}((\mathbf{Q}/\mathbf{Z})(r)[1/p]) & \longrightarrow & \mathcal{W} \\ \downarrow & & \downarrow \\ \mathrm{Ind}_{G(L/W)}^{G(L/K)}((\mathbf{Q}/\mathbf{Z})(r)[1/p] \tilde{\oplus} \mathbf{Z}[1/p]) & \longrightarrow & \tilde{\mathcal{W}} \end{array}$$

so that we immediately obtain the following result from Proposition 3.1.13.

Proposition 3.1.15
The $\mathbf{Z}[G(L/K)]$-module, $\tilde{\mathcal{W}}$, of §3.1.14 is cohomologically trivial.

3.1.16 On $(\mathbf{Q}/\mathbf{Z})(r)[1/p]$ F_0 acts by multiplication by q^d where $q = v^r$, $v = |\mathcal{O}_K/\pi_K|$. Hence, if $a_i \in (\mathbf{Q}/\mathbf{Z})(r)[1/p]$, then

$$(1 - \hat{F})(a_1, \ldots, a_d) = (a_1 - q^d a_d, a_2 - a_1, \ldots, a_d - a_{d-1}).$$

Now consider the isomorphism of §3.1.12

$$\phi : \oplus_{i=1}^{d}(\mathbf{Q}/\mathbf{Z})(r)[1/p] \xrightarrow{\cong} \mathrm{Ind}_{G(L/W)}^{G(L/K)}((\mathbf{Q}/\mathbf{Z})(r)[1/p])$$

where, on the right, $G(L/W)$ acts trivially on $(\mathbf{Q}/\mathbf{Z})(r)[1/p]$ and

$$\phi(a_1, \ldots, a_d) = (q^{1-d}a_1, q^{2-d}a_2, \ldots, q^{-1}a_{d-1}, a_d) = \sum_{i=0}^{d-1} g^i \otimes q^{-i} a_{d-i}.$$

Hence $\hat{F}$ translates to $\phi\hat{F}\phi^{-1}$, which is given by

$$\begin{aligned}
&\phi(\hat{F}(\phi^{-1}(\textstyle\sum_{i=0}^{d-1} g^i \otimes b_{d-i})))\\
&= \phi(\hat{F}(\phi^{-1}(b_1, \ldots, b_d)))\\
&= \phi(\hat{F}(q^{d-1}b_1, q^{d-2}b_2, \ldots, qb_{d-1}, b_d))\\
&= \phi(q^d b_d, q^{d-1}b_1, q^{d-2}b_2, \ldots, qb_{d-1})\\
&= (qb_d, qb_1, \ldots, qb_{d-1})\\
&= q(1 \otimes b_{d-1} + g \otimes b_{d-2} + \ldots + g^{d-1} \otimes b_d)\\
&= qg^{d-1}(\textstyle\sum_{i=0}^{d-1} g^i \otimes b_{d-i})\\
&= qg^{-1}(\textstyle\sum_{i=0}^{d-1} g^i \otimes b_{d-i}).
\end{aligned}$$

Define

$$\hat{F} : \mathrm{Ind}_{G(L/W)}^{G(L/K)}((\mathbf{Q}/\mathbf{Z})(r)[1/p]\tilde{\oplus}\mathbf{Z}[1/p]) \longrightarrow \mathrm{Ind}_{G(L/W)}^{G(L/K)}((\mathbf{Q}/\mathbf{Z})(r)[1/p]\tilde{\oplus}\mathbf{Z}[1/p])$$

by the formula ($b \in (\mathbf{Q}/\mathbf{Z})(r)[1/p]$, $m \in \mathbf{Z}[1/p]$)

$$\hat{F}(g^i \otimes (b, m)) = g^{i-1} \otimes (qb, v^{r-1}m).$$

If $u \in \mathbf{Z}$ satisfies $uv \equiv 1$ (modulo t) then $g^{-i}ag^i(0,1) = a^{u^i}(0,1) = (u^i\xi_t, 1) \in (\mathbf{Q}/\mathbf{Z})(r)[1/p]\tilde{\oplus}\mathbf{Z}[1/p]$, writing the first coordinate additively, as usual. Here a and g are as in §3.1.5. Then $\hat{F}$ is a $\mathbf{Z}[G(L/K)]$-module homomorphism since

$$\begin{aligned}
\hat{F}(a(g^i \otimes (0,1))) &= \hat{F}(g^i \otimes a^{u^i}(0,1))\\
&= \hat{F}(g^i \otimes (u^i\xi_t, 1))\\
&= g^{i-1} \otimes (qu^i\xi_t, v^{r-1})
\end{aligned}$$

while

$$\begin{aligned}
a(\hat{F}(g^i \otimes (0,1))) &= a(g^{i-1} \otimes (0, v^{r-1}))\\
&= g^{i-1} \otimes a^{u^i}(0, v^{r-1})\\
&= g^{i-1} \otimes (v^{r-1}u^{i-1}\xi_t, v^{r-1})
\end{aligned}$$

and these are equal since $v^{r-1}u^{i-1} \equiv quu^{i-1} \equiv qu^i$ (modulo t). Also

$$\hat{F}(g(g^i \otimes (0,1))) = \hat{F}(g^{i+1} \otimes (0,1)) = g^i \otimes (0, v^{r-1})$$

while

$$g(\hat{F}(g^i \otimes (0,1))) = g(g^{i-1} \otimes (0, v^{r-1})) = g^i \otimes (0, v^{r-1}).$$

Theorem 3.1.17

In the notation of §3.1.5 and §3.1.16, there is a 2-extension of $\mathbf{Z}[G(L/K)]$-*modules of the form*

$$K_{2r-1}(L) \xrightarrow{\Delta} \tilde{\oplus}_{i=1}^{d} K_{2r-1}(L_0) \xrightarrow{1-\hat{F}} \tilde{\mathcal{W}} \xrightarrow{(\pi,\pi_2)} {}_{p^\infty}K_{2r-2}(L) \oplus K_{2r-3}(\mathcal{O}_L/\pi_L).$$

Note that the right-hand module is isomorphic to $\mathrm{Tors}(K_{2r-2}(L))$.

Proof. We have merely to evaluate the kernel and cokernel of the map

$$1 - \hat{F} : \tilde{\oplus}_{i=1}^{d} K_{2r-1}(L_0) \longrightarrow \tilde{\mathcal{W}}$$

where $\hat{F}$ was defined in §3.1.16. However, as abelian groups there are isomorphisms of the form

$$\tilde{\oplus}_{i=1}^{d} K_{2r-1}(L_0) \cong (\oplus_{i=1}^{d} K_{2r-1}(L_0)) \oplus (\oplus_{i=1}^{d} \mathbf{Z}[1/p])$$

and

$$\tilde{\mathcal{W}} \cong \mathcal{W} \oplus (\oplus_{i=1}^{d} \mathbf{Z}[1/p]).$$

Furthermore the resulting map on the quotients of these direct sums by the submodules given by the first summands has the form

$$\lambda : \oplus_{i=1}^{d} \mathbf{Z}[1/p] \longrightarrow \oplus_{i=1}^{d} \mathbf{Z}[1/p]$$

and is given by

$$\lambda(m_1, \ldots, m_d) = (m_1 - v^{r-1} m_d, \ldots, m_d - v^{r-1} m_{d-1}).$$

If $(m_1, \ldots, m_d) \in \mathrm{Ker}(\lambda)$ then $m_1 = v^{r-1} m_d, m_2 = v^{r-1} m_1, \ldots$ and $m_d(1 - v^{(r-1)d}) = 0$ so that λ is injective. Also we can solve the equation $\lambda(m_1, \ldots, m_d) = (x_1, \ldots, x_d)$ by choosing $x_d, x_1, \ldots, x_{d-2}$ successively but the equations are consistent if and only if

$$m_d(1 - v^{(r-1)d}) = v^{(r-1)(d-1)} x_1 + v^{(r-1)(d-2)} x_2 + \ldots + v^{r-1} x_{d-1} + x_d.$$

Therefore, if π_2 is defined to depend only on the coordinates in the second summand and to be given by

$$\pi_2(x_1, x_2, \ldots, x_d) = v^{(r-1)(d-1)} x_1 + v^{(r-1)(d-2)} x_2 + \ldots + v^{r-1} x_{d-1} + x_d$$

then $\mathrm{Ker}(\pi_2) = Im(\lambda)$, which proves the exactness of the sequence once we recall from [115] that

$$K_{2r-3}(\mathcal{O}_L/\pi_L) \cong \mathbf{Z}/(q^d - 1) \cong \mathbf{Z}/(v^{(r-1)d} - 1).$$

Also, as Galois modules, we have

$$K_{2r-3}(\mathcal{O}_L/\pi_L) \cong (\mathbf{Q}/\mathbf{Z})(r-1)[1/p]^{\Omega_{\mathcal{O}_L/\pi_L}}.$$

Therefore it remains to verify that

$$G(L/K) \quad \text{acts on} \quad \mathbf{Z}/(v^{(r-1)d} - 1) \cong K_{2r-3}(\mathcal{O}_L/\pi_L)$$

via the quotient to $G(W/K)$ whose generator acts by "multiplication" by v^{r-1}, where $v = |\mathcal{O}_K/\pi_K|$. Certainly, $G(L/E)$ acts trivially on $\tilde{\oplus}_{i=1}^{d} K_{2r-1}(L_0)$ and $\tilde{\mathcal{W}}$.

Also the action by $a \in G(E/W)$ is trivial on the second coordinates in the direct sums. Finally the generator, $g \in G(W/K)$, acts via

$$\begin{aligned} g(x_1,\ldots,x_d) &= g(\sum_{i=0}^{d-1} g^i \otimes x_{d-i}) \\ &= \sum_{i=0}^{d-1} g^{i+1} \otimes x_{d-i} \\ &= (x_2,x_3,\ldots,x_d,x_1) \end{aligned}$$

and

$$\begin{aligned} & x_1 + v^{r-1}x_d + v^{2(r-1)}x_{d-1} + \ldots + v^{(r-1)(d-1)}x_2 \\ & \quad \equiv v^{r-1}(v^{(r-1)(d-1)}x_1 + v^{(r-1)(d-2)}x_2 + \ldots + v^{r-1}x_{d-1} + x_d) \\ & \quad (\text{modulo}\, v^{d(r-1)} - 1) \end{aligned}$$

so that g acts on $\mathbf{Z}/(v^d - 1)$ by multiplication by v^{r-1}, as required. □

3.1.18 *Properties of the local fundamental class*

We sketch the important properties on the canoncial 2-extensions of Theorem 3.1.17. The proofs are simple adaptations of those to be found in ([134] §4). Let us consider the effect of replacing ξ_t by ξ_t^b, with $HCF(b,t) = 1$, in the construction of the 2-extension of Theorem 3.1.17. Although the class of the 2-extension in $\mathrm{Ext}^2_{\mathbf{Z}[G(L/K)]}(\mathrm{Tors}(K_{2r-2}(L)), K_{2r-1}(L))$ may possibly depend on the choice of ξ_t, the following result shows that the isomorphism class, which lies in the quotient

$$\mathrm{Ext}^2_{\mathbf{Z}[G(L/K)]}(\mathrm{Tors}(K_{2r-2}(L)), K_{2r-1}(L))/\simeq$$

by the action of

$$Aut_{G(L/K)}(\mathrm{Tors}(K_{2r-2}(L))) \times Aut_{G(L/K)}(K_{2r-1}(L)),$$

is independent of the choice of ξ_t.

Proposition 3.1.19 *(cf. [134] §4.1)*

In the notation of Theorem 3.1.17 and §3.1.18 there is a commutative diagram of $\mathbf{Z}[G(L/K)]$*-isomorphisms of the form*

$$\begin{array}{ccccccc} K_{2r-1}(L) & \xrightarrow{\Delta} & \tilde{\oplus}_{i=1}^{d} K_{2r-1}(L_0) & \xrightarrow{1-\hat{F}} & \tilde{\mathcal{W}} & \longrightarrow & \mathrm{Tors}(K_{2r-2}(L)) \\ \downarrow \phi_b & & \downarrow \tilde{\Phi}_b & & \downarrow \tilde{\Phi}_b & & \downarrow 1 \\ K_{2r-1}(L) & \xrightarrow{\Delta} & \tilde{\oplus}_{i=1}^{d} K_{2r-1}(L_0) & \xrightarrow{1-\hat{F}} & \tilde{\mathcal{W}} & \longrightarrow & \mathrm{Tors}(K_{2r-2}(L)) \end{array}$$

in which the upper sequence is that of Theorem 3.1.17 defined using ξ_t *while the lower sequence is that defined using* ξ_t^b.

There is a similar isomorphism of 2-extensions of Theorem 3.1.17 corresponding to different choices of the generators, a and g, of $G(E/K)$ *in §3.1.5.*

Following the argument of ([134] Theorem 4.6) establishes the passage-to-quotients part of the following result and the passage-to-subgroups is relatively straightforward.

Theorem 3.1.20

Let L/K be a Galois extension of local fields with residue characteristic p and group, $G(L/K)$.

(i) *If $G(L/M) \lhd G(L/K)$ is a normal subgroup then the $G(L/M)$-invariant/coinvariant sequence of §3.1.2 associated to the fundamental 2-extension of Theorem 3.1.17*

$$K_{2r-1}(L) \xrightarrow{\Delta} \tilde{\oplus}_{i=1}^{d} K_{2r-1}(L_0) \xrightarrow{1-\hat{F}} \tilde{\mathcal{W}}_L \xrightarrow{(\pi,\pi_2)} \mathrm{Tors}(K_{2r-2}(L))$$

is isomorphic to the fundamental 2-extension associated to M/K

$$K_{2r-1}(M) \xrightarrow{\Delta} \tilde{\oplus}_{i=1}^{d} K_{2r-1}(M_0) \xrightarrow{1-\hat{F}} \tilde{\mathcal{W}}_M \xrightarrow{(\pi,\pi_2)} \mathrm{Tors}(K_{2r-2}(M)).$$

(ii) *If $G(L/E) \subseteq G(L/K)$, the restriction to $G(L/E)$ of the canonical 2-extension of Theorem 3.1.17 associated to L/K is isomorphic to the canonical 2-extension associated to L/E. (Here by an "isomorphism" of 2-extensions we mean a map which is an isomorphism on the ends, rather than being the identity map, as in an equivalence of 2-extensions.)*

Example 3.1.21 *The local fundamental classes in characteristic p*

Let L/K be a Galois extension of local fields of residue characteristic p and Galois group, $G(L/K)$. As in §3.1.5, set $W = L \cap L_0$ so that W/K is the maximal unramified subextension of L/K. Hence L/W is totally ramified, $\overline{K} = \mathbf{F}_v$ $\overline{W} = \overline{L} = \mathbf{F}_{v^d}$. In fact, there are isomorphisms of the form $K \cong \mathbf{F}_v((X))$, the field of fractions of $\mathbf{F}_v[[X]]$, and $W = \mathbf{F}_{v^d}((X)) \subseteq L = \mathbf{F}_{v^d}((Y))$ for some X, Y. Therefore we may choose $g \in G(L/K)$ of order d which maps to the Frobenius automorphism in $G(\overline{L}/\overline{K})$.

For each positive integer, m, set $L(m) = \mathbf{F}_{v^{dm}}L$ so that $\overline{L(m)} = \mathbf{F}_{v^{dm}}$ and there is an extension of the form

$$G(L(m)/L) \longrightarrow G(L(m)/K) \xrightarrow{\pi_m} G(L/K)$$

in which the g's may be chosen to satisfy $\pi_m(g) = g$. The kernel of π_m, $G(L(m)/L)$, is isomorphic to $G(\mathbf{F}_{v^{dm}}/\mathbf{F}_{v^d})$ which is cyclic of order m generated by g^d. If L_0 is the maximal unramified extension of L, as in §3.1.2, then L_0/K is equal to the limit of the extensions, $L(m)/K$.

For $r \geq 1$, $K_{2r}(L(m))$ and $K_{2r+1}(L(m))$ have no p-torsion [60] and so the results of [148] imply that the tame symbol

$$\delta : K_{2r}(L(m)) \longrightarrow K_{2r-1}(\overline{L(m)}) = K_{2r-1}(\mathbf{F}_{v^{dm}}) \cong \mathbf{F}^*_{v^{dmr}}$$

and the map induced by the inclusion of the field of constants

$$\mathbf{F}^*_{v^{dm(r+1)}} \cong K_{2r+1}(\mathbf{F}_{v^{dm}}) = K_{2r+1}(\overline{L(m)}) \longrightarrow K_{2r+1}(L(m))$$

each have uniquely divisible kernel and cokernel. Since $G(L/K)$ is finite we have canonical cohomology isomorphisms, for $r \geq 1$,

$$H^i(G(L(m)/K); K_{2r}(L(m))) \cong H^i(G(L(m)/K); \mathbf{F}^*_{v^{dmr}}),$$
$$H^i(G(L(m)/K); K_{2r+1}(L(m))) \cong H^i(G(L(m)/K); \mathbf{F}^*_{v^{dm(r+1)}})$$

for all $i > 0$ and

$$\mathrm{Ext}^2_{\mathbf{Z}[G(L(m)/K)]}(K_{2r}(L(m)), K_{2r+1}(L(m)))$$
$$\cong \mathrm{Ext}^2_{\mathbf{Z}[G(L(m)/K)]}(\mathbf{F}^*_{v^{dmr}}, \mathbf{F}^*_{v^{dm(r+1)}}).$$

Incidentally, for any one-dimensional local field, L, the indecomposable K-group $K_j^{\mathrm{ind}}(L)$ [80] differs from $K_j(L)$ only by uniquely divisible groups when $j \geq 3$ by [15].

Now we shall use these observations to simplify considerably the construction of the local K-theory fundamental classes in the characteristic p case. First we need a little notation.

Let $\mu_\infty[1/p]$ denote the group of roots of unity of order prime to p. This group is $\mathbf{Q}/\mathbf{Z}[1/p]$ written *multiplicatively.* Let $\mu_\infty[1/p]\tilde{\oplus}\mathbf{Z}$ denote the $\mathbf{Z}[G(L/W)]$-module given by the direct sum of $\mu_\infty[1/p]$ with the integers where $G(L/W)$ acts trivially on $\mu_\infty[1/p]$. Suppose that $[G(L/W) : G_1(L/W)] = t$, as in §3.1.5, then t divides $v^d - 1$ and so is prime to p. We let $G(L/W)$ act on $\mu_\infty[1/p]\tilde{\oplus}\mathbf{Z}$ via the quotient map to $G(L/W)/G_1(L/W) \cong \mathbf{Z}/t = \langle a \rangle$ where $a(1,1) = (\xi_t, 1)$ for some primitive t-th root of unity, $\xi_t \in \mu_\infty[1/p]$. Note that, since $\mathbf{Z}$ is written *additively,* the first coordinate of $(1,1)$ is the trivial element of $\mu_\infty[1/p]$ while the second coordinate is the integer, 1. Hence we have the induced $\mathbf{Z}[G(L(m)/K)]$-module, $\mathrm{Ind}_{G(L/W)}^{G(L(m)/K)}(\mu_\infty[1/p]\tilde{\oplus}\mathbf{Z})$. This is a submodule of the module appearing in the bottom left corner of the diagrams of §3.1.12 and §3.1.14, the only difference being that $\mathbf{Z}[1/p]$ has been replaced by the integers, $\mathbf{Z}$.

Theorem 3.1.22

Let $L(m)/K$ be as in §3.1.21. Then

(i) $\mathrm{Ind}_{G(L/W)}^{G(L(m)/K)}(\mu_\infty[1/p]\tilde{\oplus}\mathbf{Z})$ *is a cohomologically trivial* $\mathbf{Z}[G(L(m)/K)]$-*module*

and

(ii) *there is a 2-extension of* $\mathbf{Z}[G(L(m)/K)]$-*modules of the form*

$$\mathbf{F}^*_{v^{dm(r+1)}} \longrightarrow \mathrm{Ind}_{G(L/W)}^{G(L(m)/K)}(\mu_\infty[1/p]\tilde{\oplus}\mathbf{Z})$$
$$\xrightarrow{\hat{F}-1} \mathrm{Ind}_{G(L/W)}^{G(L(m)/K)}(\mu_\infty[1/p]\tilde{\oplus}\mathbf{Z}) \longrightarrow \mathbf{F}^*_{v^{dmr}}$$

where

$$\hat{F}(g^i \otimes (\eta, m)) = g^{i-1} \otimes (\eta^{v^{r+1}}, mv^r)$$

whose class in $\mathrm{Ext}^2_{\mathbf{Z}[G(L(m)/K)]}(\mathbf{F}^*_{v^{dmr}}, \mathbf{F}^*_{v^{dm(r+1)}})$ *corresponds under the isomorphism of §3.1.21 to the* K_{2r}/K_{2r+1} *local fundamental class of Theorem 3.1.17.*

Proof. Part (i) is similar to Proposition 3.1.11.

As mentioned in §3.1.21, the module $\mathrm{Ind}_{G(L/W)}^{G(L(m)/K)}(\mu_\infty[1/p]\tilde{\oplus}\mathbf{Z})$ is a submodule of both $\tilde{\oplus}_{i=1}^d K_3(L_0)$ and $\tilde{\mathcal{W}}$. Furthermore the homomorphism

$$\tilde{\oplus}_{i=1}^d K_3(L_0) \xrightarrow{1-\hat{F}} \tilde{\mathcal{W}}$$

restricts to the homomorphism $1 - \hat{F}$ where $\hat{F}$ is defined by the formula of Theorem 3.1.20. Hence we obtain a homomorphism of 2-extensions of $\mathbf{Z}[G(L(m)/K)]$-modules. It is straightforward to verify that this homomorphism induces (up to a sign) the canonical maps of §3.1.21 between the modules at the two ends. □

Remark 3.1.23 In §4 we shall use the economical form of the K-theory local fundamental class in characteristic $p > 0$ to calculate the associated Euler characteristics lying in $\tilde{K}_0(\mathbf{Z}[G(L/K)])$ when $r = 1$. The advantage of being in the positive characteristic situation is that one does not have to evaluate a regulator as in the characteristic zero case (see Theorem 4.1.24 and Theorem 3.2.2).

3.2 The higher K-theory invariants $\Omega_s(L/K, 2)$

3.2.1 Let L/K be a Galois extension of number fields. The second Chinburg invariant ($\Omega_0(L/K, 2)$ in our notation) is constructed from a projective $\mathbf{Z}[G(L/K)]$-module, X, and the classical fundamental classes of local class field theory (see [131] Chapter 7). In a similar manner, one may use the fundamental classes of Theorem 3.1.17 in place of the classical one to construct Galois module structure invariants

$$\Omega_{r-1}(L/K, 2) \in \mathcal{CL}(\mathbf{Z}[G(L/K)])$$

lying in the class-group of $\mathbf{Z}[G(L/K)]$. These invariants measure the Galois module structure of algebraic K-groups in dimensions $2r - 1$ and $2r - 2$ and are expected to be connected with the values of the L-functions at $s = 1 - r$ (or, equivalently, $s = r$). This expectation is based upon clues such as the appearance of these special L-function values in the examples of ([131] Theorem 7.4.60; [132] §7.2; [36], [38], [138], [139]).

In the construction of the original Chinburg invariant in ([131] §7.2) – denoted there by $\Omega(L/K, 2)$ rather than $\Omega_0(L/K, 2)$) – one chooses a locally free, hence projective, module X in the following manner.

For each finite prime, $P \triangleleft \mathcal{O}_K$, choose a prime lying over it, $Q \triangleleft \mathcal{O}_L$. Therefore the decomposition group of Q is isomorphic to $G(L_Q/K_P)$. We shall say that P

is *tame* if L_Q/K_P is tamely ramified (i.e. $G_1(L_Q/K_P) = \{1\}$) and that P is *wild* otherwise. By a theorem of E. Noether, if P is tame then $\mathcal{O}_{L_Q}$ is a free $\mathcal{O}_{K_P}[G(L_Q/K_P)]$-module of rank one. Therefore we may choose an adèle

$$(a_P) = a \in \prod_P \mathcal{O}_{L_Q}$$

(the product being taken over finite primes of K with Q being the chosen prime over P) such that

(i) $a_P \in \mathcal{O}_{L_Q}$, $K_P[G(L_Q/K_P)]a_P = L_Q$ for each P and

(ii) $\mathcal{O}_{K_P}[G(L_Q/K_P)]a_P = \mathcal{O}_{L_Q}$ for each tame P.

There are isomorphisms of the form

$$L \otimes_K K_P \cong \prod_{R|P} L_R \cong \mathrm{Ind}_{G(L_Q/K_P)}^{G(L/K)}(L_Q)$$

and

$$\mathcal{O}_L \otimes_{\mathcal{O}_K} \mathcal{O}_{K_P} \cong \prod_{R|P} \mathcal{O}_{L_R} \cong \mathrm{Ind}_{G(L_Q/K_P)}^{G(L/K)}(\mathcal{O}_{L_Q})$$

so that

(i) $K_P[G(L/K)]a_P \cong \prod_{R|P} L_R$ for each P and

(ii) $\mathcal{O}_{K_P}[G(L/K)]a_P \cong \prod_{R|P} \mathcal{O}_{L_R}$ for each tame P.

We shall abbreviate $\mathcal{O}_{K_P}[G(L_Q/K_P)]a_P$ to X_Q and set $X = \mathcal{O}_K[G(L/K)]a$. This is to be interpreted as meaning that X is the intersection of L with the product of its P-completions, X_P, where both are considered as subgroups of the adèles. Hence X is a locally free $\mathcal{O}_K[G(L/K)]$-module whose P-completion is

$$X_P = \mathcal{O}_{K_P}[G(L/K)]a_P = \mathrm{Ind}_{G(L_Q/K_P)}^{G(L/K)}(X_Q).$$

In addition, we shall assume henceforth (replacing X by mX for a suitable integer, $m \in \mathbf{Z}$, if necessary) that the Q-adic exponential defines an isomorphism

$$exp : X_Q \xrightarrow{\cong} 1 + X_Q \subset \mathcal{O}_{L_Q}^*$$

for all wild L_Q/K_P. This assumption is only needed in the construction of the invariant $\Omega_0(L/K, 2)$. Since X is locally free, it is cohomologically trivial and so also is X_Q for each Q. Hence X defines a class, $[X] \in \mathcal{CL}(\mathbf{Z}[G(L/K)])$.

From Theorem 3.1.17 we have a canonical class in

$$\mathrm{Ext}^2_{\mathbf{Z}[G(L_Q/K_P)]}(\mathrm{Tors}(K_{2r-2}(L_Q)), K_{2r-1}(L_Q)).$$

From the isomorphism $K_{2r-1}(L_Q) \cong K_{2r-1}(\mathcal{O}_{L_Q})$ we have a canonical map

$$\phi_{2r-1} : K_{2r-1}(L) \longrightarrow K_{2r-1}(\mathcal{O}_{L_Q}) \longrightarrow \varprojlim_s K_{2r-1}(\mathcal{O}_{L_Q}/(\pi_{L_Q}^s)) = K_{2r-1}^{top}(L_Q).$$

The target group, $K_{2r-1}^{top}(L_Q)$, is a finitely generated $\mathbf{Z}_p[G(L_Q/K_P)]$-module and ϕ_{2r-1} induces an isomorphism in p-adic cohomology [162]. In addition, we have

$$\rho_{2r-1} : K_{2r-1}(L_Q) \cong K_{2r-1}(\mathcal{O}_{L_Q}) \longrightarrow K_{2r-1}(\mathcal{O}_{L_Q}/\pi_{L_Q})$$

induced by reduction modulo π_{L_Q}, where $\mathcal{O}_{L_Q}/(\pi_{L_Q})$ is the residue field of L_Q.

The map

$$(\phi_{2r-1}, \rho_{2r-1}) : K_{2r-1}(L_Q) \longrightarrow K_{2r-1}^{top}(L_Q) \oplus K_{2r-1}(\mathcal{O}_{L_Q}/(\pi_{L_Q}))$$

induces isomorphisms in Tate cohomology in all dimensions [148]. Hence, pushing out via $(\phi_{2r-1}, \rho_{2r-1})$, we obtain a canonical 2-extension of the form

$$K_{2r-1}^{top}(L_Q) \oplus K_3(\mathcal{O}_{L_Q}/(\pi_{L_Q})) \longrightarrow A \longrightarrow B \longrightarrow \mathrm{Tors}(K_{2r-2}(L_Q))$$

in which A and B may be chosen to be cohomologically trivial, with B finitely generated over $\mathbf{Z}[G(L_Q/K_P)]$ and the p-adic part of A finitely generated over $\mathbf{Z}_p[G(L_Q/K_P)]$.

Next we choose our favourite regulator. A regulator is a Galois-equivariant isomorphism of the form

$$K_{2r-1}^{top}(L_Q) \otimes \mathbf{Q}_p \cong L_Q$$

where p is the residue characteristic. There are many to choose from and several are described in ([134] §5), which may be derived (sometimes a little work is needed) from ([11], [13], [16], [46], [51], [55], [66], [67], [73], [90], [99], [100], [141], [143], [162]). For example, we shall settle for the regulator given by the syntomic Chern classes. More precisely, we have canonical isomorphisms, given by étale and syntomic Chern classes, of the form

$$ch_{2r-1}^{\acute{e}t} : K_{2r-1}^{top}(L_Q) \xrightarrow{\cong} H_{\acute{e}t}^1(L_Q; \mathbf{Z}_p(r))$$

and

$$K_{2r-1}^{top}(L_Q) \otimes \mathbf{Q}_p \cong H_{\acute{e}t}^1(L_Q; \mathbf{Q}_p(r)) \xleftarrow{f_M} H^1(\mathcal{O}_{L_Q}; S_\infty(r)_{\mathbf{Q}_p}) \xrightarrow{\alpha} L_Q$$

such that, if $L_Q/\mathbf{Q}_p$ is unramified, $K_{2r-1}^{top}(L_Q)/\mathrm{Tors}$ is identified with the ring of integers, $\mathcal{O}_{L_Q} \subset L_Q$ for almost all p (there are errors in the formulae [90] and[91] but the calculations are correct for almost all primes.)

Let X denote the finitely generated, locally free $\mathbf{Z}_p[G(L/K)]$-module introduced previously, as used in the construction of the second Chinburg invariant, $\Omega_0(L/K, 2)$, ([131] §7.2). By construction, $X_Q \subseteq \mathcal{O}_{L_Q}$ with equality if L/K is tamely ramified at $Q \lhd \mathcal{O}_L$. We may assume that X_Q has been lifted to a p-adic lattice

$$X_Q \subseteq K_{2r-1}^{top}(L_Q)$$

mapping isomorphically to $X_Q \subseteq \mathcal{O}_{L_Q}$ under the syntomic regulator map. Dividing out by X_Q yields a 2-extension of $\mathbf{Z}[G(L_Q/K_P)]$-modules of the form

$$(K^{top}_{2r-1}(L_Q)/X_Q) \oplus K_{2r-1}(\mathcal{O}_L/Q) \longrightarrow A/X_Q \longrightarrow B \longrightarrow \mathrm{Tors}(K_{2r-2}(L_Q))$$

in which the two middle modules are finitely generated and cohomologically trivial $\mathbf{Z}[G(L_Q/K_P)]$-modules. Hence we may define an Euler characteristic

$$\Omega_{r-1}(L_Q/K_P, X_Q) = [A/X_Q] - [B] \in \mathcal{CL}(\mathbf{Z}[G(L_Q/K_P)]).$$

We define, for $r \geq 2$,

$$\Omega_{r-1}(L/K,2) \\ = [X] + \textstyle\sum_P \mathrm{Ind}^{G(L/K)}_{G(L_Q/K_P)}(\Omega_{r-1}(L_Q/K_P, X_Q)) \in \mathcal{CL}(\mathbf{Z}[G(L/K)]).$$

In the sum Q is a chosen prime over $P \lhd \mathcal{O}_K$. As explained in ([134] §5), only finitely many of the local invariants in the sum are non-zero (at least, this is true for the syntomic regulator) and $\Omega_{r-1}(L/K,2)$ is independent of all choices involved in the construction.

The properties of these invariants are summarised in the following result, whose proof is analogous to that of ([134] Theorem 5.3), which is the special case when $r = 2$.

Theorem 3.2.2

Let L/K be a Galois extension of number fields. Then, for $r \geq 2$, in the formal expression of §3.2.1

$$\Omega_{r-1}(L/K,2) \\ = [X] + \textstyle\sum_P \mathrm{Ind}^{G(L/K)}_{G(L_Q/K_P)}(\Omega_{r-1}(L_Q/K_P, X_Q)) \in \mathcal{CL}(\mathbf{Z}[G(L/K)])$$

(the sum is over $P \lhd \mathcal{O}_K$) the terms, $\Omega_{r-1}(L_Q/K_P, X_Q)$, are zero for almost all p when $L_Q/\mathbf{Q}_p$ is unramified. In fact, the class-group element defined by this expression is independent of the choices of X and of all the choices involved in the construction of the local fundamental 2-extensions associated to the $\{L_Q/K_P\}$.

In addition, $\Omega_{r-1}(L/K,2)$ satisfies the following naturality properties with respect to an inclusion of a subgroup, $G(L/N) \subseteq G(L/K)$:

(i) *the induced map*

$$\mathrm{Res}^{G(L/K)}_{G(L/N)} : \mathcal{CL}(\mathbf{Z}[G(L/K)]) \longrightarrow \mathcal{CL}(\mathbf{Z}[G(L/N)])$$

satisfies

$$\mathrm{Res}^{G(L/K)}_{G(L/N)}(\Omega_{r-1}(L/K,2)) = \Omega_{r-1}(L/N,2)$$

and

(ii) *if $G(L/N) \lhd G(L/K)$ the induced map*

$$\mathrm{Inf}_{L/N} : \mathcal{CL}(\mathbf{Z}[G(L/K)]) \longrightarrow \mathcal{CL}(\mathbf{Z}[G(N/K)])$$

satisfies

$$\mathrm{Inf}_{L/N}(\Omega_{r-1}(L/K,2)) = \Omega_{r-1}(N/K,2).$$

3.3 Two-dimensional thoughts

3.3.1 *K-theory of two-dimensional local fields*

In this section we shall take a brief look at what happens when we apply the method of §3.1 for the construction of higher K-theory local fundamental classes to the case of two-dimensional local fields. Recall that the method consisted of manufacturing a 2-extension of Galois modules (Theorem 3.1.3) and then modifying the modules in order to kill their cohomology. Accordingly, here we are going to calculate the corresponding cohomology groups to be killed in the two-dimensional case.

Let F/E be a finite Galois extension of two-dimensional local fields of characteristic zero with Galois group, $G(F/E)$, whose residue field extension is a totally ramified Galois extension, L/K, of one-dimensional local fields. There are two cases.

Case A: $\mathrm{char}(K) = 0$
In this case ([54] p. 257) $F = L((X))$ and $E = K((Y))$, where L/K is a Galois extension of p-adic local fields, so that the elements of $L((X))$ are formal series, $\sum_{n_0}^{\infty} a_i X^i$, with $a_i \in L$.

Case B: $\mathrm{char}(K) = p$
In this case ([54] p. 257) we shall only consider the situation[1] $F = L_1\{\{X\}\}$ and $E = K_1\{\{Y\}\}$, where L/K is a Galois extension of p-adic local fields and the residue fields of K_1, L_1 are K, L, respectively.

3.3.2 Let $F_0 = \bigcup_{(HCF(p,l)=1} F(\xi_l)$ where ξ_l is a primitive l-th root of unity (F_0 is called F^{pur} in ([54] p. 259)). Hence $K_0 \cap L = K$ and $E_0 \cap F = E$. Let $L(s)$ be as in §3.1.2.

In Case A we have $F = L((X))$ and $F(s) = L(s)((X))$ and we have

$$K_2(F) \cong D_F \oplus K_2(L) \oplus K_1(L),$$
$$K_2(F(s)) \cong D_{F(s)} \oplus K_2(L(s)) \oplus K_1(L(s))$$

where D_F and $D_{F(s)}$ are uniquely divisible ([15] p. 415).

Hence, by an argument with the transfer, the map

$$K_2(F) \longrightarrow K_2(F(s))^{G(F(s)/F)}$$

is an isomorphism from D_F to $D_{F(s)}^{G(F(s)/F)}$. On the other summands the map is the sum of the natural maps on $K_2(L)$ and $K_1(L) \cong L^*$ so that

$$\bigcup_s \mathrm{Ker}(K_2(F) \longrightarrow K_2(F(s))^{G(F(s)/F)}) = {}_{p^\infty}K_2(L)$$

[1]The general case occurs when F and E are finite extensions of the fields considered here.

and

$$\lim_{\overrightarrow{s}} \operatorname{Coker}(K_2(F) \longrightarrow K_2(F(s))^{G(F(s)/F)}) = 0.$$

Therefore (cf. Theorem 3.1.3) we have a 2-extension

$$K_3^{\text{ind}}(F) \longrightarrow K_3^{\text{ind}}(F_0) \xrightarrow{1-F} \mathcal{U} \longrightarrow {}_{p^\infty}K_2(L).$$

Therefore, in order to imitate the method of §3.1, we would need to know about the Tate cohomology groups $\hat{H}^*(G(F/E); K_3^{\text{ind}}(F_0))$ and $\hat{H}^*(G(F/E); \mathcal{U})$. In this section we shall only analyse the first of these.

Theorem 3.3.3

In Case A there are isomorphisms:

$$H^i(G(F/E); K_3^{\text{ind}}(F_0)) \cong \begin{cases} \mu_{e_1}(K_0) \oplus \mathbf{Z}/r & \text{if } i \text{ is odd,} \\ K_0^*/(K_0^*)^{e_1} & \text{if } i > 0 \text{ is even,} \\ K_3^{\text{ind}}(E_0) & \text{if } i = 0. \end{cases}$$

Here $v_F(Y) = e_1$ and $r = [G_0(L/K) : G_1(L/K)]$.

3.3.4 This result will be proved in the course of the subsequent discussion. We may replace F_0 by L_0 since the cokernel of the injection

$$K_3^{\text{ind}}(L_0) \longrightarrow K_3^{\text{ind}}(F_0)$$

is uniquely divisible ([106]; see also [80]), since L_0 is the subfield of constants. If $M = L((Y))$ then there is an embedding

$$G(F/M) \cong \mathbf{Z}/e_1 \longrightarrow (L[[X]]/XL[[X]])^* \cong L^*$$

given by sending g to $g(X)/X$.

Therefore we have a spectral sequence of the form

$$E_2^{s,t} = H^s(G(L/K); H^t(G(F/M); K_3^{\text{ind}}(F_0))) \Longrightarrow H^{s+t}(G(F/E); K_3^{\text{ind}}(F_0))$$

in which $E_2^{s,t} \cong H^s(G(L/K); H^t(G(F/M); K_3(L_0)))$ when $s + t > 0$. When $s = 0 = t$ these two groups differ by a uniquely divisible group and $E_2^{0,0} = E_\infty^{0,0}$. Since $G(F/M)$ acts trivially on $K_3(L_0)$,

$$E_2^{s,0} \cong H^s(G(L/K); K_3(L_0)) \cong \begin{cases} \mathbf{Z}/r & \text{if } s \text{ is odd,} \\ 0 & \text{if } s > 0 \text{ is even.} \end{cases}$$

Also we have an exact sequence [80]

$$\begin{aligned} 0 \longrightarrow H^1(G(F/M); K_3^{\text{ind}}(F_0)) \longrightarrow K_2(M_0) \longrightarrow K_2(F_0)^{G(F/M)} \\ \longrightarrow H^1(G(F/M); K_3^{\text{ind}}(F_0)) \longrightarrow 0 \end{aligned}$$

and

$$K_2(M_0) \cong D_{M_0} \oplus K_2(L_0) \oplus K_1(L_0),$$
$$K_2(F_0) \cong D_{F_0} \oplus K_2(L_0) \oplus K_1(L_0)$$

but, since $M_0 = L_0((Y))$ and $v_F(Y) = e_1$, the map between these groups is an isomorophism of the uniquely divisible parts, the identity on $K_2(L_0)$ and the e_1-th power map on $K_1(L_0) \cong L_0^*$. Hence

$$H^1(G(F/M); K_3^{\text{ind}}(F_0)) \cong \mu_{e_1}(L_0),$$
$$H^2(G(F/M); K_3^{\text{ind}}(F_0)) \cong L_0^*/(L_0^*)^{e_1}.$$

Note that, as $G(F/M)$ acts trivially, these are also $\text{Tor}(\mathbf{Z}/e_1, K_3^{\text{ind}}(L_0))$ and $K_3^{\text{ind}}(L_0) \otimes \mathbf{Z}/e_1$, respectively.

Therefore

$$E_2^{s,t} \cong \begin{cases} H^s(G(L/K); \mu_{e_1}(L_0)) & \text{if } t \text{ is odd,} \\ H^s(G(L/K); L_0^*/(L_0^*)^{e_1}) & \text{if } t > 0 \text{ is even.} \end{cases}$$

Note also that $\hat{H}^*(G(L/K); L_0^*) = 0$ when L/K is totally ramified so that the sequence

$$0 \longrightarrow \mu_{e_1}(L_0) \longrightarrow L_0^* \longrightarrow L_0^* \longrightarrow L_0^*/(L_0^*)^{e_1} \longrightarrow 0$$

implies that $E_2^{s+2,t-1} \cong E_2^{s,t}$ if t is odd and $s > 0$.

Next we calculate $H^1(G(F/E); K_3^{\text{ind}}(F_0))$ and $H^2(G(F/E); K_3^{\text{ind}}(F_0))$ from the exact sequence of [80]. The first group is isomorphic to the kernel of the map

$$D_{E_0} \oplus K_2(K_0) \oplus K_1(K_0) \longrightarrow (D_{F_0} \oplus K_2(L_0) \oplus K_1(L_0))^{G(F/E)}.$$

This homomorphism is an isomorphism on the uniquely divisible part, a surjection with kernel $\mathbf{Z}/r$ on the K_2-part and the e_1-th power of the inclusion on the K_1-part. Hence

$$H^1(G(F/E); K_3^{\text{ind}}(F_0)) \cong \mu_{e_1}(K_0) \oplus \mathbf{Z}/r$$

where $\mu_{e_1}(K_0) = K_0^* \cap \mu_{e_1}(L_0)$ and

$$H^2(G(F/E); K_3^{\text{ind}}(F_0)) \cong (L_0^*)^{G(F/E)} \otimes \mathbf{Z}/e_1 \cong K_0^*/(K_0^*)^{e_1},$$

as expected. Of course, $H^0(G(F/E); K_3^{\text{ind}}(F_0)) \cong K_3^{\text{ind}}(E_0)$, by [106].

Therefore we have, for $s + t = 2$, an exact sequence

$$0 \longrightarrow E_2^{1,1} \longrightarrow K_0^*/(K_0^*)^{e_1} \xrightarrow{i} (L_0^*/(L_0^*)^{e_1})^{G(L/K)} \xrightarrow{d_2} E_2^{2,1} \ldots.$$

The edge-homomorphism, i, is induced by the natural map.

However, we also have a commutative diagram, in which $\mathcal{H}$ temporarily denotes $H^1(G(L/K); L_0^*/\mu_{e_1}(L_0))$,

$$\begin{array}{ccccccc}
K_0^*/\mu_{e_1}(K_0) & \xrightarrow{e_1} & K_0^* & \longrightarrow & K_0^*/(K_0^*)^{e_1} & \longrightarrow & 0 \\
\downarrow j & & \downarrow \cong & & \downarrow i & & \downarrow \\
(L_0^*/\mu_{e_1}(L_0))^{G(L/K)} & \xrightarrow{e_1} & (L_0^*)^{G(L/K)} & \longrightarrow & (L_0^*/(L_0^*)^{e_1})^{G(L/K)} & \longrightarrow & \mathcal{H}
\end{array}$$

We have

$$H^1(G(L/K); L_0^*/\mu_{e_1}(L_0)) \cong H^2(G(L/K); \mu_{e_1}(L_0))$$

since L_0^* is a cohomologically trivial $\mathbf{Z}[G(L/K)]$-module, so d_2 must map onto $E_2^{2,1}$, by counting, since $E_2^{2,1}$ is a finite group. If $d_2 : E_2^{0,2} \longrightarrow E_2^{2,1}$ is surjective then so is $d_2 : E_2^{2s,2t} \longrightarrow E_2^{2s+2,2t-1}$ for all $s \geq 0$, $t \geq 1$ and, if $s > 0$ this must be an isomorphism, by counting. Also $E_2^{1,1} \cong H^1(G(L/K); \mu_{e_1}(L_0))$ so that the diagram, in which $\mathcal{H}'$ temporarily denotes $H^1(G(L/K); \mu_{e_1}(L_0))$,

$$\begin{array}{ccccccc}
\mu_{e_1}(K_0) & \longrightarrow & K_0^* & \longrightarrow & K_0^*/\mu_{e_1}(K_0) & \longrightarrow & 0 \\
\downarrow \cong & & \downarrow \cong & & \downarrow & & \downarrow \\
(\mu_{e_1}(L_0))^{G(L/K)} & \longrightarrow & (L_0^*)^{G(L/K)} & \longrightarrow & (L_0^*/\mu_{e_1}(L_0))^{G(L/K)} & \longrightarrow & \mathcal{H}'
\end{array}$$

shows that $E_2^{1,1}$ is indeed the kernel of i. By periodicity of period two in t and s we find that

$$E_3^{s,t} = \begin{cases} \mathrm{Ker}(d_2) & \text{if } s = 0,\ t > 0 \text{ even}, \\ E_2^{0,t} & \text{if } s = 0,\ t \text{ odd}, \\ E_2^{s,0} & \text{if } s > 0,\ t = 0, \\ 0 & \text{otherwise.} \end{cases}$$

This completes the calculation, since we have an isomorphism

$$H^i(G(F/E); K_3^{\mathrm{ind}}(F_0)) \longrightarrow H^{i+2}(G(F/E); K_3^{\mathrm{ind}}(F_0))$$

given by multiplying by the class of $1 \in K_0^*/(K_0^*)^{e_1} \cong H^2(G(F/E); K_3^{\mathrm{ind}}(F_0))$.

3.3.5 *Case B:*

In this case $F = L\{\{X\}\}$ and from ([15] p. 415)

$$K_2(F) \cong D_F \oplus K_2(\overline{L}) \oplus K_1(\overline{L})$$

and

$$K_2(F(s)) \cong D_{F(s)} \oplus K_2(\overline{L(s)}) \oplus K_1(\overline{L(s)})$$

where D_F and $D_{F(s)}$ are uniquely divisible and $\overline{L}$ denotes the residue field of L. Therefore

$$K_2(F) \longrightarrow K_2(F(s))^{G(F(s)/F)}$$

is an isomorphism on the uniquely divisible part and on the K_2 part, which is zero, and on the K_1 part. Hence we have an exact sequence of the form

$$0 \longrightarrow K_3^{\rm ind}(F) \longrightarrow K_3^{\rm ind}(F_0) \longrightarrow \mathcal{U} \longrightarrow 0.$$

What is $H^s(G(F/E); K_3^{\rm ind}(F_0))$?

This time $F = L\{\{X\}\}$, $E = K\{\{Y\}\}$, $M = L\{\{Y\}\}$ and we have an inclusion

$$G(F/M) \cong \mathbf{Z}/e_1 \longrightarrow \overline{L}^*,$$

a cyclic group of order prime to p.

If we ignore the group $E_2^{0,0} = E_\infty^{0,0}$, the spectral sequence again has the form

$$E_2^{s,t} = H^s(G(L/K); H^t(G(F/M); K_3(L_0))) \Longrightarrow H^{s+t}(G(F/E); K_3^{\rm ind}(F_0))$$

and, since $G(F/M)$ again acts trivially on $K_3(L_0)$,

$$E_2^{s,0} \cong H^s(G(L/K); K_3(L_0)) \cong \begin{cases} \mathbf{Z}/r & \text{if } s \text{ is odd,} \\ 0 & \text{if } s > 0 \text{ is even.} \end{cases}$$

However, this time we have

$$H^1(G(F/M); K_3(L_0)) \cong \mu_{e_1}(\overline{L_0})$$

and

$$H^2(G(F/M); K_3(L_0)) \cong \overline{L_0}/(\overline{L_0})^{e_1} = 0,$$

since p does not divide e_1. Therefore, since for $t \geq 1$ odd, $s > 0$ we have an isomorphism, $E_2^{s+2,t-1} \cong E_2^{s,t}$, these groups are zero when t is even and therefore when t is odd also. We are left with only $E_2^{s,0}$ which is $\mathbf{Z}/r$ or zero for $s > 0$ and

$$E_2^{0,t} \cong \begin{cases} \mu_{e_1}(\overline{L_0})^{G(L/K)} & \text{if } t \text{ is odd,} \\ 0 & \text{if } t \text{ is even.} \end{cases}$$

Arguing as before yields:

Theorem 3.3.6

In Case B

$$H^i(G(F/E); K_3^{\rm ind}(F_0)) \cong \begin{cases} \mu_{e_1}(\overline{K_0}) \oplus \mathbf{Z}/r & \textit{if } i \textit{ is odd,} \\ 0 & \textit{if } i > 0 \textit{ is even,} \\ K_3^{\rm ind}(E_0) & \textit{if } i = 0. \end{cases}$$

Here $v_F(Y) = e_1$ *and* $r = [G_0(L/K) : G_1(L/K)]$.

Question 3.3.7 Does there exist a canonical distinguished triangle of Galois modules of the form

$$K_3^{\mathrm{ind}}(F_0) \xrightarrow{f} X \longrightarrow C_f \longrightarrow K_3^{\mathrm{ind}}(F_0)[1]$$

is which f induces an isomorphism in Tate cohomology for all $H \subseteq G(F/E)$ in all dimensions?

In the case of one-dimensional local fields C_f was $K_3^{\mathrm{ind}}(L_0)\tilde{\oplus}\mathbf{Z}[1/p]$. Given X in the two-dimensional case one could attempt to build a fundamental class by using f to kill the cohomology of $K_3^{\mathrm{ind}}(F_0)$, following the procedure of §3.1.

Chapter 4

Positive Characteristic

4.1 $\Omega_1(L/K, 2)$ in the tame case

4.1.1 In this chapter we shall be concerned with the fundamental classes of Theorem 3.1.17 and Example 3.1.21 associated to K_2 and K_3 of a local field in characteristic $p > 0$. We are going to calculate the Euler characteristic, in the sense of Example 2.1.8, of the 2-extension described in Theorem 3.1.22. For the calculations of this section we shall make things easier for ourselves by considering the tamely ramified case.

These calculations are rather long and clumsy, so I shall begin with an outline of contents of this section. In ([34] and [131] §§7.1.40–7.1.56) homological data is given which is used to calculate the Euler characteristic of the classical (K_0/K_1) local fundamental class in the tamely ramified case. This is reproduced, for the characteristic p case, in §§4.1.3–4.1.6. This is modified in §§4.1.7–4.1.9 to give a surjection, $g_4 : \hat{X}(2) \longrightarrow K_3(\mathbf{F}_{v^d})$, whose kernel is a finitely generated, projective $\mathbf{Z}[G(L/K)]$-module representing minus the required K_2/K_3-Euler characteristic. In §4.1.11 we construct a free submodule, $W \subseteq \text{Ker}(g_4)$, which permits us to find a simpler representative for the Euler characteristic in Proposition 4.1.12. In §§4.1.13–4.1.15 we analyse the module structure of the representative in order to give a further simplification of the form

$$\frac{\ker(g_4 \mid X(2) \otimes Q_0)}{\mathbf{Z}[g]/(g^d - 1)\langle y \otimes (v - g + r\alpha_d)\rangle}$$

in Lemma 4.1.16.

In §4.1.17 we construct another free submodule, $W' \subseteq \text{Ker}(g_4 \mid X(2) \otimes Q_0)$ and use it in §§4.1.18–4.1.22 to construct a surjection from a free module

$$\mathbf{Z}[G(L/K)]\langle (x - y) \otimes 1\rangle \oplus W'$$
$$\xrightarrow{\rho} \frac{\ker(g_4 \mid X(2) \otimes Q_0)}{\mathbf{Z}[g]/(g^d - 1)\langle y \otimes (v - g + v(v^d - 1)w^{2d})\rangle}$$

so that we are reduced to finding the class in the class-group of the module $\mathrm{Ker}(\rho)$. In §4.1.23 we accomplish this by constructing a filtration of the form $A \subset B \subset \mathrm{Ker}(\rho)$. The results of all this computation are recapitulated in Theorem 4.1.24, which identifies a finitely generated projective $\mathbf{Z}[G(L/K)]$-module representing the Euler characteristic of the K_2/K_3 local fundamental class of Theorem 3.1.22 in the tamely ramified case. Theorem 4.1.24 is rephrased in terms of the Hom-description of the class-group (see Theorem 2.5.25) in §4.1.25. Finally, in §4.1.26 we verify that the Hom-description representative is independent of the choice of "Frobenius", $g \in G(L/K)$. In Theorem 4.1.27 we give the Hom-description of the global Galois module structure invariant, $\Omega_1(E/F, 2)$, in the tamely ramified case in characteristic p. This invariant is the characteristic p version of the invariant introduced in 3.2.1.

Now we can begin. The first part of this section is inspired by (and amounts to a simpler alternative to) ([34] Lemma 6.3 p. 370).

Suppose we are in the tame situation. That is, L/K is a tamely ramified Galois extension of local fields (of *any* characteristic for the moment) with Galois group $G(L/K)$ of the following form:

$$G(L/K) = \langle a, g \mid g^d = a^c,\ a^r = 1,\ \mathrm{gag}^{-1} = a^v \rangle$$

where $v = |\overline{K}|$, the order of the residue field, $\overline{K}$, of K. Here, if W/K is the maximal unramified subextension then $G(L/W) = \langle a \rangle$ and the image of g in $G(\overline{L}/\overline{K})$ is the Frobenius automorphism. Note that, as in ([34] p. 369), we may arrange that c is a divisor of r. When $\mathrm{char}(K) = p > 0$ we may arrange that $c = r$, since $K \cong \mathbf{F}_v((X))$ and L is a Kummer extension of $L^{\langle a \rangle} = \mathbf{F}_{v^d}((X))$.

Henceforth, therefore, we shall assume that $\mathrm{char}(K) = p > 0$ and that the Galois group is given by the semi-direct product

$$G(L/K) = \langle a, g \mid g^d = a^r = 1,\ \mathrm{gag}^{-1} = a^v \rangle = \langle g \rangle \propto \langle a \rangle.$$

Hence $r = [G_0(L/K) : G_1(L/K)] = [G(L/W) : G_1(L/W)]$ is the index which appears (denoted by t there) in Example 3.1.21 and Theorem 3.1.22.

Set $q = v^2$ and let w be an integer satisfying the congruence $vw \equiv 1$ (modulo $(q^d - 1)^2$). Here we have erred on the side of caution, in order to have ample congruences later. In particular, we observe that $vw \equiv 1$ (modulo $(v^d - 1)^2$).

Let $\mathbf{Q}/\mathbf{Z}(i)$ denote the i-th Tate twist of the roots of unity, which we shall write as a *multiplicative* abelian group. For $i = 1, 2$, set

$$\tilde{\mathrm{Ind}}(i) = \mathrm{Ind}_{G(L/W)}^{G(L/K)}(\mathbf{Q}/\mathbf{Z}(i)[1/p] \tilde{\oplus} \mathbf{Z})$$

where the action of $a \in G(L/W)$ on $\mathbf{Q}/\mathbf{Z}(i)[1/p] \tilde{\oplus} \mathbf{Z}$ is trivial on the first factor and is given by $a(1,1) = (\xi_r, 1)$ on the second factor. Here we have identified $\mathbf{Q}/\mathbf{Z}(1)[1/p]$ and $\mathbf{Q}/\mathbf{Z}(2)[1/p]$ as *groups*, choosing once and for all a primitive r-th root of unity, ξ_r. These $\mathbf{Z}[G(L/K)]$-modules are both equal to the module,

$\mathrm{Ind}_{G(L/W)}^{G(L(1)/K)}(\mu_\infty[1/p]\tilde{\oplus}\mathbf{Z})$ of Theorem 3.1.22, since $a \in G(L/W)$ acts trivially on the roots of unity.

The reason for distinguishing $\tilde{\mathrm{Ind}}(1)$ and $\tilde{\mathrm{Ind}}(2)$ is that they sit in different 2-extensions, which we shall now describe. The local K_0/K_1 fundamental class which is used in [34] is due to J.-P. Serre (see also [131] Chapter 7). It is a 2-extension of Galois modules and (see [132]) is filtered by the usual filtration of the multiplicative group of a local field so as to remain exact at each level. In particular, in the tame case, we may truncate Serre's fundamental class at level one to give a 2-extension of $\mathbf{Z}[G(L/K)]$-modules of the form

$$L^*/U_L^1 \cong \overline{L}^* \times \mathbf{Z} \longrightarrow \tilde{\mathrm{Ind}}(1) \stackrel{\tilde{F}-1}{\longrightarrow} \tilde{\mathrm{Ind}}(1) \longrightarrow \mathbf{Z}$$

where

$$\tilde{F}(g^i \otimes (\eta, m)) = g^{i-1} \otimes (\eta^v, m).$$

Now let us consider the K_2/K_3 fundamental class of Theorem 3.1.22. In this case, recalling the isomorphism of Galois modules

$$K_3(\overline{L}) \cong \mathbf{F}^*_{v^{2d}},$$

the 2-extension takes the form

$$K_3(\overline{L}) \longrightarrow \tilde{\mathrm{Ind}}(2) \stackrel{\hat{F}-1}{\longrightarrow} \tilde{\mathrm{Ind}}(2) \longrightarrow \overline{L}^* = \mathbf{F}^*_{v^d}$$

where

$$\hat{F}(g^i \otimes (\eta, m)) = g^{i-1} \otimes (\eta^q, mv).$$

We give the following verification for completeness although the part concerning $\hat{F}$ may be extracted from §3.1.16.

Lemma 4.1.2

The maps $\tilde{F}$ and $\hat{F}$ are both $\mathbf{Z}[G(L/K)]$-module homomorphisms.

Proof. Since $\mathrm{gag}^{-1} = a^v$ we have $ga^{uw}g^{-1} = (a^{uw})^v = a^{uwv} = a^u$ and so $g^i a^{uw^i} = a^u g^i$. Therefore one finds that

$$\begin{aligned} g(\tilde{F}(g^i \otimes (\eta, m))) &= gg^{i-1} \otimes (\eta^v, m) \\ &= g^i \otimes (\eta^v, m) \\ &= \tilde{F}(gg^i \otimes (\eta, m)) \end{aligned}$$

and

$$\begin{aligned} g(\hat{F}(g^i \otimes (\eta, m))) &= gg^{i-1} \otimes (\eta^q, mv) \\ &= g^i \otimes (\eta^q, mv) \\ &= \tilde{F}(gg^i \otimes (\eta, m)). \end{aligned}$$

Also

$$\begin{aligned} a(\tilde{F}(g^i \otimes (\eta, m))) &= ag^{i-1} \otimes (\eta^v, m) \\ &= g^{i-1}a^{w^{i-1}} \otimes (\eta^v, m) \\ &= g^{i-1} \otimes (\eta^v \xi_r^{mw^{i-1}}, m) \\ &= g^{i-1} \otimes (\eta^v \xi_r^{mvw^i}, m) \\ &= \tilde{F}(g^i \otimes (\eta \xi_r^{mw^i}, m)) \\ &= \tilde{F}(g^i a^{w^i} \otimes (\eta, m)) \\ &= \tilde{F}(ag^i \otimes (\eta, m)) \end{aligned}$$

and

$$\begin{aligned} a(\hat{F}(g^i \otimes (\eta, m))) &= ag^{i-1} \otimes (\eta^q, mv) \\ &= g^{i-1}a^{w^{i-1}} \otimes (\eta^q, mv) \\ &= g^{i-1} \otimes (\eta^q \xi_r^{mvw^{i-1}}, mv) \\ &= g^{i-1} \otimes (\eta^q \xi_r^{mqw^i}, mv) \\ &= \hat{F}(g^i \otimes (\eta \xi_r^{mw^i}, m)) \\ &= \hat{F}(g^i a^{w^i} \otimes (\eta, m)) \\ &= \hat{F}(ag^i \otimes (\eta, m)), \end{aligned}$$

as required. □

4.1.3 We now construct an explicit commutative diagram of $\mathbf{Z}[G(L/K)]$-modules which is essentially equivalent to that of ([34] Lemma 6.3 p. 370):

$$\begin{array}{ccccccc} X(2) & \longrightarrow & \mathbf{Z}[G(L/K)]\langle z_1, z_2\rangle & \xrightarrow{\delta_2} & \mathbf{Z}[G(L/K)]\langle z_0\rangle & \xrightarrow{\delta_1} & \mathbf{Z} \\ \downarrow{\scriptstyle f_4} & & \downarrow{\scriptstyle f_3} & & \downarrow{\scriptstyle f_2} & & \downarrow{\scriptstyle 1} \\ \overline{L}^* \times \mathbf{Z} & \longrightarrow & \widetilde{\mathrm{Ind}}(1) & \xrightarrow{\tilde{F}-1} & \widetilde{\mathrm{Ind}}(1) & \longrightarrow & \mathbf{Z} \end{array}$$

Here $X(2) = \ker(\delta_2)$, δ_1 is the augmentation and $\delta_2(z_2) = (a-1)z_0$, $\delta_2(z_1) = (g^{-1}-1)z_0$.

To define f_2, f_3, f_4 we need another element, $\gamma \in \mathbf{Q}/\mathbf{Z}(1)[1/p]$, which we shall choose in the same manner as [34]. We begin by considering $\mathbf{Q}/\mathbf{Z}(1)[1/p]$ as a subgroup of the multiplicative group of the maximal unramified extension of L. If π_L is a prime of L such that

$$a(\pi_L)/\pi_L = \xi_r \in \mathcal{O}_L^*/U_L^1 \cong \overline{L}^* \cong \mathbf{F}_{v^d}^*$$

then we choose γ to satisfy $\gamma^{v^d-1} = \xi_r$ so that γ is a primitive $(v^d-1)r$-th root of unity.

With this notation we define

$$\begin{aligned} f_2(z_0) &= g^{d-1}\otimes(1,1),\\ f_3(z_1) &= g^{-1}\otimes(\gamma^r,1)+1\otimes(\gamma^{rw},0)+g\otimes(\gamma^{rw^2},0)\\ &\quad +\ldots+g^{d-3}\otimes(\gamma^{rw^{d-2}},0)+g^{d-2}\otimes(\gamma^{rw^{d-1}},0),\\ f_3(z_2) &= \textstyle\sum_{i=0}^{d-1} g^i\otimes(\gamma^{w^i},0). \end{aligned}$$

Of course, f_4 will be induced by f_3 once we have verified the commutativity of the diagram.

Lemma 4.1.4

The diagram of §4.1.3 is commutative.

Proof. Since the augmentation of $g^i\otimes(\eta,m)\in \tilde{\mathrm{Ind}}(1)$ is equal to m, the right hand square clearly commutes.

Note that if $z\in \mathbf{F}^*_{v^d}$ then $z\mapsto z^v$ is the Frobenius automorphism with inverse given by $z\mapsto z^w$. Hence $\xi_r^{w^{d-1}}=\xi_r^v$. Similarly $\gamma^{wv}=\gamma$ since $vw\equiv 1$ (modulo $(v^d-1)r$).

For the middle square we have

$$\begin{aligned} f_2(\delta_2(z_1)) &= f_2(g^{-1}z_0-z_0)\\ &= g^{-1}(g^{d-1}\otimes(1,1))-g^{d-1}\otimes(1,1)\\ &= g^{d-2}\otimes(1,1)-g^{d-1}\otimes(1,1) \end{aligned}$$

while

$$\begin{aligned} &(\tilde{F}-1)f_3(z_1)\\ &\quad= \tilde{F}(g^{-1}\otimes(\gamma^r,1)+1\otimes(\gamma^{rw},0)+g\otimes(\gamma^{rw^2},0)\\ &\qquad +\ldots+g^{d-3}\otimes(\gamma^{rw^{d-2}},0)+g^{d-2}\otimes(\gamma^{rw^{d-1}},0))\\ &\qquad -(g^{-1}\otimes(\gamma^r,1)+1\otimes(\gamma^{rw},0)+g\otimes(\gamma^{rw^2},0)\\ &\qquad +\ldots+g^{d-3}\otimes(\gamma^{rw^{d-2}},0)+g^{d-2}\otimes(\gamma^{rw^{d-1}},0))\\ &\quad= g^{-2}\otimes(\gamma^{rv},1)+g^{-1}\otimes(\gamma^{rvw},0)+1\otimes(\gamma^{rvw^2},0)\\ &\qquad +\ldots+g^{d-4}\otimes(\gamma^{rvw^{d-2}},0)+g^{d-3}\otimes(\gamma^{rvw^{d-1}},0)\\ &\qquad -g^{-1}\otimes(\gamma^r,1)-1\otimes(\gamma^{rw},0)-g\otimes(\gamma^{rw^2},0)\\ &\qquad -\ldots-g^{d-3}\otimes(\gamma^{rw^{d-2}},0)-g^{d-2}\otimes(\gamma^{rw^{d-1}},0) \end{aligned}$$

$$
\begin{aligned}
&= g^{d-2}\otimes(\gamma^{rv},0)+g^{d-2}\otimes(1,1)+g^{-1}\otimes(\gamma^{r},0)+1\otimes(\gamma^{rw},0)\\
&\quad+\ldots+g^{d-4}\otimes(\gamma^{rw^{d-3}},0)+g^{d-3}\otimes(\gamma^{rw^{d-2}},0)\\
&\quad-g^{d-1}\otimes(1,1)-g^{-1}\otimes(\gamma^{r},0)-1\otimes(\gamma^{rw},0)-g\otimes(\gamma^{rw^2},0)\\
&\quad-\ldots-g^{d-3}\otimes(\gamma^{rw^{d-2}},0)-g^{d-2}\otimes(\gamma^{rw^{d-1}},0)\\
&= g^{d-2}\otimes(\gamma^{rv},0)+g^{d-2}\otimes(1,1)-g^{d-1}\otimes(1,1)-g^{d-2}\otimes(\gamma^{rw^{d-1}},0)\\
&= g^{d-2}\otimes(1,1)-g^{d-1}\otimes(1,1),
\end{aligned}
$$

as required, since $\gamma^{rw^{d-1}}=\gamma^{rw^{d-1}wv}=\gamma^{rw^{d}v}=\gamma^{rv}$.

Next we have

$$
\begin{aligned}
(\tilde{F}-1)(f_3(z_2)) &= \textstyle\sum_{i=0}^{d-1} g^{i-1}\otimes(\gamma^{w^iv},0)-\sum_{i=0}^{d-1} g^{i}\otimes(\gamma^{w^i},0)\\
&= g^{d-1}\otimes(\gamma^{v},0)-g^{d-1}\otimes(\gamma^{w^{d-1}},0)\\
&= g^{d-1}\otimes(\gamma^{v-w^{d-1}},0)\\
&= g^{d-1}\otimes(\gamma^{w^{d-1}(v^d-1)},0)\\
&= g^{d-1}\otimes(\xi_r^{w^{d-1}},0)
\end{aligned}
$$

while

$$
\begin{aligned}
f_2(\delta_2(z_2)) &= f_2((a-1)z_0)\\
&= ag^{d-1}\otimes(1,1)-g^{d-1}\otimes(1,1)\\
&= g^{d-1}a^{w^{d-1}}\otimes(1,1)-g^{d-1}\otimes(1,1)\\
&\quad(\text{since } vw\equiv 1 \ (\text{modulo } r))\\
&= g^{d-1}\otimes(\xi_r^{w^{d-1}},0)\\
&= (\tilde{F}-1)(f_3(z_2))
\end{aligned}
$$

as required. □

4.1.5 *Surjectivity of f_4*

Recall that the inclusion homomorphism, $\phi: M \longrightarrow \mathrm{Ind}_H^G(M)$, is given by $\phi(m)=\sum_{g\in G/H} g\otimes_H g^{-1}m$. If $y=(1+a+\ldots+a^{r-1})z_2\in X(2)$ then

$$
\begin{aligned}
f_3(y) &= (1+a+\ldots+a^{r-1})(\textstyle\sum_{i=0}^{d-1} g^i\otimes(\gamma^{w^i},0))\\
&= \textstyle\sum_{i=0}^{d-1} g^i\otimes(\gamma^{rw^i},0)\\
&= \phi(\gamma^r).
\end{aligned}
$$

However, γ^r is a primitive (v^d-1)-th root of unity so that $f_4(y)=\gamma^r$ generates $\mathbf{F}_{q^v}^*$.

Next set

$$x = (1 + g^{-1} + \ldots + g^{1-d})z_1 + (a^{-1} + a^{-2} + \ldots + a^{-c})z_2 \in X(2).$$

Then $f_4(x) \in \mathbf{F}^*_{q^v} \times \mathbf{Z}$ has an image in $\tilde{\mathrm{Ind}}(1)$ of the form $(\ldots, (\sum_{i=0}^{d-1} g^{-i})g^{d-1} \otimes 1)$, which has second coordinate equal to $\phi(1)$ so that

$$f_4(x) = (\ldots, 1) \in \mathbf{F}^*_{q^v} \times \mathbf{Z}$$

and so f_4 is surjective.

4.1.6 *The kernel of f_4*

Now we describe $\ker(f_4) \subseteq X(2)$.

Suppose that, for $m_i \in \mathbf{Z}[G(L/K)]$, $m_1 z_1 + m_2 z_2 \in \ker(f_4)$ then $0 = f_3(m_1 z_1 + m_2 z_2) \in \tilde{\mathrm{Ind}}(1)$ has second coordinate equal to $m_1 g^{d-1} \in \mathrm{Ind}_{G(L/W)}^{G(L/K)}(\mathbf{Z})$ so that, by ([34] §6.6), $m_1 = m_3(a-1)$ for some m_3. Also, in $\mathbf{Z}[G(L/K)]$, we have

$$(a-1)(g^{-1} - 1) = (g^{-1}(1 + a + \ldots + a^{v-1}) - 1)(a-1)$$

so that we set

$$f = (g^{-1}(1 + a + \ldots + a^{v-1}) - 1) \in \mathbf{Z}[G(L/K)].$$

Then

$$\begin{aligned} 0 = \delta_2(m_1 z_1 + m_2 z_2) &= m_3(a-1)\delta_2(z_1) + m_2\delta_2(z_2) \\ &= (m_3(a-1)(g^{-1} - 1) + m_2(a-1))z_0 \end{aligned}$$

so that $m_3 f + m_2$ annihilates $a - 1$ in $\mathbf{Z}[G(L/K)]$ and, by ([34] §6.6),

$$m_3 f + m_2 = m_4(1 + a + \ldots + a^{r-1}) \in \mathbf{Z}[G(L/K)](1 + a + \ldots + a^{r-1}).$$

Thus we have shown that

$$m_1 z_1 + m_2 z_2 = m_3((a-1)z_1 - f z_2) + m_4(1 + a + \ldots + a^{r-1})z_2,$$

which is the form of an arbitrary element in $\ker(f_4) \subseteq X(2)$.

Now let us calculate f_4 on $(a-1)z_1 - f z_2$ and $y = (1 + a + \ldots + a^{r-1})z_2$.

We have

$$\begin{aligned} &f_3((a-1)z_1 - f z_2) \\ &= (a-1)(g^{-1} \otimes (\gamma^r, 1) + 1 \otimes (\gamma^{rw}, 0) + g \otimes (\gamma^{rw^2}, 0) + \ldots + g^{d-3} \otimes (\gamma^{rw^{d-2}}, 0) \\ &\quad + g^{d-2} \otimes (\gamma^{rw^{d-1}}, 0)) - f(\textstyle\sum_{i=0}^{d-1} g^i \otimes (\gamma^{w^i}, 0)) \\ &= (a-1)(g^{-1} \otimes (1,1) - f(\textstyle\sum_{i=0}^{d-1} g^i \otimes (\gamma^{w^i}, 0))) \\ &= (g^{-1}a^v - g^{-1}) \otimes (1,1) - (g^{-1}(1 + a + \ldots + a^{v-1}) - 1)(\textstyle\sum_{i=0}^{d-1} g^i \otimes (\gamma^{w^i}, 0)) \\ &= g^{-1} \otimes (\xi_r^v, 0) - \textstyle\sum_{i=0}^{d-1} g^{i-1} \otimes (\gamma^{vw^i}, 0) + \textstyle\sum_{i=0}^{d-1} g^i \otimes (\gamma^{w^i}, 0) \\ &= g^{-1} \otimes (\xi_r^v \gamma^{-v} \gamma^{w^{d-1}}, 0) \\ &= g^{-1} \otimes ((\xi_r \gamma^{w^d - 1})^v, 0) \end{aligned}$$

which is trivial because $\gamma^{w^d - 1} = \xi_r^{-1}$.

On the other hand $f_3(y) = \phi(\gamma^r)$ has annihilator equal to $(g - v)$ but

$$(g - v)y = (1 + a + \ldots + a^{r-1})g((a - 1)z_1 - fz_2)$$

so that $\ker(f_4)$ is equal to the free $\mathbf{Z}[G(L/K)]$- module with basis element $t = (a - 1)z_1 - fz_2$, which is the same result as ([34] Lemma 6.8).

4.1.7 *The tame diagram for* K_2/K_3

Temporarily set $\hat{G} = G(W/K) = \langle g \rangle \cong \mathbf{Z}/d$ so that, by ([131] Proposition 7.1.36),

$$0 \longrightarrow Q_1 = \mathbf{Z}[\hat{G}] \stackrel{(v-g)}{\longrightarrow} Q_0 = \mathbf{Z}[\hat{G}] \stackrel{\pi}{\longrightarrow} \mathbf{F}^*_{v^d} \longrightarrow 0$$

is an exact sequence of $\mathbf{Z}[\hat{G}]$-modules with $\pi(g^i) = g^i(x) = x^{v^i}$ for some choice of generator, $x \in \mathbf{F}^*_{v^d}$. Set $\Lambda_{i,j} = P_i \otimes Q_j$ with the diagonal $G(L/K)$-action, where $P_1 = \mathbf{Z}[G(L/K)]\langle z_1, z_2\rangle$ and $P_0 = \mathbf{Z}[G(L/K)]\langle z_0\rangle$.

We wish to construct a commutative diagram of $\mathbf{Z}[G(L/K)]$-modules of the form

$$\begin{array}{ccccccc}
\hat{X}(2) & \longrightarrow & \Lambda_{0,1} \oplus \Lambda_{1,0} & \xrightarrow{\delta \otimes 1 \pm 1 \otimes \delta} & \Lambda_{0,0} & \xrightarrow{\epsilon} & \mathbf{F}^*_{v^d} \\
\downarrow g_4 & & \downarrow g_3 & & \downarrow g_2 & & \downarrow 1 \\
K_3(\mathbf{F}_{v^d}) & \longrightarrow & \tilde{\mathrm{Ind}}(2) & \xrightarrow{\hat{F} - 1} & \tilde{\mathrm{Ind}}(2) & \xrightarrow{\epsilon'} & \mathbf{F}^*_{v^d}
\end{array}$$

where $\hat{X}(2) = \ker(\delta \otimes 1 \pm 1 \otimes \delta)$. Here $\epsilon(z_0 \otimes g^i) = g^i(x) = x^{v^i} \in \mathbf{F}^*_{v^d}$ and $\epsilon'(1 \otimes (\xi, m)) = x^m$.

For $1 \leq i \leq d$ define

$$g_2(z_0 \otimes g^i) = 1 \otimes (1, v^i),$$

$$g_3(z_0 \otimes g^i) = \begin{cases} 0 & \text{if } i = 1, 2, \ldots, d-1, \\ \sum_{j=1}^{d} g^j \otimes (1, v^{d+1-j}) & \text{if } i = d, \end{cases}$$

$$g_3(z_1 \otimes g^i) = \begin{cases} 1 \otimes (1, v^i), & \text{if } 1 \leq i \leq d-1, \\ -\sum_{i=2}^{d} g^{i-1} \otimes (1, v^{d+1-i}), & \text{otherwise}, \end{cases}$$

$$g_3(z_2 \otimes g^i) = \sum_{j=0}^{d-1} g^{j+1} \otimes (\delta^{v^{i+d-2}w^{2j}}, 0) = \sum_{j=0}^{d-1} g^{j+1} \otimes (\delta^{v^{i+d-2-2j}}, 0)$$

where $\delta \in \mathbf{Q}/\mathbf{Z}$ satisfies $\delta^{1-w^{2d}} = \xi_r$. Recall that we have chosen w so that $qw^2 = v^2w^2 \equiv vw \equiv 1$ (modulo $(q^d - 1)r$) so that $\delta^{vw} = \delta$.

Hence

$$\epsilon' g_2(z_0 \otimes g^i) = \epsilon'(1 \otimes (1, v^i)) = x^{v^i} = \epsilon(z_0 \otimes g^i).$$

If $i = 1, 2, \ldots, d-1$ we have $z_0 \otimes g^i \in \Lambda_{0,1}$ and

$$\begin{aligned} g_2(\delta \otimes 1 + 1 \otimes \delta)(z_0 \otimes g^i) &= g_2(z_0 \otimes (v-g)g^i) \\ &= v(1 \otimes (1, v^i)) - 1 \otimes (1, v^{i+1}) \\ &= 1 \otimes (1^v, v^{i+1}) - 1 \otimes (1, v^{i+1}) \\ &= 0 \\ &= (\hat{F} - 1)g_3(z_0 \otimes g^i). \end{aligned}$$

When $i = d$ we have

$$\begin{aligned} g_2(\delta \otimes 1 + 1 \otimes \delta)(z_0 \otimes g^d) &= g_2(z_0 \otimes (vg^d - g)) \\ &= 1 \otimes (1, v^{d+1}) - 1 \otimes (1, v) \\ &= 1 \otimes (1, (v^d - 1)v) \end{aligned}$$

but

$$\begin{aligned} (\hat{F} - 1)g_3(z_0 \otimes g^d) &= (\hat{F} - 1)(\textstyle\sum_{j=1}^d g^j \otimes (1, v^{d+1-j})) \\ &= \textstyle\sum_{j=1}^d g^{j-1} \otimes (1, v^{d+2-j}) - g^j \otimes (1, v^{d+1-j}) \\ &= 1 \otimes (1, v^{d+1}) + g \otimes (1, v^d) + \ldots + g^{d-1} \otimes (1, v^2) \\ &\qquad -g \otimes (1, v^d) - g^2 \otimes (1, v^{d-1}) - \ldots - g^d \otimes (1, v) \\ &= 1 \otimes (1, v^{d+1}) - 1 \otimes (1, v) \\ &= 1 \otimes (1, (v^d - 1)v), \end{aligned}$$

as required.

Next we look at $z_1 \otimes g^i, z_2 \otimes g^i \in \Lambda_{1,0}$.

For $1 \leq i \leq d-1$ we have

$$\begin{aligned} g_2(\delta \otimes 1 \pm 1 \otimes \delta)(z_1 \otimes g^i) &= g_2((g^{-1} - 1)z_0 \otimes g^i) \\ &= g^{-1}g_2(z_0 \otimes g^{i+1}) - g_2(z_0 \otimes g^i) \\ &= g^{-1} \otimes (1, v^{i+1}) - 1 \otimes (1, v^i) \end{aligned}$$

while

$$\begin{aligned} g_2(\delta \otimes 1 \pm 1 \otimes \delta)(z_1 \otimes g^d) &= g^{-1}g_2(z_0 \otimes g) - g_2(z_0 \otimes g^d) \\ &= g^{-1} \otimes (1, v) - 1 \otimes (1, v^d). \end{aligned}$$

On the other hand, if $1 \leq i \leq d-1$,

$$\begin{aligned} (\hat{F} - 1)g_3(z_1 \otimes g^i) &= (\hat{F} - 1)(1 \otimes (1, v^i)) \\ &= g^{-1} \otimes (1, v^{i+1}) - 1 \otimes (1, v^i) \end{aligned}$$

and

$$\begin{aligned}(\hat{F}-1)g_3(z_1\otimes g^d) &= (\hat{F}-1)(-\textstyle\sum_{i=2}^d g^{i-1}\otimes(1,v^{d+1-i}))\\ &= -\textstyle\sum_{i=2}^d g^{i-2}\otimes(1,v^{d+2-i})+\sum_{i=2}^d g^{i-1}\otimes(1,v^{d+1-i})\\ &\quad -1\otimes(1,v^d)-g\otimes(1,v^{d-1})-\ldots-g^{d-2}\otimes(1,v^2)\\ &\qquad +g\otimes(1,v^{d-1})+g^2\otimes(1,v^{d-2})+\ldots+g^{d-1}\otimes(1,v)\\ &= g^{-1}\otimes(1,v)-1\otimes(1,v^d),\end{aligned}$$

as required.

Finally, for $1\leq i\leq d$, we have

$$\begin{aligned}g_2(\delta\otimes 1\pm 1\otimes\delta)(z_2\otimes g^i) &= g_2((a-1)z_0\otimes g^i)\\ &= ag_2(z_0\otimes g^i)-g_2(z_0\otimes g^i)\\ &= a\otimes(1,v^i)-1\otimes(1,v^i)\\ &= 1\otimes(\xi_r^{v^i},v^i)-1\otimes(1,v^i)\\ &= 1\otimes(\xi_r^{v^i},0)\end{aligned}$$

while

$$\begin{aligned}(\hat{F}-1)g_3(z_2\otimes g^i) &= \textstyle\sum_{j=0}^{d-1} g^j\otimes(\delta^{v^{d+i-2}w^{2j}q},0)-\sum_{j=0}^{d-1} g^{j+1}\otimes(\delta^{v^{d+i-2}w^{2j}},0)\\ &= 1\otimes(\delta^{v^{d+i-2}q},0)-1\otimes(\delta^{v^{d+i-2}w^{2d-2}},0)\\ &= 1\otimes(\delta^{v^{d+i}(1-w^{2d})},0)\\ &= 1\otimes(\xi_r^{v^{d+i}},0)\\ &= 1\otimes(\xi_r^{v^i},0),\end{aligned}$$

as required.

Lemma 4.1.8 *[34]*

Suppose that g_4 is surjective in the diagram of §4.1.7. Then $\mathrm{Ker}(g_4)$ *is a finitely generated projective* $\mathbf{Z}[G(L/K)]$*-module and the Euler characteristic of Example 2.1.8(i) and Example 2.2.7 associated to the lower 2-extension in the diagram is equal to*

$$-[\mathrm{Ker}(g_4)]\in\mathcal{CL}(\mathbf{Z}[G(L/K)]).$$

Proof. If g_4 is surjective then, since g_4 induces isomorphisms on all Tate cohomology groups, its kernel is a finitely generated, cohomologically trivial, torsion-free $\mathbf{Z}[G(L/K)]$-module. Hence it is projective. Therefore

$$0\longrightarrow\mathrm{Ker}(g_4)\longrightarrow\Lambda_{0,1}\oplus\Lambda_{1,0}\longrightarrow\Lambda_{0,0}\longrightarrow 0$$

is a perfect complex which is quasi-isomorphic to

$$0\longrightarrow\tilde{\mathrm{Ind}}(2)\xrightarrow{\hat{F}-1}\tilde{\mathrm{Ind}}(2)\longrightarrow 0.$$

Hence, by Examples 2.1.8 and 2.2.7 (bearing in mind that $\mathrm{Ker}(g_4)$ is in an even dimension), the Euler characteristic of the 2-extension is equal to

$$-[\mathrm{Ker}(g_4)] + [\Lambda_{0,1}] + [\Lambda_{1,0}] - \Lambda_{0,0} = -[\mathrm{Ker}(g_4)] \in \mathcal{CL}(\mathbf{Z}[G(L/K)]),$$

because the $\Lambda_{i,j}$'s are free. □

Proposition 4.1.9

The homomorphism, g_4, is surjective in §4.1.7.

Proof. Suppose that we have a resolution of free $\mathbf{Z}[G(L/K)]$-modules of the form

$$\ldots \longrightarrow P_2 \xrightarrow{\delta_3} P_1 \xrightarrow{\delta_2} P_0 \xrightarrow{\delta_1} \mathbf{Z} \longrightarrow 0$$

such that $\delta_3(P_2) = X(2) \subseteq P_1$. Then, in the notation of §4.1.7,

$$\Lambda_{2,0} \oplus \Lambda_{1,1} \oplus \Lambda_{0,2} \stackrel{\delta\otimes 1 \pm 1 \otimes \delta}{\longrightarrow} \Lambda_{1,0} \oplus \Lambda_{0,1} \stackrel{\delta\otimes 1 \pm 1 \otimes \delta}{\longrightarrow} \Lambda_{0,0}$$

is exact in the middle and so

$$\hat{X}(2) = (\delta \otimes 1)(\Lambda_{2,0}) + (\delta \otimes 1 - 1 \otimes \delta)(\Lambda_{1,1}),$$

since $\Lambda_{0,2} = 0$. Also

$$(\delta \otimes 1)(\Lambda_{2,0}) = X(2) \otimes Q_0 \subseteq P_1 \otimes Q_0.$$

Generators of the $\mathbf{Z}[G(L/K)]$-module, $X(2)$, are given by the elements $t = (a-1)z_1 - fz_2$, $y = (1 + a + \ldots + a^{r-1})z_2$ and $x = (1 + g^{-1} + \ldots + g^{1-d})z_1 + (a^{-1} + \ldots a^{-c})z_2$ of §4.1.5. This is seen by observing that all these elements are in $X(2)$, that $\mathbf{Z}[G(L/K)]\langle t\rangle = \ker(f_4)$ and that $f_4(y), f_4(x)$ generate $\mathbf{F}^*_{v^d} \times \mathbf{Z}$, by §4.1.5.

Now we have

$$\begin{aligned}
g_4(y \otimes g^i) &= g_3((1 + a + \ldots + a^{r-1})z_2 \otimes g^i) \\
&= (1 + a + \ldots + a^{r-1}) g_3(z_2 \otimes g^i) \\
&= (1 + a + \ldots + a^{r-1}) \textstyle\sum_{j=0}^{d-1} g^{j+1} \otimes (\delta^{v^{i+d-2}w^{2j}}, 0) \\
&= \textstyle\sum_{j=0}^{d-1} g^{j+1} \otimes (\delta^{rv^{i+d-2}w^{2j}}, 0) \\
&= \textstyle\sum_{j=0}^{d-1} g^{j+1} \otimes (\delta^{rv^{i+d}w^{2j+2}}, 0) \\
&= \phi(\delta^{rv^{i+d}}, 0)
\end{aligned}$$

so that, for $1 \leq i \leq d$,

$$g_4(y \otimes g^i) = (\delta^r)^{v^{i+d}} \in K_3(\mathbf{F}_{v^d}).$$

Note that, since δ is a primitive $r(v^{2d}-1)$-th root of unity, $(\delta^r)^{v^{i+d}}$ is a primitive $(v^{2d}-1)$-th root of unity and hence generates $K_3(\mathbf{F}_{v^d})$. Therefore g_4 is surjective, as required. □

4.1.10 *Some elements in* $\mathrm{Ker}(g_4)$

We are now going to find elements $z_{1,1}, \ldots, z_{1,d}, z_{2,1}, \ldots, z_{2,d}$ which will generate a free $\mathbf{Z}[G(L/K)]$-submodule of $\mathrm{Ker}(g_4)$. We calculate

$$\begin{aligned} g_4(x \otimes g^i) &= g_3((1+g^{-1}+\ldots+g^{1-d})z_1 \otimes g^i \\ &\quad +(a^{-1}+a^{-2}+\ldots+a^{-c})z_2 \otimes g^i) \\ &= g_3(z_1 \otimes g^i + g^{-1}(z_1 \otimes g^{i+1}) \\ &\quad +\ldots+g^{1-d}(z_1 \otimes g^{i+d-1})) \\ &\quad +(a^{-1}+a^{-2}+\ldots+a^{-c})g_3(z_2 \otimes g^i) \\ &= 1 \otimes (1, v^i) + g^{-1} \otimes (1, v^{i+1}) + \ldots + g^{-(d-i-1)} \otimes (1, v^{d-1}) \\ &\quad -g^{-(d-i)}(\textstyle\sum_{j=2}^{d} g^{j-1} \otimes (1, v^{d+1-j})) \\ &\quad +g^{-(d-i+1)} \otimes (1, v) + \ldots + g^{1-d} \otimes (1, v^{i-1}) \\ &\quad +(a^{-1}+a^{-2}+\ldots+a^{-c})(\textstyle\sum_{j=0}^{d-1} g^{j+1} \otimes (\delta^{v^{i+d-2}w^{2j}}, 0)) \\ &= 1 \otimes (1, v^i) + g^{-1} \otimes (1, v^{i+1}) + \ldots + g^{i+1-d} \otimes (1, v^{d-1}) \\ &\quad +g^{i-1-d} \otimes (1, v) + \ldots + g^{1-d} \otimes (1, v^{i-1}) - g^{i+1-d} \otimes (1, v^{d-1}) \\ &\quad -\ldots - g^{i-1} \otimes (1, v) + \textstyle\sum_{j=0}^{d-1} g^{j+1} \otimes (\delta^{cv^{i+d}w^{2j+2}}, 0) \\ &= \textstyle\sum_{j=0}^{d-1} g^{j+1} \otimes (\delta^{cv^{i+d}w^{2j+2}}, 0) \\ &= \phi(\delta^{cv^{i+d}}) \end{aligned}$$

so that, for $1 \leq i \leq d$,

$$g_4(x \otimes g^i) = (\delta^c)^{v^{i+d}} \in K_3(\mathbf{F}_{v^d}).$$

For $1 \leq i \leq d-1$,

$$\begin{aligned} & g_4(((a-1)z_1 - fz_2) \otimes g^i) \\ &= (a-1)(1 \otimes (1, v^i)) - g_3((g^{-1}(1+a+\ldots+a^{v-1}) - 1)z_2 \otimes g^i) \\ &= 1 \otimes (\xi_r^{v^i}, 0) - g^{-1}g_3((1+a+\ldots+a^{v-1})z_2 \otimes g^{i+1}) + g_3(z_2 \otimes g^i) \\ &= 1 \otimes (\xi_r^{v^i}, 0) - g^{-1}\textstyle\sum_{j=0}^{d-1} g^{j+1} \otimes (\delta^{v^{i+1+d-2}w^{2j}v}, 0) + \textstyle\sum_{j=0}^{d-1} g^{j+1} \otimes (\delta^{v^{i+d-2}w^{2j}}, 0) \\ &= 1 \otimes (\xi_r^{v^i}, 0) - \textstyle\sum_{j=0}^{d-1} g^j \otimes (\delta^{v^{i+d}w^{2j}}, 0) + \textstyle\sum_{j=0}^{d-1} g^{j+1} \otimes (\delta^{v^{i+d}w^{2j+2}}, 0) \\ &= 1 \otimes (\xi_r^{v^i}, 0) - 1 \otimes (\delta^{v^{i+d}}, 0) + g^d \otimes (\delta^{v^{i+d}w^{2d}}, 0) + \textstyle\sum_{j=1}^{d-1} g^j \otimes (\delta^{(v^{i+d}-v^{i+d})w^{2j}}, 0) \\ &= 1 \otimes (\xi_r^{v^i}, 0) - 1 \otimes (\delta^{v^{i+d}(1-w^{2d})}, 0) \\ &= 1 \otimes (\xi_r^{v^i}, 0) - 1 \otimes (\xi_r^{v^{i+d}}, 0) \\ &= 0. \end{aligned}$$

For $i = d$,

$$
\begin{aligned}
&g_4(((a-1)z_1 - fz_2)\otimes g^d) \\
&= -(a-1)\textstyle\sum_{i=2}^d g^{i-1}\otimes(1, v^{d+1-i}) - g_3((g^{-1}(1+a+\ldots+a^{v-1})-1)z_2\otimes g^d) \\
&= -\textstyle\sum_{i=2}^d g^{i-1}(a^{w^{i-1}}-1)\otimes(1,v^{d+1-i}) - g^{-1}(1+a+\ldots+a^{v-1})g_3(z_2\otimes g) + g_3(z_2\otimes g^d) \\
&= -\textstyle\sum_{i=2}^d g^{i-1}\otimes(\xi_r^{v^{d+1-i}w^{i-1}},0) \\
&\qquad -\textstyle\sum_{j=0}^{d-1} g^j\otimes(\delta^{v^{1+d-2}w^{2j}v},0) + \textstyle\sum_{j=0}^{d-1} g^{j+1}\otimes(\delta^{v^{d+d-2}w^{2j}},0) \\
&= -\textstyle\sum_{j=1}^{d-1} g^j\otimes(\xi_r^{w^{2j}},0) - \textstyle\sum_{j=0}^{d-1} g^j\otimes(\delta^{v^d w^{2j}},0) + \textstyle\sum_{j=0}^{d-1} g^{j+1}\otimes(\delta^{v^{2d}w^{2j+2}},0) \\
&= -1\otimes(\delta^{v^d},0) + 1\otimes(\delta^{v^{2d}w^{2d}},0) + \textstyle\sum_{j=1}^{d-1} g^j\otimes(\xi_r^{-w^{2j}}\delta^{v^{2d}w^{2j}-v^d w^{2j}},0) \\
&= 1\otimes(\delta^{1-v^d},0) + \textstyle\sum_{j=1}^{d-1} g^j\otimes(\delta^{(v^{2d}-1)(-w^{2j})}\delta^{v^{2d}w^{2j}-v^d w^{2j}},0) \\
&= 1\otimes(\delta^{1-v^d},0) + \textstyle\sum_{j=1}^{d-1} g^j\otimes(\delta^{(1-v^d)w^{2j}},0) \\
&= \phi(\delta^{1-v^d}).
\end{aligned}
$$

Next we show that $g_4(\delta\otimes 1 - 1\otimes\delta)(z_1\otimes g^i) = 0$ for all $1 \leq i \leq d$. This can be seen by noting that $g_3(\delta\otimes 1 - 1\otimes\delta)(z_1\otimes g^i)$ lies in $\mathrm{Ind}_{G(L/W)}^{G(L/K)}(\mathbf{Z}[1/p]) \subset \tilde{\mathrm{Ind}}(2)$ and that $\hat{F}-1$ is injective on this subgroup whereas

$$
\begin{aligned}
&(\hat{F}-1)g_3(\delta\otimes 1 - 1\otimes\delta)(z_1\otimes g^i) \\
&= g_2(\delta\otimes 1 \pm 1\otimes\delta)(\delta\otimes 1 - 1\otimes\delta)(z_1\otimes g^i) \\
&= 0.
\end{aligned}
$$

Alternatively this can be seen by direct calculation as follows:

$$
\begin{aligned}
&g_3(\delta\otimes 1 - 1\otimes\delta)(z_1\otimes g^i) \\
&= g_3((g^{-1}-1)z_0\otimes g^i - z_1\otimes(v-g)g^i) \\
&= g^{-1}g_3(z_0\otimes g^{i+1}) - g_3(z_0\otimes g^i) - g_3(vz_1\otimes g^i) + g_3(z_1\otimes g^{i+1}).
\end{aligned}
$$

Therefore, if $i = 1, 2, \ldots, d-2$ then

$$
\begin{aligned}
&g_3(\delta\otimes 1 - 1\otimes\delta)(z_1\otimes g^i) \\
&= -v(1\otimes(1,v^i)) + 1\otimes(1,v^{i+1})) \\
&= 0
\end{aligned}
$$

while

$$
\begin{aligned}
&g_3(\delta(z_1)\otimes g^{d-1} - z_1\otimes(vg^{d-1}-g^d)) \\
&= g^{-1}g_3(z_0\otimes g^d) - 1\otimes(1,v^d) - \textstyle\sum_{j=2}^d g^{j-1}\otimes(1,v^{d+1-j}) \\
&= 0
\end{aligned}
$$

and

$$\begin{aligned}&g_3(\delta(z_1)\otimes g^d - z_1\otimes(vg^d - g))\\&= g^{-1}g_3(z_0\otimes g) - g_3(z_0\otimes g^d)\\&+v\sum_{j=2}^{d} g^{j-1}\otimes(1, v^{d+1-j}) + 1\otimes(1,v)\\&= -\sum_{j=1}^{d} g^j\otimes(1, v^{d+1-j})\\&+\sum_{j=2}^{d} g^{j-1}\otimes(1, v^{d+1-(j-1)}) + 1\otimes(1,v)\\&= 0.\end{aligned}$$

Next, if $i = 1, 2, \ldots, d-1$,

$$\begin{aligned}&lg_3(\delta(z_2)\otimes g^i - z_2\otimes(v-g)g^i)\\&= g_3((a-1)z_0\otimes g^i) - \sum_{j=0}^{d-1} vg^{j+1}\otimes(\delta^{v^{i+d-2}w^{2j}},0) + \sum_{j=0}^{d-1} g^{j+1}\otimes(\delta^{v^{i+1+d-2}w^{2j}},0)\\&= \sum_{j=0}^{d-1} g^{j+1}\otimes(\delta^{v^{i+d-1}w^{2j}-v^{i+d-1}w^{2j}},0)\\&= 0.\end{aligned}$$

If $i = d$ we obtain

$$\begin{aligned}&g_3(\delta(z_2)\otimes g^d - z_2\otimes(v-g)g^d)\\&= g_3((a-1)z_0\otimes g^d) - g_3(z_2\otimes vg^d) + g_3(z_2\otimes g)\\&= \sum_{j=1}^{d} g^j a^{w^j}\otimes(1, v^{d+1-j}) - \sum_{j=1}^{d} g^j\otimes(1, v^{d+1-j})\\&-v\sum_{j=0}^{d-1} g^{j+1}\otimes(\delta^{v^{d+d-2}w^{2j}},0) + \sum_{j=0}^{d-1} g^{j+1}\otimes(\delta^{v^{1+d-2}w^{2j}},0)\\&= \sum_{j=1}^{d} g^j\otimes(\xi_r^{w^j v^{d+1-j}}, v^{d+1-j}) - \sum_{j=1}^{d} g^j\otimes(1, v^{d+1-j})\\&-\sum_{j=0}^{d-1} g^{j+1}\otimes(\delta^{v^{2d-1}w^{2j}},0) + \sum_{j=0}^{d-1} g^{j+1}\otimes(\delta^{v^{d-1}w^{2j}},0)\\&= \sum_{j=1}^{d} g^j\otimes(\xi_r^{v^{d+1}w^{2j}},0) - \sum_{j=0}^{d-1} g^{j+1}\otimes(\delta^{v^{d+1}(v^d-1)w^{2j+2}},0)\\&= \phi(\xi_r^v) - \phi(\delta^{v^{d+1}(v^d-1)}).\end{aligned}$$

Now we have $\delta^{v^{2d}-1} = \xi_r$ so that

$$g_3(\delta(z_2)\otimes g^d - z_2\otimes(v-g)g^d) = \phi(\delta^{v^{2d+1}-v-v^{2d+1}+v^{d+1}}) = \phi(\delta^{v(v^d-1)}).$$

In $P_0\otimes Q_1 \oplus P_1\otimes Q_0$ we define elements

$$z_{1,i} = (\delta\otimes 1 - 1\otimes\delta)(z_1\otimes g^i) = (g^{-1}-1)z_0\otimes g^i - z_1\otimes(v-g)g^i$$

for $1\le i\le d$ and

$$z_{2,i} = (\delta\otimes 1 - 1\otimes\delta)(z_2\otimes g^i) = (a-1)z_0\otimes g^i - z_2\otimes(v-g)g^i$$

for $1\le i\le d-1$. All these elements lie in $\mathrm{Ker}(g_4 : \hat{X}(2) \longrightarrow K_3(\mathbf{F}_{v^d}))$. Since $g_3(\delta(z_2)\otimes g^d - z_2\otimes(v-g)g^d) = \delta^{(v^d-1)v}$ and $g_4(y\otimes 1) = (\delta^r)^{v^{2d}}$, we set

$$z_{2,d} = (a-1)z_0\otimes 1 - z_2\otimes(v-g) - y\otimes\alpha_d$$

where $\alpha_d = ((v^d - 1)/r)vw^{2d}$. Therefore $z_{2,d} \in \text{Ker}(g_4 : \hat{X}(2) \longrightarrow K_3(\mathbf{F}_{v^d}))$, since $g_4(z_{2,d}) = \delta^{(v^d-1)v - rv^{2d}((v^d-1)/r)vw^{2d}}$, which is trivial.

Proposition 4.1.11

The $\mathbf{Z}[G(L/K)]$*-submodule*

$$W = \mathbf{Z}[G(L/K)]\langle z_{1,1}, \ldots, z_{1,d}, z_{2,1}, \ldots, z_{2,d}\rangle \subseteq \text{Ker}(g_4) \subseteq \hat{X}(2)$$

is free.

Proof. If $\sum_{\epsilon=1,2,\ 1\leq j\leq d} a_{\epsilon,j} z_{\epsilon,j} = 0$ for some $a_{\epsilon,j} \in \mathbf{Z}[G(L/K)]$ then, considering the $P_1 \otimes Q_0$-component and recalling that $y = (1 + a + \ldots + a^{r-1})z_2$,

$$0 = -\sum_{j=1}^{d} a_{1,j}(z_1 \otimes (v-g)g^j) - \sum_{j=1}^{d} a_{2,j}(z_2 \otimes (v-g)g^j) - a_{2,d}((1+a+\ldots+a^{r-1})z_2 \otimes \alpha_d)$$

in $P_1 \otimes Q_0$. Since

$$1 \otimes ((v-g)\cdot) : P_1 \otimes \mathbf{Z}[G(\overline{L}/\overline{K})] \longrightarrow P_1 \otimes \mathbf{Z}[G(\overline{L}/\overline{K})]$$

is an injective $\mathbf{Z}[G(L/K)]$-module homomorphism, we see that $a_{1,j} = 0$ for all $1 \leq j \leq d$. Also, if $u_1 \otimes u_2 \in P_1 \otimes Q_0$ then $(a-1)(u_1 \otimes u_2) = ((a-1)u_1) \otimes u_2$ so that $(a-1)z_{2,j} = 0$ for all $1 \leq j \leq d$. Therefore, for each j, $a_{2,j} = (1 + a + \ldots a^{r-1})a'_{2,j}$ for some $a'_{2,j} \in \mathbf{Z}[g]/(g^d - 1)$. Hence we have

$$\begin{aligned} &\textstyle\sum_{j=1}^{d}(1 + a + \ldots a^{r-1})a'_{2,j}(z_2 \otimes (v-g)g^j) \\ &= -(1 + a + \ldots a^{r-1})^2 a'_{2,d}(z_2 \otimes \alpha_d) \\ &= -(1 + a + \ldots a^{r-1})a'_{2,d}(z_2 \otimes r\alpha_d). \end{aligned}$$

Notice that, in $\mathbf{Z}[g]/(g^d - 1)$

$$v^d - 1 = v^d - g^d = (v-g)(v^{(d-1)} + v^{(d-2)}g + \ldots + g^{d-1}).$$

This means that

$$r\alpha_d = (v^d - 1)vw^{2d} = (v-g)vw^{2d}(v^{d-1} + v^{d-2}g + \ldots + g^{d-1})$$

Therefore, in the free $\mathbf{Z}[g]/(g^d - 1)$-module given by $\mathbf{Z}[g]/(g^d - 1) \otimes \mathbf{Z}[g]/(g^d - 1)$ with the diagonal action, we have

$$\sum_{j=1}^{d} a'_{2,j}(1 \otimes g^j) = -a'_{2,d}(1 \otimes (v^{d-1} + v^{d-2}g + \ldots + g^{d-1})vw^{2d}.$$

Equating coefficients of basis elements $\{1\otimes g^j,\ 1\leq j\leq d\}$ we find, in $\mathbf{Z}[g]/(g^d-1)$, that

$$\begin{aligned} a'_{2,d} &= -a'_{2,d}v^d w^{2d}, \\ a'_{2,d-1} &= -a'_{2,d}vw^{2d}, \\ a'_{2,d-2} &= -a'_{2,d}v^2w^{2d}, \\ a'_{2,d-3} &= -a'_{2,d}v^3w^{2d}, \\ \vdots \quad & \quad \vdots \\ a'_{2,1} &= -a'_{2,d}v^{d-1}w^{2d}. \end{aligned}$$

Since $v^dw^{2d} > 1$ we have $a'_{2,d} = 0$ and therefore $0 = a'_{2,d-1} = \ldots = a'_{2,1}$, as required. □

Proposition 4.1.12

In the notation of Lemma 4.1.8, Proposition 4.1.9(proof) and Proposition 4.1.11,

$$\frac{\mathrm{Ker}(g_4 \mid X(2)\otimes Q_0)}{W\cap(X(2)\otimes Q_0)}$$

is a finitely generated, cohomologically trivial $\mathbf{Z}[G(L/K)]$*-module and*

$$[\mathrm{Ker}(g_4)] = [\frac{\mathrm{Ker}(g_4 \mid X(2)\otimes Q_0)}{W\cap(X(2)\otimes Q_0)}] \in \mathcal{CL}(\mathbf{Z}[G(L/K)]).$$

Proof. From the proof of Proposition 4.1.9 we know that

$$\hat{X}(2) = (\delta\otimes 1)(\Lambda_{2,0}) + (\delta\otimes 1 - 1\otimes\delta)(\Lambda_{1,1}),$$

and that

$$(\delta\otimes 1)(\Lambda_{2,0}) = X(2)\otimes Q_0 \subseteq P_1\otimes Q_0.$$

Therefore, by construction of the generators, $z_{1,1},\ldots,z_{2,d}$ of W one sees easily that

$$\hat{X}(2) = (\delta\otimes 1)(\Lambda_{2,0}) + (\delta\otimes 1 - 1\otimes\delta)(\Lambda_{1,1}) = X(2)\otimes Q_0 + W.$$

Therefore, by the Second Isomorphism Theorem,

$$\frac{\mathrm{Ker}(g_4)}{W} \cong \frac{\mathrm{Ker}(g_4 \mid X(2)\otimes Q_0)}{W\cap(X(2)\otimes Q_0)}$$

and the result follows, by Proposition 4.1.11. □

4.1.13 *The submodule* $W\cap(X(2)\otimes Q_0)$

Next we must determine the submodule $W\cap(X(2)\otimes Q_0)\subseteq P_1\otimes Q_0$. In order that $\sum_{\epsilon=1,2,\ 1\leq j\leq d} a_{\epsilon,j}z_{\epsilon,j} \in P_1\otimes Q_0$ we must have, in $P_0\otimes Q_1$,

$$\sum_{j=1}^{d} a_{1,j}g^{-1}(z_0\otimes g^{j+1}) - a_{1,j}(z_0\otimes g^j) + a_{2,j}(a-1)(z_0\otimes g^j) = 0.$$

Since $\{z_0 \otimes g^j \mid 1 \leq j \leq d\}$ is a free $\mathbf{Z}[G(L/K)]$-basis for $P_0 \otimes Q_1$, we must have

$$\begin{aligned} a_{1,d}g^{-1} - a_{1,1} + a_{2,1}(a-1) &= 0, \\ a_{1,1}g^{-1} - a_{1,2} + a_{2,2}(a-1) &= 0, \\ a_{1,2}g^{-1} - a_{1,3} + a_{2,3}(a-1) &= 0, \\ \vdots \quad & \quad \vdots \\ a_{1,d-1}g^{-1} - a_{1,d} + a_{2,d}(a-1) &= 0. \end{aligned}$$

Hence

$$\begin{aligned} a_{1,1} &= a_{1,d}g^{-1} + a_{2,1}(a-1), \\ a_{1,2} &= a_{1,d}g^{-2} + a_{2,1}(a-1)g^{-1} + a_{2,2}(a-1), \\ a_{1,3} &= a_{1,d}g^{-3} + a_{2,1}(a-1)g^{-2} + a_{2,2}(a-1)g^{-1} + a_{2,3}(a-1), \\ \vdots \quad & \quad \vdots \\ a_{1,s} &= a_{1,d}g^{-s} + \textstyle\sum_{j=1}^{s} a_{2,j}(a-1)g^{j-s} \\ \vdots \quad & \quad \vdots \\ a_{1,d} &= a_{1,d}g^{-d} + \textstyle\sum_{j=1}^{d} a_{2,j}(a-1)g^{j-d}. \end{aligned}$$

If these equations hold then

$$\begin{aligned} &-(\textstyle\sum_{\epsilon=1,2,\ 1\leq j\leq d} a_{\epsilon,j}z_{\epsilon,j}) \\ &= (a_{1,d}g^{-1} + a_{2,1}(a-1))(z_1 \otimes (v-g)g) \\ &+(a_{1,d}g^{-2} + a_{2,1}(a-1)g^{-1} + a_{2,2}(a-1))(z_1 \otimes (v-g)g^2) \\ &+(a_{1,d}g^{-3} + a_{2,1}(a-1)g^{-2} + a_{2,2}(a-1)g^{-1} + a_{2,3}(a-1))(z_1 \otimes (v-g)g^3) \\ &\quad \vdots \quad \vdots \\ &+(a_{1,d}g^{-s} + \textstyle\sum_{j=1}^{s} a_{2,j}(a-1)g^{j-s})(z_1 \otimes (v-g)g^s) \\ &\quad \vdots \quad \vdots \\ &+(a_{1,d}g^{-d} + \textstyle\sum_{j=1}^{d} a_{2,j}(a-1)g^{j-d})(z_1 \otimes (v-g)) \\ &+a_{2,1}(z_2 \otimes (v-g)g) \\ &+a_{2,2}(z_2 \otimes (v-g)g^2) \\ &\quad \vdots \quad \vdots \\ &+a_{2,d-1}(z_2 \otimes (v-g)g^{d-1}) \\ &+a_{2,d}(z_2 \otimes (v-g) + y \otimes \alpha_d). \end{aligned}$$

This immense expression will lie in $X(2) \otimes Q_0$. This is seen by observing that $X(2) = \ker(\delta : P_1 \longrightarrow P_0)$ and therefore $X(2) \otimes Q_0 = \ker(\delta \otimes 1 : P_1 \otimes Q_0 \longrightarrow$

$P_0 \otimes Q_0$). However, the equations guarantee that this expression, which equals

$$\begin{aligned}&\textstyle\sum_{\epsilon=1,2,\ 1\leq j\leq d} a_{\epsilon,j} z_{\epsilon,j}\\ &= \textstyle\sum_{\epsilon=1,2,\ 1\leq j\leq d} a_{\epsilon,j}(\delta\otimes 1 - 1\otimes\delta)(z_\epsilon\otimes g^j) - a_{2,d}(y\otimes\alpha_d),\end{aligned}$$

lies in $P_1 \otimes Q_0$ and therefore so does

$$\sum_{\epsilon=1,2,\ 1\leq j\leq d} a_{\epsilon,j}(\delta\otimes 1 - 1\otimes\delta)(z_\epsilon\otimes g^j).$$

The differential, d, on $(P_1\otimes Q_0)\oplus(P_0\otimes Q_1)$ is equal to $\delta\otimes 1$ on the first summand so that

$$\begin{aligned}&(\delta\otimes 1)(\textstyle\sum_{\epsilon=1,2,\ 1\leq j\leq d} a_{\epsilon,j} z_{\epsilon,j})\\ &= d(\textstyle\sum_{\epsilon=1,2,\ 1\leq j\leq d} a_{\epsilon,j}(\delta\otimes 1 - 1\otimes\delta)(z_\epsilon\otimes g^j)) - d(a_{2,d}(y\otimes\alpha_d))\\ &= \textstyle\sum_{\epsilon=1,2,\ 1\leq j\leq d} a_{\epsilon,j} d(\delta\otimes 1 - 1\otimes\delta)(z_\epsilon\otimes g^j) - a_{2,d}(\delta(y)\otimes\alpha_d)\\ &= -a_{2,d}(\delta(y)\otimes\alpha_d)\\ &= -a_{2,d}((1+a+\ldots+a^{r-1})(a-1)z_0\otimes\alpha_d)\\ &= 0,\end{aligned}$$

as required.

Solutions to the equations may be found by assigning $a_{1,d}$, $a_{2,1}$, $a_{2,2}$, $\ldots$, $a_{2,d-1}$ arbitrarily in $\mathbf{Z}[G(L/K)]$ and, since $g^{-d} = a^{-c}$, choosing $a_{2,d}$ to satisfy

$$a_{2,d}(a-1) = a_{1,d}(1-a^{-c}) - \sum_{j=1}^{d-1} a_{2,j}(a-1)g^{j-d}.$$

Since $ag^{-1} = g^{-1}a^v$ we have $ag^{j-d} = g^{j-d}a^{v^{d-j}}$ so that we require

$$\begin{aligned}&a_{2,d}(a-1)\\ &= a_{1,d}a^{-c}(a^c-1) - \textstyle\sum_{j=1}^{d-1} a_{2,j}g^{j-d}(a^{v^{d-j}}-1)\\ &= [a_{1,d}a^{-c}(1+a+\ldots a^{c-1}) - \textstyle\sum_{j=1}^{d-1} a_{2,j}g^{j-d}(1+a+\ldots+a^{v^{d-j}-1})](a-1)\end{aligned}$$

for which the general solution is

$$\begin{aligned}a_{2,d} = {}& a_{1,d}a^{-c}(1+a+\ldots a^{c-1}) - \textstyle\sum_{j=1}^{d-1} a_{2,j}g^{j-d}(1+a+\ldots+a^{v^{d-j}-1})\\ &+\beta(1+a+\ldots+a^{r-1})\end{aligned}$$

for any $\beta\in\mathbf{Z}[g]/(g^d-1)\cong\mathbf{Z}[G(L/K)](1+a+\ldots+a^{r-1})$.

Hence we can list the generators for $W\cap(X(2)\otimes Q_0)$ by setting each of $a_{1,d}, a_{2,1}, a_{2,2}, \ldots, a_{2,d-1}, \beta$ equal to one and the rest to zero in the previous immense expression. The result is as follows:

Proposition 4.1.14

The $\mathbf{Z}[G(L/K)]$*-module,* $W \cap (X(2) \otimes Q_0)$ *of §4.1.13 is generated by the elements*

$$((\sum_{j=0}^{d-1} g^{-j})z_1 + (\sum_{t=1}^{c} a^{-t})z_2) \otimes (v-g) + y \otimes c\alpha_d = x \otimes (v-g) + y \otimes c\alpha_d,$$

$$(((a-1)(\sum_{j=0}^{d-1} g^{-j}) + g^{1-d}(a^{v^{d-1}}-1))z_1 + (1+g^{1-d}(\sum_{t=0}^{v^{d-1}-1} a^t))z_2) \otimes (v-g)g$$
$$-g^{1-d}y \otimes (v^{d-1}-1)\alpha_d g,$$

$$(((a-1)(\sum_{j=0}^{d-2} g^{-j}) + g^{2-d}(a^{v^{d-2}}-1))z_1 + (1+g^{2-d}(\sum_{t=0}^{v^{d-2}-1} a^t))z_2) \otimes (v-g)g^2$$
$$-(g^{2-d}y) \otimes (v^{d-2}-1)\alpha_d g^2,$$

$$\vdots \qquad \vdots$$

$$(((a-1)(\sum_{j=0}^{d-s} g^{-j}) + g^{s-d}(a^{v^{d-s}}-1))z_1 + (1+g^{s-d}(\sum_{t=0}^{v^{d-s}-1} a^t))z_2) \otimes (v-g)g^s$$
$$-g^{s-d}y \otimes (v^{d-s}-1)\alpha_d g^s,$$

$$\vdots \qquad \vdots$$

$$(((a-1)(\sum_{j=0}^{1} g^{-j}) + g^{-1}(a^{v}-1))z_1 + (1+g^{-1}(\sum_{t=0}^{v-1} a^t))z_2) \otimes (v-g)g^{d-1}$$
$$-g^{-1}y \otimes (v-1)\alpha_d g^{d-1},$$

$$y \otimes (v-g+r\alpha_d)$$

where $\alpha_d = ((v^d-1)/r)vw^{2d}$*, as in §4.1.10.*

4.1.15 Now we shall work to derive an easier representation of the module of Proposition 4.1.12. Since $X(2)$ is generated as a $\mathbf{Z}[G(L/K)]$-module by the elements

$$t = (a-1)z_1 - (g^{-1}(1+a+\ldots+a^{v-1})-1)z_2,$$
$$x = (1+g^{-1}+\ldots+g^{1-d})z_1 + (a^{-1}+a^{-2}+\ldots+a^{-c})z_2,$$
$$y = (1+a+\ldots+a^{r-1})z_2$$

it is not difficult to see that $X(2) \otimes Q_0 \subset P_1 \otimes Q_0$ is generated by

$$t \otimes 1 + y \otimes (v^d-1)w^{2d}/r,$$
$$t \otimes g, t \otimes g^2, \ldots, t \otimes g^{d-1},$$
$$x \otimes (v-g)g, x \otimes (v-g)g^2, \ldots, x \otimes (v-g)g^{d-1},$$
$$y \otimes (v-g)g, y \otimes (v-g)g^2, \ldots, y \otimes (v-g)g^{d-1},$$
$$(x-y) \otimes g \text{ and } y \otimes g.$$

These elements have been chosen so that all except $y \otimes g$ are in $\ker(g_4 : X(2) \otimes Q_0 \longrightarrow K_3(\mathbf{F}_{v^d}) \cong \mathbf{F}^*_{v^{2d}})$. From the exact sequence of $\mathbf{Z}[g]/(g^d-1)$-modules

$$0 \longrightarrow \mathbf{Z}[g]/(g^d-1) \stackrel{v^2-g}{\longrightarrow} \mathbf{Z}[g]/(g^d-1) \longrightarrow \mathbf{F}^*_{v^{2d}} \longrightarrow 0$$

one easily sees that $\ker(g_4|X(2)\otimes Q_0)$ is generated by

$$\begin{aligned}
&t\otimes 1+y\otimes (v^d-1)w^{2d}/r,\\
&t\otimes g, t\otimes g^2,\ldots,t\otimes g^{d-1},\\
&x\otimes (v-g)g, x\otimes (v-g)g^2,\ldots,x\otimes (v-g)g^{d-1},\\
&y\otimes (v-g)g, y\otimes (v-g)g^2,\ldots,y\otimes (v-g)g^{d-1},\\
&(x-y)\otimes g \text{ and } (v^2-g)(y\otimes g).
\end{aligned}$$

Define an isomorphism isomorphism

$$\lambda : P_1\otimes Q_0 \xrightarrow{\cong} P_1\otimes Q_0$$

by $\lambda(hz_b\otimes g^i) = hz_b\otimes\pi(h)^{-1}g^i$, extended by linearity. Here $b=1,2$, $h\in G(L/K)$ and $\pi : G(L/K)\longrightarrow G(\overline{L}/\overline{K})\cong \mathbf{Z}/d$ is given by $\pi(a^s g^u)=g^u$. Thus λ transports the diagonal $\mathbf{Z}[G(L/K)]$-action to the left-hand factor action, which is easier to work with.

Now let us apply λ to the list of generators for $\ker(g_4 \mid X(2)\otimes Q_0)$. We obtain

$$\begin{aligned}
&t\otimes 1+y\otimes (v^d-1)w^{2d}/r+g^{-1}(1+a+\ldots+a^{v-1})z_2\otimes(1-g),\\
&t\otimes g+g^{-1}(1+a+\ldots+a^{v-1})z_2\otimes(g-g^2),\\
&t\otimes g^2+g^{-1}(1+a+\ldots+a^{v-1})z_2\otimes(g^2-g^3),\\
&\qquad\vdots\qquad\vdots\\
&t\otimes g^{d-1}+g^{-1}(1+a+\ldots+a^{v-1})z_2\otimes(g^{d-1}-1),\\
&z_1\otimes(v-g)g+g^{-1}z_1\otimes(v-g)g^2+\ldots+g^{1-d}z_1\otimes(v-g)\\
&\qquad +(a^{-1}+a^{-2}+\ldots+a^{-r})z_2\otimes(v-g)g,\\
&z_1\otimes(v-g)g^2+g^{-1}z_1\otimes(v-g)g^3+\ldots+g^{1-d}z_1\otimes(v-g)g\\
&\qquad +(a^{-1}+a^{-2}+\ldots+a^{-r})z_2\otimes(v-g)g^2,\\
&\qquad\vdots\qquad\vdots\\
&z_1\otimes(v-g)g^{d-1}+g^{-1}z_1\otimes(v-g)+\ldots+g^{1-d}z_1\otimes(v-g)g^{d-2}\\
&\qquad +(a^{-1}+a^{-2}+\ldots+a^{-r})z_2\otimes(v-g)g^{d-1},\\
&y\otimes(v-g)g, y\otimes(v-g)g^2,\ldots,y\otimes(v-g)g^{d-1},\\
&z_1\otimes g+g^{-1}z_1\otimes g^2+\ldots+g^{1-d}z_1\otimes 1\\
&\qquad +(a^{-1}+a^{-2}+\ldots+a^{-r})z_2\otimes g-y\otimes g\\
&=z_1\otimes g+g^{-1}z_1\otimes g^2+\ldots+g^{1-d}z_1\otimes 1,\\
&v^2y\otimes g-gy\otimes 1.
\end{aligned}$$

By Proposition 4.1.12, we wish to determine the class in the class-group of the module

$$\frac{\ker(g_4\mid X(2)\otimes Q_0)}{W\cap\ker(g_4\mid X(2)\otimes Q_0)}.$$

However, the analysis of §4.1.13 shows that the interesection in the denominator is the direct sum of a free module with the submodule $\mathbf{Z}[g]/(g^d-1)\langle y\otimes(v-g+r\alpha_d)\rangle$. Note that $\lambda(y\otimes(v-g+r\alpha_d)) = y\otimes(v-g+r\alpha_d)$. Hence we have established the following result.

Lemma 4.1.16

In Proposition 4.1.12

$$[\mathrm{Ker}(g_4)] = [\frac{\mathrm{Ker}(g_4 \mid X(2)\otimes Q_0)}{W\cap(X(2)\otimes Q_0)}] = [\frac{\ker(g_4 \mid X(2)\otimes Q_0)}{\mathbf{Z}[g]/(g^d-1)\langle y\otimes(v-g+r\alpha_d)\rangle}]$$

in $\mathcal{CL}(\mathbf{Z}[G(L/K)])$.

Proposition 4.1.17

The $\mathbf{Z}[G(L/K)]$*-submodule,* $W'\subset\ker(g_4\mid X(2)\otimes Q_0)$ *given by*

$$W' = \mathbf{Z}[G(L/K)]\langle t\otimes 1 + y\otimes(v^d-1)w^{2d}/r, t\otimes g, t\otimes g^2,\ldots,t\otimes g^{d-1}\rangle$$

is free.

Proof. Suppose that $\{\alpha_i\in\mathbf{Z}[G(L/K)] \mid 0\le i\le d-1\}$ satisfy

$$0 = \alpha_0(t\otimes 1 + y\otimes(v^d-1)w^{2d}/r) + \sum_{i=1}^{d-1}\alpha_i(t\otimes g^i).$$

We first apply λ of §4.1.15 to transport this relation into the module with the left-factor action. The relation transforms to give

$$\begin{aligned}
0 \; &= \alpha_0 t\otimes 1 + \alpha_0 y\otimes(v^d-1)w^{2d}/r + \alpha_0 g^{-1}(1+a+\ldots+a^{v-1})z_2\otimes(1-g)\\
&\quad +\alpha_1 t\otimes g + \alpha_1 g^{-1}(1+a+\ldots+a^{v-1})z_2\otimes(g-g^2)\\
&\quad +\alpha_2 t\otimes g^2 + \alpha_2 g^{-1}(1+a+\ldots+a^{v-1})z_2\otimes(g^2-g^3)\\
&\qquad\qquad \vdots \qquad\qquad \vdots\\
&\quad +\alpha_{d-1} t\otimes g^{d-1} + \alpha_{d-1} g^{-1}(1+a+\ldots+a^{v-1})z_2\otimes(g^{d-1}-1).
\end{aligned}$$

Next we compare coefficients (in the left-factor) of the various powers of g to obtain:

$$\begin{aligned}
0 &= \alpha_0 t + \alpha_0((v^d-1)w^{2d}/r)y + \alpha_0 g^{-1}(1+a+\ldots+a^{v-1})z_2\\
&\qquad\qquad -\alpha_{d-1}g^{-1}(1+a+\ldots+a^{v-1})z_2,\\
0 &= \alpha_1 t + \alpha_1 g^{-1}(1+a+\ldots+a^{v-1})z_2 - \alpha_0 g^{-1}(1+a+\ldots+a^{v-1})z_2,\\
0 &= \alpha_2 t + \alpha_2 g^{-1}(1+a+\ldots+a^{v-1})z_2 - \alpha_1 g^{-1}(1+a+\ldots+a^{v-1})z_2,\\
&\qquad\qquad \vdots \qquad\qquad \vdots\\
0 &= \alpha_{d-1} t + \alpha_{d-1} g^{-1}(1+a+\ldots+a^{v-1})z_2 - \alpha_{d-2} g^{-1}(1+a+\ldots+a^{v-1})z_2.
\end{aligned}$$

Considering the coefficients of z_1 we see that $\alpha_i(a-1)=0$ for each i and therefore $\alpha_i=\beta_i(1+a+\cdots+a^{r-1})$ for $0\le i\le d-1$ and $\beta_i\in\mathbf{Z}[g]/(g^d-1)$. Substituting these values we obtain

$$\begin{aligned}
0&=-\beta_0(g^{-1}v-1)y+\beta_0(v^d-1)w^{2d}y+\beta_0g^{-1}vy-\beta_{d-1}g^{-1}vy,\\
0&=-\beta_1(g^{-1}v-1)y+\beta_1g^{-1}vy-\beta_0g^{-1}vy,\\
0&=-\beta_2(g^{-1}v-1)y+\beta_2g^{-1}vy-\beta_1g^{-1}vy,\\
&\qquad\vdots\qquad\quad\vdots\\
0&=-\beta_{d-1}(g^{-1}v-1)y+\beta_{d-1}g^{-1}vy-\beta_{d-2}g^{-1}vy.
\end{aligned}$$

Therefore

$$\beta_{d-1}y=\beta_{d-2}g^{-1}vy=\beta_{d-3}g^{-2}v^2y=\ldots=\beta_0g^{1-d}v^{d-1}y$$

and $\beta_0((v^d-1)w^{2d}+1)y=\beta_0v^dy$ or $(v^d-1)(w^{2d}-1)\beta_0y=0$. Hence $\beta_0=0$ which implies that $\beta_i=0$ for all $0\le i\le d-1$, as required. □

4.1.18 Consider the generators $(x-y)\otimes g=(1+g^{-1}+\ldots+g^{1-d})z_1\otimes g$ and the $(x-y)\otimes(v-g)g^i$'s. Since $g^{i-1}((x-y)\otimes g)=(x-y)\otimes g^i$ we see that all these elements may be generated from $(x-y)\otimes 1=g^{-1}((x-y)\otimes g)$. Also, from the equation

$$\lambda((x-y)\otimes 1)=z_1\otimes 1+g^{-1}z_1\otimes g+g^{-2}z_1\otimes g^2+\ldots+g^{1-d}z_1\otimes g^{d-1}$$

one sees that $(x-y)\otimes 1$ generates a free $\mathbf{Z}[G(L/K)]$-submodule $\ker(g_4)$ of rank one.

For W' as in Proposition 4.1.17, consider the intersection

$$\mathbf{Z}[G(L/K)]\langle(x-y)\otimes 1\rangle\bigcap W'\subset\ker(g_4\mid X(2)\otimes Q_0).$$

A relation of the form

$$\beta((x-y)\otimes 1)=\alpha_0(t\otimes 1+y\otimes(v^d-1)w^{2d}/r)+\sum_{i=1}^{d-1}\alpha_i(t\otimes g^i)$$

transforms under λ to give

$$\begin{aligned}
&\beta z_1\otimes 1+\beta g^{-1}z_1\otimes g+\beta g^{-2}z_1\otimes g^2+\ldots+\beta g^{1-d}z_1\otimes g^{d-1}\\
&=\alpha_0t\otimes 1+\alpha_0y\otimes(v^d-1)w^{2d}/r+\alpha_0g^{-1}(1+a+\ldots+a^{v-1})z_2\otimes(1-g)\\
&\qquad+\alpha_1t\otimes g+\alpha_1g^{-1}(1+a+\ldots+a^{v-1})z_2\otimes(g-g^2)\\
&\qquad+\alpha_2t\otimes g^2+\alpha_2g^{-1}(1+a+\ldots+a^{v-1})z_2\otimes(g^2-g^3)\\
&\qquad\qquad\vdots\qquad\quad\vdots\\
&\qquad+\alpha_{d-1}t\otimes g^{d-1}+\alpha_{d-1}g^{-1}(1+a+\ldots+a^{v-1})z_2\otimes(g^{d-1}-1).
\end{aligned}$$

Next we compare coefficients (in the left-factor) of the various powers of g to obtain:

$$\begin{aligned}
\beta z_1 &= \alpha_0 t + \alpha_0((v^d - 1)w^{2d}/r)y + \alpha_0 g^{-1}(1 + a + \ldots + a^{v-1})z_2 \\
&\qquad\qquad -\alpha_{d-1} g^{-1}(1 + a + \ldots + a^{v-1})z_2, \\
\beta g^{-1} z_1 &= \alpha_1 t + \alpha_1 g^{-1}(1 + a + \ldots + a^{v-1})z_2 - \alpha_0 g^{-1}(1 + a + \ldots + a^{v-1})z_2, \\
\beta g^{-2} z_1 &= \alpha_2 t + \alpha_2 g^{-1}(1 + a + \ldots + a^{v-1})z_2 - \alpha_1 g^{-1}(1 + a + \ldots + a^{v-1})z_2, \\
&\vdots \qquad \vdots \\
\beta g^{1-d} z_1 &= \alpha_{d-1} t + \alpha_{d-1} g^{-1}(1 + a + \ldots + a^{v-1})z_2 - \alpha_{d-2} g^{-1}(1 + a + \ldots + a^{v-1})z_2.
\end{aligned}$$

Considering the coefficients of z_1 we obtain

$$\begin{aligned}
\beta &= \alpha_0(a - 1), \\
\beta g^{-1} &= \alpha_1(a - 1), \\
\beta g^{-2} &= \alpha_2(a - 1), \\
&\vdots \qquad \vdots \\
\beta g^{1-d} &= \alpha_{d-1}(a - 1)
\end{aligned}$$

so that

$$\begin{aligned}
&\beta = \alpha_0(a - 1), \\
&\alpha_1(a - 1) = \beta g^{-1} = \alpha_0(a - 1)g^{-1} = \alpha_0 g^{-1}(a^v - 1), \\
&\alpha_2(a - 1) = \beta g^{-2} = \alpha_0(a - 1)g^{-2} = \alpha_0 g^{-2}(a^{v^2} - 1), \\
&\alpha_3(a - 1) = \beta g^{-3} = \alpha_0(a - 1)g^{-3} = \alpha_0 g^{-3}(a^{v^3} - 1), \\
&\qquad\qquad\qquad \vdots \qquad\quad \vdots \\
&\alpha_{d-1}(a - 1) = \beta g^{1-d} = \alpha_0(a - 1)g^{1-d} = \alpha_0 g^{1-d}(a^{v^{d-1}} - 1).
\end{aligned}$$

The general solution to these equations is

$$\begin{aligned}
&\beta = \alpha_0(a - 1), \\
&\alpha_1 = \alpha_0 g^{-1}(1 + a + \ldots + a^{v-1}) + \alpha_1'(1 + a + \ldots + a^{r-1}), \\
&\alpha_2 = \alpha_0 g^{-2}(1 + a + \ldots + a^{v^2-1}) + \alpha_2'(1 + a + \ldots + a^{r-1}), \\
&\alpha_3 = \alpha_0 g^{-3}(1 + a + \ldots + a^{v^3-1}) + \alpha_3'(1 + a + \ldots + a^{r-1}), \\
&\qquad\qquad\qquad \vdots \qquad\quad \vdots \\
&\alpha_{d-1} = \alpha_0 g^{1-d}(1 + a + \ldots + a^{v^{d-1}-1}) + \alpha_{d-1}'(1 + a + \ldots + a^{r-1})
\end{aligned}$$

where $\alpha_0 \in \mathbf{Z}[G(L/K)]$ and $\alpha_j' \in \mathbf{Z}[g]/(g^d - 1)$ for $1 \le j \le d - 1$.

From these equations, equating coefficients of z_2, we obtain:

$$\begin{aligned}
0 = {} & \alpha_0(-(g^{-1}(1+a+\ldots+a^{v-1})-1)) \\
& +\alpha_0((v^d-1)w^{2d}/r)(1+a+\ldots+a^{r-1}) \\
& +\alpha_0 g^{-1}(1+a+\ldots+a^{v-1}) \\
& -(\alpha_0 g^{1-d}(1+a+\ldots+a^{v^{d-1}-1}) \\
& +\alpha'_{d-1}(1+a+\ldots+a^{r-1}))g^{-1}(1+a+\ldots+a^{v-1}), \\
0 = {} & (\alpha_0 g^{-1}(1+a+\ldots+a^{v-1}) \\
& +\alpha'_1(1+a+\ldots+a^{r-1}))(-(g^{-1}(1+a+\ldots+a^{v-1})-1)) \\
& +(\alpha_0 g^{-1}(1+a+\ldots+a^{v-1})+\alpha'_1(1+a+\ldots+a^{r-1}))g^{-1}(1+a+\ldots+a^{v-1}) \\
& -\alpha_0 g^{-1}(1+a+\ldots+a^{v-1}), \\
0 = {} & (\alpha_0 g^{-2}(1+a+\ldots+a^{v^2-1}) \\
& +\alpha'_2(1+a+\ldots+a^{r-1}))(-(g^{-1}(1+a+\ldots+a^{v-1})-1)) \\
& +(\alpha_0 g^{-2}(1+a+\ldots+a^{v^2-1})+\alpha'_2(1+a+\ldots+a^{r-1}))g^{-1}(1+a+\ldots+a^{v-1}) \\
& -(\alpha_0 g^{-1}(1+a+\ldots+a^{v-1}) \\
& +\alpha'_1(1+a+\ldots+a^{r-1}))g^{-1}(1+a+\ldots+a^{v-1}), \\
& \qquad\qquad \vdots \qquad\qquad \vdots \\
0 = {} & (\alpha_0 g^{1-d}(1+a+\ldots+a^{v^{d-1}-1}) \\
& +\alpha'_{d-1}(1+a+\ldots+a^{r-1}))(-(g^{-1}(1+a+\ldots+a^{v-1})-1)) \\
& +(\alpha_0 g^{1-d}(1+a+\ldots+a^{v^{d-1}-1})+\alpha'_{d-1}(1+a+\ldots+a^{r-1}))g^{-1}(1+a+\ldots+a^{v-1}) \\
& -(\alpha_0 g^{2-d}(1+a+\ldots+a^{v^{d-2}-1}) \\
& +\alpha'_{d-2}(1+a+\ldots+a^{r-1}))g^{-1}(1+a+\ldots+a^{v-1}).
\end{aligned}$$

Simplifying we obtain:

$$\begin{aligned}
0 &= \alpha_0+\alpha_0((v^d-1)w^{2d}/r)(1+a+\ldots+a^{r-1}) \\
&\quad -\alpha_0(1+a^v+\ldots+a^{v^d-v})(1+a+\ldots+a^{v-1}) \\
&\quad -\alpha'_{d-1}(1+a+\ldots+a^{r-1})g^{-1}v \\
&= \alpha_0+\alpha_0((v^d-1)w^{2d}/r)(1+a+\ldots+a^{r-1}) \\
&\quad -\alpha_0(1+a^v+\ldots+a^{v^d-2}+1) \\
&\quad -\alpha'_{d-1}(1+a+\ldots+a^{r-1})g^{-1}v \\
&= \alpha_0((v^d-1)w^{2d}/r)(1+a+\ldots+a^{r-1}) \\
&\quad -\alpha_0((v^d-1)/r)(1+a+\ldots+a^{r-1}) \\
&\quad -\alpha'_{d-1}(1+a+\ldots+a^{r-1})g^{-1}v \\
&= \alpha_0((v^d-1)(w^{2d}-1)/r)(1+a+\ldots+a^{r-1}) \\
&\quad -\alpha'_{d-1}(1+a+\ldots+a^{r-1})g^{-1}v
\end{aligned}$$

and

$$
\begin{aligned}
0 &= \alpha_1'(1+a+\ldots+a^{r-1})\\
0 &= \alpha_2'(1+a+\ldots+a^{r-1})) - \alpha_1' g^{-1}v(1+a+\ldots+a^{r-1})\\
&\vdots \qquad\qquad \vdots\\
0 &= \alpha_{d-1}'(1+a+\ldots+a^{r-1}) - \alpha_{d-2}' g^{-1}v(1+a+\ldots+a^{r-1}).
\end{aligned}
$$

Since we can solve all these equations if and only if $\alpha_0 \in \mathbf{Z}[G(L/K)](a-1)$ we have shown the first part of the following:

Proposition 4.1.19

(i) *In Proposition 4.1.17*

$$\mathbf{Z}[G(L/K)](a-1)^2 \cong \mathbf{Z}[G(L/K)]\langle (x-y)\otimes 1\rangle \bigcap W' \subset \ker(g_4 \mid X(2)\otimes Q_0)$$

as a left $\mathbf{Z}[G(L/K)]$-submodule.

(ii) *An isomorphism of left $\mathbf{Z}[G(L/K)]$-modules*

$$\mathbf{Z}[G(L/K)](a-1) \xrightarrow{\cong} \mathbf{Z}[G(L/K)](a-1)^2$$

is given by sending z to $z(a-1)$.

(iii) *There are isomorphisms of left $\mathbf{Z}[G(L/K)]$-modules*

$$
\begin{aligned}
\mathbf{Z}[G(L/K)](a-1) &\cong \mathbf{Z}[G(L/K)]/(\mathbf{Z}[G(L/K)]\langle 1+a+\ldots+a^{r-1}\rangle)\\
&\cong \mathbf{Z}[G(L/K)]/(\mathbf{Z}[G(\overline{L}/\overline{K})]).
\end{aligned}
$$

(iv) *In positive dimensions the cohomology of $\mathbf{Z}[G(L/K)](a-1)^2$ is given by*

$$H^i(G(L/K);\mathbf{Z}[G(L/K)](a-1)^2) \cong \begin{cases} \mathbf{Z}/r & \text{if } i>0, \text{ odd},\\ 0 & \text{if } i>0, \text{ even}.\end{cases}$$

Proof. Part (i) was established during the discussion of §4.1.18. The map in part (ii) is clearly surjective. On the other hand, if $x = z(a-1)$ and $x(a-1) = 0$ then $x = u(1+a+\ldots+a^{r-1})$ so that

$$0 = z(a-1)(1+a+\ldots+a^{r-1}) = x(1+a+\ldots+a^{r-1}) = u(1+a+\ldots+a^{r-1})^2 = rx$$

and so $x = 0$.

Part (iii) is clear and part (iv) follows from the isomorphisms

$$
\begin{aligned}
H^i(G(L/K);\mathbf{Z}[G(L/K)]/(\mathbf{Z}[G(\overline{L}/\overline{K})])) &\cong H^{i+1}(G(L/K);\mathbf{Z}[G(\overline{L}/\overline{K})])\\
&\cong H^{i+1}(G(L/K);\mathrm{Ind}_{\langle a\rangle}^{G(L/K)}(1))\\
&\cong H^{i+1}(\langle a\rangle;\mathbf{Z}).
\end{aligned}
$$

□

4.1.20 Next we wish to calculate which elements of the submodule generated by $\{y \otimes (v-g)g^i\}$ and $(v^2-g)(y\otimes g)$ lie in $\mathbf{Z}[G(L/K)]\langle (x-y)\otimes 1\rangle + W'$.

Suppose that $w_0, w_1, \ldots, w_{d-1} \in \mathbf{Z}[g]/(g^d-1)$ and that

$$\lambda^{-1}(\sum_{i=0}^{d-1} w_i y \otimes g^i) = -\beta((x-y)\otimes 1) + \alpha_0(t\otimes 1 + y\otimes(v^d-1)w^{2d}/r) + \sum_{i=1}^{d-1}\alpha_i(t\otimes g^i).$$

As before, we compare coefficients (in the left-factor) of the various powers of g to obtain:

$$\begin{aligned}
&w_0 y + \beta z_1 \\
&= \alpha_0 t + \alpha_0((v^d-1)w^{2d}/r)y + \alpha_0 g^{-1}(1+a+\ldots+a^{v-1})z_2 \\
&\qquad\qquad\qquad -\alpha_{d-1}g^{-1}(1+a+\ldots+a^{v-1})z_2, \\
&w_1 y + \beta g^{-1} z_1 \\
&= \alpha_1 t + \alpha_1 g^{-1}(1+a+\ldots+a^{v-1})z_2 - \alpha_0 g^{-1}(1+a+\ldots+a^{v-1})z_2, \\
&w_2 y + \beta g^{-2} z_1 \\
&= \alpha_2 t + \alpha_2 g^{-1}(1+a+\ldots+a^{v-1})z_2 - \alpha_1 g^{-1}(1+a+\ldots+a^{v-1})z_2, \\
&\qquad\qquad \vdots \qquad\quad \vdots \\
&w_{d-1} y + \beta g^{1-d} z_1 \\
&= \alpha_{d-1} t + \alpha_{d-1} g^{-1}(1+a+\ldots+a^{v-1})z_2 - \alpha_{d-2} g^{-1}(1+a+\ldots+a^{v-1})z_2.
\end{aligned}$$

Since there are no z_1's appearing in y, by considering the coefficients of z_1, we obtain the same relations as before, namely:

$$\begin{aligned}
\beta &= \alpha_0(a-1), \\
\alpha_1 &= \alpha_0 g^{-1}(1+a+\ldots+a^{v-1}) + \alpha'_1(1+a+\ldots+a^{r-1}), \\
\alpha_2 &= \alpha_0 g^{-2}(1+a+\ldots+a^{v^2-1}) + \alpha'_2(1+a+\ldots+a^{r-1}), \\
\alpha_3 &= \alpha_0 g^{-3}(1+a+\ldots+a^{v^3-1}) + \alpha'_3(1+a+\ldots+a^{r-1}), \\
&\qquad \vdots \qquad\qquad \vdots \\
\alpha_{d-1} &= \alpha_0 g^{1-d}(1+a+\ldots+a^{v^{d-1}-1}) + \alpha'_{d-1}(1+a+\ldots+a^{r-1})
\end{aligned}$$

where $\alpha_0 \in \mathbf{Z}[G(L/K)]$ and $\alpha'_j \in \mathbf{Z}[g]/(g^d-1)$ for $1 \leq j \leq d-1$.

Comparing the coefficients of z_2 and simplifying as before we obtain:

$$\begin{aligned}
&w_0(1+a+\ldots+a^{r-1}) = \alpha_0((v^d-1)(w^{2d}-1)/r)(1+a+\ldots+a^{r-1}) \\
&\qquad -\alpha'_{d-1}(1+a+\ldots+a^{r-1})g^{-1}v \\
&w_1(1+a+\ldots+a^{r-1}) = \alpha'_1(1+a+\ldots+a^{r-1})
\end{aligned}$$

$$w_2(1+a+\ldots+a^{r-1}) = \alpha_2'(1+a+\ldots+a^{r-1})) - \alpha_1' g^{-1}v(1+a+\ldots+a^{r-1})$$

$$\vdots \qquad \vdots$$

$$w_{d-1}(1+a+\ldots+a^{r-1}) = \alpha_{d-1}'(1+a+\ldots+a^{r-1}) - \alpha_{d-2}' g^{-1}v(1+a+\ldots+a^{r-1}).$$

The equations have become:

$$\begin{aligned}
\beta &= \alpha_0(a-1), \\
\alpha_1 &= \alpha_0 g^{-1}(1+a+\ldots+a^{v-1}) + \alpha_1'(1+a+\ldots+a^{r-1}), \\
\alpha_2 &= \alpha_0 g^{-2}(1+a+\ldots+a^{v^2-1}) + \alpha_2'(1+a+\ldots+a^{r-1}), \\
\alpha_3 &= \alpha_0 g^{-3}(1+a+\ldots+a^{v^3-1}) + \alpha_3'(1+a+\ldots+a^{r-1}), \\
&\vdots \qquad \vdots \\
\alpha_{d-1} &= \alpha_0 g^{1-d}(1+a+\ldots+a^{v^{d-1}-1}) + \alpha_{d-1}'(1+a+\ldots+a^{r-1})
\end{aligned}$$

and

$$\begin{aligned}
w_0(1+a+\ldots+a^{r-1}) &= (\alpha_0((v^d-1)(w^{2d}-1)/r) - \alpha_{d-1}' g^{-1}v)(1+a+\ldots+a^{r-1}) \\
w_1(1+a+\ldots+a^{r-1}) &= \alpha_1'(1+a+\ldots+a^{r-1}) \\
w_2(1+a+\ldots+a^{r-1}) &= (\alpha_2' - \alpha_1' g^{-1}v)(1+a+\ldots+a^{r-1}) \\
&\vdots \qquad \vdots \\
w_{d-1}(1+a+\ldots+a^{r-1}) &= (\alpha_{d-1}' - \alpha_{d-2}' g^{-1}v)(1+a+\ldots+a^{r-1}).
\end{aligned}$$

Hence $\mathbf{Z}[G(L/K)]\langle (x-y)\otimes 1\rangle + W'$ contains, by setting in turn each one of $\alpha_0, \alpha_1', \ldots$ equal to one and the rest equal to zero,

$$\begin{aligned}
\lambda^{-1}(((v^d-1)(w^{2d}-1)/r)y\otimes 1) &= ((v^d-1)(w^{2d}-1)/r)y\otimes 1, \\
\lambda^{-1}(y\otimes g - g^{-1}vy\otimes g^2) &= y\otimes g - g^{-1}y\otimes vg, \\
\lambda^{-1}(y\otimes g^2 - g^{-1}vy\otimes g^3) &= y\otimes g^2 - g^{-1}y\otimes vg^2, \\
&\vdots \qquad \qquad \vdots \\
\lambda^{-1}(y\otimes g^{d-1} - g^{-1}vy\otimes 1) &= y\otimes g^{d-1} - g^{-1}y\otimes vg^{d-1}.
\end{aligned}$$

Next we want to know, in

$$\frac{\ker(g_4 \mid X(2)\otimes Q_0)}{\mathbf{Z}[g]/(g^d-1)\langle y\otimes(v-g+v(v^d-1)w^{2d})\rangle},$$

how much of the submodule generated by $y\otimes(v-g)g^i$ $(1\leq i\leq d-1)$ and $(v^2-g)(y\otimes 1)$ is generated by the elements listed above. We shall do this calculation, showing that everything may be generated in this manner, after applying λ to get into the left-factor-only action. Hence we are considering which $\mathbf{Z}[g]/(g^d-1)$-submodule of

$$\langle y\otimes(v-g)g, y\otimes(v-g)g^2,\ldots,y\otimes(v-g)g^{d-1},(v^2-g)y\otimes 1\rangle$$

is generated by

$$\begin{gathered}((v^d-1)(w^{2d}-1)/r)y\otimes 1, y\otimes g-g^{-1}vy\otimes g^2,\ldots\\ \ldots,y\otimes g^{d-1}-g^{-1}vy\otimes 1, y\otimes(v-g+v(v^d-1)w^{2d}).\end{gathered}$$

Firstly we observe that

$$\begin{aligned}&-v(v^d-1)(w^{2d}-1)y\otimes 1+y\otimes(v-g+v(v^d-1)w^{2d})\\ &+y\otimes g-g^{-1}vy\otimes g^2+g^{-1}v(y\otimes g^2-g^{-1}vy\otimes g^3)\\ &\qquad\vdots\qquad\quad\vdots\\ &+g^{2-d}v^{d-2}(y\otimes g^{d-1}-g^{-1}vy\otimes 1)\\ &=v(v^d-1)y\otimes 1+y\otimes v-gv^{d-1}y\otimes 1\\ &=(v^{d+1}-gv^{d-1})y\otimes 1\\ &=v^{d-1}(v^2-g)y\otimes 1.\end{aligned}$$

However we can also generate $((v^d-1)(w^{2d}-1)/r)(v^2-g)y\otimes 1$ and, since $HCF(v^{d-1},((v^d-1)(w^{2d}-1)/r))=1$ we may generate $(v^2-g)y\otimes 1$.

Next, for any $1\le i\le d-1$ we have

$$\begin{aligned}&v(y\otimes g^i-g^{-1}vy\otimes g^{i+1})-(1-g^{-1}v^2)((y\otimes g^{i+1}-g^{-1}vy\otimes g^{i+2})\\ &-(1-g^{-1}v^2)(g^{-1}vy\otimes g^{i+2}-g^{-2}v^2y\otimes g^{i+3})\\ &\qquad-(1-g^{-1}v^2)(g^{-2}v^2y\otimes g^{i+3}-g^{-3}v^3y\otimes g^{i+4})\\ &\qquad\quad\vdots\qquad\quad\vdots\\ &-(1-g^{-1}v^2)(g^{i+2-d}v^{d-i-2}y\otimes g^{d-1}-g^{i+3-d}v^{d-i-3}y\otimes 1)\\ &\qquad-(1-g^{-1}v^2)g^{i+3-d}v^{d-i-3}y\otimes 1\\ &=y\otimes(v-g)g^i.\end{aligned}$$

Hence we have generated everything we wished to, which proves the following result.

Lemma 4.1.21

In Proposition 4.1.17, the canonical homomorphism

$$\mathbf{Z}[G(L/K)]\langle(x-y)\otimes 1\rangle+W'\xrightarrow{\tilde{\rho}}\frac{\ker(g_4\mid X(2)\otimes Q_0)}{\mathbf{Z}[g]/(g^d-1)\langle y\otimes(v-g+v(v^d-1)w^{2d})\rangle}$$

is surjective.

4.1.22 Lemma 4.1.21 implies that we have a surjection, given by adding the factors in $\mathbf{Z}[G(L/K)]\langle (x-y)\otimes 1\rangle + W'$, of the form

$$\mathbf{Z}[G(L/K)]\langle (x-y)\otimes 1\rangle \oplus W' \stackrel{\rho}{\longrightarrow} \frac{\ker(g_4 \mid X(2)\otimes Q_0)}{\mathbf{Z}[g]/(g^d-1)\langle y\otimes (v-g+v(v^d-1)w^{2d})\rangle}.$$

Also the domain of ρ is free, by §4.1.18. In addition, by Proposition 4.1.19(i), we have an exact sequence of the form

$$\begin{aligned}0 \longrightarrow \mathbf{Z}[G(L/K)](a-1)^2 \longrightarrow \mathbf{Z}[G(L/K)]\langle (x-y)\otimes 1\rangle \oplus W' \\ \longrightarrow \mathbf{Z}[G(L/K)]\langle (x-y)\otimes 1\rangle + W' \longrightarrow 0.\end{aligned}$$

On the one hand this leads to the relation

$$[\mathrm{Ker}(\rho)] = [\frac{\ker(g_4 \mid X(2)\otimes Q_0)}{\mathbf{Z}[g]/(g^d-1)\langle y\otimes (v-g+v(v^d-1)w^{2d})\rangle}] = [\mathrm{Ker}(g_4)]$$

in $\mathcal{CL}(\mathbf{Z}[G(L/K)])$ and on the other it will enable us to find a filtration of the form

$$\mathbf{Z}[G(L/K)](a-1) \cong \mathbf{Z}[G(L/K)](a-1)^2 \cong A \subset B \subset \mathrm{Ker}(\rho)$$

which will lead to an explicit description of $\mathrm{Ker}(\rho)$ (see Theorem 4.1.24).

4.1.23 Next we need to know the kernel of the surjection, $\tilde{\rho}$. That is, in the left-factor-only action, we want all the sets of integers, $n_0, n_1, \ldots, n_{d-1}$, solving the equation

$$\begin{aligned}&\lambda^{-1}(\textstyle\sum_{i=0}^{d-1} n_i g^i y\otimes (v-g+v(v^d-1)w^{2d})) \\ &= -\beta((x-y)\otimes 1) + \alpha_0(t\otimes 1 + y\otimes (v^d-1)w^{2d}/r) \\ &\quad +\alpha_1(t\otimes g) + \alpha_2(t\otimes g^2) + \ldots + \alpha_{d-1}(t\otimes g^{d-1}).\end{aligned}$$

This equation is the same as that solved in §4.1.20 with

$$\begin{aligned}w_2 &= 0 = w_3 = \ldots = w_{d-1}, \\ w_0 &= (v+v(v^d-1)w^{2d})\textstyle\sum_{i=0}^{d-1} n_i g^i, \\ w_1 &= -\textstyle\sum_{i=0}^{d-1} n_i g^i.\end{aligned}$$

From §4.1.20, we must have the equations:

$$\begin{aligned}\beta &= \alpha_0(a-1), \\ \alpha_1 &= \alpha_0 g^{-1}(1+a+\ldots+a^{v-1}) + \alpha_1'(1+a+\ldots+a^{r-1}), \\ \alpha_2 &= \alpha_0 g^{-2}(1+a+\ldots+a^{v^2-1}) + \alpha_2'(1+a+\ldots+a^{r-1}), \\ \alpha_3 &= \alpha_0 g^{-3}(1+a+\ldots+a^{v^3-1}) + \alpha_3'(1+a+\ldots+a^{r-1}), \\ &\vdots \qquad\qquad \vdots \\ \alpha_{d-1} &= \alpha_0 g^{1-d}(1+a+\ldots+a^{v^{d-1}-1}) + \alpha_{d-1}'(1+a+\ldots+a^{r-1})\end{aligned}$$

and

$$\begin{aligned}
&(v+v(v^d-1)w^{2d})(\textstyle\sum_{i=0}^{d-1} n_i g^i)(1+a+\ldots+a^{r-1})\\
&\qquad =(\alpha_0((v^d-1)(w^{2d}-1)/r)-\alpha'_{d-1}g^{-1}v)(1+a+\ldots+a^{r-1})\\
&-(\textstyle\sum_{i=0}^{d-1} n_i g^i)(1+a+\ldots+a^{r-1})=\alpha'_1(1+a+\ldots+a^{r-1})\\
&0=(\alpha'_2-\alpha'_1 g^{-1}v)(1+a+\ldots+a^{r-1})\\
&\qquad\vdots\qquad\vdots\\
&0=(\alpha'_{d-1}-\alpha'_{d-2}g^{-1}v)(1+a+\ldots+a^{r-1}).
\end{aligned}$$

so that

$$\begin{aligned}
\alpha'_2(1+a+\ldots+a^{r-1})&=\alpha'_1 g^{-1}v(1+a+\ldots+a^{r-1})\\
\alpha'_3(1+a+\ldots+a^{r-1})&=\alpha'_1 g^{-2}v^2(1+a+\ldots+a^{r-1})\\
\vdots\qquad&\qquad\vdots\\
\alpha'_{d-1}(1+a+\ldots+a^{r-1})&=\alpha'_1 g^{2-d}v^{d-2}(1+a+\ldots+a^{r-1}).
\end{aligned}$$

These relations in turn imply that

$$\begin{aligned}
&\beta=\alpha_0(a-1),\\
&\alpha_1=\alpha_0 g^{-1}(1+a+\ldots+a^{v-1})+\alpha'_1(1+a+\ldots+a^{r-1}),\\
&\alpha_2=\alpha_0 g^{-2}(1+a+\ldots+a^{v^2-1})+\alpha'_1 g^{-1}v(1+a+\ldots+a^{r-1}),\\
&\alpha_3=\alpha_0 g^{-3}(1+a+\ldots+a^{v^3-1})+\alpha'_1 g^{-2}v^2(1+a+\ldots+a^{r-1}),\\
&\qquad\qquad\vdots\qquad\qquad\vdots\\
&\alpha_{d-1}=\alpha_0 g^{1-d}(1+a+\ldots+a^{v^{d-1}-1})+\alpha'_1 g^{2-d}v^{d-2}(1+a+\ldots+a^{r-1})
\end{aligned}$$

for some α_0 and α'_1.

The equation we started with also leads to the following equation:

$$\begin{aligned}
&\textstyle\sum_{i=0}^{d-1} n_i g^i y\otimes(v-g+v(v^d-1)w^{2d})\\
&=\alpha_0((v^d-1)(w^{2d}-1)/r)y\otimes 1\\
&\qquad+\alpha'_1(y\otimes g-g^{-1}vy\otimes g^2)\\
&\qquad+\alpha'_2(y\otimes g^2-g^{-1}vy\otimes g^3)\\
&\qquad\qquad\vdots\qquad\qquad\vdots\\
&\qquad+\alpha'_{d-1}(y\otimes g^{d-1}-g^{-1}vy\otimes 1).
\end{aligned}$$

For this we must have the following equations, in $\mathbf{Z}[G(L/K)]\langle z_2\rangle$,

$$\begin{aligned}
&(\textstyle\sum_{i=0}^{d-1} n_i g^i)(v+v(v^d-1)w^{2d})y=\alpha_0((v^d-1)(w^{2d}-1)/r)y-\alpha'_{d-1}g^{-1}vy,\\
&-(\textstyle\sum_{i=0}^{d-1} n_i g^i y)=\alpha'_1 y,
\end{aligned}$$

$$\begin{aligned}
\alpha'_2 y &= \alpha'_1 g^{-1} v y,\\
\alpha'_3 y &= \alpha'_2 g^{-1} v y = \alpha'_1 g^{-2} v^2 y,\\
&\vdots \qquad\qquad \vdots\\
\alpha'_{d-1} y &= \alpha'_{d-2} g^{-1} v y = \alpha'_1 g^{2-d} v^{d-2} y.
\end{aligned}$$

These equations give all the α'_i's in terms of the n_j's. The only restriction comes from the first equation, when rewritten as

$$\begin{aligned}
&(\textstyle\sum_{i=0}^{d-1} n_i g^i)(v + v(v^d-1)w^{2d})y\\
&\quad = \alpha_0((v^d-1)(w^{2d}-1)/r)y + (\textstyle\sum_{i=0}^{d-1} n_i g^i) g v^{d-1} y
\end{aligned}$$

or

$$\alpha_0((v^d-1)(w^{2d}-1)/r)y = (\textstyle\sum_{i=0}^{d-1} n_i g^i)(v(v^d-1)(w^{2d}-1)y + v^{d-1}(v^2-g)y).$$

Temporarily writing $\mu = (v^d-1)(w^{2d}-1)/r$ we see that μ divides

$$(\sum_{i=0}^{d-1} n_i g^i)(v^2-g) \quad \text{in} \quad \mathbf{Z}[g]/(g^d-1).$$

Hence we have integers, $n'_0, n'_1, \ldots$ such that

$$\begin{aligned}
n_{d-1} &= n_0 v^2 + \mu n'_{d-1},\\
n_{d-2} &= n_0 v^4 + \mu n'_{d-2},\\
&\vdots \qquad \vdots\\
n_1 &= n_0 v^{2d-2} + \mu n'_1,\\
n_0 &= n_0 v^{2d} + \mu n'_0.
\end{aligned}$$

If these conditions are fulfilled then

$$\sum_{i=0}^{d-1} n_i g^i = n_0(v^2 g^{d-1} + v^4 g^{d-2} + \ldots v^{2d-2} g + v^{2d}) + \mu(\sum_{i=0}^{d-1} n'_i g^i)$$

and

$$\begin{aligned}
(\textstyle\sum_{i=0}^{d-1} n_i g^i)(v^2-g) &= n_0 v^2(v^d-1) + \mu(\textstyle\sum_{i=0}^{d-1} n'_i g^i)(v^2-g)\\
&= -\mu n'_0 v^2 + \mu(\textstyle\sum_{i=0}^{d-1} n'_i g^i)(v^2-g).
\end{aligned}$$

Notice that $w^{2d}v^{2d} = 1+s(v^{2d}-1)r$ so that $\mu = ((v^d-1)/r)(w^{2d}-w^{2d}v^{2d}++s(v^{2d}-1)r) = ((v^d-1)/r)(sr-w^{2d})(v^{2d}-1)$ and therefore we may choose $n'_0, n'_1, \ldots, n'_{d-1}$ arbitrarily and thence determine $n_0, n_1, \ldots, n_{d-1}$ and subsequently α_0 and α'_1.

As we successively set equal to one each of $n'_0, n'_1, \ldots, n'_{d-1}$ and set the rest equal to zero in

$$\sum_{i=0}^{d-1} n_i g^i y \otimes (v - g + v(v^d-1)w^{2d})$$

we obtain the following generators (in the left-factor-only action) for the image in $\ker(g_4 \mid X(2) \otimes Q_0)$ of the homomorphism from $\mathbf{Z}[G(L/K)]\langle (x-y)\otimes 1\rangle \oplus W'$ which adds the two coordinates:

$$\begin{aligned}
&\tfrac{\mu}{(1-v^{2d})}(1+v^2g^{d-1}+\ldots+v^{2d-2}g)y\otimes(v-g+v(v^d-1)w^{2d}),\\
&\mu gy\otimes(v-g+v(v^d-1)w^{2d}),\\
&\mu g^2y\otimes(v-g+v(v^d-1)w^{2d}),\\
&\qquad\vdots \qquad\qquad \vdots\\
&\mu g^{d-1}y\otimes(v-g+v(v^d-1)w^{2d})
\end{aligned}$$

where $\mu = (v^d-1)(w^{2d}-1)/r$.

Now we have to unravel what is the relation of these elements to the right-hand side of the equation:

$$\begin{aligned}
&\lambda^{-1}(\textstyle\sum_{i=0}^{d-1} n_ig^iy\otimes(v-g+v(v^d-1)w^{2d}))\\
&=-\alpha_0(a-1)((x-y)\otimes 1)+\alpha_0(t\otimes 1+y\otimes(v^d-1)w^{2d}/r)\\
&+(\alpha_0g^{-1}(1+a+\ldots+a^{v-1})+\alpha_1'y)(t\otimes g)\\
&+(\alpha_0g^{-2}(1+a+\ldots+a^{v^2-1})+\alpha_1'g^{-1}vy)(t\otimes g^2)\\
&+(\alpha_0g^{-3}(1+a+\ldots+a^{v^3-1})+\alpha_1'g^{-2}v^2y)(t\otimes g^3)\\
&\qquad\qquad\qquad\qquad\vdots \qquad \vdots\\
&+(\alpha_0g^{1-d}(1+a+\ldots+a^{v^{d-1}-1})+\alpha_1'g^{2-d}v^{d-2}y)(t\otimes g^{d-1}).
\end{aligned}$$

When $n_0'=1$ and $n_1'=0=\ldots=n_{d-1}'$ then

$$\sum_{i=0}^{d-1} n_ig^i = (1+v^2g^{d-1}+\ldots+v^{2d-2}g) = \mathcal{N},$$

say. Then, from our equations,

$$\begin{aligned}
\beta &= \alpha_0(a-1),\\
\alpha_0((v^d-1)(w^{2d}-1)/r)y &= \alpha_0\mu\\
&= \mathcal{N}(v(v^d-1)(w^{2d}-1)y+v^{d-1}(v^2-g)y),\\
\alpha_1'y &= -\mathcal{N}y
\end{aligned}$$

so that

$$\alpha_1'(1+a+\ldots+a^{r-1}) = -\mathcal{N}(1+a+\ldots+a^{r-1}).$$

When $n_0'=1$ and $n_1'=0=\ldots=n_{d-1}'$ then the corresponding equation is

$$\begin{aligned}
&\lambda^{-1}(\tfrac{\mu}{(1-v^{2d})}\mathcal{N}\otimes(v-g+v(v^d-1)w^{2d}))\\
&=-\beta((x-y)\otimes 1)+\alpha_0(t\otimes 1+y\otimes(v^d-1)w^{2d}/r)+\textstyle\sum_{i=1}^{d-1}\alpha_i(t\otimes g^i)
\end{aligned}$$

$$
\begin{aligned}
&= -\alpha_0(a-1)((x-y)\otimes 1) + \alpha_0(t\otimes 1 + y\otimes (v^d-1)w^{2d}/r) \\
&+(\alpha_0 g^{-1}(1+a+\ldots+a^{v-1}) + \alpha_1'(1+a+\ldots+a^{r-1}))(t\otimes g) \\
&+(\alpha_0 g^{-2}(1+a+\ldots+a^{v^2-1}) + \alpha_1' g^{-1}v(1+a+\ldots+a^{r-1}))(t\otimes g^2) \\
&\quad\vdots \qquad\qquad \vdots \\
&+(\alpha_0 g^{1-d}(1+a+\ldots+a^{v^{d-1}-1}) + \alpha_1' g^{2-d}v^{d-2}(1+a+\ldots+a^{r-1}))(t\otimes g^{d-1}).
\end{aligned}
$$

With these equations satisfied, we are considering an element of

$$\mathbf{Z}[G(L/K)]\langle (x-y)\otimes 1\rangle \oplus W'$$

given by

$$
\begin{aligned}
Z = &(\alpha_0(a-1)((x-y)\otimes 1), \alpha_0(t\otimes 1 + y\otimes (v^d-1)w^{2d}/r)) \\
&+\textstyle\sum_{i=1}^{d-1}(\alpha_0 g^{-i}(1+a+\ldots+a^{v^i-1}) + \alpha_1' g^{1-i}v^{i-1}(1+a+\ldots+a^{r-1}))(t\otimes g^i)).
\end{aligned}
$$

Hence $Z\in \mathrm{Ker}(\rho)$ because the sum of its two coordinates in $\mathbf{Z}[G(L/K)]\langle (x-y)\otimes 1\rangle + W'$ is equal to $\lambda^{-1}(\frac{\mu}{(1-v^{2d})}\mathcal{N}y\otimes (v-g+v(v^d-1)w^{2d})$.

We may filter $\mathrm{Ker}(\rho)$ by

$$A\subset B\subset \mathrm{Ker}(\rho)$$

with $A\cong \mathbf{Z}[G(L/K)](a-1)^2 \cong \mathbf{Z}[G(L/K)](a-1)$ given by the kernel of the map which adds the two coordinates, in $\mathbf{Z}[G(L/K)]\langle (x-y)\otimes 1\rangle + W'$,

$$
\begin{aligned}
B &= \rho^{-1}(\lambda^{-1}(\mathbf{Z}[g]/(g^d-1)\langle \mu y\otimes (v-g+v(v^d-1)w^{2d})\rangle)) \\
&\cong \mathbf{Z}[G(L/K)]/\langle (a-1)\rangle
\end{aligned}
$$

and $\mathrm{Ker}(\rho)/B\cong K_3(\mathbf{F}_{v^d})$ generated by $\lambda^{-1}(\frac{\mu}{(1-v^{2d})}\mathcal{N}y\otimes (v-g+v(v^d-1)w^{2d})$. Multiplying by (v^2-g) and observing that $(v^2-g)\mathcal{N} = (v^{2d}-1)g$ we find that $(v^2-g)Z$ maps to $\lambda^{-1}(\mu g y\otimes (v-g+v(v^d-1)w^{2d})$, which generates B/A.

By the calculation made earlier in this section, we have the relation

$$\sum_{i=0}^{d-1} n_i g^i = n_0(v^2 g^{d-1}+\ldots+v^{2d}) + \mu(\sum_{i=0}^{d-1} n_i' g^i)$$

so that, if $n_0'=1$ and $n_1'=0=\ldots=n_{d-1}'$, we have (from the same calculation made earlier in this section) $\alpha_0 y = \mathcal{N}vry - v^{d-1}gy$ so we choose

$$\alpha_0 = \mathcal{N}v(1+a+\ldots+a^{r-1}) - v^{d-1}g.$$

Since $n_0'=1$ and $n_1'=0=\ldots=n_{d-1}'$, we know that Z generated $\mathrm{Ker}(\rho)/A$ as a $\mathbf{Z}[G(L/K)]/\langle (a-1)\rangle$-module and that $(v^2-g)Z$ generates B/A.

Next we need to find out what which submodule of A is generated by $(a-1)(v^2-g)Z$. In terms of the $(\beta, \alpha_0, \alpha_1')$-coordinates the middle coordinate (which determines the other two) for these elements is given by

$$\begin{aligned}(v^2-g)\alpha_0 &= (v^2-g)(\mathcal{N}v(1+a+\ldots+a^{r-1})-v^{d-1}g)\\ &= v(v^{2d}-1)g(1+a+\ldots+a^{r-1})-v^{d-1}(v^2-g)g\end{aligned}$$

and

$$(a-1)(v^2-g)\alpha_0 = -v^{d-1}(a-1)(v^2-g)g$$

in $\mathbf{Z}[G(L/K)](a-1)$.

Similarly

$$(a-1)\alpha_0 = (a-1)(-v^{d-1}g) \in A.$$

This means that if we define a homomorphism of left $\mathbf{Z}[G(L/K)]$-modules of the form

$$\Psi : \mathbf{Z}[G(L/K)] \longrightarrow \mathrm{Ker}(\rho)$$

by $\Psi(1) = \alpha_0$ then $\Psi((a-1)) = (a-1)\alpha_0 = (a-1)(-v^{d-1}g) \in A$. This discussion establishes the following result:

Theorem 4.1.24

Let L/K be a tamely ramified Galois extension of local fields in characteristic $p>0$ with group

$$G(L/K) = \langle a, g \mid g^d = a^r = 1,\ \mathrm{gag}^{-1} = a^v\rangle = \langle g\rangle \propto \langle a\rangle$$

as in §4.1.1. Then

(i) *the $\mathbf{Z}[GL/K)]$-module,* Ker(ρ) *of §4.1.22, is finitely generated and projective,*

(ii) *there is a pushout diagram of left $\mathbf{Z}[G(L/K)]$-modules of the following form:*

$$\begin{array}{ccc}\mathbf{Z}[G(L/K)](a-1) & \longrightarrow & \mathbf{Z}[G(L/K)]\\ \Big\downarrow (-v^{d-1}g) & & \Big\downarrow \Psi\\ \mathbf{Z}[G(L/K)](a-1) & \longrightarrow & \mathrm{Ker}(\rho)\end{array}$$

where $v = |\overline{K}|$,

(iii) *the Euler characteristic of the local fundamental class of Theorem 3.1.22*

$$\mathbf{F}^*_{v^{2d})} \longrightarrow \mathrm{Ind}_{G(L/W)}^{G(L/K)}(\mu_\infty[1/p]\tilde{\oplus}\mathbf{Z}) \xrightarrow{\hat{F}-1} \mathrm{Ind}_{G(L/W)}^{G(L/K)}(\mu_\infty[1/p]\tilde{\oplus}\mathbf{Z}) \longrightarrow \mathbf{F}^*_{v^d}$$

is equal to

$$-[\mathrm{Ker}(\rho)] \in \mathcal{CL}(\mathbf{Z}[G(L/K)]).$$

Proof. The discussion of §4.1.23 proves (ii) and then part (iii) follows from Lemma 4.1.8, Proposition 4.1.9 and Proposition 4.1.12. For part (i) we need only show that $\mathrm{Ker}(\rho)$ is cohomologically trivial, being clearly finitely generated and torsion-free. It is implicit in the definition of the Euler characteristic of a local fundamental class that any one of its representing modules will be cohomologically trivial.

An alternative way to see this is to observe that $HCF(v, r) = 1$ so that the left vertical homomorphism in the pushout diagram induces an isomorphism on $H^*(G(L/K); -)$, by Proposition 4.1.19(iv). □

4.1.25 *The Hom-description representative*

We conclude this section by computing the representative for $[\mathrm{Ker}(\rho)] \in \mathcal{CL}(\mathbf{Z}[G(L/K)])$ in the Hom-description of the class-group (see Theorem 2.5.25).

Since v is a unit in the rationals and in $\mathbf{Z}_p$ for all primes, p, except for the residue characteristic we may take $\alpha_0 \otimes 1$ as a basis element for $\mathrm{Ker}(\rho) \otimes \Lambda$ when $\Lambda = \mathbf{Q}$ or $\mathbf{Z}_p$ when v is not a power of p. This means that the Hom-description representative is trivial for all primes except the residue characteristic.

The same is true for the residue characteristic if $d = 1$.

If p is the residue characteristic we need a basis element for $\mathrm{Ker}(\rho) \otimes \mathbf{Z}_p$.

Choose $\beta_p \in \mathrm{Ker}(\rho) \otimes \mathbf{Z}_p$ to be the image of

$$(a - 1, 1) \in \mathbf{Z}_p[G(L/K)](a - 1) \oplus \mathbf{Z}_p[G(L/K)].$$

Then the image of β_p under the quotient map

$$\mathrm{Ker}(\rho) \otimes \mathbf{Z}_p \longrightarrow \frac{\mathrm{Ker}(\rho) \otimes \mathbf{Z}_p}{\mathbf{Z}_p[G(L/K)](a - 1)} \cong \mathbf{Z}_p[g]/(g^d - 1)$$

is equal to the generator, 1. On the other hand, in the bottom left of the pushout,

$$\begin{aligned}(a-1)^{v^d-1}\beta_p &= (a-1)^{v^d} + (a-1)^{v^d-2}(a-1)\alpha_0 \\ &= (a-1)^{v^d} + (a-1)^{v^d-1}(-v^{d-1}g) \\ &\equiv a^{v^d} - 1 \text{ (modulo } p) \\ &\equiv a - 1 \text{ (modulo } p)\end{aligned}$$

so that $(a - 1)^{v^d-1}\beta_p$ generates $\mathbf{Z}_p[G(L/K)](a - 1)$ in the bottom left and so

$$\mathrm{Ker}(\rho) \otimes \mathbf{Z}_p = \mathbf{Z}_p[G(L/K)]\beta_p.$$

When tensored with $\mathbf{Q}_p$, β_p is equal in the top right $\mathbf{Q}_p[G(L/K)]$ to $1 + (a - 1)g^{-1}v^{1-d}$ while the rational basis is 1 in the top right. In other words, $(1 + (a - 1)g^{-1}v^{1-d})(\alpha_0 \otimes 1) = \beta_p$ and so the component at primes above p of the Hom-description representative for $[\mathrm{Ker}(\rho)]$ is equal to

$$\mathrm{Det}(1 + (a - 1)g^{-1}v^{1-d}) \in \mathrm{Hom}_{\Omega_{\mathbf{Q}_p}}(R(G(L/K), O_N^*)$$

for any Galois extension, $N/\mathbf{Q}_p$, containing all $[L : K]$-th roots of unity.

Note that this means that

$$\mathrm{Det}(1+(a-1)g^{-1}v^{1-d})^{-1} \in \mathrm{Hom}_{\Omega_{\mathbf{Q}_p}}(R(G(L/K)), \mathcal{O}_N^*)$$

is the component at primes above p of the Hom-description representative for the Euler characteristic of Theorem 4.1.24.

4.1.26 *Independence of the choice of Frobenius*

Suppose that ρ is an irreducible representation of

$$G(L/K) = \{a, g \mid a^r = 1 = g^d, \ \mathrm{gag}^{-1} = a^v\}.$$

By Clifford theory, the restriction of ρ to $\langle a \rangle$ is the direct sum of conjugates of a one-dimensional representation, $\phi : \langle a \rangle \longrightarrow \mathbf{C}^*$,

$$\mathrm{Res}_{\langle a \rangle}^{G(L/K)}(\rho) = \phi \oplus g^*(\phi) \oplus \ldots (g^{t-1})^*(\phi)$$

where t is the smallest positive integer such that $(g^t)^*(\phi) = \phi$. Of course, t divides d. The representation, ϕ extends to a homomorphism, $\hat{\phi} : \langle a, g^t \rangle \longrightarrow \mathbf{C}^*$, given by the formula $\hat{\phi}(a^u g^{tw}) = \phi(a^u)$ and $\rho = \mathrm{Ind}_{\langle a, g^t \rangle}^{G(L/K)}(\hat{\phi})$.

Now we shall calculate the matrix of $1+(a-1)g^{-1}v^{1-d}$ using the basis for ρ given by $1 \otimes 1, g^{-1} \otimes 1, \ldots, g^{1-t} \otimes 1$. We have

$$\begin{aligned}
&(1+(a-1)g^{-1}v^{1-d})(g^{-s} \otimes 1) \\
&= g^{-s} \otimes 1 + (a-1)v^{1-d}g^{-s-1} \otimes 1 \\
&= g^{-s} \otimes 1 - v^{1-d}g^{-s-1} \otimes 1 + v^{1-d}g^{-s-1}g^{s+1}ag^{-s-1} \otimes 1 \\
&= g^{-s} \otimes 1 - v^{1-d}g^{-s-1} \otimes 1 + v^{1-d}g^{-s-1} \otimes \phi(g^{s+1}ag^{-s-1}) \\
&= g^{-s} \otimes 1 - v^{1-d}g^{-s-1} \otimes 1 + v^{1-d}g^{-s-1} \otimes \phi(a)^{v^{s+1}} \\
&= g^{-s} \otimes 1 + v^{1-d}(\phi(a)^{v^{s+1}} - 1)g^{-s-1} \otimes 1.
\end{aligned}$$

Hence the $t \times t$ matrix of $1+(a-1)g^{-1}v^{1-d}$ on ρ is given by

$$\begin{pmatrix}
1 & 0 & 0 & \ldots & \ldots & \frac{\phi(a)-1}{v^{d-1}} \\
\frac{\phi(a)^v-1}{v^{d-1}} & 1 & 0 & \ldots & \ldots & 0 \\
0 & \frac{\phi(a)^{v^2}-1}{v^{d-1}} & 1 & \ldots & \ldots & 0 \\
\vdots & \vdots & \vdots & \vdots & \vdots & \vdots \\
0 & 0 & 0 & \ldots & \frac{\phi(a)^{v^{t-1}}-1}{v^{d-1}} & 1
\end{pmatrix}.$$

The determinant of this matrix is equal to

$$1+(-1)^t\frac{\phi(a)-1}{v^{d-1}}\det\begin{pmatrix} \frac{\phi(a)^v-1}{v^{d-1}} & 1 & 0 & \dots & 0 \\ 0 & \frac{\phi(a)^{v^2}-1}{v^{d-1}} & 1 & \dots & 0 \\ \vdots & \vdots & \vdots & \vdots & \vdots \\ 0 & 0 & 0 & \dots & \frac{\phi(a)^{v^{t-1}}-1}{v^{d-1}} \end{pmatrix}$$

which is equal to

$$\operatorname{Det}(1+(a-1)g^{-1}v^{1-d})(\rho) = 1+v^{\dim(\rho)(1-d)}\prod_{j=0}^{t-1}(1-\phi(a)^{v^j}).$$

This expression is independent of the choice of the "Frobenius automorphism". For if we change the generator, g, to $a^i g$ of order d then we must have $1=(a^i g)^d = a^{i(v^d-1)/(v-1)}$ which means that i must be a multiple of $HCF(r, v-1)$ since a has order r which divides v^d-1. Set $z = HCF(r, v-1)$ temporarily. Suppose that a new choice of Frobenius is given by $a^{iz}g$ then the $1+(a-1)(a^{iz}g)^{-1}v^{1-d}$ acts on ρ by

$$(1+(a-1)(a^{iz}g)^{-1}v^{1-d})(g^{-s}\otimes 1) = g^{-s}\otimes 1 + v^{1-d}(\phi(a)^{v^{s+1}}-1)\phi(a)^{izv^s}g^{-s-1}\otimes 1$$

and the resulting determinant is equal to

$$\begin{aligned} &\operatorname{Det}(1+(a-1)(a^{iz}g)^{-1}v^{1-d})(\rho) \\ &= 1+v^{\dim(\rho)(1-d)}\textstyle\prod_{j=0}^{t-1}(1-\phi(a)^{v^j})\phi(a)^{izv^j} \\ &= 1+v^{\dim(\rho)(1-d)}\textstyle\prod_{j=0}^{t-1}(1-\phi(a)^{v^j}) \end{aligned}$$

because

$$\prod_{j=0}^{t-1}\phi(a)^{izv^j} = \phi(a)^{iHCF(r,v-1)(v^t-1)/(v-1)} = 1,$$

since $\phi(a)^{v^t} = \phi(a)$.

Theorem 4.1.27

Suppose that E/F is a Galois extension of global fields in characteristic $p>0$. Then, by analogy with §3.2.1, define a Galois module structure invariant by the formula

$$\Omega_1(E/F,2) = \sum_{P\lhd\mathcal{O}_F,\ \text{prime}} \operatorname{Ind}_{G(E_Q/F_P)}^{G(E/F)}(\Omega_1(G(E_Q/F_P),2)) \in \mathcal{CL}(\mathbf{Z}[G(E/F)])$$

where $\Omega_1(G(E_Q/F_P),2)\in\mathcal{CL}(\mathbf{Z}[G(E_Q/F_P)])$ is the Euler characteristic of the K_2/K_3 local fundamental class of Theorem 3.1.22.

When E/F is tamely ramified then the Hom-description representative of $\Omega_1(E/F,2)$ is trivial except at places above p. At places above p the representative is given by sending $\chi \in R(G(E/F))$ to

$$\prod_P \mathrm{Det}(1+(a_P-1)g_P^{-1}v_P^{1-d_P})^{-1}(\mathrm{Res}^{G(E/F)}_{G(E_Q/F_P)}(\chi))$$

where the product is taken over those primes of $P \lhd \mathcal{O}_F$ for which

$$G(E_Q/F_P) = \langle a_P, g_P \mid g_P^{d_P} = a^{r_P} = 1,\ g_P a_P g_P^{-1} = a_P^{v_P} \rangle$$

with $v_P = |\mathcal{O}_F/P|$ and $d \geq 2$ in the notation of §4.1.1.

Proof. This follows from the definition of $\Omega_1(E/F,2)$ together with the fact that the homomorphism

$$\mathrm{Ind}^{G(E/F)}_{G(E_Q/F_P)} : \mathcal{CL}(\mathbf{Z}[G(E_Q/F_P)]) \longrightarrow \mathcal{CL}(\mathbf{Z}[G(E/F)])$$

corresponds in the Hom-description to the map induced by restriction of representations to the decomposition group. □

Remark 4.1.28 In Theorem 4.2.19, in the tamely ramified case, we shall show that very often the elements of $\mathrm{Ext}^2_{\mathbf{Z}[G(L/K)]}(\mathbf{F}_{v^d}, \mathbf{F}_{v^{2d}})$ which induce cup-product isomorphisms on Tate cohomology all have the same Euler characteristic in $\mathcal{CL}(\mathbf{Z}[G(L/K)])$.

The obstructions to promoting our calculations from the tame case to the general case are discussed in §4.2.20.

4.2 $\mathrm{Ext}^2_{\mathbf{Z}[G(L/K)]}(\mathbf{F}^*_{v^d}, \mathbf{F}^*_{v^{2d}})$

4.2.1 As in Example 3.1.21, let L/K be a Galois extension of local fields of residue characteristic p and Galois group, $G(L/K)$. As in §3.1.5, set $W = L \cap L_0$ so that W/K is the maximal unramified subextension of L/K. Hence L/W is totally ramified, $\overline{K} = \mathbf{F}_v$, $\overline{W} = \overline{L} = \mathbf{F}_{v^d}$ and $K \cong \mathbf{F}_v((X))$, $W = \mathbf{F}_{v^d}((X)) \subseteq L = \mathbf{F}_{v^d}((Y))$ for some X, Y. Choose $g \in G(L/K)$ of order d which maps to the Frobenius automorphism in $G(\overline{L}/\overline{K})$. For each positive integer, m, set $L(m) = \mathbf{F}_{v^{dm}}L$ so that $\overline{L(m)} = \mathbf{F}_{v^{dm}}$ and there is an extension of the form

$$G(L(m)/L) \longrightarrow G(L(m)/K) \xrightarrow{\pi_m} G(L/K)$$

in which the g's may be chosen to satisfy $\pi_m(g) = g$. The kernel of π_m, $G(L(m)/L)$, is isomorphic to $G(\mathbf{F}_{v^{dm}}/\mathbf{F}_{v^d})$ which is cyclic of order m generated by g^d. If L_0 is the maximal unramified extension of L, as in §3.1.2, then L_0/K is equal to the limit of the extensions, $L(m)/K$. Set $[G_0(L/K) : G_1(L/K)] = [G(L/W) : G_1(L/W)] = r$, as in §4.1.1, then r divides $v^d - 1$ and so is prime to p. Hence r is the index which appears (denote by t there) in Example 3.1.21 and Theorem 3.1.22.

We have canonical cohomology isomorphisms (see Example 3.1.21)

$$H^i(G(L(m)/K); K_2(L(m))) \cong H^i(G(L(m)/K); \mathbf{F}^*_{v^{dm}}),$$
$$H^i(G(L(m)/K); K_3(L(m))) \cong H^i(G(L(m)/K); \mathbf{F}^*_{v^{2dm}})$$

for all $i > 0$ and

$$\mathrm{Ext}^2_{\mathbf{Z}[G(L(m)/K)]}(K_2(L(m)), K_3(L(m))) \cong \mathrm{Ext}^2_{\mathbf{Z}[G(L(m)/K)]}(\mathbf{F}^*_{v^{dm}}, \mathbf{F}^*_{v^{2dm}}).$$

In this section we are going to calculate this Ext-group and analyse those elements which possess a class-group-valued Euler characteristic in the sense of Example 2.1.8. The key to making these computations is the reduction to the tamely ramified case for which we shall retain the notation of §4.1.1. That is, if L/K is a tamely ramified then

$$G(L/K) = \langle a, g \mid g^d = a^r = 1,\ \mathrm{gag}^{-1} = a^v \rangle = \langle g \rangle \propto \langle a \rangle.$$

where $G(L/W) = \langle a \rangle$.

Proposition 4.2.2

Let L/K be a tamely ramified Galois extension of local fields in characteristic $p > 0$, as in §4.2.1.

(i) *For each $i \geq 1$ there is an exact sequence of the form*

$$0 \longrightarrow H^{2i-1}(G(L/K); K_3(\mathbf{F}_{v^d})) \longrightarrow H^{2i}(G(L/K); \mathrm{Ind}_{\langle a \rangle}^{G(L/K)}(\mathbf{Z})) \cong \mathbf{Z}/r$$
$$\stackrel{v^2-g_*}{\longrightarrow} H^{2i}(G(L/K); \mathrm{Ind}_{\langle a \rangle}^{G(L/K)}(\mathbf{Z})) \longrightarrow H^{2i}(G(L/K); K_3(\mathbf{F}_{v^d})) \longrightarrow 0.$$

(ii) *For each $i \geq 1$ there is an exact sequence of the form*

$$0 \longrightarrow H^{2i-1}(G(L/K); (\mathbf{F}_{v^d})^*) \longrightarrow H^{2i}(G(L/K); \mathrm{Ind}_{\langle a \rangle}^{G(L/K)}(\mathbf{Z})) \cong \mathbf{Z}/r$$
$$\stackrel{v-g_*}{\longrightarrow} H^{2i}(G(L/K); \mathrm{Ind}_{\langle a \rangle}^{G(L/K)}(\mathbf{Z})) \longrightarrow H^{2i}(G(L/K); (\mathbf{F}_{v^d})^*) \longrightarrow 0.$$

(iii) *In (i) and (ii) g_* equals multiplication by v^i on $H^{2i}(G(L/K); \mathrm{Ind}_{\langle a \rangle}^{G(L/K)}(\mathbf{Z}))$.*

Proof. The first sequence comes from the exact sequence

$$0 \longrightarrow \mathrm{Ind}_{\langle a \rangle}^{G(L/K)}(\mathbf{Z}) \cong \mathbf{Z}[g]/(g^d - 1) \stackrel{v^2-g}{\longrightarrow} \mathrm{Ind}_{\langle a \rangle}^{G(L/K)}(\mathbf{Z}) \longrightarrow K_3(\mathbf{F}_{v^d}) \longrightarrow 0$$

and the second comes from

$$0 \longrightarrow \mathrm{Ind}_{\langle a \rangle}^{G(L/K)}(\mathbf{Z}) \cong \mathbf{Z}[g]/(g^d - 1) \stackrel{v-g}{\longrightarrow} \mathrm{Ind}_{\langle a \rangle}^{G(L/K)}(\mathbf{Z}) \longrightarrow (\mathbf{F}_{v^d})^* \longrightarrow 0.$$

In part (iii) we claim that

$$v^i = g_* : H^{2i}(G(L/K); \mathrm{Ind}_{\langle a \rangle}^{G(L/K)}(\mathbf{Z})) \longrightarrow H^{2i}(G(L/K); \mathrm{Ind}_{\langle a \rangle}^{G(L/K)}(\mathbf{Z})).$$

This can be seen when $i = 1$ by considering the 5-term exact sequence ([130] p. 43) associated to the group extension

$$\mathbf{Z}/r \cong \langle a \rangle \longrightarrow G(L/K) \longrightarrow \langle g \rangle \cong \mathbf{Z}/d.$$

As a $\mathbf{Z}[\langle g \rangle]$-module consider $M = (\mathbf{F}_{v^d})^*$ and $M = K_3(\mathbf{F}_{v^d}) \cong (\mathbf{F}_{v^{2d}})^*$, which, by Hilbert's Theorem 90, both have trivial Tate cohomology in all dimensions. Therefore the 5-term exact sequence becomes

$$0 \longrightarrow H^1(G(L/K); M) \longrightarrow \mathrm{Hom}_{\langle g \rangle}(\langle a \rangle, M) \longrightarrow 0.$$

However, g acts as the v-th power map on $\langle a \rangle \cong \mathbf{Z}/r$ and on $(\mathbf{F}_{v^d})^*$ while it acts as the v^2-th power map on $(\mathbf{F}_{v^{2d}})^*$. Since $HCF(v, r) = 1$, this proves the result for the case $i = 1$.

For general i we first observe that the spectral sequence

$$E_2^{s,t} = H^s(\langle g \rangle; H^t(\langle a \rangle; Z)) \Longrightarrow H^{s+t}(G(L/K); \mathbf{Z})$$

has $E_2^{3,0} = 0 = E_2^{s,1}$ for all $s \geq 0$. Therefore the edge homomorphism, which may be identified with the canonical restriction map,

$$H^2(G(L/K); \mathbf{Z}) \longrightarrow H^2(\langle a \rangle; \mathbf{Z}) \cong \mathbf{Z}/r$$

is surjective. Multiplication by any element, z, mapping to a generator induces a periodicity isomorphism

$$(z \cup -) : H^n(G(L/K); \mathrm{Ind}_{\langle a \rangle}^{G(L/K)}(\mathbf{Z})) \xrightarrow{\cong} H^{n+2}(G(L/K); \mathrm{Ind}_{\langle a \rangle}^{G(L/K)}(\mathbf{Z}))$$

for all $n \geq 1$. However $g_*(z \cup w) = (vz) \cup g_*(w)$ so that the result follows for all i from the case $i = 1$. □

Remark 4.2.3 Observe that Proposition 4.2.2 yields exact Tate cohomology sequences of the form

$$0 \longrightarrow \hat{H}^{2i+1}(G(L/K); K_3(\mathbf{F}_{v^d})) \longrightarrow \hat{H}^{2i+2}(G(L/K); \mathrm{Ind}_{\langle a \rangle}^{G(L/K)}(\mathbf{Z})) \cong \mathbf{Z}/r$$
$$\xrightarrow{v^2 - v^{i+1}} \hat{H}^{2i+2}(G(L/K); \mathrm{Ind}_{\langle a \rangle}^{G(L/K)}(\mathbf{Z})) \longrightarrow \hat{H}^{2i+2}(G(L/K); K_3(\mathbf{F}_{v^d})) \longrightarrow 0$$

and

$$0 \longrightarrow \hat{H}^{2i-1}(G(L/K); (\mathbf{F}_{v^d})^*) \longrightarrow \hat{H}^{2i}(G(L/K); \mathrm{Ind}_{\langle a \rangle}^{G(L/K)}(\mathbf{Z})) \cong \mathbf{Z}/r$$
$$\xrightarrow{v - v^i} \hat{H}^{2i}(G(L/K); \mathrm{Ind}_{\langle a \rangle}^{G(L/K)}(\mathbf{Z})) \longrightarrow \hat{H}^{2i}(G(L/K); (\mathbf{F}_{v^d})^*) \longrightarrow 0$$

which is consistent with the isomorphisms for all $i \geq 1$

$$\hat{H}^{2i}(G(L/K); (\mathbf{F}_{v^d})^*) \cong \hat{H}^{2i+2}(G(L/K); K_3(\mathbf{F}_{v^d})),$$

since $HCF(r, v^2 - v^{i+1}) = HCF(r, 1 - v^{i-1}) = HCF(r, v - v^i)$.

4.2.4 *Computing* $\mathrm{Ext}^2_{\mathbf{Z}[G(L/K)]}(\mathbf{F}^*_{v^d}, \mathbf{F}^*_{v^{2d}})$

Now we shall compute $\mathrm{Ext}^2_{\mathbf{Z}[G(L/K)]}(\mathbf{F}^*_{v^d}, \mathbf{F}^*_{v^{2d}})$.

In the tamely ramified case, we have a short exact sequence of left $\mathbf{Z}[G(L/K)]$-modules given by

$$0 \longrightarrow \mathbf{Z}[g]/(g^d - 1) \stackrel{(-\cdot(v-g))}{\longrightarrow} \mathbf{Z}[g]/(g^d - 1) \longrightarrow \mathbf{F}^*_{v^d} \longrightarrow 0$$

where $a \in G(L/K)$ acts trivially. This results in a long exact sequence

$$\ldots \stackrel{(v-g)^*}{\longrightarrow} H^1(\langle a\rangle; \mathbf{F}^*_{v^{2d}}) \stackrel{\lambda_1}{\longrightarrow} \mathrm{Ext}^2_{\mathbf{Z}[G(L/K)]}(\mathbf{F}^*_{v^d}, \mathbf{F}^*_{v^{2d}})$$
$$\stackrel{\lambda_2}{\longrightarrow} H^2(\langle a\rangle; \mathbf{F}^*_{v^{2d}}) \stackrel{(v-g)^*}{\longrightarrow} \ldots \quad .$$

Also, since r divides $v^{2d} - 1$ and a acts trivially on $\mathbf{F}^*_{v^{2d}}$, we have

$$\mathrm{Ext}^i_{\mathbf{Z}[G(L/K)]}(\mathbf{Z}[g]/(g^d - 1), \mathbf{F}^*_{v^{2d}}) \cong H^i(\langle a\rangle; \mathbf{F}^*_{v^{2d}}) \cong \mathbf{Z}/r$$

for all $i \geq 1$.

If we use, as in §4.1.7,

$$\ldots \stackrel{d\otimes 1}{\longrightarrow} P_i \otimes Q_0 \stackrel{d\otimes 1}{\longrightarrow} P_{i-1} \otimes Q_0 \longrightarrow \ldots \longrightarrow Q_0 \longrightarrow 0$$

as a free $\mathbf{Z}[G(L/K)]$-resolution then the inclusion

$$P_i \otimes \mathbf{Z}[g]/(g^d - 1) = P_i \otimes Q_0 = \Lambda_{i,0} \subset \oplus_{s=0}^i \Lambda_{s,i-s}$$

induces λ_2. This inclusion also induces $X(2) \otimes Q_0 \subset \hat{X}(2)$ so that, in order to evaluate $\lambda_2([g_4])$ we must first evaluate g_4 on the generators of $X(2) \otimes Q_0$, $t \otimes g^i$, $y \otimes g^i$ and $x \otimes g^i$ $(1 \leq i \leq d)$. These values are given by

$$g_4(t \otimes g^i) = \begin{cases} 0 & \text{if } 1 \leq i \leq d-1, \\ \delta^{(1-v^d)} & \text{if } i = d, \end{cases}$$

and

$$g_4(x \otimes g^i) = g_4(y \otimes g^i) = \delta^{rv^{i+d}}$$

for $1 \leq i \leq d$. Also the cohomological change of rings isomorphism is induced by the isomorphism

$$\mathrm{Hom}_{\mathbf{Z}[G(L/K)]}(X(2) \otimes Q_0, \mathbf{F}^*_{v^{2d}}) \longrightarrow \mathrm{Hom}_{\mathbf{Z}[\langle a\rangle]}(X(2), \mathbf{F}^*_{v^{2d}})$$

which sends h to its restriction to $t \otimes 1$, $y \otimes 1$ and $x \otimes 1$. There is a commutative diagram, induced by a chain equivalence of free $\mathbf{Z}[\langle a\rangle]$-module resolutions, of the form

$$\begin{array}{ccccccc}
\mathbf{Z}\langle 1+a+\cdots+a^{r-1}\rangle & \longrightarrow & \mathbf{Z}[\langle a\rangle] & \xrightarrow{a-1} & \mathbf{Z}[\langle a\rangle] & \longrightarrow & \mathbf{Z} \\
\downarrow \tilde{h} & & \downarrow \tilde{h}_1 & & \downarrow \tilde{h}_0 & & \downarrow 1 \\
X(2) & \longrightarrow & P_1 & \longrightarrow & P_0 & \longrightarrow & \mathbf{Z}
\end{array}$$

given by $\tilde{h}_0(1) = 1$, $\tilde{h}_1(1) = z_2$. Therefore $\tilde{h}(1 + a + \ldots + a^{r-1}) = y$ and so in $\mathrm{Hom}_{\mathbf{Z}[\langle a\rangle]}(\mathbf{Z}\langle 1 + a + \ldots + a^{r-1}\rangle, \mathbf{F}^*_{v^{2d}})$ $\lambda_2([g_4])$ is represented by the map which sends the generators, $1 + a + \ldots + a^{r-1}$, to $g_4(y \otimes 1) = \delta^{rv^d}$, which is a generator of $\mathbf{F}^*_{v^{2d}}$. This implies that $\lambda_2([g_4]) \in H^i(\langle a\rangle; \mathbf{F}^*_{v^{2d}}) \cong \mathbf{Z}/r$ is a generator and so λ_2 is surjective.

Next we need to know the homomorphism g^* on $H^1(\langle a\rangle; \mathbf{F}^*_{v^{2d}})$. Our model for $H^1(\langle a\rangle; \mathbf{F}^*_{v^{2d}})$ is $\mathrm{Hom}_{\mathbf{Z}[\langle a\rangle]}(I\langle a\rangle, \mathbf{F}^*_{v^{2d}})$ where IG denotes the augmentation ideal of $\mathbf{Z}[G]$. This description comes from the projective resolution

$$\ldots \mathbf{Z}[\langle a\rangle] \xrightarrow{a-1} \mathbf{Z}[\langle a\rangle] \xrightarrow{\epsilon} \mathbf{Z} \longrightarrow 0,$$

since

$$\mathbf{Z}[\langle a\rangle](a-1) = \ker(\epsilon) = I\langle a\rangle$$

and

$$\mathrm{Hom}_{\mathbf{Z}[\langle a\rangle]}(\mathbf{Z}[\langle a\rangle], \mathbf{F}^*_{v^{2d}}) \longrightarrow \mathrm{Hom}_{\mathbf{Z}[\langle a\rangle]}(I\langle a\rangle, \mathbf{F}^*_{v^{2d}})$$

is zero.

To deduce the effect of $(-\cdot g)$ on H^1 we begin by observing that the self-map, denoted by $(-\cdot g)\otimes(-\cdot g)$, of $P_0 \otimes Q_0 = \mathbf{Z}[G(L/K)] \otimes \mathbf{Z}[g]/(g^d - 1)$ which is given by sending $x \otimes g^i$ to $xg \otimes g^{i+1}$ satisfies $(-\cdot g)(\epsilon \otimes 1) = (\epsilon \otimes 1)(-\cdot g) \otimes (-\cdot g)$. Hence we may use this map – diagonal right multiplication by g on $P_0 \otimes Q_0$ – as the start of a chain map of the free resolution of Q_0 covering $(-\cdot g)$.

The change of rings isomorphism in dimension one is induced by the isomorphism

$$\mathrm{Hom}_{\mathbf{Z}[G(L/K)]}(IG(L/K) \otimes Q_0, \mathbf{F}^*_{v^{2d}}) \cong \mathrm{Hom}_{\mathbf{Z}[\langle a\rangle]}(IG(L/K), \mathbf{F}^*_{v^{2d}})$$

which sends f to $(x \mapsto f(x \otimes 1))$. Hence $g^*(f)$ is sent to $(x \mapsto f(xg \otimes g) = g(f(g^{-1}xg \otimes 1)) = f(g^{-1}xg \otimes 1)^{v^2}$. The map induced by the inclusion of $\mathbf{Z}[\langle a\rangle]$ into $\mathbf{Z}[G(L/K)]$ also induces an isomorphism on cohomology so that the effect of g^* on $\mathrm{Hom}_{\mathbf{Z}[\langle a\rangle]}(I\langle a\rangle, \mathbf{F}^*_{v^{2d}})$ sends a map, h, given by $h(a-1) = u$ to

$$\begin{aligned} g^*(h)(a-1) &= h(a^w - 1)^{v^2} = h((1 + a + \ldots + a^{w-1})(a-1))^{v^2} \\ &= h(a-1)^{wv^2} = h(a-1)^v. \end{aligned}$$

Hence $g^* = v$ and $(v - g)^* = 0$ on $H^1(\langle a\rangle; \mathbf{F}^*_{v^{2d}})$. Therefore, in the tamely ramified case, we have a short exact sequence of the form

$$0 \longrightarrow \mathbf{Z}/r \xrightarrow{\lambda_1} \mathrm{Ext}^2_{\mathbf{Z}[G(L/K)]}(\mathbf{F}^*_{v^d}, \mathbf{F}^*_{v^{2d}}) \xrightarrow{\lambda_2} \mathbf{Z}/r \longrightarrow 0.$$

Theorem 4.2.5

Let L/K be any Galois extension of local fields in characteristic $p > 0$, as in §4.2.1. Then there is an isomorphism of the form

$$\mathrm{Ext}^2_{\mathbf{Z}[G(L/K)]}(\mathbf{F}^*_{v^d}, \mathbf{F}^*_{v^{2d}}) \cong \mathbf{Z}/r \oplus \mathbf{Z}/r.$$

Proof. We begin by showing that the short exact sequence of §4.2.4 splits in the tame case.

Since $\lambda_2([g_4]) \in \mathbf{Z}/r$ is a generator, it suffices to show that

$$g_4^r : \hat{X}(2) \longrightarrow \mathbf{F}^*_{v^{2d}}$$

extends to a $\mathbf{Z}[G(L/K)]$-module homomorphism

$$f : (P_0 \otimes Q_1) \oplus (P_1 \otimes Q_0) \longrightarrow \mathbf{F}^*_{v^{2d}}.$$

Define f to be trivial on $z_0 \otimes g^i$ and $z_1 \otimes g^i$ and set $f(z_2 \otimes g^i) = \delta^{rv^{i+d}}$ for all $1 \leq i \leq d$. Then

$$\begin{aligned} f(y \otimes g^i) &= f((1 + a + \ldots + a^{r-1})z_2 \otimes g^i) \\ &= (1 + a + \ldots + a^{r-1})f(z_2 \otimes g^i) \\ &= \delta^{r^2 v^{i+d}} \\ &= g_4(y \otimes g^i)^r. \end{aligned}$$

Similarly

$$\begin{aligned} f(x \otimes g^i) &= f((1 + g^{-1} + \ldots + g^{1-d})z_1 \otimes g^i + (a^{-1} + \ldots + a^{-r})z_2 \otimes g^i) \\ &= \delta^{r^2 v^{i+d}} \\ &= g_4(x \otimes g^i)^r. \end{aligned}$$

Also, if $1 \leq i \leq d - 1$,

$$\begin{aligned} f(t \otimes g^i) &= f((a - 1)z_1 \otimes g^i - (g^{-1}(1 + a + \ldots + a^{v-1}) - 1)z_2 \otimes g^i) \\ &= f(z_2 \otimes g^i - g^{-1}(1 + a + \ldots + a^{v-1})z_2 \otimes g^i) \\ &= f(z_2 \otimes g^i) - g^{-1}f((1 + a + \ldots + a^{v-1})z_2 \otimes g^{i+1}) \\ &= \delta^{rv^{i+d} - rv^{i+d+1}vw^2} \end{aligned}$$

which is trivial, as is $g_4(t \otimes g^i)^r$. When $i = d$

$$\begin{aligned} f(t \otimes 1) &= f(z_2 \otimes 1) - g^{-1}f((1 + a + \ldots + a^{v-1})z_2 \otimes g) \\ &= \delta^{rv^{2d} - rv^{1+d}vw^2} \\ &= \delta^{r(1-v^d)} \\ &= g_4(t \otimes 1)^r. \end{aligned}$$

Clearly $f((\delta \otimes 1 - 1 \otimes \delta)(z_1 \otimes g^i))$ and $g_4((\delta \otimes 1 - 1 \otimes \delta)(z_1 \otimes g^i))^r$ are both trivial.

Finally, if $1 \leq i \leq d-1$,

$$\begin{aligned} f((\delta \otimes 1 - 1 \otimes \delta)(z_2 \otimes g^i)) &= f(-z_2 \otimes vg^i + z_2 \otimes g^{i+1}) \\ &= \delta^{-rv^{i+d}v + rv^{i+1+d}} \end{aligned}$$

which is trivial, as is $g_4((\delta \otimes 1 - 1 \otimes \delta)(z_2 \otimes g^i))^r$ in this case. When $i = d$

$$\begin{aligned} f((\delta \otimes 1 - 1 \otimes \delta)(z_2 \otimes 1)) &= f(-z_2 \otimes v + z_2 \otimes g) \\ &= \delta^{-rv^{2d}v - rv^{1+d}} \\ &= \delta^{rv(v^d - 1)} \\ &= g_4((\delta \otimes 1 - 1 \otimes \delta)(z_2 \otimes 1))^r, \end{aligned}$$

which completes the proof in the tame case.

Now let L/K be *any* Galois extension of local fields in characteristic $p > 0$ with Galois group, $G(L/K)$. Let $G_1(L/K) \subset G(L/K)$ denote the first wild ramification group ([128] p. 62), which is a finite p-group. If M is the fixed field of $G_1(L/K) = G(L/M)$ then M/K is the maximal tamely ramified subextension. Since $\overline{L} = \overline{M} = \mathbf{F}_{v^d}$, we have shown that

$$\mathrm{Ext}^2_{\mathbf{Z}[G(M/K)]}(\mathbf{F}^*_{v^d}, \mathbf{F}^*_{v^{2d}}) \cong \mathbf{Z}/r \oplus \mathbf{Z}/r.$$

Hence it suffices to show that the natural map induces an isomorphism of the form

$$\mathrm{Ext}^2_{\mathbf{Z}[G(M/K)]}(\mathbf{F}^*_{v^d}, \mathbf{F}^*_{v^{2d}}) \longrightarrow \mathrm{Ext}^2_{\mathbf{Z}[G(L/K)]}(\mathbf{F}^*_{v^d}, \mathbf{F}^*_{v^{2d}})$$

is an isomorphism. However, for $F = M, L$ we have spectral sequences of the form

$$E_2^{s,t}(F/K) = H^s(G(F/K); \mathrm{Ext}^t_{\mathbf{Z}}(\mathbf{F}^*_{v^d}, \mathbf{F}^*_{v^{2d}})) \Longrightarrow \mathrm{Ext}^{s+t}_{\mathbf{Z}[G(F/K)]}(\mathbf{F}^*_{v^d}, \mathbf{F}^*_{v^{2d}})$$

and the natural map, $E_2^{s,t}(M/K) \longrightarrow E_2^{s,t}(L/K)$ is an isomorphism for all s, t. This last isomorphism follows from the Serre spectral sequence since

$$HCF(|G(L/M)|, v^d - 1) = HCF(|G(L/M)|, v^{2d} - 1) = 1. \qquad \square$$

Remark 4.2.6 In terms of 2-extensions the isomorphism

$$\mathrm{Ext}^2_{\mathbf{Z}[G(M/K)]}(\mathbf{F}^*_{v^d}, \mathbf{F}^*_{v^{2d}}) \xrightarrow{\cong} \mathrm{Ext}^2_{\mathbf{Z}[G(L/K)]}(\mathbf{F}^*_{v^d}, \mathbf{F}^*_{v^{2d}})$$

is given in the following manner. Any 2-extension of $\mathbf{Z}[G(M/K)]$-modules

$$\mathbf{F}^*_{v^{2d}} \longrightarrow A \longrightarrow B \longrightarrow \mathbf{F}^*_{v^d}$$

is also a 2-extension of $\mathbf{Z}[G(L/K)]$-modules, by inflation. On the other hand, any 2 -extension of $\mathbf{Z}[G(L/K)]$-modules

$$\mathbf{F}^*_{v^{2d}} \longrightarrow C \longrightarrow D \longrightarrow \mathbf{F}^*_{v^d}$$

is equivalent to one in which C and D are finite of order prime to p. Since $G(L/M)$ is a p-group we obtain a 2-extension of $\mathbf{Z}[G(M/K)]$-modules

$$\mathbf{F}^*_{v^{2d}} \longrightarrow C^{G(L/M)} \longrightarrow D^{G(L/M)} \longrightarrow \mathbf{F}^*_{v^d}$$

where $X^{G(L/M)}$ denotes the $G(L/M)$-fixed points of X. These two constructions define inverse isomorphisms of Ext^2-groups.

4.2.7 For each positive integer, $m \geq 1$, set $L_m = \mathbf{F}_{v^{dm}}L$, as in §4.2.1, so that

$$G(L_m/K) = \langle a, g \mid g^{dm} = 1 = a^r, \ gag^{-1} = a^v\rangle$$

and there is an extension of the form

$$G(L_m/L) \longrightarrow G(L_m/K) \xrightarrow{\pi_m} G(L/K)$$

in which $\pi_m(a) = a$ and $\pi_m(g) = g$. The kernel of π_m, $G(L_m/L)$, is isomorphic to $G(\mathbf{F}_{v^{dm}}/\mathbf{F}_{v^d})$ which is cyclic of order m generated by g^d. Therefore L_m/K is a tamely ramified extension which is just L/K when $m = 1$ and the limit of of the extensions L_m/L is equal to the maximal unramified extension, L_0/L.

Applying Theorem 4.2.5 to $L(m)/K$ we find that

$$\mathrm{Ext}^2_{\mathbf{Z}[G(L(m)/K)]}(\mathbf{F}^*_{v^d}, \mathbf{F}^*_{v^{2d}}) \cong \mathbf{Z}/r \oplus \mathbf{Z}/r,$$

which is independent of m.

The cohomology spectral sequence

$$E_2^{s,t} = H^s(G(L/K); H^t(G(L_m/L); \mathbf{F}^*_{v^{dm}})) \Longrightarrow H^{s+t}(G(L_m/K); \mathbf{F}^*_{v^{dm}})$$

has $E_2^{s,t} = 0$ for all $t > 0$, since $H^t(G(L_m/L); \mathbf{F}^*_{v^{dm}}) = 0$ for all $t > 0$ and $H^0(G(L_m/L); \mathbf{F}^*_{v^{dm}}) = \mathbf{F}^*_{v^d}$.

Therefore we have proved the following result:

Theorem 4.2.8

The edge homomorphism in the Serre spectral sequence of §4.2.7 yields an isomorphism of the form

$$H^*(G(L_m/K); \mathbf{F}^*_{v^{dm}}) \xrightarrow{\cong} H^*(G(L/K); \mathbf{F}^*_{v^d})$$

for all $m \geq 1$.

There is a similar isomorphism of the form

$$H^*(G(L_m/K); \mathbf{F}^*_{v^{2dm}}) \xrightarrow{\cong} H^*(G(L/K); \mathbf{F}^*_{v^{2d}}).$$

Proposition 4.2.9

The natural map yields an isomorphism

$$H^t(G(L/K); \mathbf{F}^*_{v^d}) \xrightarrow{\cong} \lim_{\overrightarrow{m}} H^t(G(L_m/K); \mathbf{F}^*_{v^{dm}})$$

for all $t \geq 0$.

Proof. This is clear when $t = 0$, since each cohomology group in the direct limit is isomorphic to $\mathbf{F}_v^*$ mapped isomorphically to itself under the inclusion, $i : \mathbf{F}^*_{v^{dm}} \longrightarrow \mathbf{F}^*_{v^{dms}}$.

Now consider the case when $t > 0$. There is an extension of the form

$$G(L_{ms}/L_m) \longrightarrow G(L_{ms}/K) \xrightarrow{\pi} G(L_m/K)$$

and the natural maps are given by the compositions

$$H^t(G(L_m/K); \mathbf{F}^*_{v^{dm}}) \xrightarrow{\pi^*} H^t(G(L_{ms}/K); \mathbf{F}^*_{v^{dm}}) \xrightarrow{i_*} H^t(G(L_{ms}/K); \mathbf{F}^*_{v^{dms}}).$$

To see that each of these compositions is an isomorphism, consider the spectral sequences:

$$E_2^{v,w} = H^v(G(L_m/K); H^w(\{1\}; \mathbf{F}^*_{v^{dm}})) \Longrightarrow H^{v+w}(G(L_m/K); \mathbf{F}^*_{v^{dm}}),$$
$$(E')_2^{v,w} = H^v(G(L_m/K); H^w(G(L_{ms}/L_m); \mathbf{F}^*_{v^{dm}})) \Longrightarrow H^{v+w}(G(L_{ms}/K); \mathbf{F}^*_{v^{dm}}),$$
$$(E'')_2^{v,w} = H^v(G(L_m/K); H^w(G(L_{ms}/L_m); \mathbf{F}^*_{v^{dms}})) \Longrightarrow H^{v+w}(G(L_{ms}/K); \mathbf{F}^*_{v^{dms}}).$$

On E_2-terms, $E_2^{t,0} \longrightarrow (E')_2^{t,0}$ and $(E')_2^{t,0} \longrightarrow (E'')_2^{t,0}$ are both the identity map. However, the edge homomorphisms of $E_r^{v,w}$ and $(E'')_r^{v,w}$ are both isomorphisms. Also

$$E_2^{v,w} \cong \begin{cases} 0 & w > 0, \\ E_\infty^{v,0} & w = 0 \end{cases}$$

and

$$(E'')_2^{v,w} \cong \begin{cases} 0 & w > 0, \\ (E'')_\infty^{v,0} & w = 0. \end{cases}$$

This means that $(E')_2^{v,0} \cong (E')_\infty^{v,0}$. This is seen by considering the first r for which $d_r : (E')_r^{v-r,w+r-1} \longrightarrow (E')_r^{v,w}$ is non-zero and observing that this implies $d_r : (E'')_r^{v-r,w+r-1} = 0 \longrightarrow (E'')_r^{v,w}$ is non-zero also.

Hence we have edge homomorphisms

$$H^v(G(L_m/K); \mathbf{F}^*_{v^{dm}}) \xrightarrow{\cong} E_\infty^{v,0},$$
$$H^v(G(L_{ms}/K); \mathbf{F}^*_{v^{dm}}) \longrightarrow (E')_\infty^{v,0},$$
$$H^v(G(L_{ms}/K); \mathbf{F}^*_{v^{dms}}) \xrightarrow{\cong} (E'')_\infty^{v,0}$$

which commute with the natural maps and the result follows. □

Remark 4.2.10 Usually a homomorphism such as $K_2(L_m) \longrightarrow K_2(L_{ms})$ would have a non-trivial kernel, but not here. This is because a transfer argument shows that the map is injective on the uniquely divisible part (if it exists) and the tame symbol [109] maps

$$\delta_m : K_2(L_m) \longrightarrow \mathbf{F}^*_{v^{dm}}, \ \delta_{ms} : K_2(L_{ms}) \longrightarrow \mathbf{F}^*_{v^{dms}}$$

are isomorphisms modulo uniquely divisible subgroups.

Furthermore, since L_{ms}/L_m is unramified,

$$\delta_m\{z, y\} = \delta_{ms}\{z, y\} \in \mathbf{F}^*_{v^{dms}}.$$

Next we are going to calculate the effect of cup-products with elements of $\mathrm{Ext}^2_{\mathbf{Z}[G(L/K)]}(\mathbf{F}^*_{v^d}, \mathbf{F}^*_{v^{2d}})$ on Tate cohomology. We shall need the following simple result.

Lemma 4.2.11

Suppose that

$$\begin{array}{ccccc}
A_1 & \xrightarrow{i_1} & B_1 & \xrightarrow{j_1} & C_1 \\
\downarrow{\scriptstyle \alpha_1} & & \downarrow{\scriptstyle \beta_1} & & \downarrow{\scriptstyle \gamma_1} \\
A_2 & \xrightarrow{i_2} & B_2 & \xrightarrow{j_2} & C_2 \\
\downarrow{\scriptstyle \alpha_2} & & \downarrow{\scriptstyle \beta_2} & & \downarrow{\scriptstyle \gamma_2} \\
A_3 & \xrightarrow{i_3} & B_3 & \xrightarrow{j_3} & C_3
\end{array}$$

is a commutative diagram of $\mathbf{Z}[G]$*-module homomorphisms in which each row and column is short exact. Suppose also that the middle row is split exact. Then for all* i *the composition*

$$H^i(G; C_3) \xrightarrow{\delta} H^{i+1}(G; C_1) \xrightarrow{\delta} H^{i+2}(G; A_1)$$

is equal to (-1) *times the composition*

$$H^i(G; C_3) \xrightarrow{\delta} H^{i+1}(G; A_3) \xrightarrow{\delta} H^{i+2}(G; A_1).$$

Proof. Suppose that $k_2 : C_2 \longrightarrow B_2$ and $l_2 : B_2 \longrightarrow A_2$ are $\mathbf{Z}[G]$-module homomorphisms such that

$$j_2 \cdot k_2 = 1, \; l_2 \cdot i_2 = 1 \text{ and } 1 = k_2 \cdot j_2 + i_2 \cdot l_2.$$

Define $\lambda = \beta_2 \cdot k_2 : C_2 \longrightarrow B_3$. Then $j_3 \cdot \beta_2 \cdot k_2 \cdot \gamma_1 = \gamma_2 \cdot j_2 \cdot k_2 \cdot \gamma_1 = \gamma_2 \cdot \gamma_1 = 0$ and so we may define $\mu : C_1 \longrightarrow A_3$ by $i_3 \cdot \mu = \beta_2 \cdot k_2 \cdot \gamma_1$. In this case $j_3 \cdot \lambda = j_3 \cdot \beta_2 \cdot k_2 = \gamma_2 \cdot j_2 \cdot k_2 = \gamma_2$ and $i_3 \cdot \mu = \beta_2 \cdot k_2 \cdot \gamma_1 = \lambda \cdot \gamma_1$ from which we obtain a homomorphism of short exact sequences. Therefore the composition

$$H^i(G; C_3) \xrightarrow{\delta} H^{i+1}(G; C_1) \xrightarrow{\mu_*} H^{i+1}(G; A_3)$$

is equal to the coboundary of the bottom short exact row in the diagram.

Define $\psi = l_2 \cdot \beta_1 : B_1 \longrightarrow A_2$. To show that $\alpha_2 \cdot \psi = \mu \cdot j_1$ it suffices to verify that $i_3 \cdot \alpha_2 \cdot \psi = i_3 \cdot \mu \cdot j_1$. However

$$\begin{aligned} i_3 \cdot \alpha_2 \cdot \psi &= i_3 \cdot \alpha_2 \cdot l_2 \cdot \beta_1 \\ &= \beta_2 \cdot i_2 \cdot l_2 \cdot \beta_1 \\ &= \beta_2 \cdot (1 - k_2 \cdot j_2) \cdot \beta_1 \\ &= \beta_2 \cdot \beta_1 - \beta_2 \cdot k_2 \cdot j_2 \cdot \beta_1 \\ &= -\beta_2 \cdot k_2 \cdot j_2 \cdot \beta_1 \\ &= -\beta_2 \cdot k_2 \cdot \gamma_1 \cdot j_1 \\ &= -i_3 \cdot \mu \cdot j_1, \end{aligned}$$

as required. Also $\psi \cdot i_1 = l_2 \cdot \beta_1 \cdot i_1 = l_2 \cdot i_2 \cdot \alpha_1 = \alpha_1$. These relations yield a homomorphism of short exact sequences showing that the composition

$$H^{i+1}(G; C_1) \xrightarrow{-\mu_*} H^{i+1}(G; A_3) \xrightarrow{\delta} H^{i+2}(G; A_1)$$

is equal to the coboundary associated to the upper row of the diagram, which completes the proof. □

4.2.12 *Cup-products with 2-extensions*

By Theorem 4.2.5, we have a split short exact sequence

$$0 \to H^1(\langle a \rangle; \mathbf{F}^*_{v^{2d}}) \xrightarrow{\lambda_1} \mathrm{Ext}^2_{\mathbf{Z}[G(L/K)]}(\mathbf{F}^*_{v^d}, \mathbf{F}^*_{v^{2d}}) \xrightarrow{\lambda_2} H^2(\langle a \rangle; \mathbf{F}^*_{v^{2d}}) \to 0$$

both extreme groups being isomorphic to $\mathbf{Z}/r$. Since $\lambda_2([g_4])$ generates the right-hand group we may take the class of $g_4 : \hat{X}(2) \longrightarrow \mathbf{F}^*_{v^{2d}}$ as one generator of order r. The cup-product with $[g_4]$ is an isomorphism. In fact, since $[g_4]$ is by definition represented by the K_2/K_3 local fundamental class, this cup-product is the canonical isomorphism.

For the other generator we shall take the image under λ_1 of any generator, $[h] \in H^1(\langle a \rangle; \mathbf{F}^*_{v^{2d}}) \cong \mathbf{Z}/r$.

Proposition 4.2.13

In the notation of §4.2.12, if $[l] = \lambda_1[h]$ *then*

$$([l] \bigcup -) : \hat{H}^i(G(L/K); \mathbf{F}^*_{v^d}) \longrightarrow \hat{H}^{i+2}(G(L/K); \mathbf{F}^*_{v^{2d}})$$

is zero if i is even and if i is odd is given by multiplication by $\alpha \cdot (r/HCF(r, v^{i-1} - 1)) \cdot (v^d - 1)/r)$ *for some integer,* α.

In particular, cup-product with $[l]$ is zero if r^2 divides $v^d - 1$.

Proof. Consider the following commutative diagram of $\mathbf{Z}[G(L/K)]$-module homomorphisms.

$$\begin{array}{ccccc}
X(2)\otimes Q_0 & \xrightarrow{i_1} & \mathrm{Ker}(\beta_2) & \xrightarrow{j_1} & \hat{X}(2) \\
\downarrow{\scriptstyle \alpha_1} & & \downarrow{\scriptstyle \beta_1} & & \downarrow{\scriptstyle \gamma_1} \\
P_1' & \xrightarrow{i_2} & P_1'\oplus P_1'' & \xrightarrow{j_2} & P_1'' \\
\downarrow{\scriptstyle \alpha_2} & & \downarrow{\scriptstyle \beta_2} & & \downarrow{\scriptstyle \gamma_2} \\
P_0' & \xrightarrow{i_3} & P_0'\oplus P_0'' & \xrightarrow{j_3} & P_0'' \\
\downarrow{\scriptstyle \alpha_3} & & \downarrow{\scriptstyle \beta_3} & & \downarrow{\scriptstyle \gamma_3} \\
Q_0 & \xrightarrow{v-g} & Q_0 & \longrightarrow & \mathbf{F}^*_{v^d}
\end{array}$$

Here $P_1' = \mathbf{Z}[G(L/K)]\langle z_1, z_2\rangle \otimes Q_0$, $P_0' = \mathbf{Z}[G(L/K)]\langle z_0\rangle \otimes Q_0$ and $X(2) = \mathrm{Ker}(\delta_2)$ in the notation of §4.1.3. In this diagram the two middle horizontal short exact sequences are split and the P_i', P_i'''s are projective modules. The diagram has vertical columns given by 2-extensions and the central 2-extension is made from the other columns by the familiar inductive construction of homological algebra. In particular, we may apply Lemma 4.2.11 twice to show that the coboundary of the bottom row followed by cup-product by the left-hand 2-extension is equal to cup-product by the right-hand 2-extension followed by the coboundary of the top row. If $Q_0 = \mathbf{Z}[g]/(g^d - 1)$, as before, then this means that the compositions

$$H^i(G(L/K); \mathbf{F}^*_{v^d}) \xrightarrow{\delta} H^{i+1}(G(L/K); Q_0) \xrightarrow{\cong} H^{i+2}(G(L/K); \mathrm{Ker}(\alpha_3))$$

and

$$H^i(G(L/K); \mathbf{F}^*_{v^d}) \xrightarrow{\cong} H^{i+1}(G(L/K); \mathrm{Ker}(\gamma_3)) \xrightarrow{-\delta} H^{i+2}(G(L/K); \mathrm{Ker}(\alpha_3))$$

are equal.

We begin by noting that it is sufficient to prove the result for ordinary group cohomology in dimensions greater than zero where it coincides with Tate cohomology.

If $[h]$ is represented by a cocycle, $h : P_1' \longrightarrow \mathbf{F}^*_{v^{2d}}$, then $[l]$ is represented by $l : \hat{X}(2) \longrightarrow \mathbf{F}^*_{v^{2d}}$ whose restriction to $\mathrm{Ker}(\beta_1)$ satisfies

$$(l \mid \mathrm{Ker}(\beta_1)) = h' \cdot \beta_1 : \mathrm{Ker}(\beta_2) \longrightarrow P_1' \oplus P_1'' \longrightarrow \mathbf{F}^*_{v^{2d}}$$

where h' is an extension of h. For $z \in H^{2i}(G(L/K); \mathbf{F}^*_{v^d})$ the cup-pduct, $[l] \bigcup z \in H^{2i+2}(G(L/K); \mathbf{F}^*_{v^{2d}})$ is equal to the image under l_* of the cup-product, $[l'] \bigcup z \in H^{2i+2}(G(L/K); \hat{X}(2))$, associated with the 2-extension given by the right-hand column in the diagram. However, by Proposition 4.2.2(proof), there exists $w \in H^{2i}(G(L/K); Q_0)$ mapping to z. Therefore $[l'] \bigcup z$ is equal to the image of $[l''] \bigcup w$, where $[l'']$ is the 2-extension given by the middle column. Hence we must show that

$$(l \mid \mathrm{Ker}(\beta_1))_*([l''] \bigcup w) = h'_*((\beta_1)_*([l''] \bigcup w)) = 0.$$

This is true because the finitely generated free module, $P_1' \oplus P_1''$, has the form $\mathrm{Ind}_{\{1\}}^{G(L/K)}(N)$ for some N so that

$$(\beta_1)_*([l''] \bigcup w) \in H^{2i+2}(G(L/K); P_1' \oplus P_1'') \cong H^{2i+2}(\{1\}; N) = 0.$$

Now consider the commutative diagram

$$\begin{array}{ccccc}
Q_0 & \xrightarrow{\ i_1\ } & \mathrm{Ker}(\beta_2) & \xrightarrow{\ j_1\ } & \hat{X}(2) \\
\downarrow{\scriptstyle \alpha_1} & & \downarrow{\scriptstyle \beta_1} & & \downarrow{\scriptstyle \gamma_1} \\
P_1' & \xrightarrow{\ i_2\ } & P_1' \oplus P_1'' & \xrightarrow{\ j_2\ } & P_1'' \\
\downarrow{\scriptstyle \alpha_2} & & \downarrow{\scriptstyle \beta_2} & & \downarrow{\scriptstyle \gamma_2} \\
\mathrm{Ker}(\alpha_3) & \xrightarrow{\ i_3\ } & \mathrm{Ker}(\beta_3) & \xrightarrow{\ j_3\ } & \mathrm{Ker}(\gamma_3)
\end{array}$$

from which we form a short exact sequence

$$0 \longrightarrow Y \longrightarrow \mathrm{Ker}(\beta_3) \oplus P_1'' \stackrel{j_3+\gamma_2}{\longrightarrow} \mathrm{Ker}(\gamma_3) \longrightarrow 0.$$

The left and right projections from $\mathrm{Ker}(\beta_3) \oplus P_1''$ map Y to $\mathrm{Ker}(\alpha_3)$ and $\hat{X}(2)$, respectively, which yields a commutative diagram of short exact sequences of the following form.

$$\begin{array}{ccccc}
\hat{X}(2) & \xrightarrow{\gamma_1} & P_1'' & \xrightarrow{\gamma_2} & \mathrm{Ker}(\gamma_3) \\
\uparrow & & \uparrow & & \uparrow 1 \\
Y & \longrightarrow & \mathrm{Ker}(\beta_3) \oplus P_1'' & \longrightarrow & \mathrm{Ker}(\gamma_3) \\
\downarrow & & \downarrow & & \downarrow 1 \\
\mathrm{Ker}(\alpha_3) & \xrightarrow{i_3} & \mathrm{Ker}(\beta_3) & \xrightarrow{j_3} & \mathrm{Ker}(\gamma_3)
\end{array}$$

Note also that $\mathrm{Ker}(\alpha_3)$ and $\hat{X}(2)$ are canonical submodules of Y. Furthermore the map from the middle row to the bottom row induces isomorphisms on $H^i(G(L/K); -)$ for all $i > 1$ since $H^j(G(L/K); P_1'') = 0$ for $j \geq 1$.

We have homomorphisms h, h', l on $P_1', P_1' \oplus P_1''$ and $\hat{X}(2)$, respectively. Also h factorises as $h = \tilde{h} \cdot \alpha_2 : P_1' \longrightarrow \mathrm{Ker}(\alpha_3) \longrightarrow \mathbf{F}^*_{v^{2d}}$. These homomorphisms enable us to construct a $\mathbf{Z}[G(L/K)]$-module homomorphism

$$F : Y \longrightarrow \mathbf{F}^*_{v^{2d}}$$

extending $\tilde{h}$ on $\mathrm{Ker}(\alpha_3)$ and l on $\hat{X}(2)$. For if $(z, p_1'') \in Y \subset \mathrm{Ker}(\beta_3) \oplus P_1''$ we may choose $(p', p'') \in P_1' \oplus P_1''$ such that $\beta_3(p', p'') = z$. Therefore $p_1'' - p'' \in \hat{X}(2) \subset P_1''$ and we define

$$F(z, p_1'') = h'(p', p'') + l(p_1'' - p'') \in \mathbf{F}^*_{v^{2d}}.$$

This is well defined because if (q', q'') is another pre-image of z then

$$h'(p', p'') - h'(q', q'') = h'(p' - q', p'' - q'') = l(p'' - q'') = l(p_1'' - q'') - l(p_1'' - p'').$$

This shows that the following three compositions are equal

$$H^i(G(L/K); \mathrm{Ker}(\gamma_3)) \longrightarrow H^{i+1}(G(L/K); \mathrm{Ker}(\alpha_3)) \xrightarrow{h'_*} H^{i+1}(G(L/K); \mathbf{F}^*_{v^{2d}}),$$

$$H^i(G(L/K); \mathrm{Ker}(\gamma_3)) \longrightarrow H^{i+1}(G(L/K); Y) \xrightarrow{F_*} H^{i+1}(G(L/K); \mathbf{F}^*_{v^{2d}}),$$

and

$$H^i(G(L/K); \mathrm{Ker}(\gamma_3)) \longrightarrow H^{i+1}(G(L/K); \hat{X}(2)) \xrightarrow{l_*} H^{i+1}(G(L/K); \mathbf{F}^*_{v^{2d}}).$$

Take $u \in H^{2i-1}(G(L/K); \mathbf{F}^*_{v^d})$ and apply the coboundary of the bottom row of the main diagram to obtain $\delta(u) \in H^{2i}(G(L/K); Q_0)$ and then apply another coboundary to obtain $u' = \delta(\delta(u)) \in H^{2i+1}(G(L/K); \mathrm{Ker}(\alpha_3)$. By Lemma 4.2.11, $-u'$ is equal, u_1, to the image of u under the composition

$$H^{2i-1}(G(L/K); \mathbf{F}^*_{v^d}) \to H^{2i}(G(L/K); \mathrm{Ker}(\gamma_3)) \to H^{2i+1}(G(L/K); \mathrm{Ker}(\alpha_3)).$$

Applying h'_* we obtain

$$-h'_*(u') = h'_*(u_1) = l_*(\delta(\delta(u)) = [l] \bigcup u.$$

However, in the composition

$$H^{2i-1}(G(L/K); \mathbf{F}^*_{v^d}) \longrightarrow H^{2i}(G(L/K); Q_0)$$
$$\xrightarrow{\cong} H^{2i+1}(G(L/K); \mathrm{Ker}(\alpha_3)) \xrightarrow{h'_*} H^{2i+1}(G(L/K); \mathbf{F}^*_{v^{2d}})$$

the first map injects $\mathrm{Tors}(\mathbf{Z}/r, \mathbf{Z}/HCF(r, v^{i-1}-1))$ into $\mathbf{Z}/r$ and h'_* is divisible by $(v^d - 1)/r$, as will be shown in Lemma 4.2.14. Hence the cup-product with $[l]$ will be divisible by r if r^2 divides $v^d - 1$ which ensures the triviality of the cup-product

$$\mathrm{Tors}(\mathbf{Z}/r, \mathbf{Z}/HCF(r, v^{i-1} - 1)) \xrightarrow{([l]\bigcup -)} \mathrm{Tors}(\mathbf{Z}/r, \mathbf{Z}/HCF(r, v^{i-1} - 1)).$$

Lemma 4.2.14 *(Evaluation of h'_*)*
In the situation of Proposition 4.2.13 h'_* is divisible by $(v^d - 1)/r$.

Proof. We wish to determine (up to units will suffice) the map induced on

$$H^{2i+1}(G(L/K); -) \quad \text{by} \quad h : \mathrm{Ker}(\alpha_3) \longrightarrow \mathbf{F}^*_{v^{2d}}.$$

We begin by describing $\mathrm{Ker}(\alpha_3)$. If

$$X(2) \longrightarrow \mathbf{Z}[G(L/K)]\langle z_1, z_2\rangle \xrightarrow{\delta_2} \mathbf{Z}[G(L/K)]\langle z_0\rangle \xrightarrow{\delta_1} \mathbf{Z}$$

is as in §4.1.7, we can tensor with the torsion-free module Q_0 to obtain the start of a free $\mathbf{Z}[G(L/K)]$-resolution of Q_0

$$X(2) \otimes Q_0 \longrightarrow P'_1 \xrightarrow{\alpha_2} P'_0 \xrightarrow{\alpha_3} Q_0$$

with $\alpha_2 = \delta_2 \otimes 1$ and $\alpha_3 = \delta_1 \otimes 1$. Hence

$$\mathrm{Ker}(\alpha_3) = I[G(L/K)] \otimes Q_0 \cong \mathrm{Ind}_{\langle a\rangle}^{G(L/K)}(I[G(L/K)])$$

where $I[G(L/K)]$ denotes the augmentation ideal. The last isomorphism sends $(g_1 - 1) \otimes g^j \in I[G(L/K)] \otimes Q_0$ to $g^j \otimes g^{-j}(g_1 - 1) \in \mathrm{Ind}_{\langle a\rangle}^{G(L/K)}(I[G(L/K)])$ for $g_1 \in G(L/K)$. Hence $h((g_1 - 1) \otimes g^j) = g^j h(g^{-j}(g_1 - 1) \otimes 1)$ and h is defined completely by the $\mathbf{Z}[\langle a\rangle]$-module homomorphism

$$h : I[G(L/K)] \longrightarrow \mathbf{F}^*_{v^{2d}}$$

given by $(g_1 - 1) \mapsto h((g_1 - 1) \otimes 1)$. One such homomorphism, h, is given by $h((a^i - 1) \otimes 1) = \xi_r^i$ while $h((g^t a^i - 1) \otimes 1) = 1$ for all $t \neq 0$.

Now suppose that

$$\ldots \longrightarrow R_2 \longrightarrow R_1 \longrightarrow R_0 \longrightarrow \mathbf{Z}$$

is a free $\mathbf{Z}[G(L/K)]$-module resolution and consider a cocycle homomorphism

$$f \in \mathrm{Hom}_{\mathbf{Z}[G(L/K)]}(R_{2i+1}, \mathrm{Ker}(\alpha_3)) \cong \mathrm{Hom}_{\mathbf{Z}[\langle a\rangle]}(R_{2i+1}, I[G(L/K)]).$$

If $f(y) = \sum_{i=0}^{d-1} f_i(y) \otimes g^i \in I[G(L/K)] \otimes Q_0$ then $f_i(y) = g(f_{i-1}(g^{-1}y)$ and f is uniquely determined by the $\mathbf{Z}[\langle a\rangle]$-module homomorphism, $R_{2i+1} \longrightarrow I[G(L/K)]$ given by $y \mapsto f_0(y)$. This recipe induces the isomorphism

$$H^{2i+1}(G(L/K); \mathrm{Ker}(\alpha_3)) \cong H^{2i+1}(\langle a\rangle; I[G(L/K)]).$$

Therefore, $h_*([f]) \in H^{2i+1}(\langle a\rangle; \mathbf{F}^*_{v^{2d}})$ is represented by the $\mathbf{Z}[\langle a\rangle]$-module homomorphism, $y \mapsto h(f_0(y) \otimes 1)$. On the other hand

$$h(f(y)) = \sum_{i=0}^{d-1} h(f_i(y) \otimes g^i) \in \mathbf{F}^*_{v^{2d}}.$$

However, $f_i(y) = g^i(f_0(g^{-i}y))$ so that

$$h(f(y)) = \sum_{i=0}^{d-1} h(g^i(f_0(g^{-i}y)) \otimes g^i) = \sum_{i=0}^{d-1} g^i(h(f_0(g^{-i}y)) \otimes 1)$$

which is $\sum_{i=0}^{d-1}(g^i)^*(y \mapsto h(f_0(y) \otimes 1))$. Since g^* acts on $H^1(\langle a\rangle; \mathbf{F}^*_{v^{2d}})$ like multiplication by v we have

$$h(f(y)) = \sum_{i=0}^{d-1} v^i(y \mapsto h(f_0(y) \otimes 1)) = (v^d - 1)/(v-1)(y \mapsto h(f_0(y) \otimes 1)).$$

This shows that the image of a generator under the composition

$$\begin{aligned} H^1(G(L/K); \mathrm{Ker}(\alpha_3)) \cong \mathbf{Z}/r \xrightarrow{h_*} H^1(G(L/K); \mathbf{F}^*_{v^{2d}}) \cong \mathbf{Z}/HCF(r, v-1) \\ \longrightarrow H^1(\langle a\rangle; \mathbf{F}^*_{v^{2d}}) \cong \mathbf{Z}/r \end{aligned}$$

is divisible by $(v^d - 1)/(v-1)$. From the spectral sequence one sees easily that the second map is injective so that in $\mathbf{Z}/HCF(r, v-1)$, for some integer α,

$$h_*(1) = m\frac{(v^d-1)/(v-1)}{r/(HCF(r, v-1)} = \alpha\frac{(v^d-1)}{r}HCF(r, v-1)/(v-1).$$

However, $(v-1)/HCF(r, v-1)$ is a unit modulo r so that $h_*(1)$ is a multiple of $\frac{(v^d-1)}{r}$ as claimed in Proposition 4.2.13.

Finally we want to get from $i = 1$ to the general case.

We observe that the iterated cup-product with the element of Proposition 4.2.2(proof), $z \in H^2(G(L/K); \mathbf{Z})$ which restricts to a generator of $H^2(\langle a\rangle; \mathbf{Z})$, yields an isomorphism $H^1(G(L/K); \mathrm{Ker}(\alpha_3)) \cong H^{2i+1}(G(L/K); \mathrm{Ker}(\alpha_3))$ so that

$$h_* : H^{2i+1}(G(L/K); \mathrm{Ker}(\alpha_3)) \longrightarrow H^{2i+1}(G(L/K); \mathbf{F}^*_{v^{2d}})$$

is divisible by $\frac{(v^d-1)}{r} \cdot \frac{HCF(r,v^{i-1}-1)}{HCF(r,v-1)}$ for all i. The homomorphism

$$\begin{aligned} H^{2i-1}(G(L/K);\mathbf{F}^*_{v^d}) \cong \mathbf{Z}/HCF(r,v^{i-1}-1) \longrightarrow H^{2i}(G(L/K);\mathrm{Ker}(\gamma_3)) \\ \longrightarrow H^{2i+1}(G(L/K);\mathrm{Ker}(\alpha_3)) \cong \mathbf{Z}/r \end{aligned}$$

sends a generator to a multiple of $r/HCF(r,v^{i-1}-1)$ which then maps via h_* to a multiple of

$$r/HCF(r,v^{i-1}-1)\frac{(v^d-1)}{r} \cdot \frac{HCF(r,v^{i-1}-1)}{HCF(r,v-1)} = \frac{(v^d-1)}{HCF(r,v-1)},$$

as required. □

Corollary 4.2.15

In §4.2.12, suppose that $b, c \in \mathbf{Z}/r$. Then, if r^2 divides $v^d - 1$,

$$((b[g_4]+c\lambda_1[h])\bigcup -) = ((b[g_4])\bigcup -) : \hat{H}^i(G(L/K);\mathbf{F}^*_{v^d}) \to \hat{H}^{i+2}(G(L/K);\mathbf{F}^*_{v^{2d}})$$

and is an isomorphism if and only if $b \in (\mathbf{Z}/r)^*$.

4.2.16 *The other Euler characteristics*

In Theorem 4.2.5 I proved that there is an isomorphism of the form

$$\mathrm{Ext}^2_{\mathbf{Z}[G(L/K)]}(\mathbf{F}^*_{v^d}, \mathbf{F}^*_{v^{2d}}) \cong \mathbf{Z}/r \oplus \mathbf{Z}/r.$$

As in §4.2.12, we may assume that the generator of the second summand is given by the canonical fundamental class and the first summand is the image of

$$\begin{aligned} \lambda_1 : H^1(\langle a\rangle;\mathbf{F}^*_{v^{2d}}) \cong \mathrm{Ext}^1_{\mathbf{Z}[G(L/K)]}(\mathbf{Z}[g]/(g^d-1), \mathbf{F}^*_{v^{2d}}) \\ \longrightarrow \mathrm{Ext}^2_{\mathbf{Z}[G(L/K)]}(\mathbf{F}^*_{v^d}, \mathbf{F}^*_{v^{2d}}). \end{aligned}$$

If r^2 divides (v^d-1) Corollary 4.2.15 implies that the cup-product, $(\lambda_1[h]\bigcup -)$, is zero for all h's.

Suppose that $[h] \in H^1(\langle a\rangle;\mathbf{F}^*_{v^{2d}})$ is a generator and that the cup-product by $\lambda_1[h]$ is zero on cohomology. Then, for $b \in (\mathbf{Z}/r)^*$ and $c \in \mathbf{Z}/r$ we would like to calculate the class-group-valued Euler characteristic of Example 2.1.8 associated to

$$b[g_4] + c\lambda_1[h] \in \mathrm{Ext}^2_{\mathbf{Z}[G(L/K)]}(\mathbf{F}^*_{v^d}, \mathbf{F}^*_{v^{2d}}).$$

We begin with $[h] \in H^1(\langle a\rangle;\mathbf{F}^*_{v^{2d}}) \cong \mathbf{Z}/r$, a generator, represented in terms of the projective resolution

$$\ldots \xrightarrow{N} \mathbf{Z}[\langle a\rangle] = \tilde{Q}_1 \xrightarrow{a-1} \mathbf{Z}[\langle a\rangle] = \tilde{Q}_0 \longrightarrow \mathbf{Z}$$

by $h : \tilde{Q}_1 \longrightarrow \mathbf{F}^*_{v^{2d}}$ is given by $h(1) = \xi_r \in \mathbf{F}^*_{v^{2d}}$. However

$$H^1(\langle a\rangle;\mathbf{F}^*_{v^{2d}}) \cong \mathrm{Ext}^1_{\mathbf{Z}[G(L/K)]}(\mathbf{Z}[g]/(g^d-1), \mathbf{F}^*_{v^{2d}})$$

and this Ext-group is computed via the resolution of left modules

$$\ldots \longrightarrow \mathbf{Z}[G(L/K)] = \overline{Q}_1 \stackrel{(-\cdot(a-1))}{\longrightarrow} \mathbf{Z}[G(L/K)] = \overline{Q}_0 \longrightarrow \mathbf{Z}[g]/(g^d - 1)$$

and h corresponds to the left $\mathbf{Z}[G(L/K)]$-module homomorphism, $h : \overline{Q}_1 \longrightarrow \mathbf{F}^*_{v^{2d}}$, given by $h(1) = \xi_r \in \mathbf{F}^*_{v^{2d}}$. Now λ_1 is the coboundary in the long exact sequence obtained by applying $\mathrm{Ext}^i_{\mathbf{Z}[G(L/K)]}(-, \mathbf{F}^*_{v^{2d}})$ to

$$0 \longrightarrow \mathbf{Z}[g]/(g^d - 1) \stackrel{(v-g)}{\longrightarrow} \mathbf{Z}[g]/(g^d - 1) \longrightarrow \mathbf{F}^*_{v^d} \longrightarrow 0.$$

To calculate the coboundary we must first build a commutative diagram of $\mathbf{Z}[G(L/K)]$-resolutions of the following form.

$$\begin{array}{ccccc}
\overline{Q}_1 & \longrightarrow & \overline{Q}_1 \oplus \Lambda_{1,0} \oplus \Lambda_{0,1} & \longrightarrow & \Lambda_{1,0} \oplus \Lambda_{0,1} \\
\downarrow {\scriptstyle (-\cdot(a-1))} & & \downarrow {\scriptstyle D} & & \downarrow \\
\overline{Q}_0 & \longrightarrow & \overline{Q}_0 \oplus \Lambda_{0,0} & \longrightarrow & \Lambda_{0,0} \\
\downarrow & & \downarrow {\scriptstyle E} & & \downarrow \\
\mathbf{Z}[g]/(g^d - 1) & \longrightarrow & \mathbf{Z}[g]/(g^d - 1) & \longrightarrow & \mathbf{F}^*_{v^d}
\end{array}$$

Next we want to define $\mathbf{Z}[G(L/K)]$-module homomorphisms D, E and

$$C : \overline{Q}_2 \oplus (\Lambda_{2,0} \oplus \Lambda_{1,1} \oplus \Lambda_{0,2}) \longrightarrow \overline{Q}_1 \oplus \Lambda_{1,0} \oplus \Lambda_{0,1}$$

so that C, D, E comprise the start of a projective resolution of $\mathbf{Z}[g]/(g^d - 1)$ fitting into the middle of the above commutative diagram. Recall that $Q_2 = 0 = \Lambda_{0,2}$.

Define

$$E : \overline{Q}_0 \oplus (P_0 \otimes Q_0) \longrightarrow \mathbf{Z}[g]/(g^d - 1)$$

by

$$E(1, 0) = v - g, \;\; E(z_0 \otimes g^i) = g^i \;\; (0 \leq i \leq d - 1).$$

This homomorphism makes the diagram commute because

$$g^i(x) = x^{v^i} = \epsilon(z_0 \otimes g^i) \in \mathbf{F}^*_{v^d}$$

where ϵ is defined in §4.1.7.

Next we must define the homomorphism

$$D : \overline{Q}_1 \oplus (P_1 \otimes Q_0) \oplus (P_0 \otimes Q_1) \longrightarrow \overline{Q}_0 \oplus (P_0 \otimes Q_0)$$

and on $\overline{Q}_1 = \mathbf{Z}[G(L/K)]$ we must have

$$D(\alpha, 0, 0) = (\alpha \cdot (a-1), 0) \in \overline{Q}_0 \oplus (P_0 \otimes Q_0).$$

The elements $(0, z_1 \otimes g^i, 0)$, $0, z_2 \otimes g^i, 0)$ and $(0, 0, z_0 \otimes g^i)$ have images in

$$\Lambda_{0,0} = P_0 \otimes Q_0 = \mathbf{Z}[G(L/K)]\langle z_0 \rangle \otimes \mathbf{Z}[g]/(g^d - 1)$$

equal, respectively, to

$$\begin{aligned}
&(a-1)(z_0 \otimes g^i) \in \mathrm{Ker}(E), \\
&g^{-1}(z_0 \otimes g^{i+1}) - z_0 \otimes g^i \in \mathrm{Ker}(E) \quad \text{and} \\
&z_0 \otimes (v - g)g^i.
\end{aligned}$$

The last element does not lie in $\mathrm{Ker}(E)$ but

$$(-g^i, z_0 \otimes (v-g)g^i) \in \mathrm{Ker}(E)$$

so that we define D by the formulae:

$$\begin{aligned}
D(0, z_1 \otimes g^i, 0) &= (0, (a-1)(z_0 \otimes g^i)), \\
D(0, z_2 \otimes g^i, 0) &= (0, g^{-1}(z_0 \otimes g^{i+1}) - z_0 \otimes g^i), \\
D(0, 0, z_0 \otimes g^i) &= (-g^i, z_0 \otimes (v-g)g^i).
\end{aligned}$$

We also need to define C and on $\overline{Q}_2 = \mathbf{Z}[G(L/K)]$ we are obliged to set

$$C(\alpha) = \alpha \cdot (1 + a + \ldots + a^{r-1}) \in \overline{Q}_1.$$

On $\Lambda_{2,0} = P_2 \otimes Q_0$ we may define

$$C : P_2 \otimes Q_0 \xrightarrow{d \otimes 1} P_1 \otimes Q_0$$

since $D = d \otimes 1$ on $P_1 \otimes Q_0$ so that $DC = 0$ on $P_2 \otimes Q_0$. On $\Lambda_{1,1} = P_1 \otimes Q_1$ we have

$$d(z_1 \otimes g^i) = (-z_1 \otimes (v-g)g^i, (a-1)(z_0 \otimes g^i)) \in \Lambda_{1,0} \oplus \Lambda_{0,1}$$

Hence if we set

$$C(z_1 \otimes g^i) = (W, -z_1 \otimes (v-g)g^i, (a-1)(z_0 \otimes g^i)) \in \overline{Q}_1 \oplus \Lambda_{1,0} \oplus \Lambda_{0,1}$$

we must have

$$\begin{aligned}
0 &= D(C(z_1 \otimes g^i)) \\
&= (W(a-1) - (a-1)g^i, -(a-1)z_0 \otimes (v-g)g^i + (a-1)(z_0 \otimes (v-g)g^i)
\end{aligned}$$

so we must have

$$W(a-1) = g^i(a^{w^i} - 1) = g^i(a^{w^i-1} + a^{w^i-2} + \ldots + a + 1)(a-1)$$

where, as in §4.1.1, w is an integer satisfying the congruence $vw \equiv 1$ (modulo $(q^d-1)^2$. Therefore we set

$$\begin{aligned} C(z_1 \otimes g^i) = (&g^i(a^{w^i-1} + a^{w^i-2} + \ldots + a + 1), \\ &-z_1 \otimes (v-g)g^i, (a-1)(z_0 \otimes g^i)) \in \overline{Q}_1 \oplus \Lambda_{1,0} \oplus \Lambda_{0,1}. \end{aligned}$$

On $\Lambda_{1,1} = P_1 \otimes Q_1$ we have

$$d(z_2 \otimes g^i) = (-z_2 \otimes (v-g)g^i, g^{-1}(z_0 \otimes g^{i+1}) - z_0 \otimes g^i) \in \Lambda_{1,0} \oplus \Lambda_{0,1}$$

Hence if we set

$$C(z_2 \otimes g^i) = (V, -z_2 \otimes (v-g)g^i, g^{-1}(z_0 \otimes g^{i+1}) - z_0 \otimes g^i)) \in \overline{Q}_1 \oplus \Lambda_{1,0} \oplus \Lambda_{0,1}$$

we must have

$$\begin{aligned} 0 &= D(C(z_2 \otimes g^i)) \\ &= (V(a-1) - g^{-1}g^{i+1} + g^i, -((g^{-1}-1)z_0) \otimes (v-g)g^i \\ &\qquad + g^{-1}(z_0 \otimes (v-g)g^{i+1}) - z_0 \otimes (v-g)g^i) \\ &= (V(a-1), 0). \end{aligned}$$

Therefore we set

$$C(z_2 \otimes g^i) = (0, -z_2 \otimes (v-g)g^i, g^{-1}(z_0 \otimes g^{i+1}) - z_0 \otimes g^i) \in \overline{Q}_1 \oplus \Lambda_{1,0} \oplus \Lambda_{0,1}.$$

Since $Q_2 = 0$ we do not have to define C on $\Lambda_{0,2} = 0$!

Now to represent $\lambda_1[h]$ where $h : \overline{Q}_1 \longrightarrow \mathbf{F}^*_{v^{2d}}$ is characterised by $h(1) = \xi_r$. First we extend h to $\tilde{h} : \overline{Q}_1 \oplus \Lambda_{1,0} \oplus \Lambda_{0,1} \longrightarrow \mathbf{F}^*_{v^{2d}}$ as $\tilde{h}(1, \alpha, \beta) = \xi_r$ for all $(\alpha, \beta) \in \Lambda_{1,0} \oplus \Lambda_{0,1}$. Then we calculate the composition

$$\tilde{h} \cdot C : \overline{Q}_2 \oplus \Lambda_{2,0} \oplus \Lambda_{1,1} \longrightarrow \overline{Q}_1 \oplus \Lambda_{1,0} \oplus \Lambda_{0,1} \longrightarrow \mathbf{F}^*_{v^{2d}}.$$

For $u \in \mathbf{Z}[G(L/K)]$ this satisfies

$$\tilde{h}(C(u,0,0)) = \tilde{h}(u(1 + a + \ldots + a^{r-1})) = u(\xi_r^r),$$

which is trivial. The composition

$$\tilde{h} \cdot C : P_2 \otimes Q_0 = \Lambda_{2,0} \xrightarrow{d \otimes 1} P_1 \otimes Q_0 = \Lambda_{1,0} \longrightarrow \mathbf{F}^*_{v^{2d}}$$

is also trivial and so is $\tilde{h}(C(0,0,z_2 \otimes g^i))$. However

$$\begin{aligned} \tilde{h}(C(0,0,z_1 \otimes g^i)) &= \tilde{h}(g^i(a^{w^i-1} + a^{w^i-2} + \ldots + a + 1)) \\ &= g^i(\xi_r^{w^i}) \\ &= \xi_r^{v^i w^i} \\ &= \xi_r. \end{aligned}$$

The composition

$$\tilde{h} \cdot C : \overline{Q}_2 \oplus \Lambda_{2,0} \oplus \Lambda_{1,1} \longrightarrow \overline{Q}_1 \oplus \Lambda_{1,0} \oplus \Lambda_{0,1} \longrightarrow \mathbf{F}^*_{v^{2d}}$$

factorises through the quotient map $\overline{Q}_1 \oplus \Lambda_{1,0} \oplus \Lambda_{0,1} \longrightarrow \Lambda_{1,0} \oplus \Lambda_{0,1}$ to give a cocycle $\overline{h} : \Lambda_{1,0} \oplus \Lambda_{0,1} \longrightarrow \mathbf{F}^*_{v^{2d}}$ which represents $\lambda_1[h]$. Explicitly

$$\overline{h}(\Lambda_{2,0}) = 0 = \overline{h}(0.z_2 \otimes g^i)$$

while $\overline{h}(0, z_1 \otimes g^i) = \xi_r$ for $0 \leq i \leq d-1$. Furthermore, being a cocycle, $\overline{h}$ induces

$$\hat{h} : \hat{X}(2) = (\delta \otimes 1)(\Lambda_{2,0}) + (\delta \otimes 1 - 1 \otimes \delta)(\Lambda_{1,1}) \longrightarrow \mathbf{F}^*_{v^{2d}}.$$

On $(\delta \otimes 1)(\Lambda_{2,0}) = X(2) \otimes Q_0$ $\hat{h}$ is trivial. Also $\hat{h}(\delta \otimes 1 - 1 \otimes \delta)(z_2 \otimes g^i) = 0$ while $\hat{h}(\delta \otimes 1 - 1 \otimes \delta)(z_1 \otimes g^i) = \xi_r$ for $0 \leq i \leq d-1$. This means that, if g_4 is as in §4.1.7 then the "sum"

$$\hat{g}_4 = g_4 + c \cdot \hat{h} : \hat{X}(2) \longrightarrow \mathbf{F}^*_{v^{2d}}$$

is surjective because

$$g_4 + c \cdot \hat{h} = g_4 : X(2) \otimes Q_0 \longrightarrow \mathbf{F}^*_{v^{2d}}$$

is surjective.

Since we are assuming that $g_4 + c \cdot h$ induces cup-product isomorphisms in Tate cohomology, the Euler characteristic associated to the class $[g_4] + c \cdot \lambda_1[h]$ is represented by the projective module which is the kernel of $\hat{g}_4$.

Let us remind ourselves how the Euler characteristic of $[g_4]$ was computed. In $\mathrm{Ker}(g_4)$ a free submodule

$$W = \mathbf{Z}[G(L/K)]\langle z_{1,1}, \ldots, z_{1,d}, z_{2,1}, \ldots, z_{2,d}\rangle \subseteq \mathrm{Ker}(g_4) \subseteq \hat{X}(2)$$

was constructed (see Propositions 4.1.11 and 4.1.12) which satisfied

$$\mathrm{Ker}(g_4) = W + (\mathrm{Ker}(g_4) \mid X(2) \otimes Q_0).$$

In this case the Euler characteristic is represented by the class of the module

$$\frac{\mathrm{Ker}(g_4)}{W} \cong \frac{(\mathrm{Ker}(g_4) \mid X(2) \otimes Q_0)}{W \bigcap (\mathrm{Ker}(g_4) \mid X(2) \otimes Q_0)}.$$

Since $\hat{h}(\delta \otimes 1 - 1 \otimes \delta)(z_2 \otimes g^i) = 0$ the formulae of §4.1.10 show that $z_{2,1}, z_{2,2}, \ldots, z_{2,d} \in \mathrm{Ker}(\hat{g}_4)$.

Since the formulae are about to become rather complicated let us simplify matters by considering henceforth only the example in which $c = 1$; that is, $\hat{g}_4 = g_4 + \hat{h}$. In this case we find that $\hat{g}_4(z_{1,i}) = \xi_r = \hat{g}_4(A(y \otimes 1)) = g_4(A(y \otimes 1))$ for some integer $A = a(v^{2d} - 1)$ with $HCF(a, r) = 1$ and $\delta^{rdA} = \xi_r \in \mathbf{F}^*_{v^{2d}}$. Therefore we may define elements, $\overline{z}_{1,i}, \overline{z}_{2,i} \in \mathrm{Ker}(\hat{g}_4)$ by the formulae

$$\overline{z}_{1,i} = z_{1,i} - Ay \otimes 1, \qquad \overline{z}_{2,i} = z_{2,i}$$

for $1 \leq i \leq d$ and $z_{1,i}, z_{2,i}$ as in Proposition 4.1.11.

Lemma 4.2.17 *(c.f. Proposition 4.1.11)*

In the notation of §4.1.16 suppose that $\hat{g}_4 = g_4 + c \cdot \hat{h}$. Then the $\mathbf{Z}[G(L/K)]$-submodule

$$\overline{W} = \mathbf{Z}[G(L/K)]\langle \overline{z}_{1,1}, \ldots, \overline{z}_{1,d}, \overline{z}_{2,1}, \ldots, \overline{z}_{2,d}\rangle \subseteq \mathrm{Ker}(\hat{g}_4) \subseteq \hat{X}(2)$$

is free.

Proof. If $\sum_{\epsilon=1,2,\ 1\leq j\leq d} a_{\epsilon,j}\overline{z}_{\epsilon,j} = 0$ for some $a_{\epsilon,j} \in \mathbf{Z}[G(L/K)]$ then, considering the $P_1 \otimes Q_0$-component and recalling that $y = (1 + a + \ldots + a^{r-1})z_2$,

$$\begin{aligned} 0 &= -\textstyle\sum_{j=1}^d a_{1,j}(z_1 \otimes (v-g)g^j) - \sum_{j=1}^d a_{2,j}(z_2 \otimes (v-g)g^j) \\ &\quad +(-A(\textstyle\sum_{1\leq j\leq d} a_{1,j}) + a_{2,d})((1 + a + \ldots + a^{r-1})z_2 \otimes \alpha_d) \end{aligned}$$

in $P_1 \otimes Q_0$. Since

$$1 \otimes ((v-g)\cdot) : P_1 \otimes \mathbf{Z}[G(\overline{L}/\overline{K})] \longrightarrow P_1 \otimes \mathbf{Z}[G(\overline{L}/\overline{K})]$$

is an injective $\mathbf{Z}[G(L/K)]$-module homomorphism, we see that $a_{1,j} = 0$ for all $1 \leq j \leq d$. The relation therefore reduces to

$$0 = \sum_{1\leq j\leq d} a_{2,j}\overline{z}_{2,j} = \sum_{1\leq j\leq d} a_{2,j}z_{2,j}$$

which implies that each $a_{2,j} = 0$ by the argument of Proposition 4.1.11. □

4.2.18 Lemma 4.2.17 implies that the Euler characteristic of $[g_4] + \lambda_1[h]$ is represented by the class of the module

$$\frac{\mathrm{Ker}(\hat{g}_4)}{\overline{W}} \cong \frac{(\mathrm{Ker}(\hat{g}_4) \mid X(2) \otimes Q_0)}{\overline{W} \bigcap (\mathrm{Ker}(\hat{g}_4) \mid X(2) \otimes Q_0)} = \frac{(\mathrm{Ker}(g_4) \mid X(2) \otimes Q_0)}{\overline{W} \bigcap (\mathrm{Ker}(g_4) \mid X(2) \otimes Q_0)}.$$

Next, imitating §4.1.13, we must determine the submodule $\overline{W} \cap (X(2) \otimes Q_0) \subseteq P_1 \otimes Q_0$. In order that $\sum_{\epsilon=1,2,\ 1\leq j\leq d} a_{\epsilon,j}\overline{z}_{\epsilon,j} \in P_1 \otimes Q_0$ we must have, in $P_0 \otimes Q_1$,

$$\sum_{j=1}^d a_{1,j}g^{-1}(z_0 \otimes g^{j+1}) - a_{1,j}(z_0 \otimes g^j) + a_{2,j}(a-1)(z_0 \otimes g^j) = 0.$$

Since $\{z_0 \otimes g^j \mid 1 \leq j \leq d\}$ is a free $\mathbf{Z}[G(L/K)]$-basis for $P_0 \otimes Q_1$, we must have

$$\begin{aligned} a_{1,d}g^{-1} - a_{1,1} + a_{2,1}(a-1) &= 0, \\ a_{1,1}g^{-1} - a_{1,2} + a_{2,2}(a-1) &= 0, \\ a_{1,2}g^{-1} - a_{1,3} + a_{2,3}(a-1) &= 0, \\ \vdots \qquad \vdots \qquad & \\ a_{1,d-1}g^{-1} - a_{1,d} + a_{2,d}(a-1) &= 0. \end{aligned}$$

Hence

$$
\begin{aligned}
a_{1,1} &= a_{1,d}g^{-1} + a_{2,1}(a-1),\\
a_{1,2} &= a_{1,d}g^{-2} + a_{2,1}(a-1)g^{-1} + a_{2,2}(a-1),\\
a_{1,3} &= a_{1,d}g^{-3} + a_{2,1}(a-1)g^{-2} + a_{2,2}(a-1)g^{-1} + a_{2,3}(a-1),\\
\vdots \quad & \quad \vdots\\
a_{1,s} &= a_{1,d}g^{-s} + \textstyle\sum_{j=1}^{s} a_{2,j}(a-1)g^{j-s}\\
\vdots \quad & \quad \vdots\\
a_{1,d} &= a_{1,d}g^{-d} + \textstyle\sum_{j=1}^{d} a_{2,j}(a-1)g^{j-d}.
\end{aligned}
$$

If these equations hold then, since $(a-1)y = 0$,

$$
\begin{aligned}
&-(\textstyle\sum_{\epsilon=1,2,\ 1\leq j\leq d} a_{\epsilon,j}\overline{z}_{\epsilon,j})\\
&= (a_{1,d}g^{-1} + a_{2,1}(a-1))(z_1 \otimes (v-g)g) + Aa_{1,d}g^{-1}(y\otimes 1)\\
&+(a_{1,d}g^{-2} + a_{2,1}(a-1)g^{-1} + Aa_{1,d}g^{-2}(y\otimes 1) + a_{2,2}(a-1))(z_1\otimes (v-g)g^2)\\
&+(a_{1,d}g^{-3} + a_{2,1}(a-1)g^{-2} + a_{2,2}(a-1)g^{-1} + a_{2,3}(a-1))(z_1\otimes (v-g)g^3)\\
&\qquad +Aa_{1,d}g^{-3}(y\otimes 1)\\
&\quad \vdots \quad \vdots\\
&\quad \vdots \quad \vdots\\
&\quad +(a_{1,d}g^{-s} + \textstyle\sum_{j=1}^{s} a_{2,j}(a-1)g^{j-s})(z_1\otimes (v-g)g^s) + Aa_{1,d}g^{-s}(y\otimes 1)\\
&\qquad \vdots \quad \vdots\\
&\quad +(a_{1,d}g^{-d} + \textstyle\sum_{j=1}^{d} a_{2,j}(a-1)g^{j-d})(z_1\otimes (v-g))\\
&\quad +a_{2,1}(z_2\otimes (v-g)g)\\
&\quad +a_{2,2}(z_2\otimes (v-g)g^2)\\
&\qquad \vdots \quad \vdots\\
&\quad +a_{2,d-1}(z_2\otimes (v-g)g^{d-1})\\
&\quad +a_{2,d}(z_2\otimes (v-g) + y\otimes \alpha_d) + Aa_{1,d}g^{-d}(y\otimes 1).
\end{aligned}
$$

As in §4.1.13, this immense expression will lie in $X(2)\otimes Q_0$.
This expression differs from that of §4.1.13 by the addition of the terms

$$Aa_{1,d}(1+g+g^2+\ldots+g^{d-1})(y\otimes 1)$$

which is an element of $X(2)\otimes Q_0$ mapping under $\hat{g}_4$ to

$$a_{1,d}(\xi_r^{1+v^2+\ldots+v^{2d-2}}) = a_{1,d}(\xi_r^{(v^{2d}-1))/v^2-1)}) \in \mathbf{F}^*_{v^{2d}}.$$

This element will be trivial if r divides $(v^{2d}-1))/v^2-1)$.

A similar argument holds when

$$\hat{g}_4 = g_4 + c \cdot \hat{h} : \hat{X}(2) \longrightarrow \mathbf{F}^*_{v^{2d}}$$

for any c. Hence we have proved the following result.

Theorem 4.2.19

As in §4.2.7, let L/K be a tamely ramified Galois extension of local fields in characteristic $p > 0$. In the notation of §4.2.18, if r divides $(v^{2d} - 1)/(v^2 - 1)$ and r^2 divides $v^d - 1$ then, in the class-group $\mathcal{CL}(\mathbf{Z}[G(L/K)])$, the Euler characteristics of all the elements

$$b \cdot [g_4] + c \cdot \lambda_1[h] \in \mathrm{Ext}^2_{\mathbf{Z}[G(L/K)]}(\mathbf{F}^*_{v^d}, \mathbf{F}^*_{v^{2d}})$$

with $b \in (\mathbf{Z}/r)^*$ are defined and coincide. By Corollary 4.2.15, these are all the elements of

$$\mathrm{Ext}^2_{\mathbf{Z}[G(L/K)]}(\mathbf{F}^*_{v^d}, \mathbf{F}^*_{v^{2d}})$$

which possess an Euler characteristic defined in $\mathcal{CL}(\mathbf{Z}[G(L/K)])$.

Proof. The preceding discussion shows that

$$\overline{W} \bigcap (\mathrm{Ker}(\hat{g}_4) \mid X(2) \otimes Q_0) = W \bigcap (\mathrm{Ker}(\hat{g}_4) \mid X(2) \otimes Q_0)$$

if r divides $(v^{2d} - 1))/v^2 - 1)$. Since $\hat{g}_4$ equal g_4 on $X(2) \otimes Q_0$, this proves the first part for all values of c when $b \equiv 1$ (modulo r). However every element possessing an Euler characteristic has the form $b([g_4] + (c/b) \cdot \lambda_1[h])$ where $b \in (\mathbf{Z}/r)^*$. Since we may find a unit, $b' \in \mathbf{Z}/(v^{2d} - 1)$ which equals b (modulo r) there is an isomorphism of $\mathbf{F}^*_{v^{2d}}$ given by the b'-th power map, whose induces map on $\mathrm{Ext}^2_{\mathbf{Z}[G(L/K)]}(\mathbf{F}^*_{v^d}, \mathbf{F}^*_{v^{2d}})$ sends $[g_4] + (c/b) \cdot \lambda_1[h]$ to $b \cdot [g_4] + c \cdot \lambda_1[h]$). Being induced by an automorphism of $\mathbf{F}^*_{v^{2d}}$, this map does not alter the Euler characteristic. □

4.2.20 Let L/K be any Galois extension of local fields in characteristic $p > 0$, as in §4.2.1. Then, from Theorem 4.2.19, one might expect that all the Euler characteristics, $\chi(z) \in \mathcal{CL}(\mathbf{Z}[G(L/K)])$, associated to elements

$$z \in \mathrm{Ext}^2_{\mathbf{Z}[G(L/K)]}(\mathbf{F}^*_{v^d}, \mathbf{F}^*_{v^{2d}}) \cong \mathbf{Z}/r \oplus \mathbf{Z}/r$$

coincide.

In the tamely ramified case one might attempt to prove this by descent from $L(m)/K$ satisfying the conditions of Theorem 4.2.19. I shall now explain why this does not work – the constructions are worth noting since they will be used later (see §4.3.6 et seq). Notice that if the condition that $r = [G_0(L/K) : G_1(L/K)]$ divides $(v^{2d} - 1)/(v^2 - 1)$ and r^2 divides $v^d - 1$is not satisfied by L/K then it will be satisfied by $L(m)/K$ providing $\mathbf{F}_{v^{dm}}$ is chosen to contain a primitive r^t-th root of unity for $t >> 0$. Fix such an m.

For any $z \in \mathrm{Ext}^2_{\mathbf{Z}[G(L/K)]}(\mathbf{F}^*_{v^d}, \mathbf{F}^*_{v^{2d}})$ suppose that we can construct a commutative diagram of 2-extensions of $\mathbf{Z}[G(L(m)/K)]$-modules of the form

$$\begin{array}{ccccccc}
\mathbf{F}^*_{v^{2d}} & \longrightarrow & A^{G(L(m)/L)} & \longrightarrow & B^{G(L(m)/L)} & \longrightarrow & \mathbf{F}^*_{v^d} \\
\downarrow & & \downarrow & & \downarrow & & \downarrow \\
\mathbf{F}^*_{v^{2dm}} & \longrightarrow & A & \longrightarrow & B & \longrightarrow & \mathbf{F}^*_{v^{dm}}
\end{array}$$

in which all the modules in the lower 2-extension are cohomologically trivial $\mathbf{Z}[G(L(m)/L)]$-modules, the vertical maps are inclusions and the upper 2-extension represents z, when considered as a 2-extension of $\mathbf{Z}[G(L/K)]$-modules. Assume the existence of such a diagram. If W is a $\mathbf{Z}[G(L(m)/K)]$-module which is cohomologically trivial as a $\mathbf{Z}[G(L(m)/L)]$-module then the composition of maps induced by inflation and inclusion

$$\begin{aligned} H^i(G(L/K); W^{G(L(m)/L)}) &\longrightarrow H^i(G(L(m)/K); W^{G(L(m)/L)}) \\ &\longrightarrow H^i(G(L(m)/K); W) \end{aligned}$$

is an isomorphism for all $i \geq 0$, because it is the edge homomorphism of the spectral sequence

$$E_2^{s,t} = H^s(G(L/K); H^t(G(L(m)/L); W)) \Longrightarrow H^{s+t}(G(L(m)/K); W).$$

Since the cup-product with a 2-extension is equal to the composition of the two cohomological coboundary maps obtained by splitting the 2-extension into two short exact sequences, the commutative diagram identifies cup-product with z in the upper row on $G(L/K)$-cohomology with cup-product with the class of the lower 2-extension, z' say, on $G(L(m)/K)$-cohomology. Hence z possesses an Euler characteristic, $\chi(z) \in \mathcal{CL}(\mathbf{Z}[G(L/K)])$, if and only if z' possesses one also. In fact, $\chi(z') = [A] - [B]$ and $\chi(z) = [A^{G(L(m)/L)}] - [B^{G(L(m)/L)}]$, where A and B can be assumed to be cohomologically trivial $\mathbf{Z}[G(L(m)/K)]$-modules. However, the canonical homomorphism from $\mathcal{CL}(\mathbf{Z}[G(L(m)/K)])$ to $\mathcal{CL}(\mathbf{Z}[G(L/K)])$ is induced by sending a cohomologically trivial module to its $G(L(m)/L)$-fixed points. Hence $\chi(z')$ maps to $\chi(z)$, which proves that $\chi(z)$ is independent of z since $\chi(z')$ is independent of z'.

Now let us consider for which z's the above construction will work. To construct the commutative diagram we begin with a projective $\mathbf{Z}[G(L(m)/K)]$-module of $\mathbf{F}^*_{v^{dm}}$

$$\ldots \xrightarrow{d_1} P_1 \xrightarrow{d_0} P_0 \xrightarrow{\epsilon} \mathbf{F}^*_{v^{dm}}.$$

Since all the modules are cohomologically trivial as $\mathbf{Z}[G(L(m)/L)]$-modules, this resolution remains exact upon taking either $G(L(m)/L)$-invariants or coinvariants.

There is a canonical quotient chain map onto the coinvariant complex

$$\ldots \xrightarrow{d_1} (P_1)_{G(L(m)/L)} \xrightarrow{d_0} (P_0)_{G(L(m)/L)} \xrightarrow{\epsilon} \mathbf{F}^*_{v^d}.$$

If $Y = \mathrm{Ker}(d_0)$ then $Y_{G(L(m)/L)} = \mathrm{Ker}(d_0 : (P_1)_{G(L(m)/L)} \longrightarrow (P_0)_{G(L(m)/L)})$. Since the coinvariant sequence is a $\mathbf{Z}[G(L/K)]$-module resolution, the 2-extension, z, is constructed by a push-out via a homomorphism, $f : Y_{G(L(m)/L)} \longrightarrow \mathbf{F}^*_{v^{2d}}$. Since the norm from $\mathbf{F}^*_{v^{2dm}}$ to $\mathbf{F}^*_{v^{2d}}$ is surjective and P_2 is projective, the composition

$$P_2 \longrightarrow (P_2)_{G(L(m)/L)} \longrightarrow Y_{G(L(m)/L)} \xrightarrow{f} \mathbf{F}^*_{v^{2d}}$$

lifts to $\tilde{f} : P_2 \longrightarrow \mathbf{F}^*_{v^{2dm}}$. Unfortunately, $\tilde{f}$ may be chosen to factorise through $\hat{f} : Y \longrightarrow \mathbf{F}^*_{v^{2dm}}$ if and only if z lies in the image of the homomorphism

$$\mathrm{Ext}^2_{\mathbf{Z}[G(L(m)/K)]}(\mathbf{F}^*_{v^{dm}}, \mathbf{F}^*_{v^{2dm}}) \longrightarrow \mathrm{Ext}^2_{\mathbf{Z}[G(L/K)]}(\mathbf{F}^*_{v^d}, \mathbf{F}^*_{v^{2d}})$$

which sends a 2-extension to its $G(L(m)/L)$-coinvariants. If such an $\hat{f}$ exists then pushing out the 2-extension

$$Y \longrightarrow P_1 \longrightarrow P_0 \longrightarrow \mathbf{F}^*_{v^{dm}}$$

by $\hat{f}$ yields the lower 2-extension in the diagram, whose 2-extension of $G(L(m)/L)$-invariants is isomorphic to z, since the norm gives an isomorphism between the associated $G(L(m)/L)$-invariants and $G(L(m)/L)$-coinvariants 2-extensions. The $G(L(m)/L)$-coinvariants homomorphism is equal to the composition

$$\begin{array}{l}\mathrm{Ext}^2_{\mathbf{Z}[G(L(m)/K)]}(\mathbf{F}^*_{v^{dm}}, \mathbf{F}^*_{v^{2dm}}) \xrightarrow{N_*} \mathrm{Ext}^2_{\mathbf{Z}[G(L(m)/K)]}(\mathbf{F}^*_{v^{dm}}, \mathbf{F}^*_{v^{2d}}) \\ \qquad \xrightarrow{(N^*)^{-1}} \mathrm{Ext}^2_{\mathbf{Z}[G(L/K)]}(\mathbf{F}^*_{v^d}, \mathbf{F}^*_{v^{2d}}).\end{array}$$

The fact that N^* is an isomorphism is an easy change-of-rings argument. The homomorphism induced by N_* between the split short exact sequences of §4.2.4

$$0 \longrightarrow \mathbf{Z}/r \xrightarrow{\lambda_1} \mathrm{Ext}^2_{\mathbf{Z}[G(L(u)/K)]}(\mathbf{F}^*_{v^{du}}, \mathbf{F}^*_{v^{2du}}) \xrightarrow{\lambda_2} \mathbf{Z}/r \longrightarrow 0,$$

for $u = m$ and $u = 1$ is an isomorphism on the right-hand $\mathbf{Z}/r$ but zero on the left-hand one! This means that the image of the $G(L(m)/L)$-coinvariants homomorphism consists only of multiples of the canonical fundamental class of Theorem 3.1.22.

The hope of reducing the general case, L/K, to the case of its maximal tamely ramified subextension, M/K, runs into the further difficulty that the inflation homomorphism (see Example 2.1.8(ii))

$$\oplus_{q \text{ prime}, \ q \neq p} K_0(\mathcal{C}_{fin,q,G(M/K)}) \longrightarrow \oplus_{q \text{ prime}, \ q \neq p} K_0(\mathcal{C}_{fin,q,G(L/K)}),$$

defined because $G(L/M)$ is a p-group, does not extend to the class-groups. Since $\mathbf{F}^*_{v^u}$ has order prime to p the Euler characteristics of the 2-extensions of this section can be lifted to lie in $\oplus_{q \text{ prime}, \ q \neq p} K_0(\mathcal{C}_{fin,q,G(L/K)})$ but emulating the calculations of Theorem 4.1.24 in this lifted context seems to be very difficult.

4.3 Connections with motivic complexes

4.3.1 *Background and motivation*

Let X be a regular Noetherian scheme [70]. In what follows we shall not need such generality because we shall only be interested in the case $X = \mathrm{Spec}(L)$, the spectrum of a field. In ([6],[97],[98]) the existence is conjectured of complexes of sheaves on X, $\{\Gamma(n,X)\}_{n\geq 0}$, satisfying a range of axioms. The complex $\Gamma(n,X)$ is known as the *motivic-cohomology complex* of weight n and its hypercohomology is called *motivic cohomology*. In [6] these are sheaves for the Zariski cohomology. However, being interested in Galois coverings, we shall describe only the axioms for the étale site, following ([97],[98]).

(0) $\Gamma(0,X) = \underline{\underline{\mathbf{Z}}}$, $\quad \Gamma(1,X) = \mathbf{G}_m[-1]$.

(1) For $n \geq 1$, $\Gamma(n,X)$ is acyclic outside the interval $[1,n]$.

(2) If α_* assigns to an étale sheaf the associated Zariski sheaf then

$$R^{n+1}\alpha_*\Gamma(n,X) = 0.$$

(3) Let q be a positive integer prime to all residue characteristics of X. Then, in the derived category, there exists a diistinguished triangle of the form

$$\Gamma(n,X) \stackrel{q}{\longrightarrow} \Gamma(n,X) \longrightarrow \mu_q^{\otimes n} \longrightarrow \Gamma(n,X)[1].$$

(4) There are pairings of the form

$$\Gamma(n_1,X) \stackrel{L}{\otimes} \Gamma(n_2,X) \longrightarrow \Gamma(n_1+n_2,X)$$

satisfying the usual properties. Here $\stackrel{L}{\otimes}$ denotes the tensor product in the derived category.

(5) The cohomology sheaves $\mathcal{H}^i(X;\Gamma(n,X))$ are isomorphic to the étale sheaves, $gr^n_\gamma \underline{\underline{K}}^{ét}_{2n-i}$ up to torsion involving primes, $p \leq n-1$. Here gr^n_γ denotes the associated graded of the K-theory presheaf.

(6) The Zariski sheaf $R^n\alpha_*\Gamma(n,X)$ is isomorphic to the sheaf of Milnor K-groups, $\underline{\underline{K}}^M_n(X)$ (c.f. [110]).

In ([97] [98]) a candidate for the motivic-cohomology complex of weight 2 was constructed and many of the axioms verified. In particular, we shall need the fact that $\Gamma(2,X)$, satisfies the axioms (save possibly for questions of 2-torsion) when $X = \mathrm{Spec}(L)$, the spectrum of a field L. In the notation of [97] we write

$$\Gamma(2,X) \ : \quad C_{2,1}(X) \stackrel{\phi_{2,X}}{\longrightarrow} C_{2,2}(X).$$

Consider the case when $X = \mathrm{Spec}(L)$. In this case an étale sheaf on X is equivalent to a Galois module; that is, an abelian group equipped with a continuous action by the absolute Galois group of L, Ω_L. The weight two motivic complex on Spec(L) is constructed in [97] as the direct limit over Galois extensions $L \subset E \subset L^{sep}$ of natural exact sequence of the form

$$\Gamma(2,E) \ : \quad 0 \longrightarrow K_3^{\mathrm{ind}}(E) \longrightarrow C_{2,1}(E) \stackrel{\phi_{2,E}}{\longrightarrow} C_{2,2}(E) \longrightarrow K_2(E) \longrightarrow 0.$$

Here L^{sep} is a fixed separable extension of L. Here $K_3^{\mathrm{ind}}(E)$ denotes the indecomposable K-group in dimension three ([96], [106]). In other words, $\Gamma(2,L)$ denotes a complex of continuous Ω_L-modules whereas $\Gamma(2,L)$ denotes a complex of abelian groups whose construction is given in §4.3.4.

We shall be interested in the complexes at the *finite level*; that is, in the complexes $\Gamma(2,L)$ as modules over the finite Galois groups, $G(L/K)$, rather than the Galois modules over Ω_L. Our interest arises from the fact that, if L/K is a finite Galois extension with Galois group $G(L/K)$, then this sequence is a 2-extension of $\mathbf{Z}[G(L/K)]$-modules representing an element

$$[\Gamma(2,L)] \in \mathrm{Ext}^2_{\mathbf{Z}[G(L/K)]}(K_2(L), K_3^{\mathrm{ind}}(L))$$

similar to that of Theorem 3.1.7 (when $r = 2$). In fact, when L/K is a Galois extension of local fields it is known that $K_3^{\mathrm{ind}}(L) = K_3(L)$ [15] so that the 2-extension $[\Gamma(2,L)] \in \mathrm{Ext}^2_{\mathbf{Z}[G(L/K)]}(K_2(L), K_3(L))$. In Tate cohomology there is an associated family of cup-product homomorphisms of the form

$$([\Gamma(2,L)] \bigcup -) : \hat{H}^i(G(L/K); K_2(L)) \longrightarrow \hat{H}^{i+2}(G(L/K); K_3^{\mathrm{ind}}(L)).$$

In the case of Theorem 3.1.7 the corresponding cup-products are isomorphism in all dimensions.

These similarities suggest the following question.

Question 4.3.2 In 4.3.1, if L/K is a Galois extension of local fields, is

$$([\Gamma(2,L)] \bigcup -) : \hat{H}^i(G(L/K); K_2(L)) \longrightarrow \hat{H}^{i+2}(G(L/K); K_3(L))$$

an isomorphism for all i?

The main result of this section (Theorem 4.3.7; see Theorem 4.3.6 for the tamely ramified case), giving partial evidence for an affirmative answer, is this cup-product yields an isomorphism of the form

$$\hat{H}^i(G(L/K); K_2(L) \otimes \mathbf{Z}[1/2]) \stackrel{\cong}{\longrightarrow} \hat{H}^{i+2}(G(L/K); K_3(L) \otimes \mathbf{Z}[1/2])$$

when L/K is any Galois extension of local fields in characteristic $p > 0$. In fact, we shall prove that this cup-product coincides with the cup-product by the $K_2 - K_3$ local fundamental class.

By other methods Michael Spiess has recently shown that Question 4.3.2 has an affirmative answer for all local fields (see Remark 4.3.8(ii)).

4.3.3 By contrast with the local fundamental classes, the motivic cohomology complexes $\Gamma(n, X)$ are far from unique. They live in the derived category, where they are difficult to identify [10]. Therefore this section should be thought of as a first attempt towards a characterisation by adding some Galois descent properties to axioms (0)–(6). Also the higher K-theory local fundamental classes, by analogy with the global case ([36] [38] [37] [39]), are expected to be connected to special values of L-functions via their Chinburg invariants ([134] [136]) and so are the motivic complexes [6] so it is reasonable to attempt to compare the two.

Here is an outline of the section. In §4.3.4 we recall an explicit construction of the weight two motivic cohomology complex of a field, L. Our construction is equivalent to that of [97] although we have based it on the localisation sequence of Theorem 2.4.6 so as to be able to compute the coboundaries more easily. In §4.3.5 we shall use the results of Theorem 3.1.22, Theorem 4.2.8 and Proposition 4.2.9 in the tamely ramified case to compare the cup-product with the fundamental class for L with that for the separable closure, L_0, doing the same for $\Gamma(2, L)$ and proving that the two cup-products coincide for L_0. This will answer Question 4.3.2 in the tamely ramified case (Theorem 4.3.5). The unique p-divisibility of $K_2(L)$ and $K_3(L)$ for a local field of characteristic p makes it easy to reduce Question 4.3.2 from the general case to the tamely ramified case, which is accomplished in Theorem 4.2.6.

4.3.4 *The construction of* $\Gamma(2, L)$

In [97] (see also [98]) Lichtenbaum constructs the motivic complex, $\Gamma(2, L)$, for a field L by taking a finite set of elements

$$\underline{b} = \{b_1, b_2, \ldots, b_n \in L^* - \{1\}\ \}$$

and taking the direct limit of the $\underline{b}$'s of the long exact algebraic K-theory sequence

$$\ldots \xrightarrow{\phi_{n+1,\underline{b}}} K_n(X, Z) \longrightarrow K_n(X - Y_{\underline{b}}, Z) \xrightarrow{\delta} K_{n-1}(Y_{\underline{b}}) \xrightarrow{\phi_{n,\underline{b}}} \ldots$$

where $X = \mathrm{Spec}(L[T])$, $Z = \mathrm{Spec}(L[T]/(T(T-1)))$ and ($\underline{b} = \{b_1, \ldots, b_n\}$) $Y_{\underline{b}} = \mathrm{Spec}(L[T]/(\prod_{i=1}^n (T - b_i)))$.

One has an isomorphism of the form $K_n(X, Z) \cong K_{n+1}(L)$ and if we set

$$S_{\underline{b}} = \{T - b_1, \ldots, T - b_n \in L[T]\}$$

then $X - Y_{\underline{b}} = \mathrm{Spec}(L[T]S_{\underline{b}}^{-1})$. In [97] it is shown that $\mathrm{im}(\lim_{\underset{\underline{b}}{\rightarrow}} \phi_{4,\underline{b}}) \subseteq K_3(L)$ consists precisely of the decomposable elements (i.e. the iterated cup-products of elements in $K_1(L) \cong L^*$.

In low dimensions there results a functorial exact sequence of the form

$$0 \longrightarrow K_3^{\mathrm{ind}}(L) \longrightarrow C_{2,1}(L) \xrightarrow{\phi_{2,L}} C_{2,2}(L) \longrightarrow K_2(L) \longrightarrow 0.$$

We may describe this sequence more explicitly, replacing the localisation sequence of the triad, $X, Z, Y_{\underline{b}}$, with the localisation sequence of $L[T]S_{\underline{b}}^{-1}$. It is straightforward to verify that the sequence we are about to construct is isomorphic to $\Gamma(2, L)$ of [97].

The long exact localisation sequence of $L[T]S_{\underline{b}}^{-1}$ splits, yielding isomorphisms of the form

$$(\pi_1, \pi_2) : K_n(L[T]S_{\underline{b}}^{-1}) \xrightarrow{\cong} K_n(L) \oplus (\oplus_{b_i \in \underline{b}} K_{n-1}(L))$$

in which π_1 is induced by setting T equal to zero and π_2 is the coboundary of the (split) localisation sequence.

Now consider the long exact K-theory sequence of the pair, $(X - Y_{\underline{b}}, Z)$,

$$\ldots \longrightarrow K_n(L[T]S_{\underline{b}}^{-1}) \xrightarrow{\lambda} K_n(L[T]/(T(T-1))) \xrightarrow{\delta} K_{n-1}(X - Y_{\underline{b}}, Z) \longrightarrow \ldots$$

in which λ is induced by the canonical ring homomorphism. There is an isomorphism of the form

$$(\lambda_1, \lambda_2) : K_n(L[T]/(T(T-1))) \xrightarrow{\cong} K_n(L) \oplus K_n(L)$$

in which λ_1 is induced by setting T equal to zero and λ_2 by setting T equal to one.

Hence λ_2 induces an isomorphism of the form

$$\lambda_2' : \frac{K_n(L[T]/(T(T-1)))}{\mathrm{im}(\lambda)} \xrightarrow{\cong} \frac{K_n(L)}{\lambda(\oplus_{b_i \in \underline{b}} K_{n-1}(L)) \cap K_n(L))}.$$

Therefore δ induces an injection of

$$\frac{K_3(L[T]/(T(T-1)))}{\mathrm{im}(\lambda)} \subseteq K_2(X - Y_{\underline{b}}, Z)$$

which, in the limit over $\underline{b}$, becomes the injection

$$K_3^{\mathrm{ind}}(L) \longrightarrow C_{2,1}(L).$$

Hence the long exact sequence of $(X - Y_{\underline{b}}, Z)$ yields the following exact sequence in low dimensions

$$0 \longrightarrow K_3(Z)/\mathrm{im}(\lambda) \longrightarrow K_2(X - Y_{\underline{b}}, Z) \longrightarrow K_2(L) \oplus (\oplus_{b_i \in \underline{b}} K_1(L))$$
$$\xrightarrow{(1, \phi_{2,\underline{b}}')} K_2(L) \oplus K_2(L) \longrightarrow \ldots.$$

Dividing out by the left-hand summand, $K_2(L)$ in the third and fourth groups and taking the limit over $\underline{b}$ yields the sequence $\Gamma(2, L)$.

The motivic cohomology complex $\Gamma(1, L)$ is equivalent to a short exact sequence of the form

$$0 \longrightarrow C_{1,1}(L) \longrightarrow C_{1,2}(L) \xrightarrow{\phi} K_1(L) \cong L^* \longrightarrow 0$$

of ([97] Proposition 2.4). As in the construction of $\Gamma(2,L)$ in §4.3.4 this sequence is made by taking $\underline{b} = \{b_1, \ldots, b_n \in L^* - \{1\}\}$, considering the localisation sequence in low dimensions and taking a direct limit. This time we have an isomorphism of the form

$$(\pi_1, \pi_2) : K_1(L[T]S_{\underline{b}}^{-1}) \xrightarrow{\cong} K_1(L) \oplus (\oplus_{b_i \in \underline{b}} K_0(L))$$

such that the limit over $\underline{b}$ satisfies

$$C_{1,2}(L) = \lim_{\overrightarrow{\underline{b}}} \frac{K_1(L[T]S_{\underline{b}}^{-1})}{K_1(L)} \cong \mathbf{Z}[\mathbf{P}^1_L - \{0,1,\infty\}].$$

Here a basis element in the projective line minus three points, $[a,b] \in \mathbf{P}^1_L - \{0,1,\infty\}$, corresponds to $T - (a/b) \in K_1(L(T)) \cong L(T)^*$. The homomorphism ϕ corresponds to evaluation at $T = 1$ so that we have

$$\phi(\sum_i n_i[a_i, b_i]) = \prod_i (1 - (a_i/b_i))^{n_i} \in L^*.$$

When $X = \mathrm{Spec}(L)$ axiom (4) of §4.3.1 follows from the pairing of the localisation sequence with $K_*(L)$. This gives rise to an important commutative diagram of the form

$$\begin{array}{ccccccc}
\mathrm{Tor}(L^*, L^*) & \longrightarrow & L^* \otimes C_{1,1}(L) & \longrightarrow & L^* \otimes C_{1,2}(L) & \xrightarrow{1 \otimes \phi} & L^* \otimes L^* \\
\downarrow \rho_L & & \downarrow & & \downarrow \mu_L & & \downarrow \phi_L \\
K_3(L) & \longrightarrow & C_{2,1}(L) & \longrightarrow & C_{2,2}(L) & \longrightarrow & K_2(L)
\end{array}$$

Observe that

$$\begin{aligned}
C_{2,2}(L) &= \lim_{\overrightarrow{\underline{b}}} \frac{K_2(L[T]S_{\underline{b}}^{-1})}{K_2(L)} \\
&= \lim_{\overrightarrow{\underline{b}}} (\oplus_{b_i \in \underline{b}} K_1(L)) \\
&\cong \oplus_{[a,b] \in \mathbf{P}^1_L - \{0,1,\infty\}} L^*.
\end{aligned}$$

This implies that the cup-product, μ_L, in the diagram is an isomorphism.

In §4 it will be convenient to replace μ_L and $1 \otimes \phi$ in the previous diagram by their compositions with the isomorphism $1 \otimes t$ where t is the automorphism of $\mathbf{Z}[\mathbf{P}^1_L - \{0,1,\infty\}]$ induced by $t[a,b] = [b-a, b]$ on the projective line. In this case $(1 \otimes \phi \cdot t)(z \otimes [a,b]) = z \otimes (a/b)$.

4.3.5 *The cup-product in the tame case*

We resume the notation of §4.2.1 so that L/K is a Galois extension of local fields of residue characteristic p and Galois group, $G(L/K)$ with W/K the maximal unramified subextension of L/K. Hence L/W is totally ramified, $\overline{K} = \mathbf{F}_v$, $\overline{W} =$

$\overline{L} = \mathbf{F}_{v^d}$ and $K \cong \mathbf{F}_v((X))$, $W = \mathbf{F}_{v^d}((X)) \subseteq L = \mathbf{F}_{v^d}((Y))$ for some X, Y. For each positive integer, $m \geq 1$, set $L(m) = \mathbf{F}_{v^{dm}}L$ so that $\overline{L(m)} = \mathbf{F}_{v^{dm}}$ and there is an extension of the form

$$G(L(m)/L) \longrightarrow G(L(m)/K) \xrightarrow{\pi_m} G(L/K).$$

Here

$$G(L_m/K) = \langle a, g \mid g^{dm} = 1 = a^r,\ \mathrm{gag}^{-1} = a^v \rangle$$

and $\pi_m(a) = a$, $\pi_m(g) = g$.

The kernel of π_m, $G(L_m/L)$, is isomorphic to $G(\mathbf{F}_{v^{dm}}/\mathbf{F}_{v^d})$ which is cyclic of order m generated by g^d. If L_0 is the maximal unramified extension of L then L_0/K is equal to the limit of the extensions, L_m/K.

The remainder of this section is devoted to showing that the cup-product with $\Gamma(2, L)$ in Question 4.3.2 is an isomorphism when L/K is a tamely ramified extension in characteristic $p > 0$.

For each m dividing m' we have a canonical commutative diagram of 2-extensions

$$\begin{array}{ccccccc}
K_3(L_m) & \longrightarrow & C_{2,1}(L_m) & \longrightarrow & C_{2,2}(L_m) & \longrightarrow & K_2(L_m) \\
\downarrow & & \downarrow & & \downarrow & & \downarrow \\
K_3(L_{m'}) & \longrightarrow & C_{2,1}(L_{m'}) & \longrightarrow & C_{2,2}(L_{m'}) & \longrightarrow & K_2(L_{m'})
\end{array}$$

In addition, since $L_{m'}/L_m$ is unramified, the right-hand vertical map composed with the tame symbol, $\delta_{L_{m'}}$ of Example 3.1.21, is equal to the tame symbol, δ_{L_m}, followed by the inclusion of $\mathbf{F}^*_{v^{dm}}$ into $\mathbf{F}^*_{v^{dm'}}$. Similarly the left-hand vertical map fits into a commutative diagram with the canonical maps $K_3(\mathbf{F}_{v^{dm}}) \longrightarrow K_3(L_m)$ and $K_3(\mathbf{F}_{v^{dm'}}) \longrightarrow K_3(L_{m'})$ of Example 3.1.21.

Therefore, by Theorem 4.2.8 and Proposition 4.2.9, the cup-product (given by the composition of two coboundaries)

$$([\Gamma(2, L)] \bigcup -) : H^i(G(L/K); K_2(L)) \longrightarrow H^{i+2}(G(L/K); K_3(L))$$

may be identified with the direct limit over m of the cup-products by the 2-extensions corresponding to L_m

$$\lim_{\vec{m}} H^i(G(L_m/K); K_2(L_m)) \longrightarrow \lim_{\vec{m}} H^{i+2}(G(L_m/K); K_3(L_m)).$$

This, in turn, may be identified with

$$\lim_{\vec{m}} H^i(G(L_m/K); \mathbf{F}^*_{v^{dm}}) \longrightarrow \lim_{\vec{m}} H^{i+2}(G(L_m/K); K_3(\mathbf{F}_{v^{dm}})).$$

Let $L_0 = \bigcup_m L_m$ denote the maximal unramified extension of L. Hence $K_i(L_0) \cong \lim_{\overrightarrow{m}} K_i(L_m)$. Consider the 2-extension of $\mathbf{Z}[G(L_0/K)]$-modules obtained by taking the limit over m

$$\Gamma(2, L_0): \ \ K_3(L_0) \longrightarrow C_{2,1}(L_0) \longrightarrow C_{2,2}(L_0) \longrightarrow K_2(L_0).$$

The cup-product with the class of this 2-extension

$$H^i(G(L_0/K); K_2(L_0)) \longrightarrow H^{i+2}(G(L_0/K); K_3(L_0))$$

is equal to the direct limit of the cup-products associated to the L_m's. This is the cup-product which we shall show to be an isomorphism for all $i > 0$ (equivalently, for all i in Tate cohomology).

For this we shall need some homomorphisms which relate this 2-extensions to those of Theorem 3.1.22.

For each m in some cofinal set of integers we choose $x_m \in \mathbf{F}^*_{v^{dm}}$, a generator, such that $x_m = x_{m'}^{(v^{dm'}-1)/(v^{dm}-1)}$ whenever $m|m'$. Hence we have a summand, $L^*_m \otimes \mathbf{Z}\{x_m, x_m^v, \ldots, x_m^{v^{dm-1}}\}$, of $L^*_m \otimes C_{1,2}(L_m)$. Furthermore, if $m|m'$, there is a commutative diagram of natural maps of the following form.

$$\begin{array}{ccc}
L^*_m \otimes \mathbf{Z}\{x_m, x_m^v, \ldots, x_m^{v^{dm-1}}\} & \longrightarrow & L^*_{m'} \otimes \mathbf{Z}\{x_{m'}, x_{m'}^v, \ldots, x_{m'}^{v^{dm'-1}}\} \\
\downarrow & & \downarrow \\
L^*_m \otimes C_{1,2}(L_m) & \longrightarrow & L^*_{m'} \otimes C_{1,2}(L_{m'}) \\
\downarrow \cong & & \downarrow \cong \\
C_{2,2}(L_m) & \longrightarrow & C_{2,2}(L_{m'})
\end{array}$$

We may define an isomorphism of $\mathbf{Z}[G(L_m/K)]$-modules of the form

$$\mu_2 : \mathrm{Ind}_{\langle a \rangle}^{G(L_m/K)}(L^*_m/U^1_{L_m}) \xrightarrow{\cong} L^*_m/U^1_{L_m} \otimes \mathbf{Z}\{x_m, x_m^v, \ldots, x_m^{v^{dm-1}}\}$$

by

$$\mu_2(g^i \otimes (\xi, t)) = \xi^{v^i}\pi_L^t \otimes [x_m^{v^i}]$$

for $0 \leq i \leq dm - 1$. Here $(\xi, t) \in \mathbf{F}^*_{v^{dm}} \times \mathbf{Z}$.

This is equivariant because

$$\begin{aligned}
g(\mu_2(g^i \otimes (\xi, m))) &= g(\xi^{v^i}\pi_L^m \otimes [x^{v^i}]) \\
&= \xi^{v^{i+1}}\pi_L^m \otimes [x^{v^{i+1}}] \\
&= \mu_2(g^{i+1} \otimes (\xi, m)).
\end{aligned}$$

Here we have chosen the prime, π_L to lie in $L^{\langle g\rangle}$. This is accomplished in the following manner. The local field, $L^{\langle g\rangle}$, must have the form $\mathbf{F}_v((Z))$ for some prime Z because the residue field is isomorphic to the field of constants in a local field of characteristic $p > 0$. Hence the ramification index of $L^{\langle g\rangle}/K$ is equal to $r = [L^{\langle g\rangle} : K]$. Since the ramification degree of L/K is also r, Z must be a prime in L, too.

Note that μ_2 does not seem to be available if $L_m^*/U_{L_m}^1$ is replaced by L_m^*.

In addition, we have $\mathbf{Z}[G(L_m/K)]$-module homomorphism of the form

$$\mu_1 : \mathrm{Ind}_{\langle a\rangle}^{G(L_m/K)}(L_m^*/U_{L_m}^1) \longrightarrow L_m^*/U_{L_m}^1 \otimes C_{1,1}(L_m)$$

given by

$$\mu_1(g^i \otimes (\xi, t)) = \xi^{v^i}\pi_L^t \otimes (v[x_m^{v^{i-1}}] - [x_m^{v^i}])$$

and if

$$\hat{F} : \mathrm{Ind}_{\langle a\rangle}^{G(L_m/K)}(L_m^*/U_{L_m}^1) \longrightarrow \mathrm{Ind}_{\langle a\rangle}^{G(L_m/K)}(L_m^*/U_{L_m}^1)$$

is given by

$$\hat{F}(g^i \otimes (\xi, t)) = g^{i-1} \otimes (\xi^{v^2}, tv)$$

then $\hat{F}$ is a homomorphism of $\mathbf{Z}[G(L_m/K)]$-modules and we have a commutative diagram of the following form

$$\begin{array}{ccc} \mathrm{Ind}_{\langle a\rangle}^{G(L_m/K)}(L_m^*/U_{L_m}^1) & \xrightarrow{\hat{F}-1} & \mathrm{Ind}_{\langle a\rangle}^{G(L_m/K)}(L_m^*/U_{L_m}^1) \\ \downarrow{\scriptstyle \mu_1} & & \downarrow{\scriptstyle \mu_2} \\ L_m^*/U_{L_m}^1 \otimes C_{1,1}(L_m) & \xrightarrow{1\otimes d} & L_m^*/U_{L_m}^1 \otimes C_{1,2}(L_m) \end{array}$$

because

$$\begin{aligned} \mu_2((\hat{F}-1)(g^i \otimes (\xi, m))) &= \mu_2(g^{i-1} \otimes (\xi^{v^2}, mv) - g^i \otimes (\xi, m)) \\ &= \xi^{v^{i+1}}\pi_L^{mv} \otimes [x^{v^{i-1}}] - \xi^{v^i}\pi_L^m \otimes [x^{v^i}] \\ &= \xi^{v^i}\pi_L^m \otimes (v[x^{v^{i-1}}] - [x^{v^i}]) \\ &= (1 \otimes d)\mu_1(g^i \otimes (\xi, m)). \end{aligned}$$

The composition

$$L_m^* \otimes C_{1,2}(L_m) \stackrel{1\otimes\phi\cdot t}{\longrightarrow} L_m^* \otimes L_m^* \stackrel{\phi_{L_m}}{\longrightarrow} K_2(L_m) \stackrel{\delta_{L_m}}{\longrightarrow} \mathbf{F}_{v^{dm}}^*$$

sends $z\otimes[\alpha, \beta]$ to $(\alpha/\beta)^{v_{L_m}(z)}z^{-v_{L_m}(\alpha/\beta)} \in (\mathcal{O}_{L_m}/(\pi_L))^* \cong \mathbf{F}_{v^{dm}}^*$ and so factorises through a homomorphism of the form

$$L_m^*/U_{L_m}^1 \otimes C_{1,2}(L_m) \longrightarrow \mathbf{F}_{v^{dm}}^*.$$

The composition

$$\pi_m : \mathrm{Ind}_{\langle a\rangle}^{G(L_m/K)}(L_m^*/U_{L_m}^1) \xrightarrow{\mu_2} L_m^*/U_{L_m}^1 \otimes C_{1,2}(L_m) \longrightarrow \mathbf{F}_{v^{dm}}^*$$

sends $g^i \otimes (\xi, s)$ to $\xi^{v^i}\pi_L^s \otimes [x_m^{v^i}]$ and thence to $x_m^{v^i s} = g^i(x_m^s)$, which is surjective.

Taking the direct limit over m

$$\lim_{\vec{m}} \mathrm{Ind}_{\langle a\rangle}^{G(L_m/K)}(L_m^*/U_{L_m}^1) \cong \mathrm{Ind}_{\langle a\rangle}^{G(L_{nr}/K)}(\mu_\infty[1/p]\tilde{\oplus}\mathbf{Z})$$

and $\hat{F}$ commutes with the limit to give

$$\hat{F} - 1 : \mathrm{Ind}_{\langle a\rangle}^{G(L_0/K)}(\mu_\infty[1/p]\tilde{\oplus}\mathbf{Z}) \longrightarrow \mathrm{Ind}_{\langle a\rangle}^{G(L_0/K)}(\mu_\infty[1/p]\tilde{\oplus}\mathbf{Z}).$$

This homomorphism is the limit of the homomorphism in the middle of the canonical 2-extension of Theorem 3.1.22(ii). Write $\overline{\mathbf{F}}_{v^d}$ for the algebraic closure of $\mathbf{F}_{v^d}$. Therefore, in the limit, we obtain a 2-extension of $\mathbf{Z}[G(L_0/K)]$-modules of the form

$$K_3(\overline{\mathbf{F}}_{v^{dm}}) \longrightarrow \mathrm{Ind}_{\langle a\rangle}^{G(L_0/K)}(\mu_\infty[1/p]\tilde{\oplus}\mathbf{Z})$$
$$\xrightarrow{\hat{F}-1} \mathrm{Ind}_{\langle a\rangle}^{G(L_0/K)}(\mu_\infty[1/p]\tilde{\oplus}\mathbf{Z}) \longrightarrow \overline{\mathbf{F}}_{v^{dm}}^*$$

in which the middle two modules are cohomologically trivial. Therefore the cup-product in Tate cohomology

$$\hat{H}^i(G(L_0/K); \overline{\mathbf{F}}_{v^{dm}}^*) \longrightarrow \hat{H}^{i+2}(G(L_0/K); K_3(\overline{\mathbf{F}}_{v^{dm}}))$$

is an isomorphism for all i.

The commutative diagram of 2-extensions

$$\begin{array}{ccccccc}
K_3(\mathbf{F}_{v^d}) & \longrightarrow & \tilde{\mathrm{Ind}}(2) & \xrightarrow{\hat{F}-1} & \tilde{\mathrm{Ind}}(2) & \longrightarrow & \mathbf{F}_{v^d}^* \\
\downarrow & & \downarrow & & \downarrow & & \downarrow \\
K_3(\overline{\mathbf{F}}_{v^d}) & \longrightarrow & \tilde{\mathrm{Ind}}(2)_\infty & \xrightarrow{\hat{F}-1} & \tilde{\mathrm{Ind}}(2)_\infty & \longrightarrow & \overline{\mathbf{F}}_{v^d}^*
\end{array}$$

in which

$$\tilde{\mathrm{Ind}}(2) = \mathrm{Ind}_{\langle a\rangle}^{G(L/K)}(\mathbf{Q}/\mathbf{Z}(2)[1/p]\tilde{\oplus}\mathbf{Z})$$

and

$$\tilde{\mathrm{Ind}}(2)_\infty = \mathrm{Ind}_{\langle a\rangle}^{G(L_0/K)}(\mathbf{Q}/\mathbf{Z}(2)[1/p]\tilde{\oplus}\mathbf{Z})$$

shows that the cup-product isomorphisms

$$\hat{H}^i(G(L/K); \mathbf{F}_{v^d}^*) \longrightarrow \hat{H}^{i+2}(G(L/K); K_3(\mathbf{F}_{v^d}))$$

and

$$\hat{H}^i(G(L_0/K); \overline{\mathbf{F}}^*_{v^d}) \longrightarrow \hat{H}^{i+2}(G(L_0/K); K_3(\overline{\mathbf{F}}_{v^d}))$$

may be identified by means of the natural isomorphisms of Theorem 4.2.8 and Proposition 4.2.9.

Now we must work backwards towards the cup-product induced by the 2-extension, $\Gamma(2, L)$.

We have a commutative diagram of $\mathbf{Z}[G(L_m/K)]$-module homomorphisms

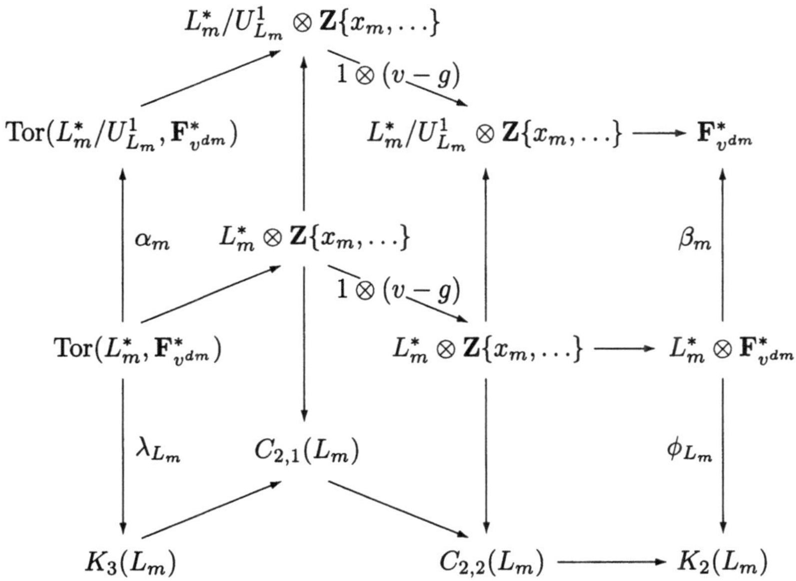

in which α_m is an isomorphism and $\lim_{\substack{\rightarrow \\ m}} \lambda_{L_m} \otimes \mathbf{Z}[1/2]$ and ϕ_{L_m} are isomorphisms modulo $\mathbf{Q}$-vector spaces. The lower part of the diagram commutes by virtue of Axiom 4 of §4.3.1 and is derived in ([97] Remark 2.6). In the upper part the middle and left-hand squares clearly commute, since the upward verticals are induced by the quotient map, $L_m^* \to L_m^*/U_{L_m}^1$. The composition

$$L_m^* \otimes \mathbf{Z}\{x_m, \ldots\} \to L_m^* \otimes \mathbf{F}^*_{v^{dm}} \stackrel{\phi_{L_m}}{\to} K_2(L_m) \stackrel{\delta_{L_m}}{\to} \mathbf{F}^*_{v^{dm}},$$

where δ_{L_m} is the tame symbol of Example 3.1.21, annihilates $U_{L_m}^1 \otimes \mathbf{Z}\{x_m, \ldots\}$ and so induces β_m to make the right upper square commute. Note that the top sequence is not exact but, by the discussion of §4.3.5, after taking the direct limit over m it becomes exact and may be identified with the limit over m of the 2-extension of Theorem 3.1.22.

Now consider the following commutative diagram in which the horizontal homomorphisms are the cup-products by the above 2-extensions taken to the limit over m.

$$\begin{array}{ccc}
\hat{H}^i(G(L_0/K); \overline{\mathbf{F}}^*_{v^{dm}}) & \xrightarrow{\cong} & \hat{H}^{i+2}(G(L_0/K); \mathrm{Tor}(L_m^*/U^1_{L_m}, \mathbf{F}^*_{v^{dm}})) \\
\Big\uparrow \lim_{\overrightarrow{m}}(\beta_m)_* & & \cong \Big\uparrow \lim_{\overrightarrow{m}}(\alpha_m)_* \\
\hat{H}^i(G(L_0/K); \lim_{\overrightarrow{m}} L_m^* \otimes \mathbf{F}^*_{v^{dm}}) & \xrightarrow{\cong} & \hat{H}^{i+2}(G(L_0/K); \mathrm{Tor}(L_m^*, \mathbf{F}^*_{v^{dm}})) \\
\Big\downarrow \lim_{\overrightarrow{m}}(\phi_{L_m})_* & & \Big\downarrow \lim_{\overrightarrow{m}}(\lambda_{L_m})_* \\
\hat{H}^i(G(L_0/K); K_2(L_0)) & \xrightarrow{\theta} & \hat{H}^{i+2}(G(L_0/K); K_3(L_0))
\end{array}$$

The upper two cup-products are isomorphisms on Tate cohomology for all i so that $\lim_{\overrightarrow{m}}(\beta_m)_*$ is also an isomorphism. On the other hand, $\lim_{\overrightarrow{m}}(\lambda_{L_m})_* \otimes \mathbf{Z}[1/2]$ is an isomorphism. Also the composition of ϕ_{L_m} with the tame symbol, δ_{L_m}, is equal to β_m. Since δ_{L_m} induces an isomorphism on cohomology so does $\lim_{\overrightarrow{m}}(\phi_{L_m})_*$. This implies that $\theta \otimes \mathbf{Z}[1/2]$ is an isomorphism for all i and, by Theorem 4.2.8 and Proposition 4.2.9, the cup-product

$$\hat{H}^i(G(L/K); K_2(L) \otimes \mathbf{Z}[1/2]) \stackrel{([\Gamma(2,L)] \cup -)}{\longrightarrow} \hat{H}^{i+2}(G(L/K); K_3(L) \otimes \mathbf{Z}[1/2])$$

is also an isomorphism for all i.

In fact, this discussion establishes the following result:

Theorem 4.3.6

Let L/K be a tamely ramified extension of local fields in characteristic p, as in §4.2.1. Then, under the canonical isomorphisms of §4.2.1, the cup-product of Question 4.3.2 is an isomorphism for all i

$$\hat{H}^i(G(L/K); K_2(L) \otimes \mathbf{Z}[1/2]) \stackrel{([\Gamma(2,L)] \cup -)}{\longrightarrow} \hat{H}^{i+2}(G(L/K); K_3(L) \otimes \mathbf{Z}[1/2])$$

which coincides with the isomorphism given by the cup-product with the K_2/K_3 local fundamental class for L/K of Theorem 3.1.22

$$\hat{H}^i(G(L/K); \mathbf{F}^*_{v^d} \otimes \mathbf{Z}[1/2])) \longrightarrow \hat{H}^{i+2}(G(L/K); \mathbf{F}^*_{v^{2d}} \otimes \mathbf{Z}[1/2]).$$

Theorem 4.3.7

Let L/K be any Galois extension of local fields in characteristic p. Then the cup-product of Question 4.3.2 is an isomorphism for all i

$$\hat{H}^i(G(L/K); K_2(L) \otimes \mathbf{Z}[1/2]) \stackrel{([\Gamma(2,L)] \cup -)}{\longrightarrow} \hat{H}^{i+2}(G(L/K); K_3(L) \otimes \mathbf{Z}[1/2])$$

which may be identified with the isomorphism given by the cup-product with the K_2/K_3 local fundamental class for L/K of Theorem 3.1.22

$$\hat{H}^i(G(L/K); \mathbf{F}^*_{v^d} \otimes \mathbf{Z}[1/2])) \longrightarrow \hat{H}^{i+2}(G(L/K); \mathbf{F}^*_{v^{2d}} \otimes \mathbf{Z}[1/2]).$$

Proof. Let L/K be any Galois extension of local fields in characteristic $p > 0$ with Galois group, $G(L/K)$. Let $G_1(L/K) \subset G(L/K)$ denote the first wild ramification group ([128] p. 62), which is a finite p-group. If M is the fixed field of $G_1(L/K) = G(L/M)$ then M/K is the maximal tamely ramified subextension. We shall apply Theorem 4.3.6 to M/K in order to prove the result for L/K.

Consider the following commutative diagram in which the rows are 2-extensions of $\mathbf{Z}[G(L/K)$-modules.

$$\begin{array}{ccccccc}
K_3(M) & \longrightarrow & C_{2,1}(M) & \longrightarrow & C_{2,2}(M) & \longrightarrow & K_2(M) \\
\downarrow & & \downarrow & & \downarrow & & \downarrow \\
K_3(L) & \longrightarrow & C_{2,1}(L) & \longrightarrow & C_{2,2}(L) & \longrightarrow & K_2(L) \\
\downarrow & & \downarrow & & \downarrow & & \downarrow \\
K_3(L) & \longrightarrow & C_{2,1}(L)[1/p] & \longrightarrow & C_{2,2}(L)[1/p] & \longrightarrow & K_2(L)
\end{array}$$

The upper two rows are given by $\Gamma(2, M)$ and $\Gamma(2, L)$. The lower row is the 2-extension obtained from $\Gamma(2, L)$ by localising to invert p. Since $K_2(L)$ and $K_3(L)$ are uniquely p-divisible [60] localisation leaves these modules unchanged. These are all 2-extensions of $\mathbf{Z}[G(L/K)]$-modules and the maps are $\mathbf{Z}[G(L/K)]$-module homomorphisms. However the action on the top row factorises through the canonical quotient, $G(L/K) \longrightarrow G(L/K)/G(L/M) \cong G(M/K)$. Therefore we obtain a commutative diagram of the following form, in which A^G denotes the subgroup of G-invariants elements of A. It is a diagram of 2-extensions because taking $G(L/M)$-fixed points is exact on p-divisible $\mathbf{Z}[G(L/K)]$-modules.

$$\begin{array}{ccccccc}
K_3(M) & \longrightarrow & C_{2,1}(M) & \longrightarrow & C_{2,2}(M) & \longrightarrow & K_2(M) \\
\downarrow & & \downarrow & & \downarrow & & \downarrow \\
K_3(L)^{G(L/M)} & \longrightarrow & C_{2,1}(L)[1/p]^{G(L/M)} & \longrightarrow & C_{2,2}(L)[1/p]^{G(L/M)} & \longrightarrow & K_2(L)^{G(L/M)}
\end{array}$$

Unique p-divisibility of $K_2(M), K_2(L), K_3(M), K_3(L)$ implies, by a transfer argument, that the left-hand and right-hand vertical maps are isomorphisms. Hence, by Theorem 4.3.6, the cup-product

$$\hat{H}^i(G(M/K); K_2(L)^{G(L/M)} \otimes \mathbf{Z}[1/2]) \to \hat{H}^{i+2}(G(M/K); K_3(L)^{G(L/M)} \otimes \mathbf{Z}[1/2])$$

is an isomorphism for all i.

For $j = 2, 3$ the Serre spectral sequence shows that $H^s(G(L/K); K_j(L)) \cong H^s(G(M/K); K_j(L)^{G(L/M)})$, since $H^t(G(L/M); K_j(L))) = 0$ for $t > 0$. This isomorphism is given by the following canonical composition

$$\begin{aligned} H^s(G(M/K); K_j(L)^{G(L/M)}) &\longrightarrow H^s(G(L/K); K_j(L)^{G(L/M)}) \\ &\longrightarrow H^s(G(L/K); K_j(L)). \end{aligned}$$

This shows that, in the three-row diagram, the cup-product on $\hat{H}^*(G(M/K); -)$ by the 2-extension on the top row may be identified with the cup-product in $\hat{H}^*(G(L/K); -)$ by the 2-extension on the bottom row and hence also on the middle row, which establishes the first part of the result.

To compare the cup-products one uses the canonical isomorphisms, for $j = 2, 3$ and $s > 0$,

$$\begin{aligned} H^s(G(L/K), K_j(\overline{L})) &\cong H^s(G(L/K), K_j(L)) \\ &\cong H^s(G(M/K), K_j(M)) \\ &\cong H^s(G(M/K), K_j(\overline{M})) \end{aligned}$$

together with the second part of Theorem 4.3.6 and the fact that the residue fields of L and M are equal. □

Remark 4.3.8 (i) There is a more sophisticated manner in which to approach Theorem 4.3.6. For any regular noetherian ring Lichtenbaum constructs [97] a natural complex of groups

$$0 \longrightarrow C_{2,1}(A) \stackrel{\phi_{2,A}}{\longrightarrow} C_{2,2}(A) \longrightarrow 0$$

generalizing the case when A is a field (see §4.3.4). One may sheafify this complex to obtain a complex of sheaves on any Grothendieck site. On the Zariski site one obtains $X \mapsto \Gamma_{zar}(2, X)$ and on the étale site one obtains $X \mapsto \Gamma(2, X)$, as in §4.3.1. One may obtain a complex of sheaves on the Zariski site by $X \mapsto \tau_{\leq 2} R\alpha_* \Gamma(2, X)$. Here α_* is the functor which associates with every sheaf on the étale site is restriction to the Zariski site. Then $\tau_{\leq 2} R\alpha_* \Gamma(2, X)$ takes an injective resolution of $\alpha_* \Gamma(2, X)$ and truncates the resulting complex of Zariski sheaves, replacing each sheaf in dimension strictly greater than two by zero ([98] pp. 37–38). In ([98] Proposition 3.1) it is shown that the Zariski complexes, $X \mapsto \Gamma_{zar}(2, X)$ and $X \mapsto \tau_{\leq 2} R\alpha_* \Gamma(2, X)$, are isomorphic up to 2-primary torsion. When $X = \mathrm{Spec}(L)$ is the spectrum of a field the sheafification in the Zariski topology just yields the complex $\Gamma(2, L)$ so that one may prove Theorems 4.3.6 and 4.3.7 by showing that $\tau_{\leq 2} R\alpha_* \Gamma(2, X)$ is quasi-isomorphic to the two-term complex given by the middle of the 2-extension of Theorem 3.1.22 when $X = \mathrm{Spec}(L(m))$. Since both complexes satisfy Galois descent in the sense of Theorem 3.1.20 this comparison can, in principle, be made "at the top". The only problem is that the complex coming from Theorem 3.1.22 depends upon a choice of a root of unity, making it canonical up to a canonical isomorphism (see Proposition 3.1.19). This means that, while $\tau_{\leq 2} R\alpha_* \Gamma(2, X)$ satisfies Galois descent for subfields of a separable closure of $L(m)$ and is easily describeable "at the top" (one just sets $X = \mathrm{Spec}(L^{\mathrm{sep}})$), fitting together the complexes coming from Theorem 3.1.22 to give the complex for $\mathrm{Spec}(L^{\mathrm{sep}})$ is rather delicate. If this could be done, the comparison of complexes would presumably not be too difficult. This would prove that the Euler characteristics of these complexes are equal in $\mathcal{CL}(\mathbf{Z}[1/2][G(L/K)])$ – a result which follows from Theorem 4.2.19 in the majority of cases.

It should not be too hard to make the comparison described above and to fill in the details which I have sketched, but for our purposes here it would require the development of too much technical machinery.

Finally, in view of the connections of the local fundamental classes arising from the study of the Galois module structure of the K-groups of one-dimensional local fields and the motivic complex, $\Gamma(2, L)$, it would be very interesting to be more ambitious and attempt to study the same questions for higher dimensional local fields in terms of the motivic complexes of ([161]; see also [56]).

(ii) In Theorem 4.3.7 we gave an affirmative answer to Question 4.3.2, after inverting 2, by showing that in characteristic $p > 0$ the cup-product with $\Gamma(2, L)$ coincides with the cup-product by the $K_2 - K_3$ local fundamental class. Recently Michael Spiess showed me another way to show that the answer to Question 4.3.2 is affirmative for any local field, without inverting 2. His method is very elegant [147] and uses results from ([81], [92], [146]).

Chapter 5

Higher K-theory of Algebraic Integers

In this chapter we shall examine invariants of the Galois module structure on the higher-dimensional algebraic K-groups of rings of algebraic integers in number fields. Special cases of these invariants were given in §§1.3.4–1.3.13. The material of this chapter originated in ([132] Chapter VII) where generalizations of the Chinburg invariant, $\Omega_0(L/K, 3)$ of [34] were constructed using algebraic K-groups of integers in dimensions two and three. Independently, Pappas (unpublished) had noticed the same construction. In [132] it was shown, in the totally real case, that the image of this new invariant in the kernel group, $D(\mathbf{Z}[G(L/K)])$ of §2.5.26, is represented in the Hom-description of Theorem 2.5.32 by the value of the Artin L-function at $s = -1$. The examples described in §§1.3.4–1.3.13 provide further evidence to connect the invariant, $\Omega_r(L/K, 3)$ constructed from K-groups in dimesions $2r$ and $2r+1$, with behaviour of the Artin L-series at (or near) $s = -r$.

To be honest, above dimension four the constructions of this chapter are made using étale cohomology sequences which are conjectured by Lichtenbaum-Quillen to be close approximations to K-theory. At the moment only the 2-adic part of the Lichtenbaum-Quillen conjecture is known as a consequence of [160] (see [123]). The construction is really a cohomological one, using a modification of the totally positive Galois cohomology of ([108] Chapter 2 §2; see also [80] p. 76, [69]), which can also be applied to cohomological motives with Galois actions (see [24] and [25]).

5.1 Positive étale cohomology

Let S be a finite set of places of a number field N which contains the set S_∞ of infinite places of N. Denote the ring of S-integers of N by $O_{N,S}$. In this section we fix a rational prime l, and we let $X_l = \mathrm{Spec}(O_{N,S}[1/l])$.

Let A be a subring of finite index in the integral closure of $\mathbf{Z}_l$ in a finite extension of $\mathbf{Q}_l$ (we shall use only the case $A = \mathbf{Z}_l$, but for other applications it may be useful to allow more general A). Let $\mathcal{M}$ be the maximal ideal of A. For $m \geq 1$

let Λ_m be the ring $A/\mathcal{M}^m$. If F is an étale sheaf on X_l, set $F_\infty = \sum_{v\in S_\infty} i_{v*}i_v^*F$ where $i_v : \mathrm{Spec}(N_v) \to X$ is the natural morphism and N_v is the completion of N at v. Let $H_l^i(F) = H^i(X_l; F)$.

Lemma 5.1.1

Suppose that F is a twisted constant constructible sheaf of Λ_m-modules on X_l. The morphism of sheaves $F \to F_\infty$ on X_l is injective. Let F_+ be the cokernel of this morphism, so that there is an exact sequence

$$0 \to F \to F_\infty \to F_+ \to 0 \tag{1}$$

of sheaves of Λ_m-modules. Then $H_l^i(F_+)$ is finite for all i, and $H_l^i(F_+) = 0$ if $i \notin \{0,1\}$. We will call $H_l^i(F_+)$ the ith positive étale cohomology group of F. If F is a flat sheaf of Λ_m-modules, then so is F_+.

Proof. We first compute the stalks of F, F_∞ and F_+ at each geometric point $u_x : \overline{x} \to X_l$ of X_l. Let M be the $\mathrm{Gal}(N^s/N)$-module defined by the restriction of F to the generic point of N. Then M is finite, since F is constructible (cf. [107] p. 156, Ex. V.1.3). Since F is twisted constant, there exist commutative diagrams

$$\begin{array}{ccc} \overline{x} & \longrightarrow & U \\ & \searrow^{u_x} & \downarrow \\ & & X \end{array} \tag{2}$$

in which $U \to X$ is étale, U is connected and one-dimensional, and $F|U$ is constant. Then $F(U)$ is the abelian group M, while

$$\begin{aligned} F_\infty(U) &= \textstyle\bigoplus_{v\in S_\infty} i_{v*}i_v^*F(U) \\ &= \textstyle\bigoplus_{v\in S_\infty} (i_v^*F)(U \times_{X_l} \mathrm{Spec}(N_v)) \\ &= \textstyle\bigoplus_{v\in S_\infty} \bigoplus_{w|v} M \end{aligned}$$

where the last inner sum on the right is over the places w over v in the function field T of U. Taking the limit over diagrams of (2) now shows the following. Let N^s be a separable closure of N. Let $v_{\overline{x}}$ be an extension to N^s of the discrete valuation defined by x on N, and let $I_{\overline{x}} \subset \mathrm{Gal}(N^s/N)$ be the inertia group of $v_{\overline{x}}$. We can take the fields T above to run through a cofinal set of subfields of $(N^s)^{I_{\overline{x}}}$. The natural morphism of sheaves $F \to F_\infty$ then gives on stalks the diagonal homomorphism

$$(F)_{\overline{x}} = M \to (F_\infty)_{\overline{x}} = \bigoplus_{v\in S_\infty} \bigoplus_{w'|v} M \tag{3}$$

where the sum on the right is over the extensions w' of v to an absolute value on $(N^s)^{I_{\overline{x}}}$. We see from this that (3) is injective with cokernel $(F_+)_{\overline{x}}$. If F is a flat sheaf of Λ_m-modules, then M is a flat Λ_m-module, and (3) shows $(F_+)_{\overline{x}}$ is as well. By ([107] p. 80), an étale sheaf of Λ_m-modules is flat if and only if its stalks are flat, so F_+ is a flat sheaf if F is.

The long exact cohomology sequence of

$$0 \to F \to F_\infty \to F_+ \to 0$$

is

$$\ldots \to H^i_l(F) \stackrel{b(i)}{\to} H^i_l(F_\infty) \stackrel{c(i)}{\to} H^i_l(F_+) \to H^{i+1}_l(F) \to \ldots. \qquad (4)$$

We shall use results of [169] and [108] to analyse (4).

Let G_{S_l} be the Galois group over N of the maximal extension N_{S_l} of N which is unramified outside S and the primes above l. The action of $\mathrm{Gal}(N^s/N)$ on M factors through G_{S_l}. By ([169] Proposition 3.3.1; see also [108] Proposition II.2.9), restriction from X_l to the generic point of X_l induces isomorphisms $H^j_l(F) = H^j(X_l; F) = H^j(G_{S_l}; M)$ for all j. Consider now

$$H^j_l(F_\infty) = \bigoplus_{v \in S_\infty} H^j_l(i_{v*}i_v^* F).$$

For $v \in S_\infty$ let $G_v \subset G_{S_l}$ be the decomposition group of a (complex) place of N_{S_l} over v. Since i_{v*} is exact ([107] Proposition II.3.14(a)),

$$H^j_l(i_{v*}i_v^* F) = H^j(\mathrm{Spec}(N_v); F) = H^j(G_v; M)$$

for all j. In this way we find that restriction from X_l to the generic point of X_l induces a commutative diagram

$$\begin{array}{ccc} H^i(G_{S_l}; M) & \to & \bigoplus_{v \in S_\infty} H^i(G_v; M) \\ \downarrow & & \downarrow \\ H^i_l(F) & \to & H^i_l(F_\infty) \end{array} \qquad (5)$$

in which the vertical arrows are isomorphisms and in which the top row results from restriction from G_{S_l} to the G_v. Since we noted earlier that M is finite, (4) and (5) imply that $H^i_l(F_+)$ is finite for all i.

By ([169] Corollary to Theorem 2.6), the top row of (5) is an isomorphism if $i \geq 3$. Hence (5) shows that $b(i)$ in (4) is an isomorphism for $i \geq 3$. Since (2.4) is exact, this shows $H^i_l(F_+) = 0$ if $i \geq 3$ and that there is an exact sequence

$$H^2(G_{S_l}; M) \stackrel{\tau}{\to} \bigoplus_{v \in S_\infty} H^2(G_v; M) \to H^2_l(F_+) \to 0. \qquad (6)$$

By ([108] Corollary I.4.16), the map τ in (6) is surjective, so $H^2_l(F_+) = 0$, which completes the proof. □

Remark 5.1.2 In view of (5), the long exact cohomology sequence (4) can be written as

$$\ldots \to H^i_l(F) \stackrel{b(i)}{\to} \bigoplus_{v \in S_\infty} H^i(G_v; M) \stackrel{c(i)}{\to} H^i_l(F_+) \to H^{i+1}_l(F) \to \ldots. \qquad (7)$$

By comparison, the compact cohomology groups $H_c^j(X;F)$ and $H_c^j(X_l;F)$ of ([108] Proposition II.2.3) lie in exact sequences

$$\ldots \to H^i(X;F) \to \bigoplus_{v\in S_\infty} H^i(G_v;M) \to H_c^{i+1}(X;F) \to H^{i+1}(X;F) \to \ldots \quad (8)$$

and

$$\ldots \to H^i(X_l;F) \to \bigoplus_{v\in S_l} H^i(N_v;M) \to H_c^{i+1}(X_l;F) \to H^{i+1}(X_l;F) \to \ldots \quad (9)$$

where $H^i(X_l;F)=H_l^i(F)$, $H^i(N_v;M) = H^i(G_v;M)$ if v is infinite, and $H^i(N_v;M)$ is local Galois cohomology for finite v. The positive Galois cohomology groups $H_+^i(N;M)$ defined in ([108] Chapter 2, §2; see also [80] p. 76) lie in an exact sequence

$$\ldots \to H^i(G_N;M) \to \bigoplus_{v\in S_\infty} H^i(G_v;M) \to H_+^{i+1}(N;M) \to H^{i+1}(G_N;M) \to \ldots \quad (10)$$

where $G_N = \mathrm{Gal}(N^{\mathrm{sep}}/N)$ for a fixed separable closure N^{sep} of N.

Since $H^i(G_{S_l};M) = H_l^i(F)$, we see from (7) and (10) that $H^i(F_+)$ results from replacing G_N in the definition of $H_+^{i+1}(N;M)$ by G_{S_l}.

We now make the following hypothesis:

Hypothesis 5.1.3 The ring A is a discrete valuation ring and N is the top field of a finite Galois extension N/K of number fields with group $G = \mathrm{Gal}(N/K)$. We suppose the finite set S of places of N is stable under the action of G and contains all of the places of N which ramify over K. We also suppose that the sheaf F on X_l is the restriction of a flat constructible locally constant sheaf of Λ_m-modules on $Y_l = \mathrm{Spec}(K \cap O_{N,S}[1/l])$, which we will also denote by F.

Proposition 5.1.4

Under Hypothesis 5.1.3, $X_l = \mathrm{Spec}(O_{N,S}[1/l])$ is an étale G-cover of $Y_l = \mathrm{Spec}(K\cap O_{N,S}[1/l])$. Define $\mathcal{R}_m = \Lambda_m[G]$ and let $H_l^\bullet(F) = H^\bullet(X_l;F)$ be the étale hypercohomology of F in the derived category, $\mathcal{D}_m$, of the homotopy category of complexes of $\mathcal{R}_m$-modules which are bounded below. Then there is a complex

$$Q^\bullet : Q_{-1} \to Q_0 \to Q_1 \quad (11)$$

of finitely generated projective $\mathcal{R}_m$-modules which is isomorphic to $H_l^\bullet(F_+)$ in $\mathcal{D}_m$, where Q_i is the degree i term of $Q^\bullet$.

Proof. Let

$$I^\bullet : I_0 \to I_1 \to I_2 \to I_3 \to \ldots \quad (12)$$

be a complex of injective sheaves of Λ_m-modules on Y_l which is quasi-isomorphic to F_+. Then

$$\Gamma(X_l, I^\bullet) : \Gamma(X_l, I_0) \to \Gamma(X_l, I_1) \to \Gamma(X_l, I_2) \to \dots \tag{13}$$

is a complex of Λ_m-modules and the action of G on X_l makes this a complex of $\mathcal{R}_m$-modules which is isomorphic to $H_l^\bullet(F_+)$ in $\mathcal{D}_m$.

Let $j : \mathrm{Spec}(K) \to Y$ be the natural morphism and fix a separable closure K^{sep} of K containing N. For each $\mathcal{R}_m$-module M, let M^s be the inflation of M from G to $\mathrm{Gal}(K^{\mathrm{sep}}/K)$, so that M^s is a $\Lambda_m[\mathrm{Gal}(K^{\mathrm{sep}}/K)]$-module. Viewing M^s as a sheaf on the étale topology of $\mathrm{Spec}(K)$, we have an étale sheaf $j_*(M^s)$ on Y_l. Since X_l is an étale G-cover of Y_l, the functor which sends M to $j_*(M^s)$ is an exact fully faithful functor from the category of $\mathcal{R}_m$-modules to that of étale sheaves of Λ_m-modules on Y. Furthermore, if J is any étale sheaf of Λ_m-modules on Y_l, $\Gamma(X_l, J)$ is an $\mathcal{R}_m$-module and $\mathrm{Hom}_{Y_l}(j_*(M^s), J) = \mathrm{Hom}_{\mathcal{R}_m}(M, \Gamma(X_l, J))$. It follows that if J is injective, then $\Gamma(X_l, J)$ is an injective $\mathcal{R}_m$-module. In particular, the terms of (13) are injective $\mathcal{R}_m$-modules.

Define $M_1 = \ker(\Gamma(X_l, I_1) \to \Gamma(X_l, I_2))$. By Lemma 5.1.1, $H^i(F_+) = 0$ if $i \geq 2$. Thus

$$\Gamma(X_l, I_0) \to M_1 \tag{14}$$

is isomorphic to $H_l^\bullet(F_+)$ in $\mathcal{D}_m$ and

$$0 \to M_1 \to \Gamma(X_l, I_1) \to \Gamma(X_l, I_2) \to \dots \tag{15}$$

is exact. We wish to show that M_1 is a cohomologically trivial G-module.

An injective $\mathcal{R}_m = \Lambda_m[G]$-module M is cohomologically trivial, since we can split the inclusion of such M into the induced G-module $M \otimes_{\Lambda_m} \mathcal{R}_m$ and induced G-modules are cohomologically trivial. Hence (15) gives a resolution of M_1 by cohomologically trivial G-modules. Therefore for each subgroup H of G, the cohomology groups of the complex

$$\Gamma(X_l, I_1)^H \to \Gamma(X_l, I_2)^H \to \dots \tag{16}$$

give the H-cohomology groups of M_1. However, since $X_l \to Y_l$ is an étale G-cover, $\Gamma(X_l, I_j)^H = \Gamma(X_l/H, I_j)$. Hence the ith cohomology group of (16) equals $H^i(X_l/H; F_+)$ and this is 0 if $i \geq 2$ by Lemma 5.1.1. Thus (16) has trivial cohomology in dimensions greater than one and we conclude that M_1 is a cohomologically trivial G-module. This proves that $H_l^\bullet(F_+)$ is isomorphic in the derived category of $\mathcal{R}_m$-modules to the complex (14) of cohomologically trivial G-modules.

We have assumed that F is a flat constructible sheaf of Λ_m-modules and so F_+ is also, by Lemma 5.1.1. Since the cohomology groups of F_+ are finite and vanish in dimensions greater than 1, the argument of ([107] Corollary VI.8.10) shows that $H_l^\bullet(F_+)$ is isomorphic in $\mathcal{D}_m$ to a complex

$$Q_{-1} \to Q_0 \to Q_1 \tag{17}$$

in which all of the Q_i are finitely generated flat Λ_m-modules and Q_i is a projective $\mathcal{R}_m$-module if $i \neq -1$. Since we have also shown $H_l^\bullet(F_+)$ isomorphic in $\mathcal{D}_m$ to a bounded complex of cohomologically trivial G-modules, the module Q_{-1} must be cohomologically trivial as a G-module. To complete the proof, it will suffice to show that Q_{-1} is a projective $\mathcal{R}_m$-module.

Let us first show that if H is a subgroup of G, then Q_{-1}^H is a free Λ_m-module. Since $X_l \to Y_l$ is an étale G-cover, we have an equality of functors $\Gamma(X_l/H, *) = \Gamma(X_l, *)^H$. Therefore $\Gamma(H, H^\bullet(X_l; F_+)) = H^\bullet(X_l/H; F_+)$, where $\Gamma(H, *)$ denotes group hypercohomology with respect to H. Because the Q_i are cohomologically trivial, the complex

$$Q_{-1}^H \to Q_0^H \to Q_1^H$$

is isomorphic to $\Gamma(H, H^\bullet(X_l; F_+))$ in the derived category $\mathcal{D}'_m$ of the homotopy category of complexes of Λ_m-modules. If $i \neq -1$ then Q_i^H is a flat Λ_m-module because Q_i is a projective $\mathcal{R}_m$-module. Again using ([107] Corollary VI.8.10), we see that $H^\bullet(X_l/H; F_+)$ is isomorphic in $\mathcal{D}'_m$ to a perfect complex of Λ_m-modules, since $H^i(X_l/H; F_+) = 0$ if $i > 1$. Thus Q_{-1}^H must have finite tor-dimension as a Λ_m-module. Since Q_{-1}^H is a finitely generated Λ_m-module, we can find a finite resolution

$$0 \to M_j \to M_{j-1} \to \cdots \to M_0 \to Q_{-1}^H \to 0 \tag{18}$$

of Q_{-1}^H by finitely generated flat Λ_m-modules. Since Λ_m is a local ring, ([107] Theorem I.2.9) shows that each of the M_i are finitely generated free modules.

We assumed in Hypothesis 5.1.3 that A is a discrete valuation ring, where $\Lambda_m = A/\mathcal{M}^m$ when $\mathcal{M}$ is the maximal ideal of A. Thus using the structure theorem for finitely generated modules over a principal ideal domain we can split the sequence (18). This proves Q_{-1}^H is a free Λ_m-module.

Let $\mathcal{M}_m$ be the unique prime ideal of Λ_m. Since Q_{-1} is a free Λ_m-module, ([129] Lemma 21 §14.4) shows that Q_{-1} is a projective $\mathcal{R}_m$-module if and only if $Q_{-1}/\mathcal{M}_m$ is a projective $\mathcal{R}_m/\mathcal{M}_m = k(\mathcal{M})[G]$-module. By ([42] Corollary 1.1), to prove the latter statement, it suffices to do so whenever G is replaced by one of its l-Sylow subgroups. Thus we can reduce to the case in which G is a finite l-group.

Since Q_{-1} is cohomologically trivial, the trace element in the group ring of G induces an isomorphism between $H_0(G; Q_{-1})$ and $H^0(G; Q_{-1}) = Q_{-1}^G$. We have shown Q_{-1}^G is free as a Λ_m-module, so we can pick a set of elements $\{q_j\}_j$ in Q_{-1} whose images in $H_0(G; Q_{-1}) = Q_{-1}/Q_{-1}^{I_G}$ form a basis over Λ_m for $H_0(G; Q_{-1})$, where I_G is the augmentation ideal in $\mathbf{Z}[G]$. As in ([128] Lemma 5 §IX.4) we now define an exact sequence

$$0 \to \mathcal{K} \to \mathcal{L} \to Q_{-1} \to 0 \tag{19}$$

in which $\mathcal{L}$ is a free $\mathcal{R}_m$-module on a basis whose elements map bijectively to the $\{q_j\}_j$ and in which $\mathcal{K}$ is the kernel of the resulting surjective $\mathcal{R}_m$-module homomorphism. By construction, $H_0(G; \mathcal{L}) \to H_0(G; Q_{-1})$ is an isomorphism, so the long exact group homology sequence of (19) shows that $H_0(G; \mathcal{K}) = H_1(G; Q_{-1})$. However, Q_{-1} is cohomologically trivial, so $H_0(G; \mathcal{K}) = 0$. Since $\mathcal{K} \to \mathcal{K}/l$ is surjective, we get a surjection $0 = H_0(G;\mathcal{K}) \to H_0(G;\mathcal{K}/l) = 0$. However, we reduced to the case in which G is an l-group. Hence ([128] Lemma 4, §IX.4) shows $\mathcal{K}/l = \{0\}$, so $\mathcal{K} = \{0\}$ and Q_{-1} is isomorphic to the free $\mathcal{R}_m$-module $\mathcal{L}$. This completes the proof of Proposition 5.1.4. □

Remark 5.1.5 In the proof of Proposition 5.1.4 the assumption that A is a discrete valuation ring was used to show that Q_{-1} is a free $\Lambda_m = A/\mathcal{M}^m$-module because it is finitely generated and of finite tor-dimension. Suppose we assumed only that A is of *finite index* in a discrete valuation ring. Then flat finitely generated Λ_m-modules are free. The argument used to prove that (18) splits would go though if one assumes Λ_m is an injective Λ_m-module. By ([95] Lemma III.8.3), $\Lambda'_m = \mathrm{Hom}_{\mathbf{Z}}(\Lambda_m, \mathbf{Q}/\mathbf{Z})$ is an injective Λ_m-module and this module has the same order as Λ_m since Λ_m is finite. Thus it would suffice for Λ'_m to be a cyclic Λ_m-module. This is always the case, for example, if $m = 1$.

Definition 5.1.6 An étale A-sheaf on X_l is a projective system $F = \{F_m\}_m$ of sheaves such that F_m is a sheaf of Λ_m-modules and the given map $F_{m+1} \to F_m$ is isomorphic to the canonical map $F_{m+1} \to F_{m+1} \otimes_{\Lambda_{m+1}} \Lambda_m$. Write F_+ for the projective system $\{(F_m)_+\}_m$. We say F is constructible (smooth or flat) if each F_m is a constructible (twisted constant constructible or flat, respectively) sheaf of Λ_m-modules.

Proposition 5.1.7

Suppose that A and S satisfy Hypothesis 5.1.3 and that $F = \{F_m\}_m$ is a flat smooth sheaf of A-modules on $Y_l = \mathrm{Spec}(K \cap O_{N,S})$. Let $\mathcal{D}$ be the derived category of the homotopy category of complexes of $A[G]$-modules which are bounded below. Define $H_l^\bullet(F) = H^\bullet(X_l; F)$ to be the étale hypercohomology in $\mathcal{D}$ of the restriction of F to X_l. Then there is a complex

$$P^\bullet : P_{-1} \xrightarrow{\delta} P_0 \to P_1 \tag{20}$$

of finitely generated projective $A[G]$-modules which is isomorphic to $H_l^\bullet(F_+)$ in $\mathcal{D}_m$, where P_i is the degree i term of $P^\bullet$.

Proof. Write $H_l^\bullet((F_m)_+)$ for a choice of three-term complex $Q^\bullet$ with the properties stated in Proposition 5.1.4 for the sheaf $(F_m)_+$. We will show that the $H_l^\bullet((F_m)_+)$ can be chosen in such a way that there are maps

$$H_l^\bullet((F_{m+1})_+) \to H_l^\bullet((F_m)_+) \tag{21}$$

inducing isomorphisms

$$H_l^\bullet((F_{m+1})_+) \otimes_{\Lambda_{m+1}} \Lambda_m \cong H_l^\bullet((F_m)_+)$$

for all m. Then as in ([107] p. 265), $H_l^\bullet(F_+) = \lim_{\overleftarrow{m}} H_l^\bullet((F_m)_+)$ will be a complex of the form (20).

The construction of (21) is the same as that of ([107] p. 264–265) with the following changes. The morphism, ψ, in ([107] Lemma VI.8.17) will not in general be a quasi-isomorphism. In the statement of ([107] Lemma 8.18) one replaces the ring A by the ring $\Lambda_{m+1}[G]$ and one replaces A_0 by $\Lambda_m[G]$. If $T^\bullet$ is a perfect complex of $\Lambda_{m+1}[G]$-modules, the complex $T^\bullet \otimes_{\Lambda_{m+1}[G]} \Lambda_m[G] = T^\bullet \otimes_{\Lambda_{m+1}} \Lambda_m$ is a perfect complex of $\Lambda_m[G]$-modules. By ([129] §14.4), if $\phi : T_1 \to T_2$ is a homomorphism between projective $\Lambda_m[G]$-modules, there is a homomorphism $\phi' : T_1' \to T_2'$ between projective $\Lambda_{m+1}[G]$-modules such that $\phi = \phi' \otimes_{\Lambda_{m+1}[G]} \Lambda_m[G]$, which implies in particular that $T_i = T_i' \otimes_{\Lambda_{m+1}[G]} \Lambda_m[G]$. The argument of ([107] Lemma VI.8.18) now applies to construct (21), as required. □

Corollary 5.1.8
With the notation of Proposition 5.1.7, the two term complex

$$M^\bullet : M_0 = P_0/\delta P_{-1} \to M_1 = P_1$$

is isomorphic to $H_l^\bullet(F_+)$ in $\mathcal{D}$ and the terms of $M^\bullet$ are finitely generated $A[G]$-modules of finite projective dimension.

5.2 The invariant $\Omega_n(N/K, 3)$

5.2.1 In this section we shall use the results of the previous section to construct an element

$$\Omega_n(N/K, 3) \in \mathcal{CL}(\mathbf{Z}[G(N/K)]) \subset K_0(\mathbf{Z}[G(N/K)])$$

for any Galois extension of number fields, N/K, with Galois group $G(N/K)$ for any integer, $n \geq 1$. The construction imitates that of the original third Chinburg invariant, $\Omega(N/K,3)$ henceforth denoted by $\Omega_0(N/K,3)$ for consistency [34]. The material of this section originates from ([36], [38]) to which the reader can refer for further details.

Let S be a finite $G(N/K)$-stable set of places of N which contains the set S_∞ of infinite places and all places which ramify over K. Resuming the notation of the previous section, the ring of S-integers of N is denoted $O_{N,S}$.

If X is a set on which $G(N/K)$ acts and R is a ring, the free R-module with basis X, $Div_R(X) = R\langle X\rangle$ is an $R[G(N/K)]$-module. Set $Y_{2n+1} = Div_{\mathbf{Z}}(S_\infty)$ if n is even and let Y_{2n+1} be the cokernel of the natural homomorphism $Div_{\mathbf{Z}}(S_\infty) \to Div_{\mathbf{Z}}(\Sigma_\infty)$ if n is odd, where Σ_∞ is the set of infinite embeddings.

For each rational prime l, set $X_l = \mathrm{Spec}(O_{N,S}[1/l])$. Let $A = \mathbf{Z}_l$ and let $F_l = \{F_{l,m}\}_m$ be a smooth flat $\mathbf{Z}_l$–sheaf on X_l, where $F_{l,m}$ is a flat locally constant constructible sheaf of $\mathbf{Z}_l/l^m$-modules. We set

$$(F_l)_\infty = \sum_{v \in S_\infty} i_{v*} i_v^* F_l \quad \text{where} \quad i_v : \mathrm{Spec}(N_v) \to X$$

is the natural morphism and N_v is the completion of N at v. The sheaf $(F_l)_+ = \lim_{\overleftarrow{m}}(F_{l,m})_+$ lies in an exact sequence of $\mathbf{Z}_l$-sheaves

$$0 \to F \to F_\infty \to F_+ \to 0 \tag{1}$$

formed by the inverse limit of the sequences (1) of Lemma 5.1.1.

Let $H_l^i(F_l) = H^i(X_l; F_l)$. If $F = \{F_l\}_l$ is a collection of sheaves F_l as above, one for each rational prime l, define $H_\infty^i(F) = \prod_l H_l^i(F_l)$. For $r \geq 0$, define $\mathbf{Z}(r) = \{\mathbf{Z}_l(r)\}_l$, $\mathbf{Z}(r)_\infty = \{\mathbf{Z}_l(r)_\infty\}_l$ and $\mathbf{Z}(r)_+ = \{\mathbf{Z}_l(r)_+\}_l$. Define $\hat{\mathbf{Z}}(r) = \prod_l \mathbf{Z}_l(r)$ and $\hat{\mathbf{Z}} = \hat{\mathbf{Z}}(0)$.

Let $H_l^\bullet(\mathbf{Z}_l(r)_+) = H^\bullet(X_l; \mathbf{Z}_l(r)_+)$ be the hypercohomology of $\mathbf{Z}_l(r)_+$ in the derived category $\mathcal{D}_l$ of the homotopy category of complexes of $\mathbf{Z}_l[G(N/K)]$-modules. By Corollary 5.1.8, there is a two-term complex $M^\bullet : M_l \to M_l'$ concentrated in degrees 0 and 1 whose terms are finitely generated $\mathbf{Z}_l[G(N/K)]$-modules of finite projective dimension such that there is an isomorphism $\psi : M^\bullet \to H_l^\bullet(\mathbf{Z}_l(r)_+)$ in $\mathcal{D}_l$. The choice of ψ gives rise to an exact sequence

$$0 \to H_l^0(\mathbf{Z}_l(r)_+) \to M_l \to M_l' \to H_l^1(\mathbf{Z}_l(r)_+) \to 0 \tag{2}$$

whose extension class $E_l \in \mathrm{Ext}^2_{\mathbf{Z}_l[G(N/K)]}(H_l^1(\mathbf{Z}_l(r)_+), H_l^0(\mathbf{Z}_l(r)_+))$ is independent of the choice of ψ. Since the terms of $M^\bullet$ are cohomologically trivial for $G(N/K)$, cup-product with E_l induces an isomorphism in Tate cohomology

$$\hat{H}^j(\Gamma; H_l^1(\mathbf{Z}_l(r)_+)) \to \hat{H}^{j+2}(\Gamma; H_l^0(\mathbf{Z}_l(r)_+)$$

for all subgroups $\Gamma \subseteq G(N/K)$ and all integers j. Therefore we have the following result.

Proposition 5.2.2

In §5.2.1, the cup-product with

$$E = \prod_l E_l \in \mathrm{Ext}^2_{\hat{\mathbf{Z}}[G(N/K)]}(H_\infty^1(\mathbf{Z}(r)_+), H_\infty^0(\mathbf{Z}(r)_+))$$

induces isomorphisms in Tate cohomology of the form

$$\hat{H}^j(\Gamma; H_\infty^1(\mathbf{Z}(r)_+)) \to \hat{H}^{j+2}(\Gamma; H_\infty^0(\mathbf{Z}(r)_+))$$

for all integers j and all subgroups $\Gamma \subseteq G(N/K)$.

Lemma 5.2.3

Let G_v denote the decomposition group of a place, v, of N and let $S_{\mathbf{R}}$ be the set of real places of N. Then there are natural isomorphisms of the form

$$H_\infty^i(\mathbf{Z}(r)_\infty) \cong \begin{cases} \bigoplus_{v \in S_\infty} (\hat{\mathbf{Z}}(r))^{G_v} & \text{if } i = 0, \\ \mathbf{Z}/2\langle S_{\mathbf{R}} \rangle & \text{if } i > 0 \text{ and } i + r \text{ is even,} \\ 0 & \text{otherwise.} \end{cases}$$

Here $\mathbf{Z}/2\langle S_{\mathbf{R}} \rangle$, which we shall sometimes denote by $Div_{\mathbf{Z}/2}(S_{\mathbf{R}})$, is the $\mathbf{Z}/2$-vector space with basis $S_{\mathbf{R}}$.

Proof. For all i there is a natural isomorphism

$$H^i_\infty(\mathbf{Z}(r)_\infty) \cong \bigoplus_{\infty \in S_\infty} H^i(G_v; \hat{\mathbf{Z}}(r))$$

from which the $i = 0$ case is clear. For $v \in S_\infty$ the group G_v is trivial unless v lies in the set, $S_{\mathbf{R}}$, of real places of N, in which case G_v has order two and non-trivial element equal to complex conjugation at v. Therefore we must compute

$$H^i(G_v; \hat{\mathbf{Z}}(r)) = H^i(G_v; \mathbf{Z}_2(r)) = \varprojlim_m H^i(G_v; \mathbf{Z}/2^m(r))$$

when $i > 0$ and $v \in S_{\mathbf{R}}$.

Choose $j \in \{-1, 0\}$ with $j \equiv i \bmod 2$. Cup-product with a power of a generator of $H^2(G_v; \mathbf{Z})$ induces an isomorphism $\hat{H}^j(G_v; \mathbf{Z}/2^m(r)) = H^i(G_v; \mathbf{Z}/2^m(r))$ which is compatible with the projection $\mathbf{Z}/2^m(r) \to \mathbf{Z}/2^{m-1}(r)$. Suppose $j = 0$ so that i is even. Then $\hat{H}^0(G_v; \mathbf{Z}/2^m(r))$ is identified with the order two quotient group (resp. subgroup) of $\mathbf{Z}/2^m(r)$ if r is even (resp. odd). This shows that for $m > 1$, the homomorphism $\hat{H}^0(G_v; \mathbf{Z}/2^m(r)) \cong \mathbf{Z}/2 \to \hat{H}^0(G_v; \mathbf{Z}/2^{m-1}(r)) \cong \mathbf{Z}/2$ is an isomorphism if and only if r is even. Thus if $i > 0$ is even and $v \in S_{\mathbf{R}}$, we find $H^i(G_v; \mathbf{Z}_2(r)) = \mathbf{Z}/2$ if r is even and $H^i(G_v; \mathbf{Z}_2(r)) = 0$ if r is odd. Suppose now that $j = -1$, so that $i > 0$ is odd. In this case $H^{-1}(G_v; \mathbf{Z}/2^m(r))$ is identified with the order two quotient group (resp subgroup) of $\mathbf{Z}/2^m(r)$ if r is odd (resp. even). Repeating the above argument completes the proof of Lemma 5.2.3. □

Corollary 5.2.4

Let $d_i : H^i_\infty(\mathbf{Z}(r)) \to H^i_\infty(\mathbf{Z}(r)_\infty)$ denote the natural map. and suppose $r > 0$. The cohomology of the exact sequence

$$0 \to \mathbf{Z}(r) \to \mathbf{Z}(r)_\infty \to \mathbf{Z}(r)_+ \to 0$$

yields two canonical exact sequences:

$$0 \to H^0_\infty(\mathbf{Z}(r)_\infty) \to H^0_\infty(\mathbf{Z}(r)_+) \xrightarrow{\delta} \ker(d_1) \to 0 \tag{3}$$

and

$$0 \to \operatorname{coker}(d_1) \to H^1_\infty(\mathbf{Z}(r)_+) \to \ker(d_2) \to 0. \tag{4}$$

A choice of an isomorphism $\tau : \hat{\mathbf{Z}} \to \hat{\mathbf{Z}}(r)$ of $\hat{\mathbf{Z}}$-modules determines a $\mathbf{Z}[G(N/K)]$-module isomorphism of the form

$$Y_{2r+1} \otimes_{\mathbf{Z}} \hat{\mathbf{Z}} \xrightarrow{\cong} H^0_\infty(\mathbf{Z}(r)_\infty) = \sum_{v \in S_\infty} \hat{\mathbf{Z}}(r)^{G_v}$$

where Y_{2r+1} is the permutation module defined in §5.2.1. Via τ and the natural injection $\mathbf{Z} \to \hat{\mathbf{Z}}$ we may view Y_{2r+1} as a $\mathbf{Z}[G(N/K)]$-submodule of $H^0_\infty(\mathbf{Z}(r)_\infty)$.

Theorem 5.2.5 *([142],[143],[51],[82])*

Suppose $r > 1$. *Define* $K_{2r-i} = K_{2r-i}(O_{N,S})$ *for* $i = 1,2$. *There are Chern character homomorphisms* $\mathrm{ch}_i : K_{2r-i} \to H^i_\infty(\mathbf{Z}(r))$ *such that* $\ker(\mathrm{ch}_i)$ *is finite and* $\mathrm{ch}_i(K_{2r-i}) \otimes_{\mathbf{Z}} \hat{\mathbf{Z}}$ *has finite two-power order index in* $H^i_\infty(\mathbf{Z}(r))$.

Proof. Let l be a prime. Define $N' = N$ and $S' = S$ if l is odd or if $\sqrt{-1} \in O_{N,S}$. Otherwise, let $N' = N(\sqrt{-1})$ and let S' be the set of primes over S in N'. It is shown in [51] (see also [82] and [52]) that for $i \in \{1,2\}$, there is a Chern character homomorphism

$$\mathrm{ch}_{i,l,N'} : K_{2r-i}(O_{N',S'}[1/l]) \to H^i(\mathrm{Spec}(O_{N',S'}[1/l]); \mathbf{Z}_l(r))$$

which induces a surjection

$$K_{2r-i}(O_{N',S'}[1/l]) \otimes_{\mathbf{Z}} \mathbf{Z}_l \to H^i(\mathrm{Spec}(O_{N',S'}[1/l]); \mathbf{Z}_l(r)).$$

Let

$$h_l : K_{2r-i} = K_{2r-i}(O_{N,S}) \to K_{2r-i}(O_{N',S'}[1/l])$$

be the homomorphism induced by the inclusion $O_{N,S} \to O_{N',S'}[1/l]$. The natural morphism $\mathrm{Spec}(O_{N',S'}[1/l]) \to \mathrm{Spec}(O_{N,S}[1/l])$ is étale, since in all cases, N' can ramify over N only over l. Therefore we have a corestriction map

$$c_l : H^i(\mathrm{Spec}(O_{N',S'}[1/l]); \mathbf{Z}_l(r)) \to H^i(\mathrm{Spec}(O_{N,S}[1/l]); \mathbf{Z}_l(r)) = H^i_l(\mathbf{Z}_l(r)).$$

Define $\mathrm{ch}_{i,l} : K_{2r-i} \to H^i_l(\mathbf{Z}_l(r))$ to be $\mathrm{ch}_{i,l} = c_l \circ \mathrm{ch}_{i,l,N'} \circ h_l$ and let $\mathrm{ch}_i = \prod_l \mathrm{ch}_{i,l}$.

By ([17], [18], [19] and [143]), $K_{2r-i}(O_{N,S})$ (resp. $H^i_l(\mathbf{Z}_l(r))$) is finitely generated over $\mathbf{Z}$ (resp. $\mathbf{Z}_l$) and the rank $t_{r,i}$ of $K_{2r-i}(O_{N,S})$ over $\mathbf{Z}$ equals the rank of $H^i_l(\mathbf{Z}_l(r))$ over $\mathbf{Z}_l$. Suppose $l \neq 2$, then localization sequence (Corollary 2.4.7) and the calculation of the K-groups of finite fields [115] show that the natural map, h_l, induces an isomorphism

$$K_{2r-i}(O_{N,S}) \otimes_{\mathbf{Z}} \mathbf{Z}_l \to K_{2r-i}(O_{N,S}[1/l]) \otimes_{\mathbf{Z}} \mathbf{Z}_l.$$

Since $l \neq 2$ we have $N' = N$. Since $\mathrm{ch}_{i,l,N'}$ induces a surjection on tensoring with $\mathbf{Z}_l$ and c_l is the identity map, we conclude that $\mathrm{ch}_{i,l}$ induces a surjection on tensoring with $\mathbf{Z}_l$. Therefore

$$\frac{H^i_\infty(\mathbf{Z}(r))}{\mathrm{ch}_i(K_{2r-i}) \otimes_{\mathbf{Z}} \hat{\mathbf{Z}}} = \frac{H^i_2(\mathbf{Z}_2(r))}{\mathrm{ch}_{i,2}(K_{2r-i}) \otimes_{\mathbf{Z}} \mathbf{Z}_2}.$$

Since $t_{r,i}$ is the rank of $H^i_l(\mathbf{Z}_l(r))$ over $\mathbf{Z}_l$, we see $\ker(\mathrm{ch}_{i,l})$ is finite, so $\ker(\mathrm{ch}_i)$ is finite. Since $H^i_2(\mathbf{Z}_2(r))$ is finitely generated over $\mathbf{Z}_2$, it will now suffice to show that $\mathrm{ch}_{i,2}$ induces a surjection

$$\mathrm{ch}_{i,2} \otimes_{\mathbf{Z}} \mathbf{Q}_2 : K_{2r-i}(O_{N,S}) \otimes_{\mathbf{Z}} \mathbf{Q}_2 \to H^i_2(\mathbf{Q}_2(r)).$$

If $i = 2$, [17] shows $t_{r,2} = 0$ so that $\mathrm{ch}_{i,2} \otimes_{\mathbf{Z}} \mathbf{Q}_2$ is trivially an isomorphism. Suppose now that $i = 1$. By [143] the inclusion $O_{N,S} \to N$ induces an isomorphism $K_{2r-1}(O_{N,S}) \otimes_{\mathbf{Z}} \mathbf{Q} \to K_{2r-1}(N) \otimes \mathbf{Q}$. By [18] the inclusion $N \to N'$ identifies $K_{2r-1}(N) \otimes_{\mathbf{Z}} \mathbf{Q}$ with $(K_{2r-1}(N') \otimes_{\mathbf{Z}} \mathbf{Q})^{G(N'/N)}$. The results [51] cited in the first part of the proof imply that for all l, $\mathrm{ch}_{1,l,N'}$ induces a $G(N'/N)$-equivariant surjection

$$K_{2r-1}(N') \otimes_{\mathbf{Z}} \mathbf{Q}_l = K_{2r-1}(O_{N',S'}[1/l]) \otimes_{\mathbf{Z}} \mathbf{Q}_l \to H^1(\mathrm{Spec}(O_{N',S'}[1/l]); \mathbf{Q}_l(r)).$$

This implies that $((\mathrm{ch}_{1,2,N'} \circ h_2)(K_{2r-1})) \otimes_{\mathbf{Z}} \mathbf{Q}_2$ is equal to the $\mathbf{Q}_2$ vector space $W = H^1(\mathrm{Spec}(O_{N',S'}[1/2]); \mathbf{Q}_2(r))^{G(N'/N)}$. The restriction map

$$H^1(\mathrm{Spec}(O_{N,S}[1/2]); \mathbf{Z}_2(r)) \to H^1(\mathrm{Spec}(O_{N',S'}[1/2]); \mathbf{Z}_2(r))$$

has image in W and the composition of restriction with the corestriction map c_2 is multiplication by $[N' : N]$. Hence c_2 maps W onto $H^1(\mathrm{Spec}(O_{N,S}[1/2]); \mathbf{Q}_2(r))$ and we conclude that $\mathrm{ch}_{1,2} \otimes_{\mathbf{Z}} \mathbf{Q}_2 = (c_2 \circ \mathrm{ch}_{1,2,N'} \circ h_2) \otimes_{\mathbf{Z}} \mathbf{Q}_2$ is surjective, which completes the proof. □

Corollary 5.2.6
If $r > 1$ the $\mathbf{Z}[G(N/K)]$-module $H^1_\infty(\mathbf{Z}(r)_+)$ is finite.

Proof. The groups $K_{2r-2}(O_{N,S})$ and $K_{2r-2}(O_{N',S'})$ are finitely generated [116] and hence finite, by ([17], [18], [19]). Hence $H^2_\infty(\mathbf{Z}(r))$ is finite by Theorem 5.2.5. Thus $\ker(d_2)$ is finite, and $\mathrm{coker}(d_1)$ is finite by Lemma 5.2.3. Therefore the exact sequence (4) shows that $H^1_\infty(\mathbf{Z}(r)_+)$ is finite also. □

In defining $\Omega_{r-1}(N/K, 3)$ we shall use $H^1_\infty(\mathbf{Z}(r)_+)$ as a substitute for $K_{2r-2}(\mathcal{O}_{N,S})$ (see Proposition 5.2.10). The objective of the following series of lemmas is to manufacture a substitute for $K_{2r-1}(\mathcal{O}_{N,S})$.

Lemma 5.2.7
Let G be a finite group. For all $\mathbf{Z}[G]$-modules, M, the inclusion of $\mathbf{Z}$ into $\hat{\mathbf{Z}}$ induces an exact sequence

$$0 \to M \to M \otimes_{\mathbf{Z}} \hat{\mathbf{Z}} \to M \otimes_{\mathbf{Z}} (\hat{\mathbf{Z}}/\mathbf{Z}) \to 0 \tag{5}$$

in which $M \otimes_{\mathbf{Z}} (\hat{\mathbf{Z}}/\mathbf{Z})$ is a rational vector space. Suppose M and N are finitely generated $\mathbf{Z}[G]$-modules. Then the natural map

$$\hat{\mathbf{Z}} \otimes_{\mathbf{Z}} \mathrm{Ext}^i_{\mathbf{Z}[G]}(M, N) \to \mathrm{Ext}^i_{\hat{\mathbf{Z}}[G]}(\hat{\mathbf{Z}} \otimes_{\mathbf{Z}} M, \hat{\mathbf{Z}} \otimes_{\mathbf{Z}} N) \tag{6}$$

is an isomorphism for all i. If $i > 0$, these groups are finite, and we have an induced isomorphism

$$\mathrm{Ext}^i_{\mathbf{Z}[G]}(M, N) = \mathrm{Ext}^i_{\hat{\mathbf{Z}}[G]}(\hat{\mathbf{Z}} \otimes_{\mathbf{Z}} M, \hat{\mathbf{Z}} \otimes_{\mathbf{Z}} N). \tag{7}$$

Proof. By the Snake Lemma [95], multiplication by a non-zero $m \in \mathbf{Z}$ defines an isomorphism of $\hat{\mathbf{Z}}/\mathbf{Z}$. Hence $\hat{\mathbf{Z}}/\mathbf{Z}$ is a rational vector space, so that (5) is clear.

If A and B are finitely generated $\mathbf{Z}[G]$-modules, then $\mathrm{Hom}_{\hat{\mathbf{Z}}}(\hat{\mathbf{Z}} \otimes_{\mathbf{Z}} A, \hat{\mathbf{Z}} \otimes_{\mathbf{Z}} B) = \hat{\mathbf{Z}} \otimes_{\mathbf{Z}} \mathrm{Hom}_{\mathbf{Z}}(A,B)$. Hence, since $\hat{\mathbf{Z}}$ is flat over $\mathbf{Z}$, we have

$$\mathrm{Hom}_{\hat{\mathbf{Z}}[G]}(\hat{\mathbf{Z}} \otimes_{\mathbf{Z}} A, \hat{\mathbf{Z}} \otimes_{\mathbf{Z}} B) = (\hat{\mathbf{Z}} \otimes_{\mathbf{Z}} \mathrm{Hom}_{\mathbf{Z}}(A,B))^G = \hat{\mathbf{Z}} \otimes_{\mathbf{Z}} \mathrm{Hom}_{\mathbf{Z}[G]}(A,B).$$

The isomorphism (6) follows from these facts and the finiteness of the groups $\mathrm{Ext}^i_{\mathbf{Z}[G]}(M,N)$ for $i > 0$ results from the spectral sequence

$$E_2^{p,q} = H^p(G; \mathrm{Ext}^q_{\mathbf{Z}}(M,N)) \Longrightarrow \mathrm{Ext}^{p+q}_{\mathbf{Z}[G]}(M,N)$$

so that (7) follows from (6). □

Lemma 5.2.8

By Theorem 5.2.5, the $\mathbf{Z}[G(N/K)]$*-module*

$$T = H^1_\infty(\mathbf{Z}(r))/(\mathrm{ch}_1(K_{2r-1}) \otimes_{\mathbf{Z}} \hat{\mathbf{Z}}) \tag{8}$$

is finite. There is a unique G-submodule $K^*_{2r-1} \subset H^1_\infty(\mathbf{Z}(r))$ *containing*

$$\mathrm{ch}_1(K_{2r-1}) = \mathrm{ch}_1(K_{2r-i}(O_{N,S}))$$

for which the natural inclusions

$$K^*_{2r-1} \hookrightarrow H^1_\infty(\mathbf{Z}(r)) \quad \text{and} \quad \mathrm{ch}_1(K_{2r-1}) \hookrightarrow (\mathrm{ch}_1(K_{2r-1}) \otimes_{\mathbf{Z}} \hat{\mathbf{Z}})$$

induce an isomorphism

$$K^*_{2r-1}/\mathrm{ch}_1(K_{2r-1}) = T. \tag{9}$$

The module K^*_{2r-1} *is finitely generated over* $\mathbf{Z}$ *and lies in an exact sequence*

$$0 \to K_{2r-1}/\ker(\mathrm{ch}_1) \to K^*_{2r-1} \to T \to 0. \tag{10}$$

The inclusion

$$K^*_{2r-1} \subset H^1_\infty(\mathbf{Z}(r))$$

induces an isomorphism

$$K^*_{2r-1} \otimes_{\mathbf{Z}} \hat{\mathbf{Z}} = H^1_\infty(\mathbf{Z}(r)).$$

Proof. By (5) of Lemma 5.2.7, we have an exact sequence

$$0 \to \mathrm{ch}_1(K_{2r-1}) \to \mathrm{ch}_1(K_{2r-1}) \otimes_{\mathbf{Z}} \hat{\mathbf{Z}} \to \mathrm{ch}_1(K_{2r-1}) \otimes_{\mathbf{Z}} (\hat{\mathbf{Z}}/\mathbf{Z}) \to 0 \tag{11}$$

in which the right term is a rational vector space.

By Theorem 5.2.5, we also have an exact sequence

$$0 \to \mathrm{ch}_1(K_{2r-1}) \otimes_{\mathbf{Z}} \hat{\mathbf{Z}} \to H^1_\infty(\mathbf{Z}(r)) \to T \to 0 \tag{12}$$

in which T is finite. This sequence defines an extension class

$$e \in \mathrm{Ext}^1_{\hat{\mathbf{Z}}[G(N/K)]}(T, \mathrm{ch}_1(K_{2r-1}) \otimes_{\mathbf{Z}} \hat{\mathbf{Z}}).$$

Since T is finite, $T = T \otimes_{\mathbf{Z}} \hat{\mathbf{Z}}$. Thus, since K_{2r-1} is finitely generated, Lemma 5.2.7 shows

$$\mathrm{Ext}^1_{\mathbf{Z}[G(N/K)]}(T, \mathrm{ch}_1(K_{2r-1})) = \mathrm{Ext}^1_{\hat{\mathbf{Z}}[G(N/K)]}(T, \mathrm{ch}_1(K_{2r-1}) \otimes_{\mathbf{Z}} \hat{\mathbf{Z}}).$$

Thus e is the image of a class in

$$\mathrm{Ext}^1_{\mathbf{Z}[G(N/K)]}(T, \mathrm{ch}_1(K_{2r-1})),$$

so there is a $\mathbf{Z}[G(N/K)]$-module, K^*_{2r-1}, fitting into a diagram with exact rows of the form

$$\begin{array}{ccccccccc} 0 & \to & \mathrm{ch}_1(K_{2r-1}) & \to & K^*_{2r-1} & \to & T & \to & 0 \\ & & \downarrow & & \downarrow & & \downarrow & & \\ 0 & \to & \mathrm{ch}_1(K_{2r-1}) \otimes_{\mathbf{Z}} \hat{\mathbf{Z}} & \to & H^1_\infty(\mathbf{Z}(r)) & \to & T & \to & 0 \end{array} \tag{13}$$

The left vertical arrow in this diagram is injective and arises from the inclusion $\mathbf{Z} \hookrightarrow \hat{\mathbf{Z}}$. The right vertical arrow is the identity map. Hence the middle vertical arrow is injective by the Snake Lemma, so we may view K^*_{2r-1} as a $\mathbf{Z}[G(N/K)]$-submodule of $H^1_\infty(\mathbf{Z}(r))$. Clearly K^*_{2r-1} is finitely generated over $\mathbf{Z}$, since K_{2r-1} and T are, and top row of (13) implies (9) and (10).

Since $\hat{\mathbf{Z}}$ is flat over $\mathbf{Z}$ and T is finite, tensoring the top row of (13) with $\hat{\mathbf{Z}}$ over $\mathbf{Z}$ shows that $K^*_{2r-1} \otimes_{\mathbf{Z}} \hat{\mathbf{Z}} = H^1_\infty(\mathbf{Z}(r))$.

It remains to show that if $K^{\#}_{2r-1}$ is another $\mathbf{Z}[G(N/K)]$-submodule of $H^1_\infty(\mathbf{Z}(r))$ with the properties stated in Lemma 5.2.8 then $K^{\#}_{2r-1} = K^*_{2r-1}$. Suppose $t \in K^{\#}_{2r-1}$. Then there is an element $t' \in K^*_{2r-1}$ such that t and t' have the same image in $T = H^1_\infty(\mathbf{Z}(r))/(\mathrm{ch}_1(K_{2r-1}) \otimes_{\mathbf{Z}} \hat{\mathbf{Z}})$. Thus $t' - t \in \mathrm{ch}_1(K_{2r-1}) \otimes_{\mathbf{Z}} \hat{\mathbf{Z}}$. However, we have assumed that the inclusions $K^*_{2r-1} \hookrightarrow H^1_\infty(\mathbf{Z}(r))$ and $K^*_{2r-1} \hookrightarrow H^1_\infty(\mathbf{Z}(r))$ induce isomorphisms

$$K^{\#}_{2r-1}/\,\mathrm{ch}_1(K_{2r-1}) = T = K^*_{2r-1}/\,\mathrm{ch}_1(K_{2r-1}).$$

Because T is finite, this implies $|T| \cdot (t - t') \in \mathrm{ch}_1(K_{2r-1})$.

Thus $t' - t \in \mathrm{ch}_1(K_{2r-1}) \otimes_{\mathbf{Z}} \hat{\mathbf{Z}}$ has image of finite order in $(\mathrm{ch}_1(K_{2r-1}) \otimes_{\mathbf{Z}} \hat{\mathbf{Z}})/\,\mathrm{ch}_1(K_{2r-1})$. However, Lemma 5.2.7 shows $(\mathrm{ch}_1(K_{2r-1}) \otimes_{\mathbf{Z}} \hat{\mathbf{Z}})/\,\mathrm{ch}_1(K_{2r-1})$ is a rational vector space. Thus, in fact, $t' - t \in \mathrm{ch}_1(K_{2r-1})$ and it follows that $K^{\#}_{2r-1} \subset K^*_{2r-1}$. By symmetry, $K^{\#}_{2r-1} = K^*_{2r-1}$. □

Proposition 5.2.9

Choose an isomorphism $\tau : \hat{\mathbf{Z}} \to \hat{\mathbf{Z}}(r)$ of $\hat{\mathbf{Z}}$-modules. Then there is a finitely generated submodule K'_{2r-1} of $H^0_\infty(\mathbf{Z}(r)_+)$ for which there is a commutative diagram

$$\begin{array}{ccccccccc} 0 & \to & Y_{2r+1} & \to & K'_{2r-1} & \xrightarrow{\delta} & K^*_{2r-1} \cap \ker(d_1) & \to & 0 \\ & & \downarrow \tau_* & & \downarrow & & \downarrow & & \\ 0 & \to & H^0_\infty(\mathbf{Z}(r)_\infty) & \to & H^0_\infty(\mathbf{Z}(r)_+) & \xrightarrow{\delta} & \ker(d_1) & \to & 0 \end{array} \tag{14}$$

with the following properties.

(i) *The rows of (14) are exact, and bottom row of (14) is sequence (3) of Corollary 5.2.4, where $d_1 : H^1_\infty(\mathbf{Z}(r)) \to H^1_\infty(\mathbf{Z}(r)_\infty)$ is the natural map.*

(ii) *The left vertical arrow of (14) is the inclusion induced by τ and defined in Corollary 5.2.4. The module $K^*_{2r-1} \cap \ker(d_1)$ in the upper right corner of (14) has finite index in K^*_{2r-1} and is commensurable to $\mathrm{ch}_1(K_{2r-1})$.*

(iii) *If we tensor the top row of (14) with $\hat{\mathbf{Z}}$ over $\mathbf{Z}$, the natural homomorphism from the resulting exact sequence to the bottom row of (14) is an isomorphism on all terms of the sequence.*

Finally, suppose that K''_{2r-1} is another submodule of $H^0_\infty(\mathbf{Z}(r)_+)$ for which there is a diagram (14) satisfying properties (i)–(iii). Then there is a $\mathbf{Z}[G(N/K)]$-module isomorphism from K'_{2r-1} to K''_{2r-1} which fixes both Y_{2r+1} and $K^*_{2r-1} \cap \ker(d_1)$.

Proof. The exact sequence

$$0 \to H^0_\infty(\mathbf{Z}(r)_\infty) \to H^0_\infty(\mathbf{Z}(r)_+) \xrightarrow{\delta} \ker(d_1) \to 0$$

in (3) defines an extension class $e_1 \in \mathrm{Ext}^1_{\hat{\mathbf{Z}}[G(N/K)]}(\ker(d_1), H^0_\infty(\mathbf{Z}(r)_\infty))$. The submodule $K^*_{2r-1} \subset H^1_\infty(\mathbf{Z}(r))$ constructed in Lemma 5.2.8 is finitely generated and $K^*_{2r-1} \otimes_{\mathbf{Z}} \hat{\mathbf{Z}} = H^1_\infty(\mathbf{Z}(r))$. Lemma 5.2.3 shows d_1 has finite image. It follows that $(\ker(d_1) \cap K^*_{2r-1}) \otimes_{\mathbf{Z}} \hat{\mathbf{Z}} = \ker(d_1)$ and that $\ker(d_1) \cap K^*_{2r-1}$ has finite index in K^*_{2r-1}. The module K'_{2r-1} is finitely generated over $\mathbf{Z}$, since Y_{2r+1} and K^*_{2r-1} are. Corollary 5.2.4 shows that $H^0_\infty(\mathbf{Z}(r)_\infty) = Y_{2r+1} \otimes_{\mathbf{Z}} \hat{\mathbf{Z}}$ via the isomorphism τ. Since Y_{2r+1} is finitely generated, Lemma 5.2.7 now gives

$$\begin{aligned} &\mathrm{Ext}^1_{\hat{\mathbf{Z}}[G]}(\ker(d_1), H^0_\infty(\mathbf{Z}(r)_\infty)) \\ &= \mathrm{Ext}^1_{\hat{\mathbf{Z}}[G]}((\ker(d_1) \cap K^*_{2r-1}) \otimes_{\mathbf{Z}} \hat{\mathbf{Z}}, Y_{2r+1} \otimes_{\mathbf{Z}} \hat{\mathbf{Z}}) \\ &= \mathrm{Ext}^1_{\mathbf{Z}[G]}(\ker(d_1) \cap K^*_{2r-1}, Y_{2r+1}) \end{aligned} \tag{15}$$

The fact that e_1 is the image of a class in the far right group in (15) implies there is a $\mathbf{Z}[G(N/K)]$-module K'_{2r-1} for which there is a diagram of the form (14). Since the right and left vertical homomorphisms in (14) are injective, the middle one must be as well so that we may view K'_{2r-1} as a submodule of $H^0_\infty(\mathbf{Z}(r)_+)$.

Statements (i) and (ii) of Proposition 5.2.9 follow from the above construction and Lemma 5.2.8.

Part (iii) is implied by the fact that the extension classes of the top and bottom rows of (14) are identified under the isomorphisms in (15). Finally the assertions at the end of the statement of Proposition 5.2.9 about other possible choices K''_{2r-1} for K'_{2r-1} follow from the fact that diagram (14) and the isomorphisms in (15) fix the extension class of the top row of (14). □

Proposition 5.2.10

Define $K'_{2r-2} = H^1_\infty(\mathbf{Z}(r)_+)$ and let K'_{2r-1} be as in Proposition 5.2.9. Via Corollary 5.2.7 and Proposition 5.2.9 there is an isomorphism $K'_{2r-i} \otimes_{\mathbf{Z}} \hat{\mathbf{Z}} \cong H^{i-1}_\infty(\mathbf{Z}(r)_+)$ for $i \in \{1,2\}$. Let E be the class in

$$\mathrm{Ext}^2_{\hat{\mathbf{Z}}[G(N/K)]}(H^1_\infty(\mathbf{Z}(r)_+), H^0_\infty(\mathbf{Z}(r)_+)) \cong \mathrm{Ext}^2_{\hat{\mathbf{Z}}[G(N/K)]}(K'_{2r-2} \otimes_{\mathbf{Z}} \hat{\mathbf{Z}}, K'_{2r-1} \otimes_{\mathbf{Z}} \hat{\mathbf{Z}})$$

defined in Proposition 5.2.2. Let E' be the class in $\mathrm{Ext}^2_{\mathbf{Z}[G(N/K)]}(K'_{2r-2}, K'_{2r-1})$ which maps to E under the isomorphism

$$\mathrm{Ext}^2_{\mathbf{Z}[G(N/K)]}(K'_{2r-2}, K'_{2r-1}) \cong \mathrm{Ext}^2_{\hat{\mathbf{Z}}[G(N/K)]}(K'_{2r-2} \otimes_{\mathbf{Z}} \hat{\mathbf{Z}}, K'_{2r-1} \otimes_{\mathbf{Z}} \hat{\mathbf{Z}})$$

provided by (7) of Lemma 5.2.7. Then E' is represented by an exact sequence

$$0 \to K'_{2r-1} \to A_r \to B_r \to K'_{2r-2} \to 0 \tag{16}$$

in which A_r and B_r are finitely generated $\mathbf{Z}[G(N/K)]$-modules of finite projective dimension (i.e. cohomologically trivial $\mathbf{Z}[G(N/K)]$-modules).

Proof. Recall from Proposition 5.2.2 that the cup-product with E induces isomorphisms $\hat{H}^j(\Gamma; H^1_\infty(\mathbf{Z}(r)_+)) \to \hat{H}^{j+2}(\Gamma; H^0_\infty(\mathbf{Z}(r)_+))$ in Tate cohomology for all integers j and all subgroups $\Gamma \subseteq G(N/K)$. Since the isomorphism (6) of Lemma 5.2.7 respects cup-products, we find that cup-product with E' induces an isomorphism $\hat{H}^j(\Gamma; K'_{2r-2}) \to \hat{H}^{j+2}(\Gamma; K'_{2r-1})$ for all j and Γ. One can now construct (16) in the manner used to construct the four term S-unit sequence of ([152] §II.5). □

Definition 5.2.11 For $n = r - 1 \geq 1$, define $\Omega_n(N/K, 3)$ by the formula

$$\Omega_n(N/K,3) = [A_r] - [B_r] - [Div_{\mathbf{Z}}(\Sigma_\infty)] \in \mathcal{CL}(\mathbf{Z}[G(N/K)]) \subset K_0(\mathbf{Z}[G(N/K)])$$

where A_r and B_r are the cohomologically trivial $\mathbf{Z}[G(N/K)]$-modules of Proposition 5.2.10.

The following result is proved in ([38] §IV; cf. [34] and [132] Chapter 7).

Theorem 5.2.12

For $n \geq 1$, the class $\Omega_n(N/K,3)$ of Definition 5.2.11 depends only on N/K. In particular, it is independent of the choice of an exact sequence (16), the choice of τ and K'_{2r-1} in Proposition 5.2.9 and the choice of S.

5.2.13 *The image of $\Omega_n(N/K,3)$ in $G_0(\mathbf{Z}[G(N/K)])$*

Recall from §5.2.1 the module Y_{2n+1}, which appears in Proposition 5.2.9. Namely, $Y_{2n+1} = Div_{\mathbf{Z}}(S_\infty)$ if n is even and Y_{2n+1} is the cokernel of the natural homomorphism $Div_{\mathbf{Z}}(S_\infty) \to Div_{\mathbf{Z}}(\Sigma_\infty)$ if n is odd. For $n \geq 1$ it can be shown, using the results of [17], that the element

$$\Omega_n(N/K)^{\text{Naive}} = (K_{2n+1}(O_N)) - (K_{2n}(O_N)) - (Y_{2n+1}) \in G_0(\mathbf{Z}[G(N/K)])$$

has finite order (cf. [38] §I).

One might ask the following:

Question 5.2.14 Let N/K be a Galois extension of number fields with group $G(N/K)$. Is

$$0 = \Omega_n(N/K)^{\text{Naive}} \in G_0(\mathbf{Z}[G(N/K)])$$

for $n \geq 1$?

The calculation of $\Omega_n(N/K,3)$ in $K_0(\mathbf{Z}[G(N/K)])$ is related to Question 5.2.14 in the following manner. From (16), taking $n = r-1$ as usual, the image of $\Omega_{r-1}(N/K,3)$ in $G_0(\mathbf{Z}[G(N/K)])$ is equal to $[K'_{2r-1}] - [K'_{2r-2}]$. The Lichtenbaum-Quillen conjecture (see [142]) states that $\ker(\mathrm{ch}_i)$ and

$$H^i_\infty(\mathbf{Z}(r))/(\mathrm{ch}_i(K_{2r-i}) \otimes_{\mathbf{Z}} \hat{\mathbf{Z}})$$

have 2-primary order for $i = 1, 2$. This discussion, together with the results of ([96], [105], [106], [154]) on K_2 and K_3 of fields, yields the following result.

Proposition 5.2.15 *([38] § V)*

The image of $\Omega_n(N/K,3)$ in $G_0(\mathbf{Z}[G])$ is the class $\Omega_n(N/K)^{\text{Naive}}$ of if $n = 1$ and if $n > 1$ there difference has 2-primary order provided that the Lichtenbaum-Quillen conjecture is true.

Definition 5.2.16 *The Cassou-Noguès-Fröhlich class, $W_{N/K}$*

Let N/K be a Galois extension of number fields and let

$$\rho : G(N/K) \longrightarrow GL(V)$$

be a finite-dimensional, complex Galois representation. Let $W_K(\rho)$ be the complex number, on the unit circle, given by the Artin root number of Definition 1.1.9. By Proposition 1.1.10(i),

$$W_K : R(G(N/K)) = K_0(\mathbf{C}[G(N/K)]) \longrightarrow \mathbf{C}^*$$

is a homomorphism on the additive group of the representation ring (cf. §2.5.8) of $G(N/K)$.

Representing ρ by a homomorphism into $GL_n(\mathbf{C})$ and taking its complex conjugate yields a representation, $\overline{\rho}$. Furthermore, if $\rho = \overline{\rho}$ then $W_K(\rho) \in \{\pm 1\}$

([101],[153]). The fixed subgroup of $R(G(N/K))$ under complex conjugation is generated by the complexifications of real or symplectic representations – these types of complex representation will be called orthogonal or symplectic, respectively – and $W_K(\rho) = 1$ if ρ is an orthogonal representation ([58],[130] p. 289, (3.9)(proof),[153] p. 130).

In terms of these invariants of analytic origin we may define a class

$$W_{N/K} \in \mathcal{CL}(\mathbf{Z}[G(N/K)])$$

by means of the Hom-description of Theorem 2.5.25. Let $E/\mathbf{Q}$ be a large Galois extension of the rational numbers, as in §2.5.8. The absolute Galois group, $\Omega_{\mathbf{Q}}$ of §2.5.11, acts transitively on the infinite places of E. Let v_∞ denote one of these places. Define a homomorphism

$$W'_{N/K} \in \mathrm{Hom}_{\Omega_{\mathbf{Q}}}(R(G(N/K)), J^*(E))$$

by the following formula for the v-th coordinate of $W'_{N/K}(\rho)$, where ρ is any irreducible representation:

$$W'_{N/K}(\rho)_v = \begin{cases} 1 & \text{if } v \text{ is finite}, \\ 1 & \text{if } \rho \text{ is not symplectic}, \\ \alpha^{-1}(W_K(\alpha(\rho))) & \text{if } \rho \text{ symplectic}, \ v = \alpha^{-1}(v_\infty), \\ & \qquad \alpha \in \Omega_{\mathbf{Q}}. \end{cases}$$

By construction, $W'_{N/K}(\rho)$ is $\Omega_{\mathbf{Q}}$-equivariant and therefore represents a class, $W_{N/K}$, in $\mathcal{CL}(\mathbf{Z}[G(N/K)])$. However, we must show that the construction of $W_{L/K}$ is independent of the choice of v_∞. In the tamely ramified case it is possible to define $W_{L/K}$ without making a choice of v_∞ and it was in this form that the construction of $W_{N/K}$ was introduced in [30]. The general definition originates in ([34]; see also [35] p. 18).

Proposition 5.2.17 *The class, $W_{N/K}$, of Definition 5.2.16 is independent of the choice of v_∞.*

Proof. If ρ is a representation of $G(N/K)$ then, for each prime $P \lhd \mathcal{O}_K$ with $Q \lhd \mathcal{O}_L$ above it, we may form the Artin conductor ideal ([101] p. 14; see also §1.1.4)

$$f(\rho, N/K) \lhd \mathcal{O}_K$$

and write $Nf(\rho)$ for its absolute norm, $Nf(\rho) = |\mathcal{O}_K/f(\rho, N/K)|$.

Write $\sqrt{(Nf(\rho))}$ for the positive square-root of $Nf(\rho)$. We may define

$$h \in \mathrm{Hom}_{\Omega_{\mathbf{Q}}}(R(G(N/K)), J^*(E))$$

by replacing

$$\alpha^{-1}(W_K(\alpha(\rho))) \quad \text{with} \quad \alpha^{-1}(\sqrt{(Nf(\alpha(\rho)))})$$

in the definition of $W'_{N/K}(\rho)_v$. The resulting class is equal to one at all finite places and is totally positive on symplectic representations at all real infinite places. By ([155] p. 9; see also [35] p. 19 (2.6.1)(proof)) h is a determinant and represents the trivial element in the Hom-description of $\mathcal{CL}(\mathbf{Z}[G(N/K)])$. Hence $W_{N/K}$ may also be represented by $W'_{N/K} \cdot h^{-1}$. However, when ρ is irreducible and symplectic,

$$\frac{\alpha^{-1}(W_K(\alpha(\rho)))}{\alpha^{-1}(\sqrt{(Nf(\alpha(\rho)))})} = \frac{W_K(\rho)}{\sqrt{(Nf(\rho))}}$$

([34] p. 327) so that the $\alpha^{-1}(v_\infty)$-component of this representative for $W_{L/K}$ is independent of α and hence does not depend on the choice of v_∞. □

Now we may state what is expected of $\Omega_n(N/K,3)$ and present some of the evidence.

Conjecture 5.2.18 For any Galois extension of number fields, N/K, and any $n \geq 1$

$$W_{N/K} = \Omega_n(N/K,3) \in \mathcal{CL}(\mathbf{Z}[G(N/K)]).$$

As for the image of such a relation when mapped to $G_0(\mathbf{Z}[G(N/K)])$, it is an (unpublished) result of Queyrut that the image of $W_{N/K}$ is trivial. Hence, by Proposition 5.2.15, Conjecture 5.2.18 would suggest $\Omega_n(N/K)^{\mathrm{Naive}}$ has 2-primary order in $G_0(\mathbf{Z}[G(N/K)])$, which is at least consistent with the (expected) affirmative answer to Question 5.2.14.

Actually more is true. In the special examples of Theorems 1.3.6 and 1.3.11, Conjecture 5.2.18 is precisely Conjecture 1.3.5 so that it is true in the cases of Corollary 1.3.8. Furthermore, in ([38] §VI) we showed how a version of the Lichtenbaum-Gross conjecture (see Conjecture 1.2.2) about $L^*(-n,\rho)$ – the leading term of the Taylor series for the Artin L-function of ρ at $s = n$ for $n \geq 1$ – implies the congruence

$$\Omega_n(N/K,3) \equiv W_{N/K} \quad \text{modulo } D(\mathbf{Z}[G(N/K)])$$

where $D(\mathbf{Z}[G(N/K)])$ is the kernel subgroup of §2.5.26.

In ([38] §VII) we constructed the analogue of $\Omega_n(N/K,3)$ for a Galois extension of global function fields in characteristic $p > 0$ and proved that

$$\Omega_n(N/K,3) = W_{N/K}$$

in those circumstances, using the cohomological interpretation of L-functions.

Notice that $W_{N/K} = 0$ whenever $G(N/K)$ has no irreducible symplectic representations. This happens, for example, when $[N : K]$ is odd or $G(N/K)$ is abelian. In particular, if $K = \mathbf{Q}$ and $N = \mathbf{Q}(\xi_m)$, the cyclotomic field, Conjecture 5.2.18 predicts that $\Omega_n(\mathbf{Q}(\xi_m)/\mathbf{Q},3) = 0$. Similar remarks apply to all abelian extensions, $L/\mathbf{Q}$. Any such extension embeds into some $\mathbf{Q}(\xi_m)/\mathbf{Q}$ and the invariant, $\Omega_n(N/K,3)$, satisfies Galois descent so that the truth of the conjecture for $\mathbf{Q}(\xi_m)/\mathbf{Q}$ implies it for the subextension, $L/\mathbf{Q}$. In ([25], [26]) the construction of $\Omega_n(N/K,3)$ is generalized to cohomological motives with Galois actions ([25] Theorem 3.2; [26] Theorem 4.1; see also [27] Theorem 3.1 where $\Omega_n(N/K,3)$ is denoted by $\Omega_n(N/K)$). Furthermore it is shown in [27] that the motivic cyclotomic invariant vanishes away from $p=2$. In other words we have the following result:

Theorem 5.2.19 *([27]; see also §5.6.13)*
Let $L/\mathbf{Q}$ be an abelian extension of number fields. Then, for each $n \geq 1$,

$$0 = \Omega_n(L/\mathbf{Q},3) = W_{L/\mathbf{Q}} \in \mathcal{CL}(\mathbf{Z}[1/2][G(L/\mathbf{Q})]).$$

5.3 A closer look at $\Omega_1(L/K,3)$

In this section we shall give an (arguably) more explicit and elementary construction of $\Omega_1(L/K,3)$ than that of Definition 5.2.11, which has proved useful for computational purposes (see [37], [39]). In §5.4 we shall outline a proof that the construction we are about to give does indeed yield $\Omega_1(L/K,3)$.

5.3.1 *Double cosets and Archimedean places* Let L/K be a Galois extension of number fields and let $E/\mathbf{Q}$ be a large Galois extension of number fields such that $L \subset E$ and E is totally complex. Let c denote complex conjugation. Let Ω_L denote the absolute Galois group, $\Omega_L = G(\mathbf{Q}^{\mathrm{sep}}/L)$, where $\mathbf{Q}^{\mathrm{sep}}$ is a separable closure of $\mathbf{Q}$, the rationals.

Let $w_\infty : L \longrightarrow E \longrightarrow \mathbf{Q}^{\mathrm{sep}}$ be a fixed embedding which restricts to a real embedding, $v_\infty : K \longrightarrow E^{\langle c\rangle} \longrightarrow (\mathbf{Q}^{\mathrm{sep}})^{\langle c\rangle}$.

Proposition 5.3.2
In the notation of §5.3.1, there is an isomorphism of $\mathbf{Z}[G(L/K)]$-modules of the form

$$(\mathrm{Ind}_{\langle c\rangle}^{\Omega_{\mathbf{Q}}}(K_3^{\mathrm{ind}}(\mathbf{Q}^{\mathrm{sep}})))^{\Omega_L} \cong \oplus_{w\in S_\infty(L)} K_3^{\mathrm{ind}}(L_w)$$

where L_w is the completion of L at the Archimedean place, w.

Lemma 5.3.3
In the notation of §5.3.1:

(i) *There is a bijection between the set of double cosets, $\Omega_L\backslash\Omega_{\mathbf{Q}}/\langle c\rangle$, and the set of Archimedean places of L, $S_\infty(L)$.*

(ii) *The intersection, $\Omega_L \cap \langle \mathrm{gcg}^{-1}\rangle$, is trivial if $(w_\infty)g$ is a complex place and has order two if $(w_\infty)g$ is real.*

(iii) *In the first case of part (ii), if $(v_\infty)g$ is a real place of K then $\mathrm{gcg}^{-1} \in \Omega_K$ and its image in $G(L/K) \cong \Omega_K/\Omega_L$ is the decomposition group, $H_g = G(L_{(w_\infty)g}/K_{(v_\infty)g})$.*

Proof. For part (i), assigning to $g \in \Omega_{\mathbf{Q}}$ the embedding $(w_\infty)g : L \xrightarrow{v_\infty} \mathbf{Q}^{\mathrm{sep}} \xrightarrow{g} \mathbf{Q}^{\mathrm{sep}}$ defines a bijection between embeddings of L and $\Omega_L\backslash\Omega_{\mathbf{Q}}$. The set of embeddings, $\{(w_\infty)g, (w_\infty)gc\}$, corresponds to an Archimedean place of L, since the completions of $(w_\infty)g$ and $(w_\infty)gc$ coincide. Hence assigning the double coset $\Omega_L g\langle c\rangle$ to this Archimedean place defines a bijection between $\Omega_L\backslash\Omega_{\mathbf{Q}}/\langle c\rangle$ and $S_\infty(L)$.

For part (ii), if $(w_\infty)g$ is a complex place then gcg^{-1} does not belong to Ω_L and $\Omega_L \cap \langle \mathrm{gcg}^{-1}\rangle = \{1\}$. If $(w_\infty)g$ is real then $\Omega_L \cap \langle \mathrm{gcg}^{-1}\rangle = \langle \mathrm{gcg}^{-1}\rangle$ is of order two.

Part (iii) is clear. □

Lemma 5.3.4

In the notation of §5.3.1, suppose that M is a $\mathbf{Z}[G(E/\mathbf{Q})]$-module.

(i) *There is a double coset isomorphism of $\mathbf{Z}[G(E/L)]$-modules of the form*

$$\oplus_{g\in G(E/L)\backslash G(E/\mathbf{Q})/\langle c\rangle} \mathrm{Ind}_{G(E/L)\cap\langle \mathrm{gcg}^{-1}\rangle}^{G(E/L)}((g^{-1})^*(M)) \xrightarrow{\cong} \mathrm{Res}_{G(E/L)}^{G(E/\mathbf{Q})}(\mathrm{Ind}_{\langle c\rangle}^{G(E/\mathbf{Q})}(M)).$$

(ii) *Also, taking $G(E/L)$-fixed points, there is an isomorphism of $\mathbf{Z}[G(L/K)]$-modules*

$$\oplus_{g\in G(E/L)\backslash G(E/\mathbf{Q})/\langle c\rangle}((g^{-1})^*(M))^{G(E/L)\cap\langle \mathrm{gcg}^{-1}\rangle} \xrightarrow{\cong} (\mathrm{Ind}_{\langle c\rangle}^{G(E/\mathbf{Q})}(M))^{G(E/L)}.$$

Here, if $G(E/L)\cap\langle \mathrm{gcg}^{-1}\rangle = \langle \mathrm{gcg}^{-1}\rangle$, $(g^{-1})^*(M)$ denotes M with a new action whereby gcg^{-1} acts on m to send it to $c(m)$. If $G(E/L)\cap\langle \mathrm{gcg}^{-1}\rangle$ is trivial then $(g^{-1})^*(M)$ is M and the action is, of course, trivial.

Proof. In part (i) the isomorphism is given by the well-known Double Coset isomorphism (see [131] Theorem 1.2.40, for example) which sends

$$u \otimes_{G(E/L)\cap\langle \mathrm{gcg}^{-1}\rangle} m \in \mathrm{Ind}_{G(E/L)\cap\langle \mathrm{gcg}^{-1}\rangle}^{G(E/L)}((g^{-1})^*(M))$$

to $ug \otimes_{\langle c\rangle} m \in \mathrm{Ind}_{\langle c\rangle}^{G(E/\mathbf{Q})}(M)$. Hence generators of the $G(E/L)g\langle c\rangle$-component of the $G(E/L)$-fixed points are given by

$$\sum_{u\in G(E/L)/(G(E/L)\cap\langle \mathrm{gcg}^{-1}\rangle)} u \otimes_{G(E/L)\cap\langle \mathrm{gcg}^{-1}\rangle} m$$

where $m \in (g^{-1})^*(M))^{G(E/L)\cap\langle \mathrm{gcg}^{-1}\rangle}$. In $\mathrm{Ind}_{\langle c\rangle}^{G(E/\mathbf{Q})}(M)$ such a generator corresponds to

$$\sum_{u\in G(E/L)/(G(E/L)\cap\langle \mathrm{gcg}^{-1}\rangle)} ug \otimes_{\langle c\rangle} m.$$

It remains to verify that this yields an isomorphism of $\mathbf{Z}[G(L/K)]$-modules as claimed in part (ii). We may lift $y \in G(L/K) \cong G(E/K)/G(E/L)$ to $y \in G(E/K)$ and then the action of y on $\sum_{u\in G(E/L)/(G(E/L)\cap\langle gcg^{-1}\rangle)} ug \otimes_{\langle c\rangle} m$ sends it to

$$\begin{aligned}&\textstyle\sum_{u\in G(E/L)/(G(E/L)\cap\langle gcg^{-1}\rangle)} yug \otimes_{\langle c\rangle} m\\ &= \textstyle\sum_{v\in G(E/L)/(G(E/L)\cap\langle ygcg^{-1}y^{-1}\rangle)} uyg \otimes_{\langle c\rangle} m,\end{aligned}$$

since $G(E/L) \lhd G(E/K)$. Therefore the element $m \in ((g^{-1})^*(M))^{G(E/L)\cap\langle gcg^{-1}\rangle}$ is mapped by y to $m \in ((g^{-1})^*(M))^{G(E/L)\cap\langle ygcg^{-1}y^{-1}\rangle}$, as required. □

5.3.5 *Proof of Proposition 5.3.2*

Setting $M = K_3^{\mathrm{ind}}(E)$ in Lemma 5.3.4 and passing to the limit over $E/\mathbf{Q}$ we obtain an isomorphism of $\mathbf{Z}[G(L/K)]$-modules of the form

$$(\mathrm{Ind}_{\langle c\rangle}^{\Omega_{\mathbf{Q}}}(K_3^{\mathrm{ind}}(\mathbf{Q}^{\mathrm{sep}})))^{\Omega_L} \cong \oplus_{w\in S_\infty(L)} K_3^{\mathrm{ind}}(L_w)$$

where L_w is the completion of L at the Archimedean place, w. Here we have used the isomorphism,

$$(K_3^{\mathrm{ind}}(E))^{G(E/L)\cap\langle gcg^{-1}\rangle} \cong K_3^{\mathrm{ind}}(E^{G(E/L)\cap\langle gcg^{-1}\rangle}),$$

together with Lemma 5.3.3(iii). □

5.3.6 *Towards the 2-extension*

Let M be a $\mathbf{Z}[G(E/\mathbf{Q})]$-module. Define an injective $\mathbf{Z}[G(E/\mathbf{Q})]$-homomorphism

$$\phi : M \longrightarrow \mathrm{Ind}_{\langle c\rangle}^{G(E/\mathbf{Q})}(M) \cong \mathbf{Z}[G(E/\mathbf{Q})] \otimes_{\mathbf{Z}[\langle c\rangle]} M$$

by $\phi(m) = \sum_{h\in G(E/\mathbf{Q})/\langle c\rangle} h \otimes_{\langle c\rangle} h^{-1}(m)$. Set $M_+ = \mathrm{coker}(\phi)$ so that there is a short exact sequence of $\mathbf{Z}[G(E/\mathbf{Q})]$-modules of the form

$$0 \longrightarrow M \longrightarrow \mathrm{Ind}_{\langle c\rangle}^{\Omega_{\mathbf{Q}}}(M) \longrightarrow M_+ \longrightarrow 0.$$

At several points in this section we shall make crucial use of the deep results of ([96], [104], [105], [106]) concerning the connection between algebraic K-groups of fields and cohomology as well as their consequences derived in [80]. The reader is referred to ([63] §18) for a description and comparison of the different approaches to these results. In particular we shall need the following canonical isomorphisms:

$$\begin{aligned}&K_3^{\mathrm{ind}}(\mathbf{Q}^{\mathrm{sep}})^{\Omega_L} \cong K_3^{\mathrm{ind}}(L) \qquad ([96],[106]),\\ &H^1(L; K_3^{\mathrm{ind}}(\mathbf{Q}^{\mathrm{sep}})) \cong K_2(L),\ H^1(L_w; K_3^{\mathrm{ind}}(L_w^{\mathrm{sep}})) \cong K_2(L_w). \quad [80]\end{aligned}$$

Proposition 5.3.7

In the notation of §5.3.6, there is a 2-extension of $\mathbf{Z}[G(L/K)]$-modules of the form

$$K_3^{\mathrm{ind}}(L) \longrightarrow \oplus_{w\in S_\infty(L)} K_3^{\mathrm{ind}}(L_w) \longrightarrow (K_3^{\mathrm{ind}}(\mathbf{Q}^{\mathrm{sep}})_+)^{\Omega_L} \longrightarrow K_2'(L).$$

Here, as in Definition 5.2.11, $K_2'(L) = \ker(K_2(L) \longrightarrow \oplus_{w\in S_\infty(L)} K_2(L_w))$.

Proof. Setting $M = K_3^{\mathrm{ind}}(E)$ and taking the limit over $E/\mathbf{Q}$ we obtain a short exact sequence of continuous $\Omega_{\mathbf{Q}}$-modules of the form

$$0 \longrightarrow K_3^{\mathrm{ind}}(\mathbf{Q}^{\mathrm{sep}}) \longrightarrow \mathrm{Ind}_{\langle c\rangle}^{\Omega_{\mathbf{Q}}}(K_3^{\mathrm{ind}}(\mathbf{Q}^{\mathrm{sep}})) \longrightarrow K_3^{\mathrm{ind}}(\mathbf{Q}^{\mathrm{sep}})_+ \longrightarrow 0.$$

Applying $H^*(L;-)$ to this short exact sequence yields, in dimensions 0 and 1, the following 2-extension

$$K_3^{\mathrm{ind}}(L) \longrightarrow \oplus_{w\in S_\infty(L)} K_3^{\mathrm{ind}}(L_w) \longrightarrow (K_3^{\mathrm{ind}}(\mathbf{Q}^{\mathrm{sep}})_+)^{\Omega_L} \longrightarrow K_2'(L).$$

Here we have used Proposition 5.3.2 to identify the second module and the cohomological isomorphisms of §5.3.6 to complete the proof. □

Remark 5.3.8

(i) As explained in [80], in Proposition 5.3.7 $(K_3^{\mathrm{ind}}(\mathbf{Q}^{\mathrm{sep}})_+)^{\Omega_L}$ is a cohomologically trivial $\mathbf{Z}[G(L/K)]$-module. However the module, $\oplus_{w\in S_\infty(L)} K_3^{\mathrm{ind}}(L_w)$, is cohomologially trivial if and only if L/K is unramified at infinity. This is one of the reasons for all the modifications which are to follow. A further reason for needing to modify the 2-extension of Proposition 5.3.7 is that the $K_2'(L)$ is not finitely generated and so we shall replace it below by $K_2'(\mathcal{O}_{L,S})$.

(ii) Notice that, by §5.3.5, as Ω_L-modules,

$$\mathrm{Ind}_{\langle c\rangle}^{\Omega_{\mathbf{Q}}}(K_3^{\mathrm{ind}}(\mathbf{Q}^{\mathrm{sep}})) \cong \oplus_{w\in S_\infty(L)} i_{w,*} i_w^* K_3^{\mathrm{ind}}(\mathbf{Q}^{\mathrm{sep}}) = (K_3^{\mathrm{ind}}(\mathbf{Q}^{\mathrm{sep}}))_\infty$$

in the notation of §5.2.1.

Proposition 5.3.9

Let S be a finite $G(L/K)$-stable set of places of L containing $S_\infty(L)$ and all places which ramify over K such as, for example, the set S of §5.2.1. Then the natural homomorphisms

$$K_2'(\mathcal{O}_{L,S}) \longrightarrow K_2'(L) \ \textit{and}\ K_3^{\mathrm{ind}}(\mathcal{O}_{L,S}) \xrightarrow{\cong} K_3^{\mathrm{ind}}(L)$$

induce an isomorphism of the form

$$\mathrm{Ext}^2_{\mathbf{Z}[G(L/K)]}(K_2'(L), K_3^{\mathrm{ind}}(L)) \xrightarrow{\cong} \mathrm{Ext}^2_{\mathbf{Z}[G(L/K)]}(K_2'(\mathcal{O}_{L,S}), K_3^{\mathrm{ind}}(\mathcal{O}_{L,S})).$$

Proof. The homomorphism, $K_2(\mathcal{O}_{L,S}) \longrightarrow \oplus_{w\in S_\infty(L)} K_2(L_w)$, is surjective onto the torsion (with uniquely divisible cokernel) so that, from the long exact localisation sequence for K-groups (see Corollary 2.4.7), we obtain a short exact localisation sequence of the form

$$0 \longrightarrow K_2'(\mathcal{O}_{L,S}) \longrightarrow K_2'(L) \longrightarrow \oplus_{P\notin S,\ P\ finite}(\mathcal{O}_L/P)^* \longrightarrow 0.$$

However, if $R \lhd \mathcal{O}_K$ is unramified in L/K then

$$\oplus_{P|R}\ (\mathcal{O}_L/P)^* \cong \mathrm{Ind}_{G(L_{P_0}/K_R)}^{G(L/K)}((\mathcal{O}_L/P_0)^*)$$

for some $P_0 \lhd \mathcal{O}_L$ lying over R. In addition, since L_{P_0}/K_R is unramified, there is a $\mathbf{Z}[G(L_{P_0}/K_R)]$-resolution ([131] §7.3.39)

$$0 \longrightarrow \mathbf{Z}[G(L_{P_0}/K_R)] \longrightarrow \mathbf{Z}[G(L_{P_0}/K_R)] \longrightarrow (\mathcal{O}_L/P_0)^* \longrightarrow 0.$$

Therefore $\mathrm{Ext}^i_{\mathbf{Z}[G(L/K)]}(\oplus_{P \notin S,\ P\ finite}(\mathcal{O}_L/P)^*, M) = 0$ for all M and all $i \geq 1$. The result follows easily from the long exact sequence of Ext-groups. □

Corollary 5.3.10

Pulling back the 2-extension of Proposition 5.3.9 via the homomorphism, $K_2'(\mathcal{O}_{L,S}) \longrightarrow K_2'(L)$, yields a 2-extension of $\mathbf{Z}[G(L/K)]$-modules of the form

$$K_3^{\mathrm{ind}}(\mathcal{O}_{L,S}) \longrightarrow \oplus_{v \in S_\infty(L)} K_3^{\mathrm{ind}}(L_v) \longrightarrow C \longrightarrow K_2'(\mathcal{O}_{L,S})$$

in which C is cohomologically trivial.

Proof. It suffices to notice that, by Remark 5.3.8(i), C is the pull-back of a cohomologically trivial module via a homomorphism which induces isomorphisms on all Tate cohomology groups. □

5.3.11 *Modification for ramification at infinity*

When L/K is unramified at infinity then $\oplus_{w \in S_\infty(L)} K_3^{\mathrm{ind}}(L_w)$ is also cohomologically trivial and in this case $\Omega_1(L/K, 3) \in \mathcal{CL}(\mathbf{Z}[G(L/K)])$ was defined in ([132] Chapter 7) to be equal to the Euler characteristic, in the sense of Example 2.1.8(i), of a representative of this 2-extension in which the central modules are finitely generated and cohomologically trivial. From the localisation sequence it is straightforward to show that the resulting Euler characteristic is independent of S ([131] §7.1.3). We shall now describe how to modify this 2-extension in the case of ramification at infinity.

Let $S'_\infty(K)$ denote the set of real places of K which become complex in L. For each $v \in S'_\infty(K)$, choose $w = w(v)$ to be a place of L over v with decomposition group, $G_w = G(L_{w(v)}/K_v) = \{1, \tau_w\}$ of order two. The short exact sequence

$$0 \longrightarrow (\mathbf{Q}/\mathbf{Z})(2) \longrightarrow K_3^{\mathrm{ind}}(L_w) \longrightarrow K_3^{\mathrm{ind}}(L_w) \otimes \mathbf{Q} \longrightarrow 0$$

shows that there exists a unique element of order two, $(-1)_w \in K_3^{\mathrm{ind}}(L_w)$, which is fixed by τ_w. Define a $\mathbf{Z}[G_w]$-module, $\tilde{K}_3^{\mathrm{ind}}(L_w)$, in the following manner. The underlying abelian group of $\tilde{K}_3^{\mathrm{ind}}(L_w)$ is $K_3(L_w) \oplus \mathbf{Z}$. If we denote by $a\tilde{\oplus}b$ the element of $\tilde{K}_3^{\mathrm{ind}}(L_w)$ corresponding to $a \oplus b \in K_3^{\mathrm{ind}}(L_w) \oplus \mathbf{Z}$ then the action of τ_w on $\tilde{K}_3^{\mathrm{ind}}(L_w)$ is given by $\tau_w(a\tilde{\oplus}b) = (\tau_w(a) \cdot (-1)_w^b)\tilde{\oplus}b$.

The following result is straightforward.

Lemma 5.3.12

In the notation of §5.3.11, $\tilde{K}_3^{\mathrm{ind}}(L_w)$ is a cohomologically trivial G_w-module.

The lower row of the following commutative diagram is the 2-extension whose Euler characteristic will define $\Omega_1(L/K,3) \in \mathcal{CL}(\mathbf{Z}[G(L/K)])$ in general.

Proposition 5.3.13

With the notation of Corollary 5.3.10 and Lemma 5.3.12, there exists a canonical commutative diagram of 2-extensions of $\mathbf{Z}[G(L/K)]$-modules in which $\bigoplus_w \tilde{K}_3^{\text{ind}}(L_w)$ and $C \bigoplus T_2$ are cohomologically trivial and $\tilde{K}_2'(O_{L,S})$ is a finitely generated extension of $K_2'(O_{L,S})$.

$$\begin{array}{ccccccc}
K_3^{\text{ind}}(L) & \longrightarrow & \bigoplus_w K_3^{\text{ind}}(L_w) & \xrightarrow{\delta} & C & \longrightarrow & K_2'(O_{L,S}) \\
\downarrow & & \downarrow & & \downarrow & & \downarrow \\
K_3^{\text{ind}}(L) & \longrightarrow & \bigoplus_w \tilde{K}_3^{\text{ind}}(L_w) & \xrightarrow{\tilde{\delta}_w} & C \bigoplus T_2 & \longrightarrow & \tilde{K}_2'(O_{L,S})
\end{array}$$

Proof. The proof of the proposition will consist of the construction of a sequence of diagrams (1)–(5). The required 2-extension will be found as the middle row of diagram (5).

Consider once more the 2-extension of $\mathbf{Z}[G(L/K)]$-modules

$$K_3^{\text{ind}}(\mathcal{O}_{L,S}) \longrightarrow \oplus_{w \in S_\infty(L)} K_3^{\text{ind}}(L_w) \xrightarrow{\delta} C \longrightarrow K_2'(\mathcal{O}_{L,S}).$$

Suppose that $v \in S'_\infty(K)$ and $w = w(v)$, as in §5.3.11. Since $(1+\tau_w)((-1)_w) = 0$, $(1+\tau_w)(\delta((-1)_w)) = 0$ in C. Therefore, because C is cohomologically trivial, there is an $\alpha_w \in C$ such that $(1-\tau_w)(\alpha_w) = \delta((-1)_w)$.

We now construct a commutative diagram of $\mathbf{Z}[G_w]$-modules of the form

$$\begin{array}{ccccc}
K_3(L_w) & \xrightarrow{\delta} & C & \longrightarrow & K_2'(O_{L,S}) \\
\downarrow & & \downarrow & & \downarrow \\
\tilde{K}_3^{\text{ind}}(L_w) & \xrightarrow{\tilde{\delta}_w} & C \bigoplus \mathbf{Z}[G_w] & \longrightarrow & \tilde{K}_2'(O_{L,S})_w \\
\downarrow & & \downarrow & & \downarrow \\
\mathbf{Z} & \longrightarrow & \mathbf{Z}[G_w] & \longrightarrow & \mathbf{Z}_{w,-}
\end{array} \qquad (1)$$

with the following properties. The vertical columns are short exact and the terms of the second column map surjectively onto those of the third column. The morphisms $C \longrightarrow C \bigoplus \mathbf{Z}[G_w]$ and $C \bigoplus \mathbf{Z}[G_w] \longrightarrow \mathbf{Z}[G_w]$ in the middle column are defined by $(c \mapsto c \oplus 0)$ and $(c \oplus a \mapsto a)$, respectively. The morphism $\tilde{\delta}_w : \tilde{K}_3^{\text{ind}}(L_w) = K_3^{\text{ind}}(L_w) \tilde{\oplus} \mathbf{Z} \longrightarrow C \bigoplus \mathbf{Z}[G_w]$ is defined by $\tilde{\delta}(a \tilde{\oplus} 0) = \delta(a) \oplus 0$ and $\tilde{\delta}(0 \tilde{\oplus} 1) = \alpha_w \oplus (1+\tau_w)$. The module $\tilde{K}_2'(O_{L,S})_w$ is defined to be the cokernel of $\tilde{\delta}_w$, while $\mathbf{Z}_{w,-}$ is the G_w-module with underlying abelian group $\mathbf{Z}$ on which τ_w acts by multiplication by -1. To check that one has a commutative diagram in (1) it suffices to check that

$$\begin{array}{ccc} K_3^{\mathrm{ind}}(L_w) & \xrightarrow{\delta} & C \\ \downarrow & & \downarrow \\ \tilde{K}_3^{\mathrm{ind}}(L_w) & \xrightarrow{\tilde{\delta}_w} & C \bigoplus \mathbf{Z}[G_w] \end{array} \qquad (2)$$

is a commutative square of $\mathbf{Z}[G_w]$-module homomorphisms. This follows from the formulae

$$(1 - \tau_w)(0\tilde{\oplus}1) = (-1)_w\tilde{\oplus}0 \in \tilde{K}_3^{\mathrm{ind}}(L_w)$$

and

$$(1 - \tau_w)\ \tilde{\delta}(0\tilde{\oplus}1) = (1 - \tau_w)(\alpha_w \oplus (1 + \tau_w)) = \delta(-1)_w \oplus 0 \in C \oplus \mathbf{Z}[G_w].$$

It will be convenient to define versions of the diagram (1) when $w = w(v)$ and v is an arbitrary infinite place of F for which purpose we must pause in order to introduce some temporary notation.

Definition 5.3.14 Let $S_1(K)$ denote the set of complex places of K and let $S_2(K)$ denote the set of real places of K which do not ramify in L. Thus $S_\infty(K)$ is the disjoint union, $S_1(K) \cup S_2(K) \cup S'_\infty(K)$.

Suppose $v \in S_\infty(K)$ and $w = w(v)$, as in §5.3.11. Set $\tilde{K}_3^{\mathrm{ind}}(L_w) = K_3^{\mathrm{ind}}(L_w)$ if $v \in S_1(K) \cup S_2(K)$. Let $T_{1,w} = \mathbf{Z}$ if $v \in S'_\infty(K)$ and let $T_{1,w} = 0$ otherwise. Define $T_{2,w} = \mathbf{Z}[G_w]$ if $v \in S'_\infty(K) \cup S_1(K)$ and $T_{2,w} = 0$ otherwise. Finally, let $T_{3,w} = \mathbf{Z}_{w,-}$ if $v \in S'_\infty(K)$, $T_{3,w} = 0$ if $v \in S_2(K)$ and $T_{3,w} = \mathbf{Z}[G_w]$ if $v \in S_1(K)$.

5.3.15 *The proof of Proposition 5.3.13 continued*

For all infinite places v of K and $w = w(v)$, we now have a commutative diagram of $\mathbf{Z}[G_w]$-modules

$$\begin{array}{ccccc} K_3^{\mathrm{ind}}(L_w) & \xrightarrow{\delta} & C & \longrightarrow & K'_2(O_{L,S}) \\ \downarrow & & \downarrow & & \downarrow \\ \tilde{K}_3^{\mathrm{ind}}(L_w) & \xrightarrow{\tilde{\delta}_w} & C \bigoplus T_{2,w} & \longrightarrow & \tilde{K}'_2(O_{L,S})_w \\ \downarrow & & \downarrow & & \downarrow \\ T_{1,w} & \longrightarrow & T_{2,w} & \longrightarrow & T_{3,w} \end{array} \qquad (3)$$

of the following kind. The vertical sequences are short exact. If $v \in S'_\infty(K)$ then (3) is diagram (1). If $v \in S_1(K) \cup S_2(K)$ then G_w and $T_{1,w}$ are trivial and (3) is obtained from the 2-extension of (1) by push-out via the inclusion morphism, $C \longrightarrow \tilde{C} = C \oplus T_{2,w}$, where $T_{2,w} = T_{3,w}$ in this case. Once again the maps from the middle to the right-hand column are all surjective.

Inducing up from G_w to $G(L/K)$ and using the fact that C and $K'_2(O_{L,S})$ are $\mathbf{Z}[G(L/K)]$-modules, we may use (3) to construct a diagram

$$\begin{array}{ccccc}
\mathrm{Ind}_{G_w}^{G(L/K)} K_3^{\mathrm{ind}}(L_w) & \xrightarrow{\delta} & C & \longrightarrow & K_2'(O_{L,S}) \\
\downarrow & & \downarrow & & \downarrow \\
\mathrm{Ind}_{G_w}^{G(L/K)} \tilde{K}_3^{\mathrm{ind}}(L_w) & \xrightarrow{\tilde{\delta}_w} & C \bigoplus \mathrm{Ind}_{G_w}^{G(L/K)} T_{2,w} & \longrightarrow & \tilde{K}_2'(O_{L,S})_{w,0} \\
\downarrow & & \downarrow & & \downarrow \\
\mathrm{Ind}_{G_w}^{G(L/K)} T_{1,w} & \longrightarrow & \mathrm{Ind}_{G_w}^{G(L/K)} T_{2,w} & \longrightarrow & \mathrm{Ind}_{G_w}^{G(L/K)} T_{3,w}
\end{array} \qquad (4)$$

whose columns are short exact and in which the terms of the second column map surjectively onto those of the third.

We now identify

$$\mathrm{Ind}_{G_w}^{G(L/K)} K_3^{\mathrm{ind}}(L_w) \qquad \text{with} \qquad \oplus_{w'|v} K_3^{\mathrm{ind}}(L_{w'}),$$

and define

$$\oplus_{w'|v} \tilde{K}_3^{\mathrm{ind}}(L_w) \qquad \text{to be} \qquad \mathrm{Ind}_{G_w}^{G(L/K)} \tilde{K}_3^{\mathrm{ind}}(L_w).$$

Using (4) to push-out the 2-extension of Corollary 5.3.10 yields a diagram of the following form in which the columns are short exact and the rows are exact 2-extensions.

$$\begin{array}{ccccccc}
K_3^{\mathrm{ind}}(L) & \longrightarrow & \bigoplus_w K_3^{\mathrm{ind}}(L_w) & \xrightarrow{\delta} & C & \longrightarrow & K_2'(O_{L,S}) \\
\downarrow & & \downarrow & & \downarrow & & \downarrow \\
K_3^{\mathrm{ind}}(L) & \longrightarrow & \bigoplus_w \tilde{K}_3^{\mathrm{ind}}(L_w) & \xrightarrow{\tilde{\delta}_w} & C \bigoplus T_2 & \longrightarrow & \tilde{K}_2'(O_{L,S}) \\
\downarrow & & \downarrow & & \downarrow & & \downarrow \\
0 & \longrightarrow & T_1 & \longrightarrow & T_2 & \longrightarrow & T_3
\end{array} \qquad (5)$$

For $i = 1, 2, 3$ one has

$$T_i = \bigoplus_{v \in S_\infty(K)} \mathrm{Ind}_{G_{w(v)}}^{G(L/K)} T_{i,w(v)}$$

where the sum is over the infinite places v of K. The module $\tilde{K}_2'(O_{L,S})$ is defined to be the cokernel of $\tilde{\delta}$.

Since $\tilde{K}_3^{\mathrm{ind}}(L_{w(v)})$ and $T_{2,w}$ are cohomologically trivial $\mathbf{Z}[G_{w(v)}]$-modules, the modules $\oplus_w \tilde{K}_3^{\mathrm{ind}}(L_w)$ and $C \oplus T_2$ in the middle row of (5) are cohomologically trivial $\mathbf{Z}[G(L/K)]$-modules, which completes the proof of Proposition 5.3.13. □

5.3.16 Since $K_3^{\mathrm{ind}}(L)$ and $\tilde{K}_2'(O_{L,S})$ are finitely generated $\mathbf{Z}[G(L/K)]$-modules, we may construct a commutative diagram representing an equivalence of 2-extensions

$$\begin{array}{ccccccc}
K_3^{\mathrm{ind}}(L) & \longrightarrow & A_1 & \longrightarrow & B_1 & \longrightarrow & \tilde{K}_2'(O_{L,S}) \\
\downarrow{\scriptstyle 1} & & \downarrow & & \downarrow & & \downarrow{\scriptstyle 1} \\
K_3^{\mathrm{ind}}(L) & \longrightarrow & \bigoplus_w \tilde{K}_3^{\mathrm{ind}}(L_w) & \xrightarrow{\tilde{\delta}} & C \bigoplus T_2 & \longrightarrow & \tilde{K}_2'(O_{L,S})
\end{array} \qquad (6)$$

in which A_1 and B_1 are finitely generated, cohomologically trivial $\mathbf{Z}[G(L/K)]$-modules.

Theorem 5.3.17

(i) *For A_1 and B_1 as in diagram (6) of §5.3.16, temporarily write $\Omega(L/K)$ for the the Euler characteristic, $[A_1] - [B_1] \in K_0(\mathbf{Z}[G(L/K)])$. Then*

$$\Omega(L/K) = [A_1] - [B_1] \in \mathcal{CL}(\mathbf{Z}[G(L/K)]) \subset K_0(\mathbf{Z}[G(L/K)]).$$

(ii) *Also $\Omega(L/K)$ depends only on the extension L/K. In particular, $\Omega(L/K)$ does not depend on the choice of places S in §5.3.6, on the $w(v)$ chosen for $v \in S_\infty(K)$, on the $\alpha_{w(v)} \in C$ for $v \in S'_\infty(K)$ used to construct the diagrams (2) and (5) or on the choice of a top row of diagram (6).*

Proof. For part (i), it is clear that $[A_1]-[B_1] \in K_0(\mathbf{Z}[G(L/K)])$ and, by definition, $\mathcal{CL}(\mathbf{Z}[G(L/K)])$ is equal to the kernel of the rank homomorphism,

$$K_0(\mathbf{Z}[G(L/K)]) \longrightarrow \mathbf{Z}.$$

Firstly we remark that the rank of $T_{3,w(v)}$ over $\mathbf{Z}$ is 1 if $w(v)$ is complex and is 0 if $w(v)$ is real. Let $r_2(L)$ be the number of complex places of L. Since $K_2'(O_{L,S})$ is finite, T_3 and $\tilde{K}_2'(O_{L,S})$ both have rank $r_2(L)$ over $\mathbf{Z}$. By a result of Borel, $K_3^{\text{ind}}(L)$ also has rank $r_2(L)$ and the exactness of the first row of (6) shows $\Omega(L/K)$ has rank 0 and therefore lies in $\mathcal{CL}(\mathbf{Z}[G(L/K)]) \subset K_0(\mathbf{Z}[G(L/K)])$.

For part (ii), we now consider, in reverse order, the dependence of $\Omega(L/K)$ upon the choices listed in the statement of Theorem 5.3.17.

It is well known that if the bottom row of (6) is fixed and the top row is changed, the class $[A_1]-[B_1]$ in $K_0(\mathbf{Z}[G(L/K)])$ is unaltered (see [132] Proposition 2.2.2 p. 47, for example).

Suppose now that we fix S and the choice of the $w(v)$ for v an infinite place of K. We must show that $\Omega(L/K)$ does not depend on the choice of the elements $\alpha_{w(v)} \in C$ for $v \in S'_\infty(K)$ which were used in the construction of (2) and (5). Fix $v \in S'_\infty(K)$ and let $w = w(v)$. Suppose α'_w is another element of C which satisfies the defining condition for α_w, namely that $(1-\tau_w)\alpha_w = (1-\tau_w)\alpha'_w = \delta(-1)_w$ where, as in §5.3.11, $\tau_w \in G_w$ is complex conjugation at w and $\delta : K_3^{\text{ind}}(L_w) \longrightarrow C$ is the homomorphism in the 2-extension of §5.3.15. Then $(1-\tau_w)(\alpha_w - \alpha'_w) = 0$ and, since C is cohomologically trivial, there is an element, $\gamma_w \in C$, such that $(1+\tau_w)\gamma_w = \alpha_w - \alpha'_w$. Define a $\mathbf{Z}[G_w]$-automorphism $\pi_w : C \oplus \mathbf{Z}[G_w] \longrightarrow C \oplus \mathbf{Z}[G_w]$ by $\pi_w(c \oplus 0) = c \oplus 0$ and $\pi_w(0 \oplus 1) = \gamma_w \oplus 1$. We claim that the diagram

$$\begin{array}{ccc}
\tilde{K}_3^{\text{ind}}(L_w) = K_3^{\text{ind}}(L_w)\tilde{\oplus}\mathbf{Z} & \xrightarrow{\tilde{\delta}'_w} & C \oplus \mathbf{Z}[G_w] \\
\Big\downarrow & & \Big\downarrow \pi_w \\
\tilde{K}_3^{\text{ind}}(L_w) = K_3^{\text{ind}}(L_w)\tilde{\oplus}\mathbf{Z} & \xrightarrow{\tilde{\delta}_w} & C \oplus \mathbf{Z}[G_w]
\end{array} \tag{7}$$

commutes, where the morphisms are described in the following manner. The left vertical morphism is the identity. The bottom row is the morphism defined in the middle row of (1). Thus $\tilde{\delta}_w(a\tilde{\oplus}0) = \delta(a) \oplus 0$ and $\tilde{\delta}_w(0\tilde{\oplus}1) = \alpha_w \oplus (1+\tau_w)$. The morphism $\tilde{\delta}'_w$ in the top row of (7) is defined in the same way as $\tilde{\delta}_w$ except that α_w is replaced by α'_w in the definition. The commutativity of (7) follows from

$$\begin{aligned}
\pi_w(\tilde{\delta}'_w(0\tilde{\oplus}1)) &= \pi_w(\alpha'_w \oplus (1+\tau_w)) \\
&= (\alpha'_w \oplus 0) \ + \ (1+\tau_w)\pi_w(0 \oplus 1) \\
&= (\alpha'_w \oplus 0) \ + \ (1+\tau_w)(\gamma_w \oplus 1) \\
&= (\alpha'_w \oplus 0) \ + \ ((\alpha_w - \alpha'_w) \oplus (1+\tau_w)) \\
&= \alpha_w \oplus (1+\tau_w) \\
&= \tilde{\delta}_w(0\tilde{\oplus}1).
\end{aligned}$$

For all infinite places v of K we now have a diagram of $\mathbf{Z}[G_{w(v)}]$-modules

$$\begin{array}{ccc}
\tilde{K}_3^{\mathrm{ind}}(L_{w(v)}) & \xrightarrow{\tilde{\delta}'_{w(v)}} & C \oplus T_{2,w(v)} \\
\downarrow & & \downarrow{\scriptstyle \pi_{w(v)}} \\
\tilde{K}_3^{\mathrm{ind}}(L_{w(v)}) & \xrightarrow{\tilde{\delta}_{w(v)}} & C \oplus T_{2,w(v)}
\end{array} \tag{8}$$

which specialises to (7) if $v \in S'_\infty(K)$ and in which the vertical homomorphisms are the identity otherwise. Fitting together the inductions of these diagrams from $G_{w(v)}$ to $G(L/K)$ yields a commutative diagram

$$\begin{array}{ccccccc}
K_3^{\mathrm{ind}}(L) & \longrightarrow & \bigoplus_w \tilde{K}_3^{\mathrm{ind}}(L_w) & \xrightarrow{\tilde{\delta}'} & C \bigoplus T_2 & \longrightarrow & \tilde{K}'_2(O_{L,S}) \\
\downarrow{\scriptstyle 1} & & \downarrow & & \downarrow{\scriptstyle \pi} & & \downarrow{\scriptstyle 1} \\
K_3^{\mathrm{ind}}(L) & \longrightarrow & \bigoplus_w \tilde{K}_3^{\mathrm{ind}}(L_w) & \xrightarrow{\tilde{\delta}} & C \bigoplus T_2 & \longrightarrow & \tilde{K}'_2(O_{L,S})
\end{array} \tag{9}$$

in which the top row is the sequence resulting from using $\alpha'_{w(v)}$ instead of $\alpha_{w(v)}$ for $v \in S'_\infty(K)$. All of the vertical morphisms in (9) are $\mathbf{Z}[G(L/K)]$-isomorphisms.

Suppose now that we use the top row of (9) to compute $\Omega(L/K)$. Thus we choose a diagram

$$\begin{array}{ccccccc}
K_3^{\mathrm{ind}}(L) & \longrightarrow & A'_1 & \longrightarrow & B'_1 & \longrightarrow & \tilde{K}'_2(O_{L,S}) \\
\downarrow{\scriptstyle 1} & & \downarrow & & \downarrow & & \downarrow{\scriptstyle 1} \\
K_3^{\mathrm{ind}}(L) & \longrightarrow & \bigoplus_w \tilde{K}_3^{\mathrm{ind}}(L_w) & \xrightarrow{\tilde{\delta}'} & C \bigoplus T_2 & \longrightarrow & \tilde{K}'_2(O_{L,S})
\end{array} \tag{10}$$

in which A'_1 and B'_1 are finitely generated and cohomologically trivial $\mathbf{Z}[G(L/K)]$-modules.

Composing the vertical morphisms in (10) with the vertical isomorphisms in (9) gives a diagram of the form (6). This shows that we can take the top rows of

(6) and (10) to be the same sequence. We conclude that $\Omega(E/F)$ does not depend on the choice of the $\alpha_{w(v)}$ for $v \in S'_\infty(K)$.

We must now show that $\Omega(L/K)$ does not depend on the choice of the $w(v)$ for v an infinite place of F. Suppose v is fixed, $\sigma \in G$ and that $\sigma w(v) = w(v)'$ is another place of L over v. If $v \in S'_\infty(K)$, let $\alpha_{w(v)'} = \sigma\alpha_{w(v)}$ in C. Via the identification of $\mathrm{Ind}_{G_{w(v)}}^{G(L/K)} K_3^{\mathrm{ind}}(L_{w(v)})$ with $\oplus_{w'|v} K_3^{\mathrm{ind}}(L_{w'})$, the choice of $w(v)$ (together with $\alpha_{w(v)}$ if $v \in S'_\infty(K)$) leads to a diagram (5) which is isomorphic to the one resulting from the choice of $w(v)'$ (together with $\alpha'_{w(v)}$ if $v \in S'_\infty(K)$). Thus the invariant, $\Omega(L/K)$, is independent of the choice of the $w(v)$, since we have already shown it is independent of the choice of the $\alpha_{w(v)}$ for $v \in S'_\infty(K)$.

The last point is to check that $\Omega(L/K)$ is independent of the choice of the $G(L/K)$-stable set of places S of L which contains the ramified primes of L. The argument for this uses the localisation sequence (see Corollary 2.4.7) and is given in ([132] §7.1), so the proof of Theorem 5.3.17 is complete. □

Corollary 5.3.18

In Theorem 5.3.17, if L/K is unramified at infinity then $\Omega(L/K)$ is equal to the Euler characteristic of the 2-extension of Corollary 5.3.10

$$K_3^{\mathrm{ind}}(\mathcal{O}_{L,S}) \longrightarrow \oplus_{w\in S_\infty(L)} K_3^{\mathrm{ind}}(L_w) \xrightarrow{\delta} C \longrightarrow K_2'(\mathcal{O}_{L,S}).$$

Proof. By construction, we may choose a commutative diagram of 2-extensions of the form

$$\begin{array}{ccccccc}
K_3^{\mathrm{ind}}(L) & \longrightarrow & A_2 & \longrightarrow & B_2 & \longrightarrow & K_2'(O_{L,S}) \\
\downarrow{\scriptstyle 1} & & \downarrow & & \downarrow & & \downarrow{\scriptstyle i} \\
K_3^{\mathrm{ind}}(L) & \longrightarrow & A_1 & \longrightarrow & B_1 & \longrightarrow & \tilde{K}_2'(O_{L,S})
\end{array}$$

in which A_1, B_1, A_2, B_2 are finitely generated, cohomologically trivial $\mathbf{Z}[G(L/K)]$-modules with A_1, B_1 as in diagram (6). The homomorphism, i, is the injection of diagram (5). Hence the cokernel of i is equal to T_3, which is free in the absence of ramification at infinity. Hence there is an exact sequence of the form

$$0 \longrightarrow A_2 \longrightarrow A_1 \oplus B_2 \longrightarrow B_1 \longrightarrow T_3 \longrightarrow 0.$$

Therefore, in $\mathcal{CL}(\mathbf{Z}[G(L/K)])$, we have

$$\begin{aligned}
\Omega_1(L/K,3) &= [A_1] - [B_1] \\
&= [A_2] - [B_2] + [T_3] \\
&= [A_2] - [B_2],
\end{aligned}$$

as required. □

5.4 Comparing the two definitions

5.4.1 Let L/K be a Galois extension of number fields with group, $G(L/K)$. As in §5.3.1, let $E/\mathbf{Q}$ be a large Galois extension of number fields such that $L \subset E$ and E is totally complex. Let c denote complex conjugation. Let M be a $\mathbf{Z}[G(E/\mathbf{Q})]$-module and consider the short exact sequence

$$0 \longrightarrow M \xrightarrow{\phi} \mathrm{Ind}_{\langle c\rangle}^{G(E/\mathbf{Q})}(M) \longrightarrow M_+ \longrightarrow 0$$

where ϕ is as in §5.3.6. The resulting $G(E/L)$-cohomology sequence yields a 2-extension of the form

$$M^{G(E/L)} \longrightarrow \mathrm{Ind}_{\langle c\rangle}^{G(E/\mathbf{Q})}(M)^{G(E/L)} \longrightarrow M_+^{G(E/L)} \longrightarrow \mathrm{Ker}(\phi_*)$$

where $\mathrm{Ker}(\phi_*) = \ker(H^1(G(E/L); M) \longrightarrow H^1(G(E/L); \mathrm{Ind}_{\langle c\rangle}^{G(E/\mathbf{Q})}(M)))$.

Let $\{B_iG(E/\mathbf{Q}), d_i\}$ denote the bar resolution of $G(E/\mathbf{Q})$ ([132] p. 1) so that $B_iG(E/\mathbf{Q})$ is the free $\mathbf{Z}[G(E/\mathbf{Q})]$-module on i-tuples of elements of $G(E/\mathbf{Q})$. Define a homomorphism

$$\lambda_0 : \mathrm{Ind}_{\langle c\rangle}^{G(E/\mathbf{Q})}(M) \longrightarrow \mathrm{Hom}(B_0G(E/\mathbf{Q}), M)$$

by the formula

$$\lambda_0(\sum_{h\in G(E/\mathbf{Q})/\langle c\rangle} h \otimes m_h)(h') = h'(m_{h'})$$

for $h' \in G(E/\mathbf{Q})$. This is well defined since the formula selects either $h = h'$ or $h = h'c$ and $h' \otimes m_{h'} = h'c \otimes m_{h'c}$ if and only if $m_{h'c} = cm_{h'}$ so that $h'c(m_{h'c}) = h'c(cm_{h'}) = h'(m_{h'})$.

If $v \in G(E/K)$,

$$\lambda_0(v(\sum_{h\in G(E/\mathbf{Q})/\langle c\rangle} h \otimes m_h))(h') = h'(m_{v^{-1}h'})$$

while

$$\begin{aligned} &v(\lambda_0(\textstyle\sum_{h\in G(E/\mathbf{Q})/\langle c\rangle} h \otimes m_h))(h') \\ &\quad = v(\lambda_0(\textstyle\sum_{h\in G(E/\mathbf{Q})/\langle c\rangle} h \otimes m_h)(v^{-1}h'))v(v^{-1}h'(m_{v^{-1}h'})), \end{aligned}$$

so that λ_0 is a $\mathbf{Z}[G(E/K)]$-homomorphism.

Define

$$\lambda_1 : M_+ \longrightarrow \mathrm{Hom}(B_1G(E/\mathbf{Q}), M)$$

by, $g, h' \in G(E/\mathbf{Q})$,

$$\lambda_1([\sum_{h\in G(E/\mathbf{Q})/\langle c\rangle} h \otimes m_h])(g[h']) = gh'(m_{gh'}) - g(m_g).$$

This is well defined since

$$\begin{aligned}
&\lambda_1([\textstyle\sum_{h\in G(E/\mathbf{Q})/\langle c\rangle} h\otimes m_h + \phi(a)])(g[h'])\\
&= gh'(m_{gh'}) - g(m_g) + gh'(h'^{-1}g^{-1}(a)) - g(g^{-1}(a))\\
&= gh'(m_{gh'}) - g(m_g)\\
&= \lambda_1([\textstyle\sum_{h\in G(E/\mathbf{Q})/\langle c\rangle} h\otimes m_h])(g[h']).
\end{aligned}$$

It is a $G(E/K)$-map because, if $g_1 \in G(E_+/K)$, then

$$\begin{aligned}
&\lambda_1(\textstyle\sum g_1 h\otimes m_h)(g[h'])\\
&= gh'(m_{g_1^{-1}gh'} - g(m_{g_1^{-1}g}))\\
&= g_1 g_1^{-1} gh'(m_{g_1^{-1}gh'} - g_1 g_1^{-1} g(m_{g_1^{-1}g}))\\
&= g_1(\lambda_1(\textstyle\sum h\otimes m_h)(g_1^{-1}g[h']))\\
&= g_1(\lambda_1(\textstyle\sum h\otimes m_h))(g[h']).
\end{aligned}$$

We have a commutative diagram of the following form.

$$\begin{array}{ccc}
(\mathrm{Ind}_{\langle c\rangle}^{G(E/\mathbf{Q})}(M))^{G(E/L)} & \xrightarrow{\quad\pi\quad} & M_+^{G(E/L)}\\
\downarrow{\scriptstyle\lambda_0} & & {\scriptstyle\lambda_1}\downarrow\\
\mathrm{Hom}_{G(E/L)}(B_0G(E/\mathbf{Q}), M) & \xrightarrow{d_0^*} & \mathrm{Hom}_{G(E/L)}(B_1G(E/\mathbf{Q}), M)
\end{array}$$

in which π is the canonical surjection and d_0^* is induced by the bar-resolution differential. The commutativity of this diagram follows from the formulae

$$\begin{aligned}
\lambda_1(\pi(\textstyle\sum h\otimes m_h))(g[h']) &= \lambda_1([\textstyle\sum h\otimes m_h])(g[h'])\\
&= gh'(m_{gh'}) - g(m_g)\\
&= \lambda_0(\textstyle\sum h\otimes m_h)(gh' - g)\\
&= \lambda_0(\textstyle\sum h\otimes m_h)(d_0(g[h']))\\
&= d_0^*(\lambda_0(\textstyle\sum h\otimes m_h))(g[h']).
\end{aligned}$$

The $G(E/L)$-cohomology of M may be computed from the complex

$$\mathrm{Hom}_{G(E/L)}(B_0G(E/\mathbf{Q}), M) \xrightarrow{d_0^*} \mathrm{Hom}_{G(E/L)}(B_1G(E/\mathbf{Q}), M) \xrightarrow{d_1^*} \dots$$

and the resulting isomorphism, $\rho : M^{G(E/L)} \xrightarrow{\cong} \mathrm{Ker}(d_0^*)$, is given by $(m \mapsto (g \mapsto m))$ for $g \in G(E/\mathbf{Q})$, if we identify $B_0G(E/\mathbf{Q})$ with $\mathbf{Z}[G(E/\mathbf{Q})]$. Furthermore, if $m \in M^{G(E/L)}$, the formulae

$$\begin{aligned}
\lambda_0(\phi(m))[h'] &= \lambda_0(\textstyle\sum_{h\in G(E/\mathbf{Q})/\langle c\rangle} h\otimes_{\langle c\rangle} h^{-1}(m))[h']\\
&= h'((h')^{-1}(m)\\
&= m\\
&= \rho(m)[h']
\end{aligned}$$

shows that λ_0 induces the identity map between $M^{G(E/L)}$ and $\mathrm{Ker}(d_0^*) \cong M^{G(E/L)}$.

Next we consider the map induced by λ_1 on the cokernel of π, which is isomorphic to

$$\mathrm{Ker}(\phi_*) = \ker(H^1(G(E/L); M) \longrightarrow H^1(G(E/L); \mathrm{Ind}_{\langle c\rangle}^{G(E/\mathbf{Q})}(M))).$$

The homomorphism, d_1, induces

$$d_1^* : \mathrm{Hom}_{G(E/L)}(B_1G(E/\mathbf{Q}), M) \longrightarrow \mathrm{Hom}_{G(E/L)}(B_2G(E/\mathbf{Q}), M)$$

and we observe that

$$\begin{aligned} & d_1^*(\lambda_1(\textstyle\sum_{h\in G(E/\mathbf{Q})/\langle c\rangle}))(g[h_1|h_2]) \\ &= gh_1h_2(m_{h_1h_2}) - gh_1(m_{h_1}) - gh_1h_2(m_{h_1h_2}) \\ &\qquad +g(m_g) + gh_1(m_{h_1}) - g(m_g) \\ &= 0 \end{aligned}$$

so that the cokernel of π, maps to

$$\mathrm{Ker}(d_1^*)/d_0^*(\mathrm{Hom}_{G(E/L)}(B_0G(E/\mathbf{Q}), M)) \cong H^1(G(E/L) : M).$$

Let $z \in (M_+)^{G(E/L)}$ be represented by $\sum_{h\in G(E/\mathbf{Q})/\langle c\rangle} h \otimes_{\langle c\rangle} m_h$. Therefore, if $g \in G(E/L)$, there exists $f(g) \in M$ such that $g(z) - z = \phi(f(g))$. This means that

$$\sum_{h\in G(E/\mathbf{Q})/\langle c\rangle} gh \otimes_{\langle c\rangle} m_h - h \otimes_{\langle c\rangle} m_h = \sum_h h \otimes_{\langle c\rangle} h^{-1}(f(g))$$

so that $f(g) = h(m_{g^{-1}h}) - h(m_h)$ for all h.

The coboundary map

$$\delta : (M_+)^{G(E/L)} \longrightarrow H^1(G(E/L); M)$$

is given in terms of the bar resolution for $G(E/L)$ by $\delta(z) = [f]$, the class of the 1-cocycle, f. On the other hand, we have a composition

$$(M_+)^{G(E/L)} \xrightarrow{\lambda_1} \mathrm{Hom}_{G(E/L)}(B_1G(E/\mathbf{Q}), M) \longrightarrow \mathrm{Hom}_{G(E/L)}(B_1G(E/L), M)$$

in which the second map is induced by the inclusion, $G(E/L) \subset G(E/\mathbf{Q})$. The image of z under this composite is

$$([g] \mapsto \lambda_1(z)(1[g]) = g(m_g) - m_1).$$

Since $f(g) = g(m_1) - g(m_g)$ the sum of $\delta(z)$ and $\lambda_1(z)$ is $([g] \mapsto g(m_1) - m_1)$ which is the coboundary of $m_1 \in M$. Hence

$$[\delta(z)] = -[\lambda_1(z)] \in H^1(G(E/L); M).$$

Therefore we have proved the following result:

Theorem 5.4.2

In the notation of §5.4.1, the homomorphisms λ_0 and λ_1 induce a commutative diagram of 2-extensions of $\mathbf{Z}[G(L/K)]$-modules

$$\begin{array}{ccccccc}
M^{G(E/L)} & \longrightarrow & (\mathrm{Ind}_{\langle c\rangle}^{G(E/\mathbf{Q})}(M))^{G(E/L)} & \xrightarrow{\pi} & M_+^{G(E/L)} & \longrightarrow & (H^1)' \\
\downarrow 1 & & \downarrow \lambda_0 & & \downarrow \lambda_1 & & \downarrow (-1) \\
M^{G(E/L)} & \longrightarrow & \mathrm{Hom}_{G(E/L)}(B_0G(E/\mathbf{Q}), M) & \xrightarrow{d_0^*} & \mathrm{Ker}(d_1^*) & \longrightarrow & (H^1)'
\end{array}$$

where $(H^1)' = \ker(H^1(G(E/L); M) \to H^1(G(E/L); \mathrm{Ind}_{\langle c\rangle}^{G(E/\mathbf{Q})}(M)))$.

Lemma 5.4.3

Let M be a finitely generated $\mathbf{Z}[G(L/K)]$-module. Then, in §5.4.1, the $\mathbf{Z}[G(L/K)]$-module, $\mathrm{Hom}_{G(E/L)}(B_sG(E/\mathbf{Q}), M)$ is cohomologically trivial.

Proof. There is a finitely generated, free abelian group, X, such that

$$\mathrm{Hom}_{G(E/L)}(B_sG(E/\mathbf{Q}), M) \cong \mathrm{Hom}_{G(E/L)}(\mathbf{Z}[G(E/K)], \mathrm{Hom}(X, M))$$

so that this module is an induced module of the form $\mathrm{Ind}_{G(E/L)}^{G(E/K)}(\mathrm{Hom}(X, M))$, which is cohomologically trivial.

This is seen as follows: Let V be any $G(E/\mathbf{Q})$-module and let $J \subseteq G(E/\mathbf{Q})$ be a subgroup. Then $\mathrm{Hom}(\mathbf{Z}[G(E/\mathbf{Q})], V)$ is a $\mathbf{Z}[J]$-module with action given by $j(f)(u) = j(f(j^{-1}u))$. Hence, if $d = [G(E/\mathbf{Q}) : J]$ and $z_1, \ldots, z_d$ are coset representatives for $J\backslash G(E/\mathbf{Q})$ then $\mathrm{Hom}(\mathbf{Z}[G(E/\mathbf{Q})], V)$ is the "free" module on d copies of V where an element, w, belonging to the i-th copy corresponds to the function given by

$$f(z_s) = \begin{cases} w & \text{if } s = i, \\ 0 & \text{otherwise.} \end{cases}$$

Hence

$$\mathrm{Hom}(\mathbf{Z}[G(E/\mathbf{Q})], V) \cong \oplus_1^d \mathrm{Ind}_{\{1\}}^J(V)$$

and $H^s(J; \mathrm{Hom}(\mathbf{Z}[G(E/\mathbf{Q})], V)) = 0$ if $s > 0$. From the spectral sequence associated to $\mathbf{Q} \subset K \subset F \subset L \subset E$, with L/F Galois,

$$\begin{aligned} E_2^{s,t} &= H^s(G(L/F); H^t(G(E/L); \mathrm{Hom}(\mathbf{Z}[G(E/\mathbf{Q})], V))) \\ &\Longrightarrow H^{s+t}(G(E/F); \mathrm{Hom}(\mathbf{Z}[G(E/\mathbf{Q})], V)) \end{aligned}$$

we see that $E_2^{s,t} = 0$ unless $t = 0$ and then

$$\begin{aligned} E_2^{s,0} &= H^s(G(L/F); \mathrm{Hom}_{G(E/L)}(\mathbf{Z}[G(E/\mathbf{Q})], V))) \\ &\cong H^s(G(E/F); \mathrm{Hom}(\mathbf{Z}[G(E/\mathbf{Q})], V)) \end{aligned}$$

which is zero unless $s = 0$. □

5.4.4 We shall need the case in which $M = K_3^{\mathrm{ind}}(E)$ in Theorem 5.4.2. More precisely, let $R\Gamma_s(G(E/L), K_3^{\mathrm{ind}}(E)) = \mathrm{Hom}_{G(E/L)}(B_sG(E/\mathbf{Q}), K_3^{\mathrm{ind}}(E))$ and let $R\Gamma_*(L, K_3^{\mathrm{ind}}(\mathbf{Q}^{\mathrm{sep}}))$ denote the limit of $R\Gamma_*(G(E/L), K_3^{\mathrm{ind}}(E))$ taken over Galois extensions, $\mathbf{Q} \subset E \subset \mathbf{Q}^{\mathrm{sep}}$. We have a chain complex

$$R\Gamma_0(L, K_3^{\mathrm{ind}}(\mathbf{Q}^{\mathrm{sep}})) \xrightarrow{d_0^*} R\Gamma_1(L, K_3^{\mathrm{ind}}(\mathbf{Q}^{\mathrm{sep}})) \xrightarrow{d_0^*} \ldots.$$

From Theorem 5.4.2 we obtain a quasi-isomorphism between the 2-extension of Proposition 5.3.7

$$K_3^{\mathrm{ind}}(L) \longrightarrow \oplus_{w \in S_\infty(L)} K_3^{\mathrm{ind}}(L_w) \longrightarrow (K_3^{\mathrm{ind}}(\mathbf{Q}^{\mathrm{sep}})_+)^{\Omega_L} \longrightarrow K_2'(L)$$

and

$$K_3^{\mathrm{ind}}(L) \longrightarrow R\Gamma_0(L, K_3^{\mathrm{ind}}(\mathbf{Q}^{\mathrm{sep}})) \xrightarrow{d_0^*} \mathrm{Ker}(d_1^*)' \longrightarrow K_2'(L)$$

(where $\mathrm{Ker}(d_1^*)'$ is the inverse image of $K_2'(L)$) using the natural identification of $K_2'(L)$ with

$$\mathrm{Ker}(H^1(L; \mathbf{Q}^{\mathrm{sep}}) \longrightarrow H^1(L; \mathrm{Ind}_{\langle c \rangle}^{\Omega_{\mathbf{Q}}}(K_3^{\mathrm{ind}}(\mathbf{Q}^{\mathrm{sep}})))$$

of ([80] §5). In fact, this quasi-isomorphism is almost an equivalence of 2-extensions, being the identity on $K_3^{\mathrm{ind}}(L)$ and minus the identity on $K_2'(L)$.

Now that we have the 2-extension of Proposition 5.3.7 described in terms of $R\Gamma_*(L, K_3^{\mathrm{ind}}(\mathbf{Q}^{\mathrm{sep}}))$ we may use the short exact sequence

$$0 \longrightarrow \mathbf{Z}/l^m(2) \longrightarrow K_3^{\mathrm{ind}}(\mathbf{Q}^{\mathrm{sep}}) \xrightarrow{l^m} K_3^{\mathrm{ind}}(\mathbf{Q}^{\mathrm{sep}}) \longrightarrow 0$$

to relate it to the 2-extension coming from $R\Gamma_*(-)$ applied to the mapping cone of

$$\mathbf{Z}/l^m(2) \longrightarrow \oplus_{w \in S_\infty(L)} i_{w,*} i_w^*(\mathbf{Z}/l^m(2))$$

and thence to that coming from $R\Gamma_*(-)$ applied to the mapping cone of

$$\mathbf{Z}_l(2) \longrightarrow \oplus_{w \in S_\infty(L)} i_{w,*} i_w^*(\mathbf{Z}_l(2)),$$

which is the 2-extension of Proposition 5.2.10. This comparison is carefully carried out in ([40] §4) and leads to the following result.

Theorem 5.4.5

Let L/K be a Galois extension of number fields. Then the set, S, of places of L may be chosen so that there is a natural commutative diagram of 2-extensions of $\mathbf{Z}[G(L/K)]$-modules of the form

$$\begin{array}{ccccccc}
K_3^{\mathrm{ind}}(O_{L,S})' & \longrightarrow & A & \xrightarrow{\quad a \quad} & B & \longrightarrow & K_2'(O_{L,S}) \\
\downarrow \pi & & \downarrow \pi_1 & & \downarrow \pi_2 & & \pi_3 \downarrow \cong \\
K_3^{\mathrm{ind}}(O_{L,S}) & \longrightarrow & \oplus_{w \in S_\infty(L)} K_3^{\mathrm{ind}}(L_w) & \xrightarrow{a'} & C & \longrightarrow & K_2'(O_{L,S})
\end{array}$$

such that

(i) π *is the surjection of* §1, *with kernel* $\oplus_{w\in S_\infty(L)}\mathbf{Z}$,

(ii) π_1 *and* π_2 *are surjective and the isomorphism,* π_3, *is given by* ± 1 *on each Sylow p-subgroup of the finite group,* $K_2'(\mathcal{O}_{L,S})$,

(iii) *the upper 2-extension defines the class*

$$E \in \mathrm{Ext}^2_{\mathbf{Z}[G(L/K)]}(K_2'(\mathcal{O}_{L,S}), K_3^{\mathrm{ind}}(\mathcal{O}_{L,S})')$$

whose Euler characteristic is equal to

$$\Omega_1(L/K,3) \in \mathcal{CL}(\mathbf{Z}[G(L/K)]),$$

(iv) *the lower 2-extension is that of Corollary 5.3.10, as used in the construction of the invariant* $\Omega(L/K)$ *of Theorem 5.3.17.*

We are going to use Theorem 5.4.5 to show that $\Omega_1(L/K,3) = \Omega(L/K) \in \mathcal{CL}(\mathbf{Z}[G(L/K)])$. Before proceeding to the proof we record the following corollary, which is the totally real case of this comparison.

Corollary 5.4.6

If L/K *is totally real then the invariant,* $\Omega_1(L/K,3) \in \mathcal{CL}(\mathbf{Z}[G(L/K)])$, *of Definition 5.2.11 is equal to the Euler characteristic,* $\Omega(L/K)$, *of Theorem 5.3.17.*

Proof. From the commutative diagram of Theorem 5.4.5 we may obtain a commutative diagram of equivalent 2-extensions in which the modules

$$A,\ B,\ \oplus_{w\in S_\infty(L)}K_3^{\mathrm{ind}}(L_w) \quad \text{and} \quad C$$

are replaced by finitely generated, cohomologically trivial modules

$$A_1,\ B_2,\ A_2 \quad \text{and} \quad B_2$$

repectively. This modified diagram gives rise to an exact sequence of the form

$$0 \longrightarrow \oplus_{w\in S_\infty(L)}\mathbf{Z} \longrightarrow A_1 \longrightarrow A_2 \oplus B_1 \longrightarrow B_2 \longrightarrow 0.$$

Since $\oplus_{w\in S_\infty(L)}\mathbf{Z}$ is a free module in the totally real case, in $\mathcal{CL}(\mathbf{Z}[G(L/K)])$ we obtain the equation

$$\Omega_1(L/K) = [A_1] - [B_1] = [A_2] - [B_2]$$

and the result follows from Corollary 5.3.18. □

5.4.7 For the moment let us consider the special case in which the Galois extension of number fields, L/K, has the form $E/\mathbf{Q}$ with E totally complex. Temporarily write $G = G(E/\mathbf{Q})$. In this case, if we glue the diagram of Theorem 5.4.5 to diagram (6) of §5.3.15 we obtain a diagram of the following type in which the horizontal rows are 2-extensions in which A_1, B_1, A_2, B_2 are finitely generated, cohomologically trivial $\mathbf{Z}[G]$-modules and the extreme vertical columns are short exact sequences. In our arithmetical situation the upper 2-extension will be the upper 2-extension of Theorem 5.4.5 and the lower 2-extension will be the lower 2-extension of §5.3.15(6) so that $[A_1]-[B_1] = \Omega_1(E/\mathbf{Q}, 3)$ and $[A_2]-[B_2] = \Omega(E/\mathbf{Q})$

$$\begin{array}{ccccccc}
E_+ & & & & & & \\
\downarrow & & & & & & \\
X_1 & \longrightarrow & A_1 & \longrightarrow & B_1 & \longrightarrow & Y_1 \\
\downarrow & & \downarrow & & \downarrow & & \downarrow \\
X_2 & \longrightarrow & A_2 & \longrightarrow & B_2 & \longrightarrow & Y_2 \\
 & & & & & & \downarrow \\
 & & & & & & E_-
\end{array}$$

In this diagram

$$E_\pm = \oplus_{i=1}^r \operatorname{Ind}_{H_i}^G(\mathbf{Z}_\pm)$$

where $H_1, \ldots, H_r$ are distinct subgroups of order two, each of whose generators acts by ± 1 on a copy of the integers denoted by $\mathbf{Z}_\pm$. In the arithmetical situation the H_i's will be the decomposition groups of the infinite places of E.

The exact Mayer-Vietoris sequence resulting from this diagram defines a 3-extension, in $\operatorname{Ext}^3_{\mathbf{Z}[G]}(E_-, E_+)$, whose Euler characteristic is equal to

$$[A_1] - [B_1] - [A_2] + [B_2] \in \mathcal{CL}(\mathbf{Z}[G]).$$

Therefore, in the arithmetical situation, we wish to show that this element is zero.

With this in mind, for the rest of this section we shall study the (equivalent) process of producing Euler characteristics in the class-group from elements of

$$\operatorname{Ext}^1_{\mathbf{Z}[G]}(E_-, E_+) \cong \operatorname{Ext}^3_{\mathbf{Z}[G]}(E_-, E_+)$$

with the objective of proving that they often vanish. Our main result (Theorem 5.4.20 and Corollary 5.4.21) shows that, if $N_G H_i$ is the normaliser of H_i in G, the

Euler characteristics which arise lie in the subgroup generated by the images of compositions of the form

$$K_1(\mathbf{Z}[N_G H_i]) \xrightarrow{\delta} \mathcal{CL}(\mathbf{Z}[N_G H_i]) \xrightarrow{\mathrm{Ind}_{N_G H_i}^{G}} \mathcal{CL}(\mathbf{Z}[G]).$$

As explained in Theorem 5.4.22, this suffices to show that the difference of Euler characteristics vanishes in the arithmetical setting (see Example 5.4.11) in which $G = G(E/\mathbf{Q})$ is the Galois group of a number field extension in which E is totally complex. Heuristically, this is because ("in the limit") $N_G H_i = H_i$ in this case and $\mathcal{CL}(\mathbf{Z}[H]) = 0$ when $|H| = 2$.

5.4.8 *The Hecke algebras*

Let R be a commutative ring with identity and let G be a finite group. Suppose that

$$\mathcal{T} = \{\tau_i \in G; 1 \leq i \leq r\}$$

is a set of distinct elements of order two and set $H_i = \langle \tau_i \rangle$, the subgroup of order two generated by τ_i. Let $R_{\pm}$ be a copy of R on which τ_i acts by ± 1, respectively. Then there are isomorphisms, for $1 \leq i, j \leq r$,

$$\begin{aligned}
&\mathrm{Hom}_G(\mathrm{Ind}_{H_i}^G(R_{\pm}), \mathrm{Ind}_{H_j}^G(R_{\pm})) \\
&\cong \mathrm{Hom}_{R[H_i]}(R_{\pm}, \mathrm{Ind}_{H_j}^G(R_{\pm})) \\
&\cong \mathrm{Hom}_{R[H_i]}(R_{\pm}, \oplus_{z \in H_i \backslash G / H_j} \mathrm{Ind}_{H_i \cap z H_j z^{-1}}^{H_i}(R_{\pm}))
\end{aligned}$$

where $\mathrm{Ind}_H^G(R_{\pm})$ is the induced $R[G]$-module, $R[G] \otimes_H R_{\pm}$. The second isomorphism is induced by the Double Coset isomorphism ([131] Theorem 1.2.40)

$$\mathrm{Res}_{H_i}^G \mathrm{Ind}_{Hj}^G(R_{\pm}) \cong \oplus_{z \in H_i \backslash G / H_j} \mathrm{Ind}_{H_i \cap z H_j z^{-1}}^{H_i}(R_{\pm})$$

under which $a \otimes_{H_i \cap z H_j z^{-1}} b$ on the right corresponds to $az \otimes_{H_j} b$ on the left.

Inside the Hecke algebra

$$\begin{aligned}
&End_{R[G]}(\oplus_{i=1}^r \mathrm{Ind}_{H_i}^G(R_{\pm})) \\
&\cong \mathrm{Hom}_{R[G]}(\oplus_{i=1}^r \mathrm{Ind}_{H_i}^G(R_{\pm}), \oplus_{j=1}^r \mathrm{Ind}_{H_j}^G(R_{\pm})) \\
&\cong \oplus_{i=1}^r \oplus_{j=1}^r \oplus_{z \in H_i \backslash G / H_j} \mathrm{Hom}_{R[H_i]}(R_{\pm}, \mathrm{Ind}_{H_i \cap z H_j z^{-1}}^{H_i}(R_{\pm}))
\end{aligned}$$

let $S_{R,\mathcal{T}}^{\pm}(G)$ correspond to the subgroup of summands for which $H_i = z H_j z^{-1}$. Hence, additively, if we write $J \sim K$ to indicate conjugacy,

$$S_{R,\mathcal{T}}^{\pm}(G) \cong \oplus_{H_i \sim H_j} \oplus_{z \in H_i \backslash G / H_j,\ H_i = z H_j z^{-1}} \mathrm{Hom}_R(R_{\pm}, R_{\pm}).$$

Henceforth we shall identify $\mathrm{Hom}_R(R_{\pm}, R_{\pm})$ with R by the isomorphism which sends f to $f(1)$. Write $\underline{\mu} = \{\mu(i,j,z) \in R\} \in S_{R,\mathcal{T}}^{\pm}(G)$ for the element whose

(i,j,z)-component equals to $\mu(i,j,z) \in R$. Therefore we may consider $\mu(i,j,-)$ as a function from G to R with the property that, in the case of $S^{\pm}_{R,\mathcal{T}}(G)$,

$$\mu(i,j,\tau_i z) = \pm\mu(i,j,z) = \mu(i,j,z\tau_j),$$

respectively. Hence $\underline{\mu}$ is characterised, as a homomorphism, by the fact that it sends $g \otimes_{H_i} v \in \mathrm{Ind}_{H_i}^{G}(R_{\pm})$ $(g \in G, v \in R)$ to

$$\oplus_{H_i \sim H_j} \oplus_{z \in H_i\backslash G/H_j,\ H_i = zH_jz^{-1}}\ gz \otimes_{H_j} \mu(i,j,z)v \in \oplus_{j=1}^{r} \mathrm{Ind}_{H_j}^{G}(R_{\pm}).$$

Proposition 5.4.9

The subgroup, $S^{\pm}_{R,\mathcal{T}}(G)$, is a subring of $End_{R[G]}(\oplus_{i=1}^{r} \mathrm{Ind}_{H_i}^{G}(R_{\pm}))$. In fact, the product of $\underline{\mu} = \{\mu(i,j,z)\}$ and $\underline{\lambda} = \{\lambda(s,t,w)\}$ is given by $(g \in G, v \in R)$

$$(\underline{\lambda} \cdot \underline{\mu})(g \otimes_{H_i} v) = \sum_{\substack{H_i \sim H_j,\ z \in H_i\backslash G/H_j \\ H_i = zH_jz^{-1}}} \ \sum_{\substack{H_j \sim H_k,\ w \in H_j\backslash G/H_k \\ H_j = wH_kw^{-1}}} gzw \otimes_{H_k} \mu(i,j,z)\lambda(j,k,w)v.$$

Therefore

$$\underline{\lambda} \cdot \underline{\mu} = \underline{\nu} = \{\nu(a,b,y)\}$$

where

$$\nu(i,k,y) = \sum_{\substack{H_i \sim H_j,\ z \in H_i\backslash G/H_j \\ H_i = zH_jz^{-1}}} \ \sum_{\substack{H_j \sim H_k,\ w \in H_j\backslash G/H_k \\ H_j = wH_kw^{-1}}} \mu(i,j,z)\lambda(j,k,w)$$

the sum being taken only over pairs of double cosets, H_izH_j and H_jwH_k such that $H_iyH_k = H_izwH_k$.

Proof. The homomorphism corresponding to $\underline{\mu}$ sends $g \otimes_{H_i} v$ to

$$\oplus_{H_i \sim H_j} \oplus_{z \in H_i\backslash G/H_j,\ H_i = zH_jz^{-1}}\ gz \otimes_{H_j} \mu(i,j,z)v$$

which is sent by $\underline{\lambda}$ to

$$\sum_{\substack{H_i \sim H_j,\ z \in H_i\backslash G/H_j \\ H_i = zH_jz^{-1}}} \ \sum_{\substack{H_j \sim H_k,\ w \in H_j\backslash G/H_k \\ H_j = wH_kw^{-1}}} gzw \otimes_{H_k} \mu(i,j,z)\lambda(j,k,w)v,$$

as required. □

Example 5.4.10 Let $G = Q_8 = \{x, y | x^2 = y^2,\ x^4 = 1,\ xyx^{-1} = y^{-1}\}$ denote the quaternion group of order eight. Then $H_1 = \langle x^2 \rangle$ is the unique subgroup of order two, which is central. Let $V = Q_8/\langle x^2 \rangle \cong \mathbf{Z}/2 \times \mathbf{Z}/2$ consisting of the images of $1, x, y, xy$. Then $S^{+}_{R,\langle x^2 \rangle}(Q_8) \cong R[V]$, the isomorphism being to send $\underline{\mu}$ to $\sum_{z \in V} \mu(1,1,z)z$, where we have identified $\langle x^2 \rangle\backslash Q_8/\langle x^2 \rangle = Q_8/\langle x^2 \rangle$ with V.

Similarly, $S^{-}_{R,\langle x^2 \rangle}(Q_8) \cong \mathbf{H_Z}$, the integral quaternions.

Example 5.4.11 *The arithmetical situation*

Let L/K be a Galois extension of number fields and let $E/\mathbf{Q}$ be a large Galois extension of number fields such that $L \subset E$ and E is totally complex. Let c denote complex conjugation. Let Ω_L denote the absolute Galois group, $\Omega_L = G(\mathbf{Q}^{\text{sep}}/L)$, where $\mathbf{Q}^{\text{sep}}$ is a separable closure of $\mathbf{Q}$, the rationals. Let $v_\infty : L \longrightarrow E \longrightarrow \mathbf{Q}^{\text{sep}}$ be a fixed embedding which restricts to a real embedding, $w_\infty : K \longrightarrow E^{\langle c\rangle} \longrightarrow (\mathbf{Q}^{\text{sep}})^{\langle c\rangle}$. Assigning to $g \in \Omega_{\mathbf{Q}}$ the embedding $(v_\infty)g : L \xrightarrow{v_\infty} \mathbf{Q}^{\text{sep}} \xrightarrow{g} \mathbf{Q}^{\text{sep}}$ defines a bijection between embeddings of L and $\Omega_L\backslash\Omega_{\mathbf{Q}}$. The set of embeddings, $\{(v_\infty)g, (v_\infty)gc\}$, corresponds to an Archimedean place of L, since the completions of of $(v_\infty)g$ and $(v_\infty)gc$ coincide. Hence assigning the double coset $\Omega_L g\langle c\rangle$ to this Archimedean place defines a bijection between $\Omega_L\backslash\Omega_{\mathbf{Q}}/\langle c\rangle$ and $\Sigma_\infty(L)$, the set of Archimedean places of L.

If $(v_\infty)g$ is a complex place then gcg^{-1} does not belong to Ω_L and $\Omega_L \cap \langle \text{gcg}^{-1}\rangle = \{1\}$. If $(v_\infty)g$ is real then $\Omega_L \cap \langle \text{gcg}^{-1}\rangle = \langle \text{gcg}^{-1}\rangle$ is of order two. In the first case, if $(w_\infty)g$ is a real place of K then $\text{gcg}^{-1} \in \Omega_K$ and its image in $G(L/K) \cong \Omega_K/\Omega_L$ is the decomposition group, $H_g = G(L_{(v_\infty)g}/K_{(w_\infty)g})$.

Taking $G = G(L/K)$ and $\mathcal{T} = \{H_i = \langle g_i c g_i^{-1}\rangle ; 1 \leq i \leq r\}$, the set of non-trivial decomposition groups at infinity, gives the example which is our main motivation for the study of the class-group invariant which we shall introduce in §5.4.13.

5.4.12 *Some homological algebra*

Let R, G and H_i $(1 \leq i \leq r)$ be as in §5.4.7 and let $R_\pm$ denote the $R[H_i]$-module upon which τ_i acts as multiplication by ± 1. Let $E_{R,\pm}$ denote the $R[G]$ module given by $\oplus_{i=1}^r \text{Ind}_{H_i}^G(R_\pm)$.

Taking $R = \mathbf{Z}$ or $\mathbf{Z}_2$, the integers or the 2-adic integers, we have a chain of isomorphisms of the form

$$\begin{array}{ll}
\text{Ext}^1_{R[G]}(E_{R,-}, E_{R,+}) & \cong \oplus_{i=1}^r \text{Ext}^1_{R[G]}(\text{Ind}_{H_i}^G(R_-), E_{R,+}) \\
\cong \oplus_{i=1}^r \text{Ext}^1_{R[H_i]}(R_-, E_{R,+}) & \cong \oplus_{i=1}^r \text{Ext}^2_{R[H_i]}(R_+, E_{R,+}) \\
\cong \oplus_{i=1}^r H^2(H_i; E_{R,+}) & \cong \oplus_{i=1}^r \oplus_{j=1}^r H^2(H_i; \text{Ind}_{H_j}^G(R_+))
\end{array}$$

$$\begin{aligned}
&\cong \oplus_{i=1}^r \oplus_{j=1}^r \oplus_{z\in H_i\backslash G/H_j} H^2(H_i; \text{Ind}^{H_i}_{H_i\cap zH_jz^{-1}}(R_+)) \\
&\cong \oplus_{H_i\sim H_j} \oplus_{z\in H_i\backslash G/H_j,\ H_i=zH_jz^{-1}} H^2(H_i; R_+) \\
&\cong \oplus_{H_i\sim H_j} \oplus_{z\in H_i\backslash G/H_j,\ H_i=zH_jz^{-1}} \mathbf{Z}/2 \\
&\cong S^+_{\mathbf{Z},\mathcal{T}}(G) \otimes \mathbf{Z}/2 \\
&\cong S_{\mathbf{Z}/2,\mathcal{T}}(G)
\end{aligned}$$

where we have abbreviated the isomorphic rings, $S^\pm_{\mathbf{Z}/2,\mathcal{T}}(G)$, both to $S_{\mathbf{Z}/2,\mathcal{T}}(G)$.

Let us also abbreviate $E_{\mathbf{Z},\pm}$ to $E_\pm$. Then we have the following diagram of isomorphisms.

$$\begin{array}{ccc} S^+_{\mathbf{Z},\mathcal{T}}(G)\otimes \mathbf{Z}/2 & \xrightarrow{\cong} & S^+_{\mathbf{Z}_2,\mathcal{T}}(G)\otimes \mathbf{Z}/2 \\ \cong\downarrow & & \cong\downarrow \\ \mathrm{Ext}^1_{\mathbf{Z}[G]}(E_-,E_+) & \xrightarrow{\cong} & \mathrm{Ext}^1_{\mathbf{Z}_2[G]}(E_{\mathbf{Z}_2,-},E_{\mathbf{Z}_2,+}) \end{array}$$

Unravelling the chain of isomorphisms it is not hard to see that the isomorphism

$$\begin{aligned} \mathrm{Ext}^1_{\mathbf{Z}[G]}(E_-,E_+) &\cong \oplus_{H_i\sim H_j} \oplus_{z\in H_i\backslash G/H_j,\ H_i=zH_jz^{-1}} \mathbf{Z}/2 \\ &\cong \oplus_{i=1}^r H^2(H_i;E_+) \end{aligned}$$

sends a 1-extension of $\mathbf{Z}[G]$-modules

$$E_+ \longrightarrow X \longrightarrow E_-$$

to the element whose i-th coordinate is the image of the generator of $H^1(H_i;\mathbf{Z}_-)\cong \mathbf{Z}/2$ under the composition

$$H^1(H_i;\mathbf{Z}_-) \longrightarrow H^1(H_i;E_-) \stackrel{\delta}{\longrightarrow} H^2(H_i;E_+).$$

Here the first map is induced by the H_i-map, $\mathbf{Z}_- \longrightarrow \mathrm{Ind}^G_{H_i}(\mathbf{Z}_-)$, given by sending 1 to $1\otimes_{H_i} 1$ and the second map is the coboundary associated to the long exact cohomology sequence

$$\ldots H^n(H_i;E_+) \longrightarrow H^n(H_i;X) \longrightarrow H^n(H_i;E_-) \stackrel{\delta}{\longrightarrow} H^{n+1}(H_i;E_+)\ldots$$

If $\underline{\delta} = \oplus_{H_i\sim H_j} \oplus_{z\in H_i\backslash G/H_j,\ H_i=zH_jz^{-1}} \delta(i,j,z)$ lies in $\mathrm{Ext}^1_{\mathbf{Z}[G]}(E_-,E_+)$ then $\underline{\lambda}\in S^+_{\mathbf{Z},\mathcal{T}}(G)$ gives an endomorphism of $H^2(H_i;E_+)$ which we shall now evaluate. The $\mathbf{Z}[H_i]$-resolution

$$\ldots \longrightarrow \mathbf{Z}[H_i] \stackrel{1+\tau_i}{\longrightarrow} \mathbf{Z}[H_i] \stackrel{1-\tau_i}{\longrightarrow} \mathbf{Z}[H_i] \longrightarrow \mathbf{Z} \longrightarrow 0$$

shows that the cohomology group is computed from the complex

$$\oplus_{j=1}^r \mathrm{Ind}^G_{H_j}(\mathbf{Z}) \stackrel{1+\tau_i}{\longrightarrow} \oplus_{j=1}^r \mathrm{Ind}^G_{H_j}(\mathbf{Z}) \stackrel{1-\tau_i}{\longrightarrow} \oplus_{j=1}^r \mathrm{Ind}^G_{H_j}(\mathbf{Z})$$

and that δ is represented by

$$\oplus_{H_i\sim H_j} \oplus_{z\in H_i\backslash G/H_j, H_i=zH_jz^{-1}} z\otimes_{H_j} \delta(i,j,z)$$

where $\delta(i,j,z)\in \mathbf{Z}$ now denotes any lifting of $\delta(i,j,z)\in \mathbf{Z}/2$. This element is mapped by $\underline{\lambda}$ to

$$\sum_{\substack{H_i\sim H_j,\ z\in H_i\backslash G/H_j \\ H_i=zH_jz^{-1}}} \ \sum_{\substack{H_j\sim H_k,\ w\in H_j\backslash G/H_k \\ H_j=wH_kw^{-1}}} zw\otimes_{H_k} \delta(i,j,z)\lambda(j,k,w)$$

so that

$$\underline{\lambda}_*(\underline{\delta}) \equiv \underline{\lambda}\cdot\underline{\delta} \text{ (modulo 2)}$$

where the product is that of $S_{\mathbf{Z}/2,\mathcal{T}}(G)$.

5.4.13 Identify $\oplus_{H_i \sim H_j} \oplus_{z \in H_i \backslash G/H_j,\ H_i = zH_jz^{-1}} \mathbf{Z}/2$ with $S_{\mathbf{Z}/2,\mathcal{T}}(G)$, as in §5.4.12, and denote by $S_{\mathbf{Z}/2,\mathcal{T}}(G)^*$ the multiplicative group of units. We may define a map to the class-group of the integral group-ring of G

$$\psi : S_{\mathbf{Z}/2,\mathcal{T}}(G)^* \longrightarrow \mathcal{CL}(\mathbf{Z}[G])$$

by the following procedure. Let $\underline{\delta}_0 \in S_{\mathbf{Z}/2,\mathcal{T}}(G)$ denote the element for which $\delta(i,j,z) = 0$ except for $1 \leq i = j \leq r$ and $z = 1 \in G$ when $\delta(i,i,1) = 1$. Given another element, $\underline{\mu} \in S_{\mathbf{Z}/2,\mathcal{T}}(G)^*$, we may lift $\underline{\mu}$ to $\underline{\mu}' \in S^+_{\mathbf{Z},\mathcal{T}}(G) \cong End_{\mathbf{Z}[G]}(E_+)$ and then we may form

$$\underline{\mu}'_*(\underline{\delta}_0) \in \mathrm{Ext}^1_{\mathbf{Z}[G]}(E_-, E_+).$$

This 1-extension only depends upon $\underline{\mu} \in S_{\mathbf{Z}/2,\mathcal{T}}(G)^*$. Also, since $\underline{\mu} \in S_{\mathbf{Z}/2,\mathcal{T}}(G)^*$, the homomorphism $\underline{\mu}' : E_+ \longrightarrow E_+$ will be injective, since it is an isomorphism when reduced modulo 2. The 1-extension, $\underline{\mu}'_*(\underline{\delta}_0)$, is constructed by a push-out diagram of the form

$$\begin{array}{ccccc} E_+ & \longrightarrow & E & \longrightarrow & E_- \\ \downarrow{\scriptstyle \underline{\mu}'} & & \downarrow{\scriptstyle \tilde{\underline{\mu}}} & & \downarrow{\scriptstyle 1} \\ E_+ & \longrightarrow & Y(\underline{\mu}) & \longrightarrow & E_- \end{array}$$

where $E = \oplus_{i=1}^r \mathrm{Ind}_{H_i}^G(\mathbf{Z}[H_i]) \cong \oplus_{i=1}^r \mathbf{Z}[G]$. The upper 1-extension is constructed by applying $\mathrm{Ind}_{H_i}^G(-)$ to the canonical 1-extension of $\mathbf{Z}[H_i]$-modules

$$0 \longrightarrow \mathbf{Z}_+\langle 1 + \tau_i \rangle \longrightarrow \mathbf{Z}[H_i] \longrightarrow \mathbf{Z}_- \longrightarrow 0$$

and summing over $i = 1, \ldots, r$. Therefore

$$Y(\underline{\mu}) \cong (E_+ \oplus E)/\{(\underline{\mu}'(x), x) | x \in E_+\}$$

is a finitely generated $\mathbf{Z}[G]$-module.

Since $\underline{\mu} \in S_{\mathbf{Z}/2,\mathcal{T}}(G)^*$ the map $\underline{\mu}'_*$ is an isomorphism on the $\mathbf{Z}/2$-vector space, $\hat{H}^*(J; E_+)$ (Tate cohomology), for all $J \subseteq G$. Hence $Y(\underline{\mu})$ is a finitely generated, torsion-free, cohomologically trivial $\mathbf{Z}[G]$-module and is therefore projective, defining a class

$$[Y(\underline{\mu})] - \mathrm{rank}(Y(\underline{\mu}))[\mathbf{Z}[G]] \in \mathcal{CL}(\mathbf{Z}[G]) = \tilde{K}_0(\mathbf{Z}[G])$$

in the class-group of the integral group-ring (c.f. Example 2.1.2; see also [49]II).

Set

$$\psi(\underline{\mu}) = [Y(\underline{\mu})] - \mathrm{rank}(Y(\underline{\mu}))[\mathbf{Z}[G]] \in \mathcal{CL}(\mathbf{Z}[G]).$$

Proposition 5.4.14

The map, ψ, of §5.4.13 is a homomorphism

$$\psi : S_{\mathbf{Z}/2,\mathcal{T}}(G)^* \longrightarrow \mathcal{CL}(\mathbf{Z}[G])$$

which factors through the quotient by $S_{\mathbf{Z},\mathcal{T}}(G)^*$ to give

$$\psi : \frac{S_{\mathbf{Z}/2,\mathcal{T}}(G)^*}{S^+_{\mathbf{Z},\mathcal{T}}(G)^*} \longrightarrow \mathcal{CL}(\mathbf{Z}[G]).$$

Proof. The canonical 1-extension

$$E_+ \longrightarrow E \longrightarrow E_-$$

represents the class

$$\underline{\delta}_0 \in \mathrm{Ext}^1_{\mathbf{Z}[G]}(E_-, E_+) \cong \oplus_{H_i \sim H_j} \oplus_{z \in H_i \backslash G/H_j,\ H_i = zH_jz^{-1}} \mathbf{Z}/2.$$

Also $\underline{\delta}_0$ is the identity element of the ring $S_{\mathbf{Z}/2,\mathcal{T}}(G)$.

The H_i-cohomology coboundary associated to the 1-extension

$$E_+ \longrightarrow Y(\underline{\mu}) \longrightarrow E_-$$

may, by naturality, be computed by composing

$$\underline{\mu}'_* : \hat{H}^*(H_i; E_+) \longrightarrow \hat{H}^*(H_i; E_+)$$

for $i = 1, \ldots, r$ with the coboundary associated to the canonical 1-extension. Therefore, by the discussion of §5.4.12, the $Y(\underline{\mu})$-extension corresponds to

$$\underline{\mu}'_*(\underline{\delta}_0) = \underline{\mu} \cdot \underline{\delta}_0 = \underline{\mu} \in S_{\mathbf{Z}/2,\mathcal{T}}(G)^* \subset S_{\mathbf{Z}/2,\mathcal{T}}(G).$$

Now suppose that we have $\underline{\mu}, \underline{\lambda} \in S_{\mathbf{Z}/2,\mathcal{T}}(G)^*$. We may lift these elements to $\underline{\mu}', \underline{\lambda}' \in S^+_{\mathbf{Z},\mathcal{T}}(G)$, respectively. In $\mathcal{CL}(\mathbf{Z}[G])$, since E is free,

$$[Y(\underline{\mu})] - \mathrm{rank}(Y(\underline{\mu}))[\mathbf{Z}[G]] = [\mathrm{Coker}(\tilde{\underline{\mu}})] = [\mathrm{Coker}(\underline{\mu}')]$$

since $\mathrm{Coker}(\tilde{\underline{\mu}}) \cong \mathrm{Coker}(\underline{\mu}')$ is a finite group which is also a cohomologically trivial $\mathbf{Z}[G]$-module. Therefore the short exact sequence

$$0 \longrightarrow \mathrm{Coker}(\underline{\lambda}') \longrightarrow \mathrm{Coker}(\underline{\mu}' \cdot \underline{\lambda}') \longrightarrow \mathrm{Coker}(\underline{\mu}') \longrightarrow 0$$

shows that, in $\mathcal{CL}(\mathbf{Z}[G])$,

$$\begin{aligned} \psi(\underline{\mu} \cdot \underline{\lambda}) &= [\mathrm{Coker}(\underline{\mu}' \cdot \underline{\lambda}')] \\ &= [\mathrm{Coker}(\underline{\mu}')] + [\mathrm{Coker}(\underline{\lambda}')] \\ &= \psi(\underline{\mu}) + \psi(\underline{\lambda}), \end{aligned}$$

as required.

Finally, if $\underline{\mu} \in S_{\mathbf{Z}/2,\mathcal{T}}(G)^*$ is in the image of $S^+_{\mathbf{Z},\mathcal{T}}(G)^*$ then $\underline{\mu}'$ may be chosen to be a $\mathbf{Z}[G]$-module isomorphism in which case $Y(\underline{\mu}) \cong E$ and $\psi(\underline{\mu}) = 0$. □

Now we shall develop another description of ψ. Let us begin with two examples.

Example 5.4.15 Consider the Example 5.4.10, where $G = Q_8$ and $\mathcal{T}$ consists of the centre, $\langle x^2 \rangle$. In this case

$$\frac{S_{\mathbf{Z}/2,\mathcal{T}}(Q_8)^*}{S^+_{\mathbf{Z},\mathcal{T}}(Q_8)^*} \cong (\mathbf{Z}/4)^* \cong \mathcal{CL}(\mathbf{Z}[Q_8]).$$

In fact, the homomorphism ψ is an isomorphism. In order to see this, using the results about the Hom-description of $\psi(\underline{\mu})$ which are explained in §5.4.23, we need only show that $\psi(1+x+y)$ is non-trivial. However the determinant of

$$1 + ((1+x+y)^{-1} - 1)/2)(1+x^2) \in \mathbf{Q}_2[Q_8]^*$$

is trivial on all the non-trivial irreducible representations of Q_8 while being equal to 3 (modulo 4) on the trivial one-dimensional representation. This is the Hom-description of a representative of the non-trivial element of $\mathcal{CL}(\mathbf{Z}[Q_8])$.

Example 5.4.16 The following discussion applies to any G with a normal subgroup, $\langle \tau \rangle \lhd G$, of order two. In this case

$$S^+_{R,\mathcal{T}}(G) \cong R[G/\langle \tau \rangle]$$

and the homomorphism

$$\psi : \frac{\mathbf{Z}/2[G/\langle \tau \rangle]^*}{\mathbf{Z}[G/\langle \tau \rangle]^*} \longrightarrow \mathcal{CL}(\mathbf{Z}[G])$$

may be identified, as we shall prove below, with the Mayer-Vietoris homomorphisms of the type studied in ([49] II p. 273).

If $\mu \in \mathbf{Z}/2[G/\langle \tau \rangle]^*$ lifts to $\mu' \in \mathbf{Z}_2[G/\langle \tau \rangle]^*$ then the Hom-description of $\psi(\mu)$ sends an irreducible representation, ρ, to the identity at all places if ρ does not factor through $G/\langle \tau \rangle$ - that is, if $\rho(\tau) = -1$. If $\rho(\tau) = 1$ then ρ factorises through ρ' on $G/\langle \tau \rangle$. In this case ρ is sent to 1 at all places except above 2 where it goes to $\mathrm{Det}(\mu')(\rho) \in (\mathbf{Z}_2 \otimes \mathcal{O}_E)^*$ where $E/\mathbf{Q}$ is a Galois extension containing all the $|G|$-th roots of unity.

Hence the image of ψ lies in $D(\mathbf{Z}[G]) \subseteq \mathcal{CL}(\mathbf{Z}[G])$.

When $G = Q_{2^n}$, the generalized quaternion group of order 2^n, then

$$D(\mathbf{Z}[Q_{2^n}]) \cong \mathbf{Z}/2$$

and ψ is surjective (c.f. [49] II p. 273).

Later in this section we shall see that Example 5.4.16 is typical.

5.4.17 Our first observation is that $\mathcal{T}$ is the disjoint union of G-conjugacy classes, $\mathcal{T}_s$, and that, in these circumstances, there is an isomorphism of the form

$$S^{\pm}_{R,\mathcal{T}}(G) \cong \prod_s S^{\pm}_{R,\mathcal{T}_s}(G).$$

In addition, the homomorphism, ψ, of Proposition 5.4.14 is evidently equal to the sum of the ψ's for each of the factors. Therefore we shall henceforth restrict ourselves to the case when $\mathcal{T} = \{g_i H g_i^{-1}; 1 \leq i \leq r\}$ where $r = [G : N_G H]$ is the index of the normaliser of H in G and $g_1, \ldots, g_r$ are coset representatives for $G/N_G H$. Write τ for the generator of H.

Next we observe that

$$z g_j H g_j^{-1} z^{-1} = g_i H g_i^{-1} \quad \text{if and only if} \quad g_i^{-1} z g_j \in N_G H.$$

On the other hand,

$$g_i H g_i^{-1} z g_j H g_j^{-1} = g_i H g_i^{-1} w g_j H g_j^{-1} \quad \text{if and only if} \quad g_j^{-1} z^{-1} w g_j \in H.$$

Therefore there is an obvious isomorphism of abelian groups of the form

$$S^{\pm}_{R,\mathcal{T}}(G) \cong \oplus_{1 \leq i,j \leq r}\ g_i R[N_G H]/(\tau - (\pm 1)) g_j^{-1}.$$

If we endow the right-hand side with the multiplication given on generators by

$$(g_j(wH)g_k^{-1}) \cdot (g_i(zH)g_j^{-1}) = (g_i(zwH)g_k^{-1})$$

and zero in all other cases one sees easily that this becomes an isomorphism of rings. Furthermore it is clear that the right-hand side is isomorphism to the ring, $M_r(R[N_G H]/(\tau - (\pm 1)))^{op}$, of $r \times r$ matrices with entries in the quotient, $R[N_G H]/(\tau - (\pm 1))$, of the group-ring of $N_G H$ with coefficients in R and the *opposite* multiplication.

Therefore we have proved the following result.

Proposition 5.4.18

In the notation of §5.4.8 and §5.4.17, there is an isomorphism of rings

$$S^{\pm}_{R,\mathcal{T}}(G) \cong M_r(R[N_G H]/(\tau - (\pm 1)))^{op}$$

where Λ^{op} *denotes the opposite ring of the ring* Λ.

5.4.19 Next consider the pull-back diagram

$$\begin{array}{ccc} \mathbf{Z}[H] & \longrightarrow & \mathbf{Z}_+ \\ \downarrow & & \downarrow \\ \mathbf{Z}_- & \longrightarrow & \mathbf{Z}/2 \end{array}$$

which induces up to yield the pull-back square

$$\begin{array}{ccc} \mathbf{Z}[N_GH] & \longrightarrow & \mathbf{Z}[N_GH]/(\tau-1) \\ \downarrow & & \downarrow \\ \mathbf{Z}[N_GH]/(\tau+1) & \longrightarrow & \mathbf{Z}/2[N_GH]/(\tau-1) \end{array}$$

from which we may obtain a K-theory Mayer-Vietoris sequence by the method described in ([109] §3)

$$\ldots \longrightarrow K_1(\mathbf{Z}[N_GH]/(\tau+1)) \oplus K_1(\mathbf{Z}[N_GH]/(\tau-1)) \longrightarrow$$
$$K_1(\mathbf{Z}/2[N_GH]/(\tau-1)) \xrightarrow{\delta} K_0(\mathbf{Z}[N_GH]) \longrightarrow \ldots.$$

If $\mathcal{M}_G$ is a maximal $\mathbf{Z}$-order in $\mathbf{Q}[G]$ let $D(\mathbf{Z}[G])$ denote the kernel of the canonical map of class-groups, $\mathcal{CL}(\mathbf{Z}[G]) \longrightarrow \mathcal{CL}(\mathcal{M}_G)$. The definition of $D(\mathbf{Z}[G])$ is independent of choice of maximal order. When we tensor the second Cartesian square with the rationals the bottom right corner vanishes. Therefore the maximal order of $\mathbf{Q}[N_GH]$ is isomorphic to the direct sum of the maximal orders in $\mathbf{Z}[N_GH]/(\tau \pm 1)$. Therefore we have an inclusion

$$\mathrm{im}(K_1(\mathbf{Z}/2[N_GH]/(\tau-1)) \xrightarrow{\delta} K_0(\mathbf{Z}[N_GH])) \subseteq D(\mathbf{Z}[N_GH]).$$

This is because $\delta(x)$ must vanish under the map

$$K_0(\mathbf{Z}[N_GH]) \longrightarrow K_0(\mathbf{Z}[N_GH]/(\tau+1)) \oplus K_0(\mathbf{Z}[N_GH]/(\tau-1))$$

but the corresponding map on the K-theory of maximal orders is an isomorphism. Note also that the canonical homomorphism

$$\mathrm{Ind}_{N_GH}^G : K_0(\mathbf{Z}[N_GH]) \longrightarrow K_0(\mathbf{Z}[G])$$

satisfies

$$\mathrm{Ind}_{N_GH}^G(D(\mathbf{Z}[N_GH]) \subseteq D(\mathbf{Z}[G]).$$

By Proposition 5.4.18, a unit

$$\underline{\mu} \in S_{\mathbf{Z}/2,\mathcal{T}}(G)^* = GL_1 S_{\mathbf{Z}/2,\mathcal{T}}(G)$$

defines a class

$$[\underline{\mu}] \in K_1(S_{\mathbf{Z}/2,\mathcal{T}}(G)) \cong K_1(M_r(\mathbf{Z}/2[N_GH]/(\tau-1))^{op}) \cong K_1(\mathbf{Z}/2[N_GH]/(\tau-1)),$$

since $K_1(\Lambda)$ is defined to be the abelianisation of the infinite general linear group of Λ. The second isomorphism is induced by Morita equivalence (see [119]).

Theorem 5.4.20

If $\underline{\mu} \in S_{\mathbf{Z}/2,\mathcal{T}}(G)^*$ then, when $\mathcal{T} = \{g_iHg_i^{-1}; 1 \leq i \leq r\}$ and $r = [G : N_GH]$ as in §5.4.17,

$$\psi(\underline{\mu}) = \mathrm{Ind}_{N_GH}^G(\delta([\underline{\mu}]) - r[\mathbf{Z}[N_GH]]) \in \mathcal{CL}(\mathbf{Z}[G]) \subseteq K_0(\mathbf{Z}[G]).$$

Proof. Firstly, we must find an alternative description of the $\mathbf{Z}[G]$-module, $Y(\underline{\mu})$, of §5.4.13 which is more obviously related to the boundary homomorphism, δ.

Consider the following commutative diagram of $\mathbf{Z}[G]$-modules in which the right-hand square is a pull-back which defines $X(\underline{\mu})$.

$$\begin{array}{ccccccc} E_+ \cong 2E_+ & \longrightarrow & X(\underline{\mu}) & \longrightarrow & & & E_- \\ \downarrow{\scriptstyle 1} & & \downarrow & & & & \downarrow \\ E_+ & \xrightarrow{2} & E_+ & \longrightarrow & E_+ \otimes \mathbf{Z}/2 & \xleftarrow{\underline{\mu}} & E_- \otimes \mathbf{Z}/2 \end{array}$$

In the pull-back square defining $X(\underline{\mu})$ we have identified $E_\pm \otimes \mathbf{Z}/2$ by the isomorphism induced by the unique identification of $\mathbf{Z}_\pm \otimes \mathbf{Z}/2$ so that $\underline{\mu}$ may be interpreted as an isomorphism of $\mathbf{Z}[G]$-modules in the bottom right-hand corner. Therefore $X(\underline{\mu})$ consists of pairs, $(e_+, e_-) \in E_+ \times E_-$, mapping to the same element in $E_+ \otimes \mathbf{Z}/2$. Hence $\ker(X(\underline{\mu}) \longrightarrow E_-)$ consists of all pairs, $(e_+, 0)$, in which e_+ is divisible by 2. Therefore both rows of the diagram are short exact sequences. In fact, the two row is equivalent to the 1-extension which defines $Y(\underline{\mu})$ and therefore there is an isomorphism of $\mathbf{Z}[G]$-modules of the form

$$Y(\underline{\mu}) \cong X(\underline{\mu}).$$

To see this it suffices to calculate all the compositions of §5.4.12

$$H^1(H_i; \mathbf{Z}_-) \longrightarrow H^1(H_i; E_-) \xrightarrow{\delta} H^2(H_i; E_+)$$

where δ is the coboundary associated to the upper row of the diagram. We must show that this family of homomorphisms is induced by composition with $\underline{\mu}$, in the sense of §5.4.12. However there are canonical isomorphisms of the form

$$H^1(H_i; E_-) \cong H^1(H_i; E_- \otimes \mathbf{Z}/2) \cong H^2(H_i; E_+ \otimes \mathbf{Z}/2) \cong H^2(H_i; E_+)$$

in terms of which the family of coboundary compositions for the exact sequence

$$0 \longrightarrow E_+ \xrightarrow{2} E_+ \longrightarrow E_+ \otimes \mathbf{Z}/2 \longrightarrow 0$$

is induced by composition with the identity map. Therefore the family of coboundary compositions for the exact sequence of the lower row is induced by composition with $\underline{\mu}$ and consequently the same is true for the upper row.

Next we consider the (left) $\mathbf{Z}[G]$-modules, $E_\pm = \oplus_{i=1}^r \mathrm{Ind}_{g_i H g_i^{-1}}^G(\mathbf{Z}_\pm)$. These modules are isomorphic, respectively, to $\oplus_{i=1}^r \mathrm{Ind}_H^G(\mathbf{Z}_\pm)$ by means of the homomorphism which sends $g \otimes_{g_i H g_i^{-1}} v \in \mathrm{Ind}_{g_i H g_i^{-1}}^G(\mathbf{Z}_\pm)$ to $g g_i \otimes_H v \in \mathrm{Ind}_H^G(\mathbf{Z}_\pm)$ in the i-th summand. The isomorphism of $M_r(\mathbf{Z}[N_G H]/(\tau - (\pm 1)))^{op}$ with $S^{\pm}_{\mathbf{Z},T}(G)$ sends the elementary matrix, $e_{ij}^{\lambda z}$ ($\lambda \in \mathbf{Z}$, $z \in N_G H$) to $\underline{\mu} = \{\mu(s,t,w)\}$ whose only non-zero coordinate is given by $\mu(i, j, g_i z g_j^{-1}) = \lambda$. This elementary matrix acts by sending $g \otimes_{g_i H g_i^{-1}} v$ to $g g_i z g_j^{-1} \otimes_{g_j H g_j^{-1}} v$. Translated under these isomorphisms, $e_{ij}^{\lambda z}$ acts on $g g_i \otimes_H v$ in the i-th coordinate by sending it to $g g_i z \otimes_H v$ in the j-th coordinate.

Hence there are (left) $\mathbf{Z}[G]$-module isomorphisms of the form

$$E_{\pm} \cong \mathrm{Ind}_{N_G H}^{G}(E'_{\pm})$$

where the (left) $\mathbf{Z}[N_G H]$-modules, $E'_{\pm}$ are equal to $\oplus_{i=1}^{r} \mathrm{Ind}_{H}^{N_G H}(\mathbf{Z}_{\pm})$, respectively. The elements of these modules are considered as r-tuple row vectors with entries in $\mathrm{Ind}_{H}^{N_G H}(\mathbf{Z}_{\pm})$ upon which an element of $S^{\pm}_{\mathbf{Z},\mathcal{T}}(G)$, considered as an $r \times r$ matrix via Proposition 5.4.18, acts via right matrix multiplication. The action of $\mathbf{Z}[N_G H]$ is by means of left multiplication on each coordinate.

Acting via right multiplication as described above, a unit

$$\underline{\mu} \in S_{\mathbf{Z}/2,\mathcal{T}}(G)^* \cong GL_r(\mathbf{Z}/2[N_G H]/H)$$

defines a (left) $\mathbf{Z}/2[N_G H]/H$-automorphism, $\underline{\mu} : E'_{-} \otimes \mathbf{Z}/2 \xrightarrow{\cong} E'_{-} \otimes \mathbf{Z}/2$ which defines a class

$$[\underline{\mu}] \in K_1(\mathbf{Z}/2[N_G H]/H) = GL_{\infty}(\mathbf{Z}/2[N_G H]/H)_{ab}.$$

By definition of the coboundary, δ, in the K-theory Mayer-Vietoris sequence ([109] §3)

$$\delta([\underline{\mu}]) = [X'(\underline{\mu})] \in K_0(\mathbf{Z}[N_G H])$$

where $X'(\underline{\mu})$ is defined by a pull-back in the same manner as $X(\underline{\mu})$ but with G and $E_{\pm}$ replaced by $N_G H$ and $E'_{\pm}$. From the preceding discussion, $X(\underline{\mu}) = \mathrm{Ind}_{N_G H}^{G}(X'(\underline{\mu}))$, so that in $\mathcal{CL}(\mathbf{Z}[G]) \subset K_0(\mathbf{Z}[G])$

$$\begin{aligned}
\psi(\underline{\mu}) &= Y(\underline{\mu}) - r\mathbf{Z}[G] \\
&= X(\underline{\mu}) - r\mathbf{Z}[G] \\
&= \mathrm{Ind}_{N_G H}^{G}(X'(\underline{\mu}) - r\mathbf{Z}[N_G H]) \\
&= \mathrm{Ind}_{N_G H}^{G}(\delta([\underline{\mu}]) - r\mathbf{Z}[N_G H])
\end{aligned}$$

as required. □

The following result is immediate from the discussion concerning the subgroup, $D(\mathbf{Z}[G])$, together with the fact that the class-group of the group-ring of the group of order two is trivial.

Corollary 5.4.21

For any G and $\mathcal{T}$ the homomorphism of Proposition 5.4.14

$$\psi : \frac{S_{\mathbf{Z}/2,\mathcal{T}}(G)^*}{S^{+}_{\mathbf{Z},\mathcal{T}}(G)^*} \longrightarrow \mathcal{CL}(\mathbf{Z}[G])$$

takes values in the subgroup $D(\mathbf{Z}[G])$ of Theorem 2.5.32.

Furthermore, if the $N_G H = H$ for each $H \in \mathcal{T}$ then ψ is trivial.

Theorem 5.4.22 *(see also [40])*
Let L/K be any Galois extension of number fields. Then

$$\Omega(L/K) = \Omega_1(L/K, 3) \in \mathcal{CL}(\mathbf{Z}[G(L/K)])$$

where $\Omega_1(L/K, 3)$ is the invariant of Definition 5.2.11 and $\Omega(L/K)$ is the Euler characteristic of Theorem 5.3.17.

Proof. From the discussion of §5.4.7, which explained how, if E is a totally complex field, to write $\Omega_1(E/\mathbf{Q}, 3) - \Omega(E/\mathbf{Q})$ as an Euler characteristic of an element of

$$\mathrm{Ext}^1_{\mathbf{Z}[G]}(E_-, E_+) \cong \mathrm{Ext}^3_{\mathbf{Z}[G]}(E_-, E_+)$$

of the type we have been studying.

From this we shall explain how to prove that $\Omega_1(E/\mathbf{Q}) = \Omega_1(E/\mathbf{Q}, 3)$ when E is totally complex – equality in the general case following at once, by naturality.

The idea is that, in the limit, $N_{G(E/\mathbf{Q})}\langle c\rangle/\langle c\rangle$ is trivial and so Corollary 5.4.21 should yield the result. Unfortunately, for each $E/\mathbf{Q}$, the quotient, $N_{G(E/\mathbf{Q})}\langle c\rangle/\langle c\rangle$, may be non-trivial. Therefore we have to proceed slightly differently.

Suppose $E/\mathbf{Q}$ and $M/\mathbf{Q}$ are two totally complex Galois extensions with $E \subseteq M$ and Galois groups

$$G(M/E) \lhd G(M/\mathbf{Q}) \stackrel{\pi}{\longrightarrow} G(E/\mathbf{Q}) \cong G(M/\mathbf{Q})/G(M/E).$$

Each of the 2-extensions of $\mathbf{Z}[G(M/\mathbf{Q})]$-modules, defining

$$\Omega_1(M/\mathbf{Q}) \qquad \text{and} \qquad \Omega_1(M/\mathbf{Q}, 3),$$

is natural with respect to passage to quotients. This sort of naturality means, for example, that if

$$X_1 \longrightarrow A_1 \longrightarrow B_1 \longrightarrow Y_1$$

is the 2-extension which defines $\Omega_1(M/\mathbf{Q})$ then the associated 2-extension of $\mathbf{Z}[G(E/\mathbf{Q})]$-modules

$$X_1^{G(M/E)} \longrightarrow A_1^{G(M/E)} \longrightarrow B_1^{G(M/E)} \longrightarrow (Y_1)_{G(M/E)}$$

defines $\Omega_1(E/\mathbf{Q})$. In fact, the diagram of §5.4.7 involving these 2-extensions is natural in this sense, with respect to passage to quotient groups.

The diagram of §5.4.7 relating the 2-extensions for $\Omega(M/\mathbf{Q})$ and $\Omega_1(M/\mathbf{Q}, 3)$ defined an element

$$\omega_{M/\mathbf{Q}} \in \mathrm{Ext}^3_{\mathbf{Z}[G(M/\mathbf{Q})]}(E_-, E_+) \cong \mathrm{Ext}^1_{\mathbf{Z}[G(M/\mathbf{Q})]}(E_-, E_+) \cong S_{\mathbf{Z}/2, \mathcal{T}_M}(G(M/\mathbf{Q}))$$

where $\mathcal{T}_M$ is the set of all conjugates of complex conjugation, c, in $G(M/\mathbf{Q})$. Naturality implies that these elements fit together to define

$$\omega = \lim_{\leftarrow} \omega_{M/\mathbf{Q}} \in \lim_{\leftarrow} S_{\mathbf{Z}/2, \mathcal{T}_M}(G(M/\mathbf{Q})).$$

Returning to $\mathbf{Q} \subset E \subseteq M$, the canonical homomorphism

$$S_{\mathbf{Z}/2,\mathcal{T}_M}(G(M/\mathbf{Q})) \longrightarrow S_{\mathbf{Z}/2,\mathcal{T}_E}(G(E/\mathbf{Q}))$$

is induced by considering an element of

$$S_{\mathbf{Z}/2,\mathcal{T}_M}(G(M/\mathbf{Q}))$$

as lifting to a $\mathbf{Z}[G(M/\mathbf{Q})]$-endomorphism of E_+ and sending it to the reduction modulo 2 of the induced $\mathbf{Z}[G(E/\mathbf{Q})]$-module endomorphism of $E_+^{G(M/E)}$, considered as an element of $S_{\mathbf{Z}/2,\mathcal{T}_E}(G(E/\mathbf{Q}))$.

Let $G(M/E) = \{x_1, \ldots, x_r\}$ and let $G(E/\mathbf{Q}) = \{y_1'c^\epsilon, \ldots, y_s'c^\epsilon;\ \epsilon = 0, 1\}$, where c denotes complex conjugation. Lifting each y_j' to $y_j \in G(M/\mathbf{Q})$ we may set

$$\mathcal{T}_M = \{\langle x_i y_j c y_j^{-1} x_i^{-1}\rangle \mid 1 \leq i \leq r,\ 1 \leq j \leq s\}.$$

Hence $\mathcal{T}_M$ is the set of all conjugates of $\langle c\rangle$ in $G(M/\mathbf{Q})$. By Proposition 5.4.18, we have an isomorphism of the form

$$S^+_{\mathbf{Z},\mathcal{T}_M}(G(M/\mathbf{Q})) \cong M_{rs}(\mathbf{Z}[N_{G(M/\mathbf{Q})}\langle c\rangle/\langle c\rangle])^{op}.$$

Set $H_{i,j} = \langle x_i y_j c y_j^{-1} x_i^{-1}\rangle$ so that associated pairs of suffices, (i,j), index the rows and columns of these matrices.

With this notation, the canonical homomorphism of Hecke algebras, induced by passing to $G(M/E)$-fixed points, corresponds to the the ring homomorphism

$$M_{rs}(\mathbf{Z}[N_{G(M/\mathbf{Q})}\langle c\rangle/\langle c\rangle])^{op} \longrightarrow M_s(\mathbf{Z}[N_{G(E/\mathbf{Q})}\langle c\rangle/\langle c\rangle])^{op}$$

which sends the $((i,j),(i_1,j_1))$-th entry to the (j,j_1)-th entry by the map induced by $\pi : G(M/\mathbf{Q}) \longrightarrow G(E/\mathbf{Q})$ (see §5.4.24).

However, the inverse limit

$$\mathcal{N} = \lim_{\leftarrow} N_{G(M/\mathbf{Q})}\langle c\rangle/\langle c\rangle$$

is trivial. For if not then the compositum of all the $M^{\langle c\rangle}$ would be a real closed field, F, and $F/F^{\mathcal{N}}$ would be a non-trivial Galois extension. However, real closed fields do not have any non-trivial Galois automorphisms (c.f. [95] pp. 392–398), since the roots in F of a minimum polynomial over $F^{\mathcal{N}}$ are ordered and the automorphism must preserve the ordering.

Therefore, for any given $E/\mathbf{Q}$ there exists M such that the image of the homomorphism

$$S_{\mathbf{Z}/2,\mathcal{T}_M}(G(M/\mathbf{Q})) \rightarrow S_{\mathbf{Z}/2,\mathcal{T}_E}(G(E/\mathbf{Q})) \cong M_s(\mathbf{Z}/2[N_{G(E/\mathbf{Q})}\langle c\rangle/\langle c\rangle])^{op}$$

lies in the subring of "constants", $M_s(\mathbf{Z}/2)^{op}$. The images under ψ of §5.4.13 of such elements are trivial in the class-group of the group-ring. Therefore $\Omega_1(E/\mathbf{Q}) = \Omega_1(E/\mathbf{Q}, 3)$, as claimed.

However, a general Galois extension, L/K, may be embedded in one of the form $E/\mathbf{Q}$ in which E is totally complex and E/L is Galois. Each of the invariants is natural in the sense that $\Omega(E/\mathbf{Q})$ and $\Omega_1(E/\mathbf{Q},3)$ map to $\Omega(L/K)$ and $\Omega_1(L/K,3)$, respectively, under the homomorphism

$$\mathcal{CL}(\mathbf{Z}[G(E/\mathbf{Q})]) \longrightarrow \mathcal{CL}(\mathbf{Z}[G(E/K)]) \longrightarrow \mathcal{CL}(\mathbf{Z}[G(L/K)]),$$

which completes the proof. □

5.4.23 *Representing ψ in the Hom-description*

In the construction of §5.4.13 suppose that $\underline{\mu} \in S_{\mathbf{Z}/2,\mathcal{T}}(G)^*$ and that $\underline{\mu}' \in S_{\mathbf{Z},\mathcal{T}}(G)$ is a lifting of $\underline{\mu}$. If $R = \mathbf{Z}_l$, the l-adic integers for an odd prime l, or $R = \mathbf{Q}$, the rationals, then the canonical 1-extension of $R[G]$-modules

$$E_+ \otimes R \longrightarrow E \otimes R \longrightarrow E_- \otimes R$$

is split. Hence there are isomorphisms

$$\begin{aligned} Y(\underline{\mu}) \otimes R &\cong \left(\frac{(E_+\otimes R)\oplus(E_+\otimes R)}{((\underline{\mu}'(x),x) \mid x\in E_+\otimes R)} \right) \oplus E_- \otimes R \\ &\cong (Im(\underline{\mu}' \otimes 1 : E_+ \otimes R \longrightarrow E_+ \otimes R)) \oplus (E_- \otimes R). \end{aligned}$$

Hence we may choose the $R[G]$-basis for $E \otimes R$ given by the elements $\{v_i | 1 \leq i \leq r\}$ where

$$v_i = (1/2 \otimes_{H_i} (1+\tau_i), 1/2 \otimes_{H_i} (1-\tau_i)) \in Y(\underline{\mu}) \otimes R.$$

Then

$$Y(\underline{\mu}) \otimes R = R[G]\langle v_1, \ldots, v_r \rangle.$$

Next we must consider $Y(\underline{\mu}) \otimes \mathbf{Z}_2$ as a $\mathbf{Z}_2[G]$-module. Firstly, we observe that the image of the lifting, $\underline{\mu}' \in S_{\mathbf{Z},\mathcal{T}}(G)$, lies in $S_{\mathbf{Z}_2,\mathcal{T}}(G)^*$ if $\underline{\mu} \in S_{\mathbf{Z}/2,\mathcal{T}}(G)^*$. For we can find $\underline{\lambda}_0 \in S_{\mathbf{Z},\mathcal{T}}(G)$ such that

$$\underline{\mu}' \cdot \underline{\lambda}_0 = \underline{\delta}_0 + 2^m \underline{\lambda}_1 \in S_{\mathbf{Z},\mathcal{T}}(G)$$

for some $m \geq 1$. Hence

$$\begin{aligned} &\underline{\mu}' \cdot \underline{\lambda}_0 \cdot (\underline{\delta}_0 - 2^m \underline{\lambda}_1) \\ &= \underline{\mu}' \cdot \underline{\lambda}_0 \cdot \underline{\delta}_0 - 2^m \underline{\mu}' \cdot \underline{\lambda}_0 \cdot \underline{\lambda}_1 \\ &= \underline{\mu}' \cdot \underline{\lambda}_0 - 2^m (\underline{\delta}_0 + 2^m \underline{\lambda}_1) \cdot \underline{\lambda}_1 \\ &= \underline{\delta}_0 + 2^m \underline{\lambda}_1 - 2^m \underline{\lambda}_1 - 2^{2m} \underline{\lambda}_1^2 \end{aligned}$$

and so we may find a 2-adically convergent sequence, $\underline{\lambda}_n \in S_{\mathbf{Z},\mathcal{T}}(G)$, whose limit, $\underline{\lambda} \in S_{\mathbf{Z}_2,\mathcal{T}}(G)$ satisfies $\underline{\mu}' \cdot \underline{\lambda} = \underline{\delta}_0$, the identity of $S_{\mathbf{Z}_2,\mathcal{T}}(G)$.
Consider the following commutative diagram of 1-extensions of $\mathbf{Z}_2[G]$-modules.

$$\begin{array}{ccccc}
E_+ \otimes \mathbf{Z}_2 & \longrightarrow & E \otimes \mathbf{Z}_2 & \longrightarrow & E_- \otimes \mathbf{Z}_2 \\
\underline{\mu}' \downarrow \cong & & \tilde{\underline{\mu}} \downarrow & & 1 \downarrow \\
E_+ \otimes \mathbf{Z}_2 & \longrightarrow & Y(\underline{\mu}) \otimes \mathbf{Z}_2 & \longrightarrow & E_- \otimes \mathbf{Z}_2
\end{array}$$

The elements

$$w_i = (\underline{\mu}'(1 \otimes_{H_i} (1+\tau_i)), 1 \otimes_{H_i} 1)$$

are a $\mathbf{Z}_2[G]$-module basis for

$$Y(\underline{\mu}) \otimes \mathbf{Z}_2 \cong \frac{(E_+ \otimes \mathbf{Z}_2) \oplus (E \otimes \mathbf{Z}_2)}{((\underline{\mu}'(x), x) | x \in E_+ \otimes \mathbf{Z}_2)}$$

since

$$\begin{aligned}
(1+\tau_i)w_i &= (2\underline{\mu}'(1 \otimes_{H_i} (1+\tau_i)), 1 \otimes_{H_i} (1+\tau_i)) \\
&= (\underline{\mu}'(1 \otimes_{H_i} (1+\tau_i)), 0)
\end{aligned}$$

and $\underline{\mu}'$ is 2-adically invertible while the $\{(1 \otimes_{H_i} (1+\tau_i))\}$ generate $E_+ \otimes \mathbf{Z}_2$. Hence $w_1, \ldots, w_r$ generate the free $\mathbf{Z}_2[G]$-module, $Y(\underline{\mu}) \otimes \mathbf{Z}_2$, and therefore $w_1, \ldots, w_r$ must be a $\mathbf{Z}_2[G]$-module basis.

With these bases the recipe for the Hom-description of $\psi(\underline{\mu})$ involves finding an $r \times r$ matrix, $X_2 \in GL_2(\mathbf{Q}_2[G])$, such that

$$X_2 \begin{pmatrix} v_1 \\ \vdots \\ v_r \end{pmatrix} = \begin{pmatrix} w_1 \\ \vdots \\ w_r \end{pmatrix}.$$

It will be more convenient to describe the inverse of X_2.

In $Y(\underline{\mu}) \otimes \mathbf{Q}_2$ we have

$$\begin{aligned}
v_i &= (1/2 \otimes_{H_i} (1+\tau_i), 1/2 \otimes_{H_i} (1-\tau_i)) \\
&= (1/2 \otimes_{H_i} (1+\tau_i), 1 \otimes_{H_i} 1 - 1/2 \otimes_{H_i} (1+\tau_i)) \\
&= (1/2 \otimes_{H_i} (1+\tau_i) + \underline{\mu}'(1/2 \otimes_{H_i} (1+\tau_i)), 1 \otimes_{H_i} 1)
\end{aligned}$$

so that

$$\begin{aligned}
v_i - w_i &= (((1-\underline{\mu}')/2)(1 \otimes_{H_i} (1+\tau_i)), 0) \\
&= (((1-\underline{\mu}')/2)(\underline{\mu}')^{-1}(\underline{\mu}'(1 \otimes_{H_i} (1+\tau_i))), 0) \\
&= ((1-\underline{\mu}')/2)(\underline{\mu}')^{-1}(1+\tau_i)w_i.
\end{aligned}$$

Hence

$$v_i = (1 + (((\underline{\mu}')^{-1} - 1)/2)(1+\tau_i))w_i.$$

Therefore

$$X_2^{-1} = I_r - \underline{\nu} \cdot diag(1+\tau_1, \ldots, 1+\tau_r)$$

where $diag(1+\tau_1,\ldots,1+\tau_r)$ is the map which sends v_i to $(1+\tau_i)v_i) \in E_+$ and

$$\underline{\nu} = ((\underline{\mu}')^{-1}-1)/2.$$

The (i,j)-th entry of X_2^{-1} is equal to

$$\delta_{ij} - \sum_{H_i \sim H_j} \sum_{z \in H_i\backslash G/H_j,\ H_i = zH_jz^{-1}} \nu(i,j,z)z(1+\tau_j)$$

which in turn is equal to

$$\delta_{ij} - \sum_{H_i \sim H_j} \sum_{z \in G/H_j,\ H_i = zH_jz^{-1}} \nu(i,j,z)z.$$

The first expression is derived by noting that $diag(1+\tau_1,\ldots,1+\tau_r)$ sends v_i to $1 \otimes_{H_i} (1+\tau_i)$ which is the generator of $\mathrm{Ind}_{H_i}^G(\mathbf{Z}_+)$ which is mapped by $\underline{\nu}$ to the sum of the $z \otimes_{H_j} \nu(i,j,z)(1+\tau_j) = z(1+\tau_j) \otimes_{H_j} \nu(i,j,z)$ but $\tau_i z = z\tau_j$.

By definition of the Hom-description, the representative of $\psi(\underline{\mu})$ is given by the idèlic-valued function on the (complex) representation ring of G, $R(G)$, which sends χ to $\mathrm{Det}(X_2)(\chi) = \det(\chi(X_2))$ at the primes above 2 and to 1 everywhere else.

In the spirit of Proposition 5.4.18 we may interpret the Hom-description of $\psi(\underline{\mu})$ in terms of $M_r(\mathbf{Z}_2[N_GH/H])$ when

$$\mathcal{T} = \{\langle g_i\tau g_i^{-1}\rangle \mid 1 \le i \le r = [G : N_GH]\}.$$

The element, $(\underline{\mu}')^{-1} \in S^+_{\mathbf{Z}_2,\mathcal{T}}(G)^*$, has (i,j)-th entry equal to $\sum_z \phi(i,j,g_izg_j^{-1})$ where $z \in N_GH/H$. It corresponds to a matrix

$$\underline{\phi} \in M_r(\mathbf{Z}_2[N_GH/(\tau-1)])$$

whose (i,j)-th entry is $\sum_z \phi'(i,j,z)z$ where $\phi'(i,j,z) = \phi(i,j,g(i)zg_j^{-1})$. The matrix in $M_r(\mathbf{Z}_2[N_GH])$ whose (i,j)-th entry is $\sum_z \phi'(i,j,z)z(1+\tau)$ is well defined and corresponds to $(\underline{\mu}')^{-1}diag(1+\tau_1,\ldots,1+\tau_r)$.

Since the homomorphisms

$$\mathrm{Ind}_{N_GH}^G : \mathcal{CL}(\mathbf{Z}[N_GH]) \longrightarrow \mathcal{CL}(\mathbf{Z}[G])$$

corresponds to restriction of representations under the Hom-description we see that, for $\chi \in R(G)$,

$$\mathrm{Det}(X_2)(\chi) = \mathrm{Det}(I_r - (\underline{\phi} - I_r)(1+\tau)/2)\,\mathrm{Res}_{N_GH}^G(\chi).$$

Also, if λ is an irreducible representation of N_GH then $\lambda(\tau) = \pm 1$. If $\lambda(\tau) = -1$ then $\mathrm{Det}(I_r - (\underline{\phi} - I_r)(1+\tau)/2)(\lambda) = 1$ while if $\lambda(\tau) = 1$ then $\mathrm{Det}(I_r - (\underline{\phi} - I_r)(1+\tau)/2)(\lambda) = \mathrm{Det}(\underline{\phi})(\lambda)$. Since $\underline{\phi}$ corresponds to $(\underline{\mu}')^{-1}$ we see that $\psi(\underline{\mu})$ is the image under $\mathrm{Ind}_{N_GH}^G$ of the element of $\mathcal{CL}(\mathbf{Z}[N_GH])$ whose Hom-description is trivial away from primes above 2 and there is given on irreducibles, $\lambda \in R(N_GH)$, by $\mathrm{Det}(\underline{\mu}')(\lambda)$ if $\lambda(\tau) = 1$ and is trivial otherwise, where $\underline{\mu}' \in S^+_{\mathbf{Z}_2,\mathcal{T}}(G)$ is interpreted as as an element of $GL_r(\mathbf{Z}_2[N_GH/H])$.

5.4.24 *Passing to quotients*

The following discussion is necessary to complete the details of the proof of Theorem 5.4.22.

Suppose $E/\mathbf{Q}$ and $M/\mathbf{Q}$ are two totally complex Galois extensions with $E \subseteq M$ and Galois groups

$$G(M/E) \lhd G(M/\mathbf{Q}) \xrightarrow{\pi} G(E/\mathbf{Q}) \cong G(M/\mathbf{Q})/G(M/E).$$

Let $G(M/E) = \{x_1, \ldots, x_r\}$ and let $G(E/\mathbf{Q}) = \{y_1'c^\epsilon, \ldots, y_s'c^\epsilon;\ \epsilon = 0, 1\}$, where c denotes complex conjugation. Lifting each y_j' to $y_j \in G(M/\mathbf{Q})$ we may set

$$\mathcal{T}_M = \{\langle x_i y_j c y_j^{-1} x_i^{-1}\rangle \mid 1 \leq i \leq r,\ 1 \leq j \leq s\}.$$

Hence $\mathcal{T}_M$ is the set of all conjugates of $\langle c\rangle$ in $G(M/\mathbf{Q})$. We wish to consider the isomorphism

$$S^+_{\mathbf{Z},\mathcal{T}_M}(G(M/\mathbf{Q})) \cong M_{rs}(\mathbf{Z}[N_{G(M/\mathbf{Q})}\langle c\rangle/\langle c\rangle])^{op}.$$

Set $H_{i,j} = \langle x_i y_j c y_j^{-1} x_i^{-1}\rangle$. Then

$$S^+_{\mathbf{Z},\mathcal{T}}(G(M/\mathbf{Q})) \cong \oplus_{H_{i,j} \sim H_{i_1,j_1}} \oplus_{z \in H_{i,j}\backslash G(M/\mathbf{Q})/H_{i_1,j_1},\ H_{i,j} = zH_{i_1,j_1}z^{-1}} \mathbf{Z}$$

is a sub-algebra of

$$End_{G(M/\mathbf{Q})}(\oplus_{i,j} \operatorname{Ind}_{H_{i,j}}^{G(M/\mathbf{Q})}(\mathbf{Z})).$$

From §5.4.8, the element given by an integer, n, in the $((i,j),(i_1,j_1),z)$-summand corresponds to the endomorphism which is zero except for sending $g \otimes_{H_{i,j}} 1 \in \operatorname{Ind}_{H_{i,j}}^{G(M/\mathbf{Q})}(\mathbf{Z})$ to $g \otimes_{H_{i_1,j_1}} n \in \operatorname{Ind}_{H_{i_1,j_1}}^{G(M/\mathbf{Q})}(\mathbf{Z})$.

Any $\mathbf{Z}[G(M/\mathbf{Q})]$-module endomorphism of $\oplus_{i,j} \operatorname{Ind}_{H_{i,j}}^{G(M/\mathbf{Q})}(\mathbf{Z})$ induces a $\mathbf{Z}[G(E/\mathbf{Q})]$-module endomorphism of the fixed points

$$(\oplus_{i,j} \operatorname{Ind}_{H_{i,j}}^{G(M/\mathbf{Q})}(\mathbf{Z}))^{G(M/E)}.$$

Also there is an isomorphism of $\mathbf{Z}[G(E/\mathbf{Q})]$-modules of the form

$$\operatorname{Ind}_{H_{i,j}}^{G(M/\mathbf{Q})}(\mathbf{Z}))^{G(M/E)} \cong \operatorname{Ind}_{H_j'}^{G(E/\mathbf{Q})}(\mathbf{Z})$$

sending $\sum_{i=1}^r x_r g \otimes_{H_{i,j}} 1$ to

$$\pi(g) \otimes_{H_j'} 1 \in \operatorname{Ind}_{H_j'}^{G(E/\mathbf{Q})}(\mathbf{Z}),$$

where $H_j' = \pi(H_{i,j})$.

Therefore the image of n in the $((i,j),(i_1,j_1),z)$-summand induces a map which sends $\pi(g) \otimes_{H_j'} 1$ to

$$\pi(g)\pi(z) \otimes_{H_{j_1}'} n \in \operatorname{Ind}_{H_j'}^{G(E/\mathbf{Q})}(\mathbf{Z}).$$

The element of $S^+_{\mathbf{Z},\mathcal{T}_M}(G(M/\mathbf{Q}))$ given by n in the $((i,j),(i_1,j_1),z)$-summand corresponds in $M_{rs}(\mathbf{Z}[N_{G(M/\mathbf{Q})}\langle c\rangle/\langle c\rangle])$ to

$$n(y_j^{-1}x_i^{-1}zx_{i_1}y_{j_1}) \in \mathbf{Z}[N_{G(M/\mathbf{Q})}\langle c\rangle/\langle c\rangle]$$

in the $((i,j),(i_1,j_1))$-th entry. The endomorphism induced on fixed points corresponds to $n(y_j'^{-1}\pi(z)y'_{j_1}) \in \mathbf{Z}[N_{G(E/\mathbf{Q})}\langle c\rangle/\langle c\rangle]$ in the (j,j_1)-entry. Therefore the map induced by passing to $G(M/E)$-fixed points is given by the ring homomorphism

$$M_{rs}(\mathbf{Z}[N_{G(M/\mathbf{Q})}\langle c\rangle/\langle c\rangle])^{op} \longrightarrow M_s(\mathbf{Z}[N_{G(E/\mathbf{Q})}\langle c\rangle/\langle c\rangle])^{op}$$

which sends the $((i,j),(i_1,j_1))$-th entry to the (j,j_1)-th entry by the map induced by $\pi : G(M/\mathbf{Q}) \longrightarrow G(E/\mathbf{Q})$.

5.5 Some calculations

In this section we shall explain how to prove the formulae of Theorem 1.3.6 and Theorem 1.3.11 for $\Omega_1(L/K,3)$ in the case of some families of quaternion fields. We shall see that Iwasawa theory plays a crucial role in these calculations, which presages further connections with Iwasawa theory and the results of [166] which we shall encounter in §6.

5.5.1 The quaternion extensions came in two families $N/\mathbf{Q}$'s (see Example 1.3.2) and $N_q/\mathbf{Q}$'s (see Example 1.3.3). The calculation for the second family is similar to that for the first except that the 2-primary part becomes more complicated. Therefore we shall merely describe the calculation of

$$\Omega_1(N/\mathbf{Q},3) \in \mathcal{CL}(\mathbf{Z}[G(N/\mathbf{Q})]) \cong \mathbf{Z}/4)^*$$

for the family of Example 1.3.2 and Theorem 1.3.6. In this case the biquadratic subfield of N is $E = \mathbf{Q}(\sqrt{2},\sqrt{p})$ where p is a prime satisfying $p \equiv 3$ (modulo 8). The only primes to ramify are 2 and p with only one prime of N above each, P_2 and P_p respectively, so that we shall choose $S = \{P_2, P_p\}$. The localisation sequence of Theorem 2.4.6 yields a short exact sequence of the form

$$0 \longrightarrow K_2'(\mathcal{O}_N) \longrightarrow K_2'(\mathcal{O}_{N,S}) \longrightarrow (\mathcal{O}_N/P_2)^* \oplus (\mathcal{O}_N/P_p)^* \longrightarrow 0.$$

By Corollary 5.3.18 and Corollary 5.4.6 we must evaluate the Euler characteristic of a 2-extension of finite groups

$$K_3^{\text{ind}}(\mathcal{O}_{N,S}) \longrightarrow A \longrightarrow B \longrightarrow K_2'(\mathcal{O}_{N,S})$$

having A, B cohomologically trivial and which is equivalent to the 2-extension of Corollary 5.3.10

$$K_3^{\text{ind}}(\mathcal{O}_{N,S}) \longrightarrow \oplus_{w\in S_\infty(N)}K_3^{\text{ind}}(N_w) \xrightarrow{\delta} C \longrightarrow K_2'(\mathcal{O}_{N,S}).$$

If $x, y \in Q_8$ denote the generators of Example 1.3.2, write $(\mathbf{Z}/2^k)_m$ for the cyclic group of order 2^k on which y acts trivially and x acts like multiplication by m. One finds that the Sylow 2-subgroup of $K_2'(\mathcal{O}_{N,S})$ is isomorphic to $(\mathbf{Z}/8)_3$. By ([96], [106]), the Sylow 2-subgroup of $K_3^{\mathrm{ind}}(\mathcal{O}_{N,S})$ is isomorphic to $(\mathbf{Z}/16)_9$. Hence the 2-primary part of the 2-extension has the form

$$0 \longrightarrow (\mathbf{Z}/16)_9 \longrightarrow A' \longrightarrow B' \longrightarrow (\mathbf{Z}/8)_3 \longrightarrow 0$$

where A', B' are finite, cohomologically trivial 2-groups.

The Euler characteristic of such a 2-extension may be computed by means of the following result.

Proposition 5.5.2 *([37] Corollary 4.6)*
Suppose that we have a 2-extension of $\mathbf{Z}[Q_8]$-modules of the form

$$(\mathbf{Z}/16)_9 \longrightarrow A' \longrightarrow B' \longrightarrow (\mathbf{Z}/8)_3$$

which possesses an Euler characteristic, as in Example 2.1.8.

Choose any element, $\gamma \in B'$, which maps to a generator of $(\mathbf{Z}/8)_3$. Hence there exist elements $b, b' \in A'$ mapping to $(x-3)\gamma$ and $(y-1)\gamma$ respectively. Form

$$x_5 = (xy - xy^3)b + (y + y^2 - 3xy - 3xy^2)b' \in A'$$

and

$$x_6 = (xy^2 - xy^3)b - (y^3 + 3xy^2)b' \in A'.$$

In fact x_5, x_6 lie in $(\mathbf{Z}/16)_9$ and are units (modulo 16). Forthermore, there exists an integer, z, such that

$$x_5 = x_6(-1-4z) \in (\mathbf{Z}/16)_9.$$

The value of z (modulo 4) is independent of the choices and depends only on the equivalence class of the 2-extension. The Euler characteristic, $[A'] - [B']$ of Theorem 2.2.5, is non-trivial if and only if $z \equiv 0, 1$ (modulo 4).

Using this criterion on the original 2-extension of Corollary 5.3.10 one may prove the following result.

Proposition 5.5.3 *([37] Proposition 5.3)*
Let $N/\mathbf{Q}$ be one of the quaternion extensions of Example 1.3.2 and, as in §5.4.1, let

$$0 \longrightarrow (\mathbf{Z}/16)_9 \longrightarrow A' \longrightarrow B' \longrightarrow (\mathbf{Z}/8)_3 \longrightarrow 0$$

be a 2-extension representing the 2-primary part of the 2-extension defining $\Omega_1(N/\mathbf{Q}, 3)$. Then

$$[A'] - [B'] = -1 \in \{\pm 1\} \cong C\mathcal{L}(\mathbf{Z}[G(N/\mathbf{Q})]).$$

In order to evaluate the Euler characteristic of the odd primary part of the 2-extension defining $\Omega_1(N/\mathbf{Q},3)$ we shall use the following result.

Proposition 5.5.4 *([131] Theorem 5.2.23; [35], §4.3.6, p. 44; [132] Lemma 6.3.7; [155] p. 88)*

Let X be a finite group of odd order with a Q_8-action then

$$[X] \in \mathcal{CL}(\mathbf{Z}[Q_8]) \cong (\mathbf{Z}/4)^*$$

is given by

$$[X] = \chi_+(|X_+|)(\sqrt{|X_-|}) = (\sqrt{|X_-|})(-1)^{(1/4)\log_2(|X_+|)} \in (\mathbf{Z}/4)^*$$

where $X_\pm = \{z \in X | x^2(z) = \pm z\}$ and $0 \leq (\sqrt{|X_-|})$ is the positive square root of the cardinality of X_-. Alternatively, if $z \in \mathbf{Z}$, $(-1)^{(1/4)\log_2(|X_+|)} \in (\mathbf{Z}/4)^*$ is equal to $\chi_+(z)$ where $\chi_+ : (\mathbf{Z}/8)^* \longrightarrow (\mathbf{Z}/4)^*$ is the surjective homomorphism with kernel equal to $\{\pm 1\}$.

5.5.5 *Proof of Theorem 1.3.6*

The class, $\Omega_1(N/\mathbf{Q},3)$, is defined as the Euler characteristic of a 2-extension of the form

$$0 \longrightarrow K_3^{\mathrm{ind}}(\mathcal{O}_{N,S}) \longrightarrow A \longrightarrow B \longrightarrow K_2(\mathcal{O}_{N,S})' \longrightarrow 0$$

where S is the set of primes of N over $\{2,p\}$ and A, B are finite groups. The Euler characteristic is the sum of that of the 2-primary part – non-trivial by Proposition 5.4.4 – and that of the odd primary part, which we shall denote by

$$0 \longrightarrow k_3^{\mathrm{ind}}(\mathcal{O}_{N,S}) \longrightarrow A'' \longrightarrow B'' \longrightarrow k_2(\mathcal{O}_{N,S})' \longrightarrow 0.$$

Since all these groups are finite and of odd order (and hence cohomologically trivial) we have

$$[A''] - [B''] = [k_3^{\mathrm{ind}}(\mathcal{O}_{N,S})] - [k_2(\mathcal{O}_{N,S})'] \in \mathcal{CL}(\mathbf{Z}[Q_8]) \cong (\mathbf{Z}/4)^*.$$

Now we shall apply the results of [166] (cf. [132] §7.2) to calculate

$$[A''] - [B''] \in \mathcal{CL}(\mathbf{Z}[Q_8]) \cong (\mathbf{Z}/4)^*.$$

Let $\chi_1, \chi_2 : \langle x^2 \rangle \longrightarrow \{\pm 1\}$ be given by $\chi_1(x^2) = 1$ and $\chi_2(x^2) = -1$. Let S' denote the primes of $E = N^{\langle x^2 \rangle}$ over 2 and p. Then, if X is a finite Q_8-group of odd order and if X^λ denotes the λ-eigenspace of X, we have $X_+ = X^{\chi_1}$, $X_- = X^{\chi_2}$. Therefore, by the odd primary part of the Birch-Tate conjecture proved in [166],

$$\begin{aligned} |\{k_2(\mathcal{O}_{N,S})' - k_3^{\mathrm{ind}}(\mathcal{O}_{N,S})\}_+| &= \tfrac{|k_2(\mathcal{O}_{N,S})'_+|}{|k_3^{\mathrm{ind}}(\mathcal{O}_{N,S})_+|} \\ &= \tilde{L}_{E,S'}(-1,\chi_{1)}) \\ &= \tilde{L}_{\mathbf{Q},\Sigma}(-1, \mathrm{Ind}_{\langle x^2 \rangle}^{Q_8}(1)), \end{aligned}$$

where $\Sigma = \{2, p\}$ and $\tilde{L}$ is as in Theorem 1.3.6, the odd primary part of the rational number given by the value of the L-function. Therefore we find that

$$\chi_+(|\{k_2(\mathcal{O}_{N,S})' - k_3^{\mathrm{ind}}(\mathcal{O}_{N,S})\}_+|) = \chi_+(\tilde{L}_{\mathbf{Q},\Sigma}(-1, \mathrm{Ind}_{\langle x^2 \rangle}^{Q_8}(1))).$$

Next we must decide the positive square root of $|\{k_2(\mathcal{O}_{N,S})' - k_3^{\mathrm{ind}}(\mathcal{O}_{N,S})\}_-|$. This time, by [166], we have

$$\begin{aligned}
&|\{k_2(\mathcal{O}_{N,S})' - k_3^{\mathrm{ind}}(\mathcal{O}_{N,S})\}_-| \\
&= \tilde{L}_{E,S'}(-1, \chi_2) \\
&= \tilde{L}_{\mathbf{Q},\Sigma}(-1, \mathrm{Ind}_{\langle x^2 \rangle}^{Q_8}(\chi_2)) \\
&= \tilde{L}_{\mathbf{Q},\Sigma}(-1, 2\nu(N)) \\
&= \tilde{L}_{\mathbf{Q},\Sigma}(-1, \nu(N))^2
\end{aligned}$$

where $\nu(N)$ is the unique 2-dimensional irreducible representation of Q_8. Since the Artin root number satisfies $W_{\mathbf{Q}}(\nu(N)) = 1$ [37], the sign of $\tilde{L}_{\mathbf{Q},\Sigma}(-1, \nu(N))$ is positive, which completes the calculation. □

5.6 Lifted Galois invariants

5.6.1 In this section we shall use study the lifting of the invariant

$$\Omega_1(N/K, 3) \in \mathcal{CL}(\mathbf{Z}[G(N/K)]) \subset K_0(\mathbf{Z}[G(N/K)])$$

in the localisation sequence of Example 2.1.7 and Theorem 2.4.6

$$\begin{aligned}
\ldots \to K_1(\mathbf{Z}[G(N/K)]) &\xrightarrow{f_*} K_1(\mathbf{Q}[G(N/K)]) \xrightarrow{\partial} K_0(\mathbf{Z}[G(N/K)], \mathbf{Q}) \\
&\xrightarrow{\pi} K_0(\mathbf{Z}[G(N/K)]) \xrightarrow{f_*} K_0(\mathbf{Q}[G(N/K)]).
\end{aligned}$$

That is, for any Galois extension of number fields, N/K, with Galois group $G(N/K)$ we would like to define an element

$$\hat{\Omega}_1(N/K, 3) \in K_0(\mathbf{Z}[G(N/K)], \mathbf{Q})$$

such that $\pi(\hat{\Omega}_1(N/K, 3)) = \Omega_1(N/K, 3)$.

Most of what we have to say in this section extends to the lifting of the invariants, $\Omega_n(N/K, 3)$, of Definition 5.2.11 (cf. Remark 7.2.11(iv)). On the other hand, it is only in the totally real case when $n = 1$ that I have any explicit calculations to offer, so we shall start with that.

Suppose that N/K is a Galois extension of totally real number fields. Then, by Corollary 5.4.6, the 2-extension of Corollary 5.3.10

$$K_3^{\mathrm{ind}}(\mathcal{O}_{N,S}) \longrightarrow \oplus_{w \in S_\infty(N)} K_3^{\mathrm{ind}}(N_w) \xrightarrow{\delta} C \longrightarrow K_2'(\mathcal{O}_{N,S})$$

defines an canonical element of $\mathrm{Ext}^2_{\mathbf{Z}[G(N/K)]}(K_2'(\mathcal{O}_{N,S}), K_3^{\mathrm{ind}}(N))$ possessing an Euler characteristic in the sense of Example 2.1.8(i) equal to $\Omega_1(N/K, 3)$. In fact, this 2-extension is canonical because it came from the truncation of a canonical long exact cohomology sequence. In addition, both $K_2'(\mathcal{O}_{N,S})$ and $K_3^{\mathrm{ind}}(N)$ are finite when N is totally real. Therefore we may choose an equivalent 2-extension of the form

$$K_3^{\mathrm{ind}}(\mathcal{O}_{N,S}) \longrightarrow A \xrightarrow{d} B \longrightarrow K_2'(\mathcal{O}_{N,S})$$

in which A, B are finite and cohomologically trivial. The complex

$$0 \longrightarrow A \xrightarrow{d} B \longrightarrow 0$$

is well defined up to quasi-isomorphism.

Definition 5.6.2 Suppose that N/K is a Galois extension of totally real number fields. In the notation of §5.6.1 let

$$F_* : \ 0 \longrightarrow F_k \xrightarrow{d_k} F_{k-1} \xrightarrow{d_k} \ldots \xrightarrow{d_1} F_0 \longrightarrow 0$$

be a perfect chain complex of $\mathbf{Z}[G(N/K)]$-modules which is quasi-isomorphic to

$$0 \longrightarrow A \xrightarrow{d} B \longrightarrow 0$$

(with B in degree zero). Since the homology groups of F_* are finite, we may apply the construction of Example 2.1.8(ii) to produce a rational isomorphism

$$X : \oplus_j F_{2j} \otimes \mathbf{Q} \xrightarrow{\cong} \oplus_j F_{2j+1} \otimes \mathbf{Q}$$

and a class, $[\oplus_j F_{2j}, X, \oplus_j F_{2j+1}] \in K_0(\mathbf{Z}[G(N/K)], \mathbf{Q})$ satisfying

$$\pi([\oplus_j F_{2j}, X, \oplus_j F_{2j+1}]) = \oplus_j (-1)^j [F_j] = [A] - [B] \in K_0(\mathbf{Z}[G(N/K)]).$$

Define $\hat{\Omega}_1(N/K, 3) = [\oplus_j F_{2j}, X, \oplus_j F_{2j+1}]$, which is well defined by Proposition 2.5.35. This means that, because the 2-extension of §5.6.1 is natural with respect to passage to subgroups and quotient groups the same it true for $\hat{\Omega}_1(N/K, 3)$.

It should be noted that $\hat{\Omega}_1(N/K, 3)$, as I have defined it, depends on the set of primes S. I imagine that it is not too difficult to modify this lifted invariant so as to be independent of S. After all the lifted invariants of ([23], [68], [121]) are independent of S. However, I am concerned here to give a few worked examples for which purpose being independent of S will not be crucial.

Example 5.6.3 Suppose that N/K is equal to the quaternion extension, $N_q/\mathbf{Q}$ of Example 1.3.3. Here q is either a prime satisfying $q \equiv 17$ (modulo 96) or $q = 0$, meaning $N_0 = N = \mathbf{Q}(\sqrt{2}, \sqrt{3}, \beta)$ of Example 1.3.3. These fields all have $E = \mathbf{Q}(\sqrt{2}, \sqrt{3})$ as their biquadratic subfield and $\Omega_1(N_q/\mathbf{Q}, 3)$ was calculated in Theorem 1.3.6 and Theorem 1.3.11 (see also §5.5.5).

As explained in Example 2.5.36, there are isomorphisms

$$\begin{aligned}&K_0(\mathbf{Z}[G(N_q/\mathbf{Q})],\mathbf{Q})\\ &\cong \mathrm{TorsK}_0(\mathbf{Z}[G(N_q/\mathbf{Q})],\mathbf{Q}) \oplus (\oplus_{p \text{ prime}} \ \mathrm{Hom}(R(G(N_q/\mathbf{Q})),\mathbf{Z}) \quad \text{and}\\ &\mathrm{TorsK}_0(\mathbf{Z}[G(N_q/\mathbf{Q})],\mathbf{Q})\\ &\cong \mathrm{TorsK}_0(\mathbf{Z}[G(E/\mathbf{Q})],\mathbf{Q}) \oplus \mathcal{CL}(\mathbf{Z}[G(N_q/\mathbf{Q})]) \cong \mathbf{Z}/2 \oplus \mathbf{Z}/2.\end{aligned}$$

This means that, with the exception of the valuation component at each prime for the 2-dimensional irreducible representation of $G(N_q/\mathbf{Q}) \cong Q_8$, we can determine $\hat{\Omega}_1(N_q/\mathbf{Q},3)$ by calculating $\hat{\Omega}_1(E/\mathbf{Q},3)$ (see Remark 5.6.11).

Example 5.6.4 We wish to compute the possibilities for $\hat{\Omega}_1(E/\mathbf{Q},3)$ where

$$G(E/\mathbf{Q}) = \langle x, y \mid x^2 = y^2, xy = yx\rangle \cong Q_8^{ab}$$

acts according to the formulae

$$x(\sqrt{2}) = -\sqrt{2}, \quad x(\sqrt{3}) = \sqrt{3}, \quad y(\sqrt{2}) = \sqrt{2} \quad \text{and} \quad y(\sqrt{3}) = -\sqrt{3}.$$

Write $\mathbf{Z}/2^s_a$ for the cyclic group, $\mathbf{Z}/2^s$, on which $G(E/\mathbf{Q})$ acts via $x(m) = am$ and $y(m) = m$. Then we have an isomorphism

$$K_3^{\mathrm{ind}}(E) \cong (\mu(L) \otimes \mu(L))^{G(L/E)}$$

where L/E is a Galois extension; this group is independent of L provided that L contains enough roots of unity. In fact, $K_3^{\mathrm{ind}}(E) \cong \mathbf{Z}/48$ as is seen in the following manner. Let $\xi_n = e^{2\pi/n}$ then for any prime, $p > 3$, $\mathbf{Q}(\xi_{p^n})/\mathbf{Q}$ and $E/\mathbf{Q}$ are linearly disjoint so that we may find a Galois automorphism fixing E and sending $\xi_{p^n} \otimes \xi_{p^n}$ to any power prime to p. This shows that the only torsion in $K_3^{\mathrm{ind}}(E)$ must be 2-primary or 3-primary. From [37] we know that E is the biquadratic subfield of a quaternion field, $N/\mathbf{Q}$, for which $K_3^{\mathrm{ind}}(N) \otimes \mathbf{Z}_2 \cong \mathbf{Z}/16$ on which the lifts of x and y from $G(E/\mathbf{Q}) \cong Q_8^{ab}$ to Q_8 act as multiplication by 9 and 1, respectively. This is fixed by $x^2 \in Q_8$ so that

$$K_3^{\mathrm{ind}}(E) \otimes \mathbf{Z}_2 \cong (K_3^{\mathrm{ind}}(N) \otimes \mathbf{Z}_2)^{\langle x^2 \rangle} \cong (\mathbf{Z}/16)_9.$$

Clearly $K_3^{\mathrm{ind}}(E)$ contains 3-torsion, since $\xi_3^{\pm 1} \otimes \xi_3^{\pm 1} = \xi_3 \otimes \xi_3$. Also there exist Galois automorphisms fixing E and sending ξ_{3^n} to $(\xi_{3^n})^{1+3s}$ for any integer, s. This raises $\xi_{3^n} \otimes \xi_{3^n}$ to the $(1+3s)^2$-th power so that varying s shows there is no higher 3-primary torsion in $K_3^{\mathrm{ind}}(E)$. The preceding discussion also shows that the $G(E/\mathbf{Q})$-action on $\xi_3 \otimes \xi_3$ is trivial.

Now let $\mathcal{O}_E$ denote the ring of algebraic integers of E. By the Birch-Tate conjecture, which holds for E [37]

$$1 = \zeta_E(-1) = \frac{|K_2(\mathcal{O}_E)|}{|K_3^{\mathrm{ind}}(E)|}.$$

Therefore, if

$$K_2'(\mathcal{O}_E) = \mathrm{Ker}(K_2(\mathcal{O}_E) \longrightarrow \oplus_{v \ real \ place} K_2(\mathbf{R}))$$

then $K_2'(\mathcal{O}_E) \cong \mathbf{Z}/3$ because the image of the map to the real places is an abelian elementary 2-group of rank four.

Let $S = \{P_2, P_3\}$ where P_j is the unique prime of $\mathcal{O}_E$ above $j = 2, 3$. From the localisation sequence one finds [37] that

$$K_2'(\mathcal{O}_{E,S}) = K_2'(\mathcal{O}_E) \oplus \mathbf{F}_9^* \cong \mathbf{Z}/3 \oplus (\mathbf{Z}/8)_3$$

Incidentally, the action of $G(E/\mathbf{Q})$ on the 3-torsion of $K_2'(\mathcal{O}_{E,S})$ must be non-trivial. This is because, by naturality of the 2-extension under passage to quotients, the 2-group $K_2'(\mathbf{Z}[1/2, 1/3])$ is equal to the $G(E/\mathbf{Q})$-coinvariants of $K_2'(\mathcal{O}_{E,S})$. In fact, $K_2'(\mathcal{O}_F)$ is trivial for $F = \mathbf{Q}(\sqrt{2}), \mathbf{Q}(\sqrt{3})$ and of order three for $F = \mathbf{Q}(\sqrt{6})$. The same is true for the 3-torsion in $K_2'(\mathcal{O}_{F,S'})$ where S' consists of the two primes above 2 and 3. However, there is an isomorphism of the form

$$K_2'(\mathcal{O}_{E,S})_{G(E/F)} \cong K_2'(\mathcal{O}_{F,S'}).$$

Therefore we must have

$$K_2'(\mathcal{O}_{E,S}) = K_2'(\mathcal{O}_E) \oplus \mathbf{F}_9^* \cong (\mathbf{Z}/3)_{-1,-1} \oplus (\mathbf{Z}/8)_3$$

in order to have the correct co-invariants. Here $(\mathbf{Z}/3)_{a,b}$ means $\mathbf{Z}/3$ on which x and y both act as multiplication by a and b, respectively.

Next we observe that

$$\mathrm{Ext}^2_{\mathbf{Z}[G(E/\mathbf{Q})]}(K_2'(\mathcal{O}_{E,S}), K_3^{\mathrm{ind}}(E)) \cong \mathrm{Ext}^2_{\mathbf{Z}[G(E/\mathbf{Q})]}((\mathbf{Z}/8)_3, (\mathbf{Z}/16)_9)$$

since $\mathrm{Ext}^2_{\mathbf{Z}[G(E/\mathbf{Q})]}((\mathbf{Z}/3)_{-1,-1}, M) = 0 = \mathrm{Ext}^2_{\mathbf{Z}[G(E/\mathbf{Q})]}(M, (\mathbf{Z}/3)_{1,1})$ for any M.

Theorem 5.6.5

With the notation introduced above,

$$\mathrm{Ext}^2_{\mathbf{Z}[G(E/\mathbf{Q})]}((\mathbf{Z}/8)_3, (\mathbf{Z}/16)_9) \cong \mathbf{Z}/2 \oplus \mathbf{Z}/2.$$

Proof. Consider the following commutative diagram of $\mathbf{Z}[G(E/\mathbf{Q})]$-module homomorphisms.

$$\begin{array}{ccccccc}
\mathbf{Z}[G(E/\mathbf{Q})] & \xrightarrow{y-1} & \mathbf{Z}[G(E/\mathbf{Q})] & \xrightarrow{y+1} & \mathbf{Z}[G(E/\mathbf{Q})] & \xrightarrow{y-1} & \mathbf{Z}[G(E/\mathbf{Q})] \\
\downarrow{\scriptstyle x-3} & & \downarrow{\scriptstyle x-3} & & \downarrow{\scriptstyle x-3} & & \downarrow{\scriptstyle x-3} \\
\mathbf{Z}[G(E/\mathbf{Q})] & \xrightarrow{y-1} & \mathbf{Z}[G(E/\mathbf{Q})] & \xrightarrow{y+1} & \mathbf{Z}[G(E/\mathbf{Q})] & \xrightarrow{y-1} & \mathbf{Z}[G(E/\mathbf{Q})]
\end{array}$$

The vertical maps are each injective, since $(x+3)(x-3) = -8$, and have cokernel isomorphic to $\mathrm{Ind}_{\langle x\rangle}^{G(E/\mathbf{Q})}((\mathbf{Z}/8)_3)$ because

$$0 \longrightarrow \mathbf{Z}[\langle x\rangle] \xrightarrow{x-3} \mathbf{Z}[\langle x\rangle] \longrightarrow (\mathbf{Z}/8)_3 \longrightarrow 0$$

is exact. Then the sequence

$$\ldots \xrightarrow{y-1} \mathrm{Ind}_{\langle x\rangle}^{G(E/\mathbf{Q})}((\mathbf{Z}/8)_3) \xrightarrow{y+1} \mathrm{Ind}_{\langle x\rangle}^{G(E/\mathbf{Q})}((\mathbf{Z}/8)_3)$$
$$\xrightarrow{y-1} \mathrm{Ind}_{\langle x\rangle}^{G(E/\mathbf{Q})}((\mathbf{Z}/8)_3) \longrightarrow (\mathbf{Z}/8)_{3,1}) \longrightarrow 0$$

is exact. This is because the kernel of the surjection

$$\mathrm{Ind}_{\langle x\rangle}^{G(E/\mathbf{Q})}((\mathbf{Z}/8)_3) \longrightarrow (\mathbf{Z}/8)_{3,\pm 1}$$

is equal to $(\mathbf{Z}/8)_{3,\mp 1}$ generated by $(y \mp 1) \otimes 1$.

Therefore we obtain a free $\mathbf{Z}[G(E/\mathbf{Q})]$-modules resolution of the form

$$\ldots \longrightarrow \mathbf{Z}[G(E/\mathbf{Q})]\langle a_{0,3}, a_{1,2}\rangle \xrightarrow{d} \mathbf{Z}[G(E/\mathbf{Q})]\langle a_{0,2}, a_{1,1}\rangle \xrightarrow{d}$$
$$\mathbf{Z}[G(E/\mathbf{Q})]\langle a_{0,1}, a_{1,0}\rangle \xrightarrow{d} \mathbf{Z}[G(E/\mathbf{Q})]\langle a_{0,0}\rangle \xrightarrow{\epsilon} (\mathbf{Z}/8)_3 \longrightarrow 0$$

where

$$\begin{aligned} d(a_{0,1}) &= (y-1)a_{0,0}, & d(a_{1,0}) &= (x-3)a_{0,0}, \\ d(a_{1,1}) &= (y-1)a_{1,0} - (x-3)a_{0,1}, & d(a_{0,2}) &= (y+1)a_{0,1}, \\ d(a_{1,2}) &= (x-3)a_{0,2} + (y+1)a_{1,1}, & d(a_{0,3}) &= (y-1)a_{0,2} \qquad \text{etc.} \end{aligned}$$

Take a $\mathbf{Z}[G(E/\mathbf{Q})]$-module homomorphism

$$f : \mathbf{Z}[G(E/\mathbf{Q})]\langle a_{0,2}, a_{1,1}\rangle \longrightarrow (\mathbf{Z}/16)_9$$

given by $f(a_{0,2}) \equiv \alpha$, $f(a_{1,1}) \equiv \beta$ (modulo 16). To be a 2-cocycle we must have $0 \equiv f(d(a_{0,3})) \equiv f((y-1)a_{0,2}) \equiv (y-1)f(a_{0,2})$, which is automatic, and $0 \equiv f(d(a_{1,2})) \equiv f((x-3)a_{0,2} + (y+1)a_{1,1}) \equiv 6\alpha + 2\beta$ whose most general solution is

$$f(a_{0,2}) \equiv \alpha, \ f(a_{1,1}) \equiv -3\alpha + 8t.$$

Now consider the 2-coboundaries. Suppose that

$$h : \mathbf{Z}[G(E/\mathbf{Q})]\langle a_{0,1}, a_{1,0}\rangle \longrightarrow (\mathbf{Z}/16)_9$$

is given by $h(a_{1,0}) \equiv \gamma$, $h(a_{0,1}) \equiv \delta$ (modulo 16). Then $d(h)(a_{0,2}) \equiv h((y+1)a_{0,1}) \equiv 2\delta$ and $d(h)(a_{1,1}) \equiv h((y-1)a_{1,0} - (x-3)a_{0,1}) \equiv -6\delta$. Therefore we may change α by any even integer and so the distinct possibilities for the cohomology class of f are

α	β
0	0
0	8
1	−3
1	5

as required. □

Remark 5.6.6 Of the four possibilities for f in the table of Theorem 5.6.5(proof) only the last two can possibly corrrespond to 2-extensions which induce cup-product isomorphisms in Tate cohomology. This is because the homomorphism, f, induces a map of the form

$$f' : \mathrm{Ker}(d : \mathbf{Z}[G(E/\mathbf{Q})]\langle a_{0,1}, a_{1,0}\rangle \longrightarrow \mathbf{Z}[G(E/\mathbf{Q})]\langle a_{0,0}\rangle) \longrightarrow (\mathbf{Z}/16)_9$$

inducing isomorphisms in cohomology. However, the non-surjective f's cannot do this. From the Serre spectral sequence

$$E_2^{s,t} = H^s(\langle y\rangle; H^t(\langle x\rangle; (\mathbf{Z}/16)_9)) \Longrightarrow H^{s+t}(G(E/\mathbf{Q}); (\mathbf{Z}/8)_9)$$

we see that

$$H^s(V; (\mathbf{Z}/8)_{3,1}) \cong H^s(\langle y\rangle; \mathbf{Z}/2) \cong \mathbf{Z}/2 \quad \text{for all} \quad s \geq 0,$$

because

$$H^t(\langle x\rangle; (\mathbf{Z}/16)_9) = 0 \quad \text{for all} \quad t > 0.$$

The non-surjective f's factorise through the inclusion of $\mathbf{Z}/2$ into $(\mathbf{Z}/16)_9$. On $G(E/\mathbf{Q})$-invariants this map induces a homomorphism $\mathbf{Z}/2 \longrightarrow (\mathbf{Z}/16)_9^{G(E/\mathbf{Q})} \cong \mathbf{Z}/8$ sending 1 to 8 (modulo 16). However, 4 (modulo 16) is trivial in the Tate cohomology group, $\hat{H}^0(G(E/\mathbf{Q}); \mathbf{Z}/16)_9) \cong (\mathbf{Z}/16)_9^{G(E/\mathbf{Q})}/(1+x+y+xy)(\mathbf{Z}/16)_9)$.

Therefore, in order to construct a perfect complex of $\mathbf{Z}[G(E/\mathbf{Q})]$-modules which is quasi-isomorphic to a complex of cohomologically trivial, finite modules

$$0 \longrightarrow A \xrightarrow{d} B \longrightarrow 0$$

fitting into a 2-extension of the form

$$(\mathbf{Z}/16)_9 \longrightarrow A \xrightarrow{d} B \longrightarrow (\mathbf{Z}/8)_3$$

we have to consider two cases, in the notation of table in Theorem 5.6.5(proof):

Case A: f corresponding to $\alpha = 1, \beta = 5$.

Case B: f_1 corresponding to $\alpha = 1, \beta = -3$.

5.6.7 *Case A*

Now, from the resolution of Theorem 5.6.5, we have

$$\mathrm{Ker}(d : \mathbf{Z}[G(E/\mathbf{Q})]\langle a_{0,1}, a_{1,0}\rangle \longrightarrow \mathbf{Z}[G(E/\mathbf{Q})]\langle a_{0,0}\rangle) = \langle X, Y\rangle$$

where

$$X = (y+1)a_{0,1} = d(a_{0,2}), \quad Y = (y-1)a_{1,0} - (x-3)a_{0,1} = d(a_{1,1}).$$

Set $Z = 5X - Y$. Then f induces a homomorphism, f', from $\mathrm{Ker}(d)$ to $(\mathbf{Z}/16)_9$ satisfying $f(a_{0,2}) = f'(X), f(a_{1,1}) = f'(Y)$, whose kernel we wish to find. Certainly $\mathbf{Z}[G(E/\mathbf{Q})]\langle Z\rangle \subseteq \mathrm{Ker}(f' : \mathrm{Ker}(d) \longrightarrow (\mathbf{Z}/16)_9)$. If $(a + bx + cy + dxy)Z = 0$ then the coefficient of $a_{1,0}$ shows that $0 = (y-1)(a+bx+cy+dxy)$ and so $a = b, c = d$. The coefficient of $a_{0,1}$ then implies that

$$0 = (a(1+y)+cx(1+y))(5(y+1)+(x-3)) = (1+y)(10a+c-3a+x(10c+a-3c))$$

which implies that $a = c = 0$ and so $\mathbf{Z}[G(E/\mathbf{Q})]\langle Z\rangle$ is free of rank one. Also this free submodule has finite index in $\mathrm{Ker}(f')$.

Now suppose that

$$(a + bx)X + (u + vx + wy + zxy)Y \in \mathrm{Ker}(d) \bigcap \mathrm{Ker}(f'),$$

which is the form of an arbitary element. Then adding $(u+vx+wy+zxy)Z$ gives an element of the form $(a' + b'x)X \in \mathrm{Ker}(d) \bigcap \mathrm{Ker}(f')$ and we must have $a' + 9b' \equiv 0$ (modulo 16). Therefore this element has the form $(b'(x-9)+16t)X$. Now $(x-3)X = (x-3)(y+1)a_{0,1} = -(y+1)Y$ so that $(y+1)Z = (y+1)5X - (y+1)Y = 10X + (x-3)X = (x+7)X = (x-9+16)X$ and therefore $(7-x)(y+1)Z = 48X$. Hence, modulo $\mathbf{Z}[G(E/\mathbf{Q})]\langle Z\rangle$, the action of x on $16X$ is given by $x(16X) \equiv (x-3)(16X) \equiv -(y+1)(16Y) \equiv -(y+1)16(5X - Z) \equiv -160X \equiv -16X$ so that

$$\frac{\mathrm{Ker}(f')}{\mathbf{Z}[G(E/\mathbf{Q})]\langle Z\rangle} \cong (\mathbf{Z}/16)_9 \oplus (\mathbf{Z}/3)_{-1,1}.$$

Next we need a resolution of $(\mathbf{Z}/3)_{-1,1}$ – and the key to this is the well-known resolution of $(\mathbf{Z}/3)_{1,1}$ using Swan modules [131]:

$$0 \longrightarrow \mathbf{Z}[G(E/\mathbf{Q})] \stackrel{(3\cdot -)}{\longrightarrow} \langle 3, 1 + x + y + xy\rangle \longrightarrow (\mathbf{Z}/3)_{1,1} \longrightarrow 0.$$

The Swan module, $\langle 3, 1+x+y+xy\rangle$, is the free module on $3-(1+x)(1+y)$ since $(1+x)(1+y)(3-(1+x)(1+y)) = (3-4)(1+x)(1+y) = -(1+x)(1+y)$ and $(1-(1+x)(1+y))(3-(1+x)(1+y)) = 3-(1+x)(1+y)+(1+x)(1+y) = 3$. Similarly we have a resolution of the form

$$0 \longrightarrow \mathbf{Z}[G(E/\mathbf{Q})] \stackrel{(3\cdot -)}{\longrightarrow} \langle 3, 1 - x + y - xy\rangle \longrightarrow (\mathbf{Z}/3)_{-1,1} \longrightarrow 0.$$

Since $(1-x)(1+y)(3-(1-x)(1+y)) = (3-4)(1-x)(1+y) = -(1-x)(1+y)$ and $(1-(1-x)(1+y))(3-(1-x)(1+y)) = 3-(1-x)(1+y)+(1-x)(1+y) = 3$, we see that $\langle 3, 1-x+y-xy\rangle = \mathbf{Z}[G(E/\mathbf{Q})](3-(1-x)(1+y))$.

In other words, we have a resolution of the form

$$0 \longrightarrow \mathbf{Z}[G(E/\mathbf{Q})] \stackrel{((1-(1-x)(1+y))\cdot -)}{\longrightarrow} \mathbf{Z}[G(E/\mathbf{Q})] \longrightarrow (\mathbf{Z}/3)_{-1,1} \longrightarrow 0$$

where the right-hand homomorphism sends 1 to 1 (modulo 3).

So far, then, we have a perfect complex of the form

$$0 \longrightarrow \mathbf{Z}[G(E/\mathbf{Q})]\langle b_1\rangle \xrightarrow{d} \mathbf{Z}[G(E/\mathbf{Q})]\langle a_{0,1}, a_{1,0}\rangle \xrightarrow{d} \mathbf{Z}[G(E/\mathbf{Q})]\langle a_{0,0}\rangle \longrightarrow 0$$

where $d(b_1) = Z$ and whose non-zero homology is $H_0 = (\mathbf{Z}/8)_3, H_1 = (\mathbf{Z}/16)_9 \oplus (\mathbf{Z}/3)_{-1,1}$. Since the $(\mathbf{Z}/3)_{-1,1}$-factor is generated by $16X$ we need a homomorphism

$$\mathbf{Z}[G(E/\mathbf{Q})]\langle c\rangle \xrightarrow{d} \mathbf{Z}[G(E/\mathbf{Q})]\langle b_1, b_2\rangle \xrightarrow{d} \mathbf{Z}[G(E/\mathbf{Q})]\langle a_{0,1}, a_{1,0}\rangle$$

of the form $d(b_2) = 16X$ and $d(c) = (1-(1-x)(1+y))b_2 + \gamma b_1$ for some $\gamma \in \mathbf{Z}[G(E/\mathbf{Q})]$. To give a complex γ must satisfy

$$\begin{aligned} 0 &= (1-(1-x)(1+y))16X + \gamma Z \\ &= (1-(1-x)(1+y))16(y+1)a_{0,1} + \gamma(5X - Y) \\ &= (16-(1-x)32)(y+1)a_{0,1} + \gamma 5(y+1)a_{0,1} - \gamma Y \\ &= (32x - 16 + 5\gamma)(y+1)a_{0,1} - \gamma((y-1)a_{1,0} - (x-3)a_{0,1}). \end{aligned}$$

Therefore $\gamma = (1+y)(u+vx)$ and

$$\begin{aligned} 0 &= (y+1)(32x - 16 + 10(u+vx) + (u+vx)(x-3)) \\ &= (y+1)(32x - 16 + 10u + 10vx + ux + v - 3u - 3vx) \end{aligned}$$

so that

$$0 = -16 + 10u + v - 3u, \quad 0 = 32 + 10v + u - 3v$$

whose solution is $u = 3, v = -5$. Hence

$$d(c) = (1-(1-x)(1+y))b_2 + (1+y)(3-5x)b_1.$$

To recap, our perfect complex is

$$0 \to \mathbf{Z}[G(E/\mathbf{Q})]\langle c\rangle \xrightarrow{d_3} \mathbf{Z}[G(E/\mathbf{Q})]\langle b_1, b_2\rangle \xrightarrow{d_2} \mathbf{Z}[G(E/\mathbf{Q})]\langle a_{0,1}, a_{1,0}\rangle \xrightarrow{d_1} \mathbf{Z}[G(E/\mathbf{Q})]\langle a_{0,0}\rangle \to 0$$

with $a_{0,0}$ in degree zero.

Writing F_i for the domain of d_i we must next calculate the rational isomorphism

$$X : (F_0 \otimes \mathbf{Q}) \oplus (F_2 \otimes \mathbf{Q}) \xrightarrow{\cong} (F_1 \otimes \mathbf{Q}) \oplus (F_3 \otimes \mathbf{Q})$$

given by

$$X(w_0, w_2) = (\eta_0(w_0) + d_2(w_2), \eta_2(w_2 - \eta_1(d_2(w_2))))$$

where $\eta_i : B_i \otimes \mathbf{Q} \longrightarrow F_{i+1} \otimes \mathbf{Q}$ is a rational splitting of d_{i+1}. We may choose $\eta_0(a_{0,0}) = -((x+3)/8)a_{1,0}$ since then $d_1(\eta_0(a_{0,0})) = -((x-3)(x+3)/8)a_{0,0} =$

$a_{0,0}$. Since the free submodule $\mathbf{Z}[G(E/\mathbf{Q})]\langle Z\rangle$ has finite index in $\mathrm{Ker}(d_1)$ and $d_2(b_1) = Z$, we may set $\eta_1(Z) = b_1$. Then, since $48X = (7-x)(1+y)Z$ we have $\eta_1(16X) = ((7-x)(1+y)/3)b_1$. Finally, $d_3 : \mathbf{Z}[G(E/\mathbf{Q})]\langle c\rangle \longrightarrow B_2$ is an isomorphism so that we set $\eta_2 = d_3^{-1}$.

Now we can calculate the matrix for

$$X : \mathbf{Q}[G(E/\mathbf{Q})]\langle a_{0,0}, b_1, b_2\rangle \xrightarrow{\cong} \mathbf{Q}[G(E/\mathbf{Q})]\langle a_{1,0}, a_{0,1}, c\rangle.$$

Firstly $X(a_{0,0}) = -((x+3)/8)a_{1,0}$. Secondly $X(b_1) = d_2(b_1) = Z = 5(y+1)a_{0,1} - (y-1)a_{1,0} + (x-3)a_{0,1}$ and $X(b_2) = 16X + \eta_2(b_2 - \eta_1(16X)) = 16(y+1)a_{0,1} + d_3^{-1}(b_2 - ((7-x)(1+y)/3)b_1)$. Now $d(c) = (1-(1-x)(1+y))b_2 + (1+y)(3-5x)b_1$ and

$$\begin{aligned}
&(1-(1-x)(1+y))(-((7-x)(1+y)/3)\\
&= (-1/3)((7-x)(1+y) - (7-x)(1-x)(1+y)^2)\\
&= (-1/3)((7-x)(1+y) - (7-x-7x+1)2(1+y))\\
&= (-1/3)(1+y)(7-x-16+16x)\\
&= (1+y)(3-5x)
\end{aligned}$$

so that $d_3^{-1}(b_2 - ((7-x)(1+y)/3)b_1) = (1-(1-x)(1+y))^{-1}c = ((3-(1-x)(1+y))/3)c$. Therefore the matrix of X is given by

$$\begin{pmatrix} -(x+3)/8 & -(y-1) & 0 \\ 0 & 5(y+1)+(x-3) & 16(y+1) \\ 0 & 0 & (3-(1-x)(1+y))/3 \end{pmatrix}.$$

Hence

$$\begin{aligned}
\det(X) &= (1-5(1+y)(x+3)/8)(1-(1-x)(1+y)/3)\\
&= 1 + (1+y)(-(15x+45)/24) - (8-8x)/24) + 10(x+3)(1-x)/24)\\
&= 1 + (1/24)(1+y)(-15x-45-8+8x+20-20x)\\
&= 1 + (1+y)(-33-27x)/24\\
&= 1 - (1+y)(11+9x)/8.
\end{aligned}$$

If $\chi_1(x) = -1, \chi_1(y) = 1, \chi_2(x) = 1, \chi_2(y) = -1$, for the corresponding Hom-description representative in $K_0(\mathbf{Z}[G(E/\mathbf{Q})], \mathbf{Q})$, we shall need the following table of values:

χ	$\chi(\det(X))$
1	-4
χ_1	$1/2$
χ_2	1
$\chi_1\chi_2$	1

5.6.8 *Case B*

Set $Z_1 = 3X + Y = 8X - Z$. Then f_1 induces a homomorphism, f_1', from $\mathrm{Ker}(d)$ to $(\mathbf{Z}/16)_9$ satisfying $f_1(a_{0,2}) = f_1'(X), f_1(a_{1,1}) = f_1'(Y)$, whose kernel we wish to find. Certainly $\mathbf{Z}[G(E/\mathbf{Q})]\langle Z_1\rangle \subseteq \mathrm{Ker}(f' : \mathrm{Ker}(d) \longrightarrow (\mathbf{Z}/16)_9)$. If $(a+bx+cy+dxy)Z_1 = 0$ then the coefficient of $a_{1,0}$ shows that $0 = (y-1)(a+bx+cy+dxy)$ and so $a = b, c = d$. The coefficient of $a_{0,1}$ then implies that

$$0 = (a(1+y)+cx(1+y))(3(y+1)-(x-3)) = (1+y)(6a-c+3a+x(6c-a+3c))$$

which implies that $a = c = 0$ and so $\mathbf{Z}[G(E/\mathbf{Q})]\langle Z_1\rangle$ is free of rank one. Also this free submodule has finite index in $\mathrm{Ker}(f_1')$.

Now suppose that

$$(a+bx)X + (u+vx+wy+zxy)Y \in \mathrm{Ker}(d)\bigcap \mathrm{Ker}(f_1'),$$

which is the form of an arbitary element. Then subtracting $(u+vx+wy+zxy)Z_1$ gives an element of the form $(a'+b'x)X \in \mathrm{Ker}(d)\bigcap\mathrm{Ker}(f_1')$ and we must have $a'+9b' \equiv 0$ (modulo 16). Therefore this element has the form $(b'(x-9)+16t)X$. Now $(x-3)X = (x-3)(y+1)a_{0,1} = -(y+1)Y$ so that $(y+1)Z_1 = (y+1)3X+(y+1)Y = 6X-(x-3)X = -(x-9)X$ and therefore $(9+x)(y+1)Z_1 = 80X$. Therefore, modulo $\mathbf{Z}[V]\langle Z_1\rangle$, the action of x on $16X$ is given by $x(16X) \equiv (x-3)(16X) \equiv -(y+1)(16Y) \equiv (y+1)16(3X-Z_1) \equiv 96X \equiv 16X$ so that

$$\frac{\mathrm{Ker}(f')}{\mathbf{Z}[V]\langle Z_1\rangle} \cong (\mathbf{Z}/16)_9 \oplus (\mathbf{Z}/5)_{-1,1}.$$

Next we need a resolution of $(\mathbf{Z}/5)_{-1,1}$ – and again the key to this is the well-known resolution of $(\mathbf{Z}/5)_{1,1}$ using Swan modules:

$$0 \longrightarrow \mathbf{Z}[G(E/\mathbf{Q})] \stackrel{(5\cdot -)}{\longrightarrow} \langle 5, 1+x+y+xy\rangle \longrightarrow (\mathbf{Z}/5)_{1,1} \longrightarrow 0.$$

The Swan module, $\langle 5, 1+x+y+xy\rangle$, is the free module on $5-(1+x)(1+y)$ since $(1+x)(1+y)(5-(1+x)(1+y)) = (5-4)(1+x)(1+y) = (1+x)(1+y)$ and $(1+(1+x)(1+y))(5-(1+x)(1+y)) = 5-(1+x)(1+y)+(1+x)(1+y) = 5$. Similarly we have a resolution of the form

$$0 \longrightarrow \mathbf{Z}[G(E/\mathbf{Q})] \stackrel{(5\cdot -)}{\longrightarrow} \langle 5, 1-x+y-xy\rangle \longrightarrow (\mathbf{Z}/5)_{-1,1} \longrightarrow 0.$$

Since $(1-x)(1+y)(5-(1-x)(1+y)) = (5-4)(1-x)(1+y) = (1-x)(1+y)$ and $(1+(1-x)(1+y))(5-(1-x)(1+y)) = 5-(1-x)(1+y)+(1-x)(1+y) = 5$, we see that $\langle 5, 1-x+y-xy\rangle = \mathbf{Z}[V](5-(1-x)(1+y))$.

In other words, we have a resolution of the form

$$0 \longrightarrow \mathbf{Z}[G(E/\mathbf{Q})] \stackrel{(1+(1-x)(1+y))\cdot -)}{\longrightarrow} \mathbf{Z}[G(E/\mathbf{Q})] \longrightarrow (\mathbf{Z}/5)_{-1,1} \longrightarrow 0$$

where the right-hand homomorphism sends 1 to 1 (modulo 5).

So far, then, we have a perfect complex of the form

$$\begin{aligned} 0 \longrightarrow \mathbf{Z}[G(E/\mathbf{Q})]\langle b_1\rangle &\stackrel{d}{\longrightarrow} \mathbf{Z}[G(E/\mathbf{Q})]\langle a_{0,1}, a_{1,0}\rangle \\ &\stackrel{d}{\longrightarrow} \mathbf{Z}[G(E/\mathbf{Q})]\langle a_{0,0}\rangle \longrightarrow 0 \end{aligned}$$

where $d(b_1) = Z_1$ and whose non-zero homology is $H_0 = (\mathbf{Z}/8)_3, H_1 = (\mathbf{Z}/16)_9 \oplus (\mathbf{Z}/5)_{-1,1}$. Since the $(\mathbf{Z}/5)_{-1,1}$-factor is generated by $16X$ we need a homomorphism

$$\mathbf{Z}[G(E/\mathbf{Q})]\langle c\rangle \xrightarrow{d} \mathbf{Z}[G(E/\mathbf{Q})]\langle b_1, b_2\rangle \xrightarrow{d} \mathbf{Z}[G(E/\mathbf{Q})]\langle a_{0,1}, a_{1,0}\rangle$$

of the form $d(b_2) = 16X$ and $d(c) = (1 + (1-x)(1+y))b_2 + \gamma_1 b_1$ for some $\gamma_1 \in \mathbf{Z}[G(E/\mathbf{Q})]$. To give a complex γ_1 must satisfy

$$\begin{aligned} 0 &= (1+(1-x)(1+y))16X + \gamma_1 Z_1 \\ &= (1+(1-x)(1+y))16(y+1)a_{0,1} + \gamma_1(3X+Y) \\ &= (16+(1-x)32)(y+1)a_{0,1} + \gamma_1 3(y+1)a_{0,1} + \gamma_1 Y \\ &= (-32x+48+3\gamma_1)(y+1)a_{0,1} + \gamma_1((y-1)a_{1,0} - (x-3)a_{0,1}). \end{aligned}$$

Therefore $\gamma_1 = (1+y)(u+vx)$ and

$$\begin{aligned} 0 &= (y+1)(-32x+48+6(u+vx) - (u+vx)(x-3)) \\ &= (y+1)(-32x+48+6u+6vx - ux - v + 3u + 3vx) \end{aligned}$$

so that

$$0 = 48 + 6u - v + 3u, \quad 0 = -32 + 6v - u + 3v$$

whose solution is $u = 5, v = -3$. Hence

$$d(c) = (1+(1-x)(1+y))b_2 + (1+y)(5-3x)b_1.$$

To recap, our perfect complex is

$$\begin{aligned} 0 \to \mathbf{Z}[G(E/\mathbf{Q})]\langle c\rangle &\overset{d_3}{\to} \mathbf{Z}[G(E/\mathbf{Q})]\langle b_1, b_2\rangle \\ &\overset{d_2}{\to} \mathbf{Z}[G(E/\mathbf{Q})]\langle a_{0,1}, a_{1,0}\rangle \overset{d_1}{\to} \mathbf{Z}[G(E/\mathbf{Q})]\langle a_{0,0}\rangle \to 0 \end{aligned}$$

with $a_{0,0}$ in degree zero.

Writing F_i for the domain of d_i we must next calculate the rational isomorphism

$$X_1 : (F_0 \otimes \mathbf{Q}) \oplus (F_2 \otimes \mathbf{Q}) \xrightarrow{\cong} (F_1 \otimes \mathbf{Q}) \oplus (F_3 \otimes \mathbf{Q})$$

given by

$$X_1(w_0, w_2) = (\eta_0(w_0) + d_2(w_2), \eta_2(w_2 - \eta_1(d_2(w_2))))$$

where $\eta_i : B_i \otimes \mathbf{Q} \longrightarrow F_{i+1} \otimes \mathbf{Q}$ is a rational splitting of d_{i+1}. We may choose $\eta_0(a_{0,0}) = -((x+3)/8)a_{1,0}$ since then $d_1(\eta_0(a_{0,0})) = -((x-3)(x+3)/8)a_{0,0} = a_{0,0}$. Since the free submodule $\mathbf{Z}[G(E/\mathbf{Q})]\langle Z_1\rangle$ has finite index in $\mathrm{Ker}(d_1)$ and $d_2(b_1) = Z_1$, we may set $\eta_1(Z_1) = b_1$. Then, since $(9+x)(y+1)Z_1 = 80X$ we have $\eta_1(16X) = ((9+x)(1+y)/5)b_1$. Finally, $d_3 : \mathbf{Z}[V]\langle c\rangle \longrightarrow B_2$ is an isomorphism so that we set $\eta_2 = d_3^{-1}$.

Now we can calculate the matrix for

$$X_1 : \mathbf{Q}[G(E/\mathbf{Q})]\langle a_{0,0}, b_1, b_2\rangle \xrightarrow{\cong} \mathbf{Q}[G(E/\mathbf{Q})]\langle a_{1,0}, a_{0,1}, c\rangle.$$

Firstly $X_1(a_{0,0}) = -((x+3)/8)a_{1,0}$. Secondly $X_1(b_1) = d_2(b_1) = Z_1 = 3(y+1)a_{0,1} + (y-1)a_{1,0} - (x-3)a_{0,1}$ and $X_1(b_2) = 16X + \eta_2(b_2 - \eta_1(16X)) = 16(y+1)a_{0,1} + d_3^{-1}(b_2 - ((9+x)(1+y)/5)b_1)$. Now $d(c) = (1+(1-x)(1+y))b_2 + (1+y)(5-3x)b_1$ and

$$\begin{aligned}
&(1+(1-x)(1+y))(9+x)(1+y)/5 \\
&= (9+x)(1+y)/5 + (9+x)(1-x)(1+y)^2/5 \\
&= (1/5)((9+x)(1+y) + (9+x-9x-1)2(1+y)) \\
&= (1/5)(1+y)(9+x+16-16x) \\
&= (1+y)(5-3x)
\end{aligned}$$

so that $d_3^{-1}(b_2 - ((9+x)(1+y)/5)b_1) = (1+(1-x)(1+y))^{-1}c$. Therefore the matrix of X_1 is given by

$$\begin{pmatrix} -(x+3)/8 & (y-1) & 0 \\ 0 & 3(y+1)-(x-3) & 16(y+1) \\ 0 & 0 & (1+(1-x)(1+y))^{-1} \end{pmatrix}.$$

Hence

$$\det(X) = (-1-3(1+y)(x+3)/8)(1+(1-x)(1+y))^{-1}$$

and if $\chi_1(x) = -1, \chi_1(y) = 1, \chi_2(x) = 1, \chi_2(y) = -1$, for the corresponding Hom-description representative in $K_0(\mathbf{Z}[G(E/\mathbf{Q})], \mathbf{Q})$, we shall need the following table of values:

χ	$\chi((-1-3(1+y)(x+3)/8))$	$\chi(1+(1-x)(1+y))$	$\chi(\det(X_1))$
1	-4	1	-4
χ_1	$-5/2$	5	$-1/2$
χ_2	-1	1	-1
$\chi_1\chi_2$	-1	1	-1

.

Theorem 5.6.9
Let $E/\mathbf{Q}$ be as in Example 5.6.4 and let S be the set of two primes of E dividing 2 and 3. Then the 2-primary part of the 2-extension

$$K_3^{\text{ind}}(\mathcal{O}_{E,S}) \longrightarrow \oplus_{w\in S_\infty(E)} K_3^{\text{ind}}(E_w) \xrightarrow{\delta} C \longrightarrow K_2'(\mathcal{O}_{E,S})$$

used to define $\hat{\Omega}_1(E/\mathbf{Q}, 3)$ in Definition 5.6.2 is represented by a 2-extension of the form

$$0 \longrightarrow A \xrightarrow{d} B \longrightarrow 0$$

fitting into a 2-extension of the form

$$(\mathbf{Z}/16)_9 \longrightarrow A \xrightarrow{d} B \longrightarrow (\mathbf{Z}/8)_3$$

corresponding (only) to Case B in Remark 5.6.6.

Proof. Suppose that, as in Proposition 5.5.2,

$$(\mathbf{Z}/16)_9 \longrightarrow A' \longrightarrow B' \longrightarrow (\mathbf{Z}/8)_3$$

represents the 2-primary part of the canonical 2-extension of $\mathbf{Z}[Q_8]$-modules used in the definition of $\hat{\Omega}_1(N_0/\mathbf{Q}, 3)$ in Definition 5.6.2. By Proposition 5.5.3 we know that the Euler characteristic of this 2-extension is the non-trivial element of $\mathcal{CL}(\mathbf{Z}[Q_8])$. By Proposition 5.5.2 this means that if $\gamma \in B'$ maps to a generator of $(\mathbf{Z}/8)_3$ and $b, b' \in A'$ map to $(x-3)\gamma, (y-1)\gamma$, respectively, then

$$x_5 = (xy - xy^3)b + (y + y^2 - 3xy - 3xy^2)b', \quad x_6 = (xy^2 - xy^3)b - (y^3 + 3xy^2)b'$$

lie in $(\mathbf{Z}/16)_9$. Furthermore they are units (modulo 16) which satisfy

$$x_5 = x_6(-1 - 4z) \in (\mathbf{Z}/16)_9.$$

for some $z \equiv 0, 1$ (modulo 4). In fact, in [37], it is shown that $z = 0$.

By naturality under passage to quotients of these 2-extensions, the $\mathbf{Z}[G(E/\mathbf{Q})]$-module 2-extension in the statement of the theorem is obtained by glueing together the the $G(N_0/E)$-invariant sequence

$$(\mathbf{Z}/16)_9 \longrightarrow A'^{G(N_0/E)} \longrightarrow B'^{G}(N_0/E)$$

to the $G(N_0/E)$-coinvariant sequence

$$A'_{G(N_0/E)} \longrightarrow B'_{G(N_0/E)} \longrightarrow (\mathbf{Z}/8)_3$$

by the norm isomorphisms, $z \mapsto (1 + x^2)z$, in the middle. Therefore, in the $\mathbf{Z}[G(E/\mathbf{Q})]$-module 2-extension, if $\gamma \in B$ maps to a generator of $(\mathbf{Z}/8)_3$ and $b, b' \in A$ map to $(x-3)\gamma, (y-1)\gamma$, respectively, then

$$\begin{aligned}\tilde{x}_5 &= (xy - xy^3)b + (y + y^2 - 3xy - 3xy^2)b' = (1 + y - 3x - 3xy)b', \\ \tilde{x}_6 &= (xy^2 - xy^3)b - (y^3 + 3xy^2)b' = (x - xy(b - (y + 3x)b'\end{aligned}$$

lie in $(\mathbf{Z}/16)_9$. Furthermore they are units (modulo 16) which satisfy $\tilde{x}_5 \equiv -\tilde{x}_6$ (modulo 16). In the 2-extensions of §5.6.7 and §5.6.8 we may choose $\gamma = a_{0,0}, b = a_{1,0}, b' = a_{0,1}$. Then $\tilde{x}_5 = (1+y-3x-3xy)a_{0,1} = (1-3x)X$ and $\tilde{x}_6 = x(1-y)a_{1,0} - (y+3x)a_{0,1} = -xY - X$. In both Case A and Case B, $\tilde{x}_5 = f((1-3x)X) = f_1((1-3x)X) \equiv -26 \equiv 6$ (modulo 16). However, in Case A, $\tilde{x}_6 = f(-xY - X) \equiv -46 \equiv 2$ (modulo 16) and $(-1) \cdot 2$ is not congruent to 6 (modulo 16). On the other hand, in Case B, $\tilde{x}_6 = f_1(-xY - X) \equiv 26 \equiv -6$ (modulo 16). □

Corollary 5.6.10

Let $E/\mathbf{Q}$ be as in Example 5.6.4 and let S be the set of two primes of E dividing 2 and 3. Then, in the Hom-description of Example 2.5.37

$$\begin{aligned} & K_0(\mathbf{Z}[G(E/\mathbf{Q})], \mathbf{Q}) \\ & \cong \mathrm{TorsK}_0(\mathbf{Z}[G(E/\mathbf{Q})], \mathbf{Q}) \oplus (\oplus_{p \text{ prime}} \ \mathrm{Hom}(R(G(N/\mathbf{Q})), \mathbf{Z}) \\ & \cong \mathbf{Z}/2 \oplus (\oplus_{p \text{ prime}} \ \mathrm{Hom}(R(G(N/\mathbf{Q})), \mathbf{Z}) \end{aligned}$$

the class $\hat{\Omega}_1(E/\mathbf{Q}, 3)$ corresponds to $(0, \{f_p\}_{p \text{ prime}})$ where f_p is trivial when $p \neq 2, 3$ and f_2, f_3 are given by the following table.

χ	1	χ_1	χ_2	$\chi_1\chi_2$
$f_2(\chi)$	2	-1	0	0
$f_3(\chi)$	1	0	0	-1

Proof. Theorem 5.6.9 together with the calculation of §5.6.8 shows that the 2-primary part of the 2-extension used to define $\hat{\Omega}_1(E/\mathbf{Q}, 3)$ gives rise to an element whose Hom-representative, g, is given by $g(1) = -4, g(\chi_1) = -1/2, g(\chi_2) = -1 = g(\chi_1\chi_2)$. By Corollary 2.5.40, the component of $[g]$ in $\mathrm{TorsK}_0(\mathbf{Z}[G(E/\mathbf{Q})], \mathbf{Q}) \cong \{\pm 1\}$ is equal to $g(1 + \chi_1 + \chi_2 + \chi_1\chi_2) = (-1)^4$, which is trivial. The contribution to the function f_p is the p-adic valuation of g, which is trivial for odd primes and at 2 is given by $1 \mapsto 2, \chi_1 \mapsto -1, \chi_2 \mapsto 0, \chi_1\chi_2 \mapsto 0$.

The 2-extension used to define $\hat{\Omega}_1(E/\mathbf{Q}, 3)$ also has a 3-primary part which, by Example 5.6.4 has the form

$$(\mathbf{Z}/3)_{1,1} \longrightarrow A'' \longrightarrow B'' \longrightarrow (\mathbf{Z}/3)_{-1,-1}.$$

Since $G(E/\mathbf{Q})$ is a 2-group this 2-extension must be trivial. Hence $A'' \to B''$ must be quasi-isomorphic to the sum of a resolution for $(\mathbf{Z}/3)_{-1,-1}$ ending in dimension zero and a resolution for $(\mathbf{Z}/3)_{1,1}$ ending in dimension one. From Example 5.6.7 such resolutions are given by

$$0 \longrightarrow \mathbf{Z}[G(E/\mathbf{Q})] \stackrel{((1-(1+x)(1+y))\cdot -)}{\longrightarrow} \mathbf{Z}[G(E/\mathbf{Q})] \longrightarrow (\mathbf{Z}/3)_{1,1} \longrightarrow 0$$

and

$$0 \longrightarrow \mathbf{Z}[G(E/\mathbf{Q})] \stackrel{((1-(1-x)(1-y))\cdot -)}{\longrightarrow} \mathbf{Z}[G(E/\mathbf{Q})] \longrightarrow (\mathbf{Z}/3)_{-1,-1} \longrightarrow 0$$

where the right-hand homomorphisms send 1 to 1 (modulo 3). The rational isomorphism corresponding to this has determinant

$$\det(X) = \frac{(1 - (1 + x)(1 + y))}{(1 - (1 - x)(1 - y))} \in \mathbf{Q}[G(E/\mathbf{Q})]^*$$

which correspond in the Hom-description to the function, g', given by

$$g'(1) = -3, g'(\chi_2) = 1 = g'(\chi_2), g'(\chi_1\chi_2) = -1/3.$$

Again the component of $[g']$ in $\mathrm{TorsK}_0(\mathbf{Z}[G(E/\mathbf{Q})], \mathbf{Q})$ is trivial and the contribution to the f_p's is trivial except at 3 where we obtain $1 \mapsto 1, \chi_1 \mapsto 0, \chi_2 \mapsto 0, \chi_1\chi_2 \mapsto -1$, which completes the proof. □

Remark 5.6.11 As explained in Example 5.6.3, there is an isomorphism of the form

$$\begin{aligned}&K_0(\mathbf{Z}[G(N_q/\mathbf{Q})],\mathbf{Q})\\ &\cong \mathrm{TorsK}_0(\mathbf{Z}[G(N_q/\mathbf{Q})],\mathbf{Q}) \oplus (\oplus_{p \text{ prime}} \mathrm{Hom}(R(G(N_q/\mathbf{Q})),\mathbf{Z})\\ &\cong \mathrm{TorsK}_0(\mathbf{Z}[G(E/\mathbf{Q})],\mathbf{Q}) \oplus \mathcal{CL}(\mathbf{Z}[G(N_q/\mathbf{Q})])\\ &\qquad \oplus(\oplus_{p \text{ prime}} \mathrm{Hom}(R(G(N_q/\mathbf{Q})),\mathbf{Z}).\end{aligned}$$

By naturality, $\hat{\Omega}_1(N_q/\mathbf{Q},3)$ will correspond to the element having first component trivial, second component equal to the formula of Theorem 1.3.11 and the function, f_p, at the prime p in the last component will be given on the one-dimensional representations $1,\chi_1,\chi_2,\chi_1\chi_2$ by the formulae of Corollary 5.6.10.

There remains the problem of determining $f_p(\nu)$, where ν is the two-dimensional irreducible of $G(N_q/\mathbf{Q}) \cong Q_8$. The conjectures of [23] contain a prediction for these values.

5.6.12 *Lifting $\Omega_1(N/K,3)$ in general*

Now let us turn to the problem of constructing $\hat{\Omega}_1(N/K,3)$ for an arbitrary Galois extension of number fields. Definition 5.6.2 accomplishes this in the totally real case. Since $\Omega_1(N/K,3)$ is natural with respect passage to subgroups and quotient groups, it will suffice to consider the special case in which the Galois extension of number fields, N/K, has the form $E/\mathbf{Q}$ with E totally complex. As explained in §5.4.7, if we glue the diagram of Theorem 5.4.5 to diagram (6) of §5.3.15 we obtain a commutative diagram of the form

$$\begin{array}{ccccccc}
E_+ & & & & & & \\
\downarrow & & & & & & \\
K_3^{\mathrm{ind}}(\mathcal{O}_{E,S})' & \longrightarrow & A_1 & \longrightarrow & B_1 & \longrightarrow & K_2'(\mathcal{O}_{E,S}) \\
\downarrow & & \downarrow & & \downarrow & & \downarrow \\
K_3^{\mathrm{ind}}(E) & \longrightarrow & A_2 & \longrightarrow & B_2 & \longrightarrow & \tilde{K}_2'(\mathcal{O}_{E,S}) \\
 & & & & & & \downarrow \\
 & & & & & & E_-
\end{array}$$

in which the horizontal rows are 2-extensions in which A_1, B_1, A_2, B_2 are finitely generated, cohomologically trivial $\mathbf{Z}[G(E/\mathbf{Q})]$-modules and the extreme vertical columns are short exact sequences. The upper 2-extension is the upper 2-extension of Theorem 5.4.5 and the lower 2-extension is the lower 2-extension of §5.3.15(6) so that $[A_1] - [B_1] = [A_2] - [B_2] = \Omega_1(E/\mathbf{Q}, 3)$, by Theorem 5.4.22. In this diagram $E_\pm = \mathrm{Ind}_H^{G(E/\mathbf{Q})}(\mathbf{Z}_\pm)$ where H is the decomposition group one of the infinite places of E, whose generator acts by ± 1 on a copy of the integers denoted by $\mathbf{Z}_\pm$.

There are two (equivalent) ways in which we can construct $\hat{\Omega}_1(E/\mathbf{Q}, 3)$, one based on the upper and one on the lower 2-extension in the diagram.

Consider first the lower 2-extension. Since $K_2'(\mathcal{O}_{E,S})$ is finite we have a rational isomorphism of the form $\tilde{K}_2'(\mathcal{O}_{E,S}) \otimes \mathbf{Q} \cong \mathbf{Q}[G(E/\mathbf{Q})] \otimes_{\mathbf{Z}[H]} \mathbf{Z}_-$. On the other hand, the Borel regulator ([17], [18], [19]) gives an isomorphism of real representations of the form $K_3^{\mathrm{ind}}(E) \otimes \mathbf{R} \cong \mathbf{R}[G(E/\mathbf{Q})] \otimes_{\mathbf{Z}[H]} \mathbf{Z}_-$, which implies that there is a rational isomorphism of the form $K_3^{\mathrm{ind}}(E) \otimes \mathbf{Q} \cong \mathbf{Q}[G(E/\mathbf{Q})] \otimes_{\mathbf{Z}[H]} \mathbf{Z}_-$. This rational isomorphism is not unique and a choice must be made - a similar situation occurs in the construction of the Stark regulator (see [152]). Having made this choice we have a $\mathbf{Q}[G(E/\mathbf{Q})]$-module isomorphism

$$\tilde{K}_2'(\mathcal{O}_{E,S}) \otimes \mathbf{Q} \cong K_3^{\mathrm{ind}}(E) \otimes \mathbf{Q}$$

between the odd and even-dimensional homology of the chain complex

$$0 \longrightarrow A_2 \longrightarrow B_2 \longrightarrow 0.$$

Choosing a quasi-isomorphic perfect chain complex of $\mathbf{Z}[G(E/\mathbf{Q})]$-modules of the form

$$F_* : \ 0 \longrightarrow F_k \xrightarrow{d_k} F_{k-1} \xrightarrow{d_k} \cdots \xrightarrow{d_1} F_0 \longrightarrow 0$$

and applying the construction of Example 2.1.8(ii), using the chosen rational homology isomorphism, yields the required lifting, $\hat{\Omega}_1(E/\mathbf{Q}, 3) \in K_0(\mathbf{Z}[G(E/\mathbf{Q})], \mathbf{Q})$, mapping, under π, to $\Omega_1(E/\mathbf{Q}, 3) \in K_0(\mathbf{Z}[G(E/\mathbf{Q})])$.

The second approach uses a $\mathbf{Q}[G(E/\mathbf{Q})]$-module isomorphism of the form

$$K_3^{\mathrm{ind}}(\mathcal{O}_{E,S})' \otimes \mathbf{Q} \cong \mathbf{Q}[G(E/\mathbf{Q})].$$

The left-hand vertical exact sequence in the diagram shows that such an isomorphism exists, since rationally the left and right ends of the 2-extension must differ by a multiple of the regular representation and the right-hand end is a finite group. Hence we may choose free $\mathbf{Z}[G(E/\mathbf{Q})]$-submodule of finite index in $K_3^{\mathrm{ind}}(\mathcal{O}_{E,S})'$, call it $\mathcal{L}$, and form the 2-extension

$$K_3^{\mathrm{ind}}(\mathcal{O}_{E,S})'/\mathcal{L} \longrightarrow A_1/\mathcal{L} \longrightarrow B_1 \longrightarrow K_2'(\mathcal{O}_{E,S}).$$

This extension has finite groups at the ends and finitely generated, cohomologically trivial modules in the middle. Hence we may apply the construction of Definition 5.6.2 to this 2-extension to define $\hat{\Omega}_1(E/\mathbf{Q}, 3)$.

Different choices of the lattice, $\mathcal{L}$, are afforded by different regulators (see [7], [8], [13], [90], [112], [118], [141], [143], [144]). The connection between the two constructions of $\hat{\Omega}_1(E/\mathbf{Q}, 3)$ come from the fact that the map of $\mathcal{L} \otimes \mathbf{Q}$ onto $K_3^{\mathrm{ind}}(E) \otimes \mathbf{Q}$ yields an isomorphism to $\mathbf{Q}[G(E/\mathbf{Q})] \otimes_{\mathbf{Z}[H]} \mathbf{Z}_- \cong K_3^{\mathrm{ind}}(E) \otimes \mathbf{Q}$.

As it stands, there are too many choices in the constructions discussed here. One might improve on the second one by choosing $\mathcal{L}$ to minimise the index, $[E_+ : E_+ \bigcap \mathcal{L}]$. I would not be surprised if this minimal value turned out to be $2^{|G(E/\mathbf{Q})|/2}$.

5.6.13 *Abelian extensions*

Let $E/\mathbf{Q}$ be an abelian Galois extension of number fields with E totally complex, as in §5.6.12, and let

$$K'_{2r-1} \longrightarrow A_r \longrightarrow B_r \longrightarrow K'_{2r-2}$$

be the associated 2-extension of $\mathbf{Z}[G(E/\mathbf{Q})]$-modules, as in Proposition 5.2.10. In this situation there is a canonical lattice, $\mathcal{L}_E \subseteq K'_{2r-1}$, generated by Beilinson's cyclotomic elements ([9], [78], [90], [91], [112]). This lattice is constructed first for some cyclotomic field, $\mathbf{Q}(\xi_m)$, containing E and then we set

$$\mathcal{L}_E = (\mathcal{L}_{\mathbf{Q}(\xi_m)})^{G(\mathbf{Q}(\xi_m)/E)}.$$

Using this lattice, as in §5.6.12, we may construct a lifted invariant for each $r \geq 2$

$$\hat{\Omega}_{r-1}(E/\mathbf{Q}, 3) \in K_0(\mathbf{Z}[G(E/\mathbf{Q}], \mathbf{Q})$$

such that $\pi(\hat{\Omega}_{r-1})(E/\mathbf{Q}, 3)) = \Omega_{r-1}(E/\mathbf{Q}, 3) \in K_0(\mathbf{Z}[G(E/\mathbf{Q}])$.

These lifted invariants are closely related to those constructed in [22], inspired by work of Bloch and Kato ([14], [85], [86], [87]). Generally the the latter invariants lie in $K_0(\mathbf{Z}[G(N/K)], \mathbf{R})$ but in the case of abelian fields they lie in $K_0(\mathbf{Z}[G(E/\mathbf{Q})], \mathbf{Q})$. Using ([53], [86], [159], [166]) this invariant is evaluated in [27], as mentioned in Conjecture 5.2.18 (see also Remark 7.2.11(iv)).

Chapter 6

The Wiles unit

In this chapter I want to introduce a conjecture (Conjecture 6.3.4) whose cryptic form would be: "The Wiles Unit is a determinant". The "Wiles Unit" unit is the p-adic unit-valued function on Galois representations given (approximately) by the ratio of the p-adic L-function to the Iwasawa polynomial. The values of this function are p-adic units by the main result of [166]. In §6.1 I give a functorial treatment of the material of [64], which results in Iwasawa polynomials even in the non-abelian case. In §6.2 we shall recall the definition of the p-adic L-function. In §6.3 we examine what determinantal functions are and how to detect them when the Galois group is an elementary abelian p-group. Then, having set up the background material, we state the conjecture and accompany it with some preliminary evidence in its favour.

6.1 The Iwasawa polynomial

6.1.1 In this section we shall define the Iwasawa polynomial, $h_\rho(T)$, of a p-adic Galois representation, ρ, and we shall derive for it all the formal properties (twisting by characters of type W, invariance under inflation and induction) which are enjoyed by the p-adic L-function (see §6.2.2). This will be used in §6.2.4 to derive (Theorem 6.2.5) the existence of the Wiles Unit polynomial for an arbitrary Galois representation of totally real number fields from the results of [166], which establish the existence for one-dimensional representations of type S.

Let E/F be a Galois extension of totally real number fields with Galois group, $G(E/F)$. Let p be a prime and let $\mathbf{Z}_p$ denote the p-adic integers. Let $\overline{\mathbf{Q}}_p$ be an algebraic closure of $\mathbf{Q}_p$, the p-adics. Let F_∞/F be the cyclotomic $\mathbf{Z}_p$-extension so that $\Gamma = G(F_\infty/F) \cong \mathbf{Z}_p$ and we choose, once and for all, a topological generator, γ, for Γ.

Let $X_{E/F}$ denote the Galois group, $G(L/EF_\infty)$, where L/EF_∞ is the maximal abelian pro-p extension in which only primes over p ramify. Hence $X_{E/F}$ is a module over the completed (with respect to the profinite topology on the Galois group) group-ring , $\mathbf{Z}_p[G(EF_\infty/F)]$, where $g \in G(EF_\infty/F)$ acts on $x \in X_{E/F}$ by $g(x) = \overline{g}x\overline{g}^{-1}$ for any lifting of g to an F-automorphism, $\overline{g}$, of L.

Now let $\mathcal{O}_K$ denote the ring of integers in a p-adic local field, $K \subset \overline{\mathbf{Q}}_p$. Suppose that U is an $\mathcal{O}_K[G(EF_\infty/F)]$-module which is free of finite rank as an $\mathcal{O}_K$-module. A primary source of such U's, by inflation, is the set of $\mathcal{O}_K[G(E/F)]$-lattices of finite rank. Let

$$\operatorname{Hom} U, X_{E/F} \otimes_{\mathbf{Z}_p} \mathcal{O}_K)$$

denote the set of $\mathcal{O}_K$-module homomorphisms, $U \longrightarrow X_{E/F} \otimes_{\mathbf{Z}_p} \mathcal{O}_K$, endowed with the (left) $\mathcal{O}_K[G(EF_\infty/F)]$-module structure given by $h(f)(u) = h(f(h^{-1}(u)))$ for $h \in G(EF_\infty/F)$. Since $X_{E/F}$ is a Noetherian $\mathcal{O}_K[G(EF_\infty/F)]$-module so is $\operatorname{Hom}(U, X_{E/F} \otimes_{\mathbf{Z}_p} \mathcal{O}_K)$.

The subgroup of $G(EF_\infty/F_\infty)$-fixed points of $\operatorname{Hom}(U, X_{E/F} \otimes_{\mathbf{Z}_p} \mathcal{O}_K)$, written

$$I_{E/F}(U) = \operatorname{Hom}_{G(EF_\infty/F_\infty)}(U, X_{E/F} \otimes_{\mathbf{Z}_p} \mathcal{O}_K),$$

is a Noetherian, torsion module over $\mathcal{O}_K[\Gamma] \cong \mathcal{O}_K[[T]]$ if $T+1$ acts like $\gamma \in G(EF_\infty/F)/G(EF_\infty/F_\infty) \cong \Gamma$. The $\overline{\mathbf{Q}}_p$-vector space, $V_{E/F} = X_{E/F} \otimes_{\mathbf{Z}_p} \overline{\mathbf{Q}}_p$, is finite-dimensional and so is

$$I_{E/F}(U) \otimes_{\mathcal{O}_K} \overline{\mathbf{Q}}_p \cong \operatorname{Hom}_{G(EF_\infty/F_\infty)}(U \otimes_{\mathcal{O}_K} \overline{\mathbf{Q}}_p, X_{E/F} \otimes_{\mathbf{Z}_p} \overline{\mathbf{Q}}_p).$$

Therefore the characteristic polynomial, $\det(tI - ((\gamma - 1) \cdot -))$, of $\gamma - 1$ acting on $I_{E/F}(U) \otimes_{\mathbf{Z}_p} \overline{\mathbf{Q}}_p$ is a monic polynomial with coefficients in $\mathcal{O}_K$. This characteristic polynomial clearly depends only on the $\overline{\mathbf{Q}}_p$-representation

$$\rho : G(EF_\infty/F) \longrightarrow Aut_{\overline{\mathbf{Q}}_p}(U \otimes_{\mathcal{O}_K} \overline{\mathbf{Q}}_p).$$

Hence we set

$$h_\rho(T) = \det(T - ((\gamma - 1) \cdot -) \mid I_{E/F}(U) \otimes_{\mathbf{Z}_p} \overline{\mathbf{Q}}_p) \in \mathcal{O}_K[T].$$

Example 6.1.2 Suppose that that $E \cap F_\infty = F$ and that ρ is one-dimensional, coming from a homomorphism of the form $\rho : G(E/F) \longrightarrow \mathcal{O}_K^*$. We have a commutative diagram (top of next page) of canonical maps of Galois groups.

Let $\gamma' \in G(EF_\infty/E)$ denote the unique element which satisfies $\pi'(i(\gamma')) = \gamma$ and set $\psi = \rho \cdot \pi : G(EF_\infty/F) \longrightarrow \overline{\mathbf{Q}}_p^*$. Then ψ is a one-dimensional representation of type S in the terminology of [64]. That is, the fixed field of the kernel of ψ is linearly disjoint from F_∞ and ψ annihilates some lift of γ (for example, γ').

There is an isomorphism of $\mathcal{O}_K[\Gamma]$-modules of the form

$$\lambda : I_{E/F}(U) \otimes_{\mathbf{Z}_p} \overline{\mathbf{Q}}_p \cong \operatorname{Hom}_{G(EF_\infty/F_\infty)}(U \otimes_{\mathcal{O}_K} \overline{\mathbf{Q}}_p, V_{E/F}) \longrightarrow V_{E/F}^\psi$$

where

$$V_{E/F}^\psi = \{v \in V_{E/F} \mid g(v) = \psi(g)v \text{ for all } g \in G(EF_\infty/F_\infty) \cong G(E/F)\}.$$

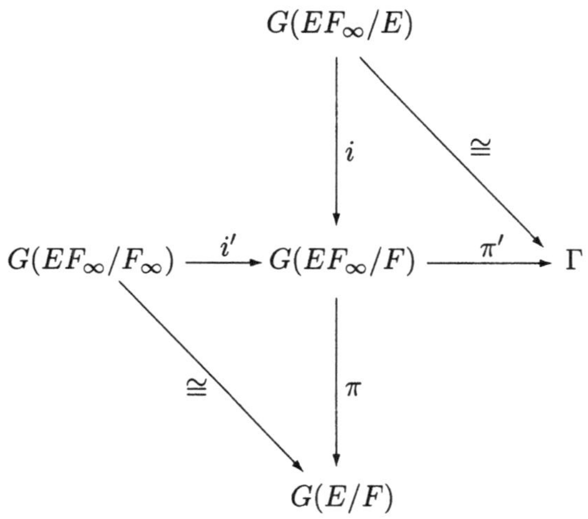

In fact, $\lambda(f) = f(1)$ is such an isomorphism, since $g(\lambda(f)) = f(g(1)) = f(\psi(1)) = \psi(1)f(1) = \psi(1)\lambda(f)$.

This means that $h_\rho(T)$ defined in §6.1.1 coincides with the $h_\rho(T)$ (or, equivalently, with $h_\psi(T)$) of ([64] §1).

In the terminology of [64] a character of type W is a one-dimensional representation of the form

$$\rho' : G(EF_\infty/F) \xrightarrow{\pi'} \Gamma \longrightarrow \overline{\mathbf{Q}}_p^*.$$

In ([64] Proposition 3) the process of twisting representations by characters of type W to obtain representations of type S is studied in relation to its effect on $h_\psi(T)$.

The following result describes the effect of twisting by characters of type W in general.

Proposition 6.1.3

Let

$$\rho' : G(EF_\infty/F) \xrightarrow{\pi'} G(EF_\infty/F)/G(EF_\infty/F_\infty) \cong \Gamma \xrightarrow{\overline{\rho}} \overline{\mathbf{Q}}_p^*$$

be a one-dimensional representation of type W in the terminology of [64]. Then, in the notation of §6.1.1, if $\rho\rho'$ is of type S

$$h_{\rho\rho'}(T) = \overline{\rho}(\gamma)^{-d} h_\rho(\overline{\rho}(\gamma)(1+T) - 1)$$

where $d = \dim_{\overline{\mathbf{Q}}_p}(I_{E/F}(U(\rho)) \otimes_{\mathbf{Z}_p} \overline{\mathbf{Q}}_p)$.

Proof. Let $U(\rho)$ and $U(\rho\rho')$ be the modules associated with ρ and $\rho\rho'$ respectively. The action of $\gamma \in \Gamma$ on $f \in I_{E/F}(U(\rho)) \otimes_{\mathcal{O}_K} \overline{\mathbf{Q}}_p$ is given by lifting γ to

$\gamma' \in G(EF_\infty/F)$, and conjugating $\gamma(f)(-) = \gamma'(f((\gamma')^{-1}(-))))$. If we identify the vector spaces, $U(\rho\rho') \otimes_{\mathcal{O}_K} \overline{\mathbf{Q}}_p$ and $U(\rho) \otimes_{\mathcal{O}_K} \overline{\mathbf{Q}}_p$, then the action of γ' on the former is equal to $\rho'(\gamma') = \overline{\rho}(\gamma)$ (these are equal because $\rho\rho'$ is of type S) times the action of γ' on the latter. Therefore, if A is a matrix which describes the action of γ on $I_{E/F}(U(\rho)) \otimes_{\mathcal{O}_K} \overline{\mathbf{Q}}_p$, with respect to some choice of basis, then $\overline{\rho}(\gamma)^{-1}A$ is the corresponding matrix on $I_{E/F}(U(\rho\rho')) \otimes_{\mathcal{O}_K} \overline{\mathbf{Q}}_p$. Computing characteristic polynomials of $\gamma - 1$, we obtain

$$\begin{aligned} h_{\rho\rho'}(T) &= \det(T+1-\overline{\rho}(\gamma)^{-1}A) \\ &= \overline{\rho}(\gamma)^{-d}\det(\overline{\rho}(\gamma)(1+T)-A) \\ &= \overline{\rho}(\gamma)^{-d}\det((\overline{\rho}(\gamma)(1+T)-1)-A+1) \\ &= \overline{\rho}(\gamma)^{-d}h_\rho(\overline{\rho}(\gamma)(1+T)-1), \end{aligned}$$

as required. □

Proposition 6.1.4

Let N/F be an intermediate Galois extension of E/F and let U be an $\mathcal{O}_K[G(NF_\infty/F)]$-module which is free of finite rank as an $\mathcal{O}_K$-module. Then there is an isomorphism of Γ-representations of the form

$$I_{E/F}(\mathrm{Inf}_{G(NF_\infty/F)}^{G(EF_\infty/F)}(U)) \otimes_{\mathcal{O}_K} \overline{\mathbf{Q}}_p \xrightarrow{\cong} I_{N/F}(U) \otimes_{\mathcal{O}_K} \overline{\mathbf{Q}}_p,$$

where Inf *denotes inflation.*

Proof. Since $G(EF_\infty/NF_\infty)$ acts trivially on $\mathrm{Inf}_{G(NF_\infty/F)}^{G(EF_\infty/F)}(U)$, we have an isomorphism

$$\begin{aligned} &\mathrm{Hom}_{G(EF_\infty/F_\infty)}(\mathrm{Inf}_{G(NF_\infty/F)}^{G(EF_\infty/F)}(U) \otimes_{\mathcal{O}_K} \overline{\mathbf{Q}}_p, V_{E/F}) \\ &\cong \mathrm{Hom}_{G(NF_\infty/F_\infty)}(U \otimes_{\mathcal{O}_K} \overline{\mathbf{Q}}_p, V_{E/F}^{G(EF_\infty/NF_\infty)}). \end{aligned}$$

However the natural map, $X_{E/F} \longrightarrow X_{N/F}$, induces an isomorphism

$$(V_{E/F})_{G(EF_\infty/NF_\infty)} \xrightarrow{\cong} V_{N/F},$$

where $(A)_G$ denotes the coinvariants of G acting on A. Also the norm, $N(v) = \sum_{g \in G(EF_\infty/NF_\infty)} g(v)$, induces an isomorphism

$$N : (V_{E/F})_{G(EF_\infty/NF_\infty)} \xrightarrow{\cong} (V_{E/F})^{G(EF_\infty/NF_\infty)}.$$

Since a lifting of $\gamma \in \Gamma$ to $G(EF_\infty/F)$ maps to a lifting of γ to $G(NF_\infty/F)$, it is clear that the resulting isomorphism commutes with the Γ-action, as required. □

Proposition 6.1.5
Let N/F be an intermediate extension of E/F and let U be an $\mathcal{O}_K[G(EF_\infty/N)]$-module which is free of finite rank as an $\mathcal{O}_K$-module. Then there is an isomorphism of $\mathcal{O}_K[\Gamma]$-modules of the form

$$I_{E/F}(\mathrm{Ind}_{G(EF_\infty/N)}^{G(EF_\infty/F)}(U)) \xrightarrow{\cong} \mathrm{Ind}_{G(N_\infty/N)}^{\Gamma}(I_{E/N}(U)),$$

where Ind *denotes induction. Here $G(N_\infty/N)$ has been identified with $G(F_\infty/N \cap F_\infty) \subseteq \Gamma$.*

Proof. We have $F \subseteq N \cap F_\infty \subseteq N \subseteq E$ and $G(N \cap F_\infty/F) \cong \mathbf{Z}/p^m$, generated by γ, since it is a finite quotient of $\mathbf{Z}_p$. Therefore $1, \gamma, \ldots, \gamma^{p^m-1}$ constitute a set of representatives of the double cosets

$$G(EF_\infty/F_\infty)\backslash G(EF_\infty/F)/G(EF_\infty/N) \equiv \Gamma/G(F_\infty/N \cap F_\infty).$$

By the Double Coset Formula [131], $\mathrm{Ind}_{G(EF_\infty/N)}^{G(EF_\infty/F)}(U)$ is equal as an $\mathcal{O}_K[G(EF_\infty/F_\infty)]$-module to

$$\begin{aligned}&\mathrm{Res}_{G(EF_\infty/F_\infty)}^{G(EF_\infty/F)}(\mathrm{Ind}_{G(EF_\infty/N)}^{G(EF_\infty/F)}(U))\\ &\cong \oplus_{0\leq i\leq p^m-1} \mathrm{Ind}_{G(EF_\infty/F_\infty)\cap\gamma^i G(EF_\infty/N)\gamma^{-i}}^{G(EF_\infty/F_\infty)}((\gamma^{-i})^*(U))\end{aligned}$$

where $(\gamma^{-i})^*(U)$ is equal to a copy of U on which $\gamma^i g \gamma^{-i}$ acts as g does on U. Hence

$$\mathrm{Res}_{G(EF_\infty/F_\infty)}^{G(EF_\infty/F)}(\mathrm{Ind}_{G(EF_\infty/N)}^{G(EF_\infty/F)}(U)) \cong \oplus_{0\leq i\leq p^m-1} \mathrm{Ind}_{G(EF_\infty/\gamma^i(N)F_\infty)}^{G(EF_\infty/F_\infty)}((\gamma^{-i})^*(U)).$$

It is sufficient to prove the proposition in the two special cases (a) $N \cap F_\infty = F$ and (b) $N \subset F_\infty$ because then we have a chain of isomorphisms of the following form:

$$\begin{aligned}I_{E/F}(\mathrm{Ind}_{G(EF_\infty/N)}^{G(EF_\infty/F)}(U)) &\cong I_{E/F}(\mathrm{Ind}_{G(EF_\infty/N\cap F_\infty)}^{G(EF_\infty/F)}(\mathrm{Ind}_{G(EF_\infty/N)}^{G(EF_\infty/N\cap F_\infty)}(U)))\\ &\cong \mathrm{Ind}_{G(F_\infty/N\cap F_\infty)}^{\Gamma}(I_{E/N\cap F_\infty}(\mathrm{Ind}_{G(EF_\infty/N)}^{G(EF_\infty/N\cap F_\infty)}(U)))\\ &\cong \mathrm{Ind}_{G(F_\infty/N\cap F_\infty)}^{\Gamma}(\mathrm{Ind}_{G(F_\infty/N\cap F_\infty)}^{G(F_\infty/N\cap F_\infty)}(I_{E/N}(U)))\\ &\cong \mathrm{Ind}_{G(F_\infty/N\cap F_\infty)}^{\Gamma}(I_{E/N}(U)).\end{aligned}$$

In case (a) we have $m = 0$ and $N_\infty = NF_\infty$ so that $EN_\infty = EF_\infty$ and $X_{E/F} = X_{E/N}$. Hence, in this case,

$$\begin{aligned}&\mathrm{Hom}_{G(EF_\infty/F_\infty)}(\mathrm{Ind}_{G(EF_\infty/N)}^{G(EF_\infty/F)}(U), X_{E/F} \otimes_{\mathbf{Z}_p} \mathcal{O}_K)\\ &\cong \mathrm{Hom}_{G(EF_\infty/F_\infty)}(\mathrm{Ind}_{G(EF_\infty/F_\infty N)}^{G(EF_\infty/F_\infty)}(U), X_{E/F} \otimes_{\mathbf{Z}_p} \mathcal{O}_K)\\ &\cong \mathrm{Hom}_{G(EF_\infty/N_\infty)}(U, X_{E/N} \otimes_{\mathbf{Z}_p} \mathcal{O}_K),\end{aligned}$$

as required.

In case (b) $N_\infty = F_\infty$, $X_{E/F} = X_{E/N}$ and $\gamma^i(N) \subset \gamma^i(F_\infty) = F_\infty$ so that

$$\begin{aligned}
&\mathrm{Hom}_{G(EF_\infty/F_\infty)}(\mathrm{Ind}_{G(EF_\infty/N)}^{G(EF_\infty/F)}(U), X_{E/F} \otimes_{\mathbf{Z}_p} \mathcal{O}_K) \\
&\cong \oplus_{0 \le i \le p^m - 1} \mathrm{Hom}_{G(EF_\infty/F_\infty)}(\mathrm{Ind}_{G(EF_\infty/\gamma^i(N)F_\infty)}^{G(EF_\infty/F_\infty)}((\gamma^{-i})^*(U)), X_{E/F} \otimes_{\mathbf{Z}_p} \mathcal{O}_K) \\
&\cong \oplus_{0 \le i \le p^m - 1} \mathrm{Hom}_{G(EN_\infty/N_\infty)}((\gamma^{-i})^*(U), X_{E/N} \otimes_{\mathbf{Z}_p} \mathcal{O}_K),
\end{aligned}$$

since $\gamma^i(N)F_\infty = F_\infty$. However, $1, \gamma, \ldots, \gamma^{p^m - 1}$ are also coset representatives for $\Gamma/G(F_\infty/N)$ and it is easy to see that

$$\begin{aligned}
&\oplus_{0 \le i \le p^m - 1} \mathrm{Hom}_{G(EN_\infty/N_\infty)}((\gamma^{-i})^*(U), X_{E/N} \otimes_{\mathbf{Z}_p} \mathcal{O}_K) \\
&\cong \mathrm{Ind}_{G(F_\infty/N)}^{\Gamma}(\mathrm{Hom}_{G(EN_\infty/N_\infty)}(U, X_{E/N} \otimes_{\mathbf{Z}_p} \mathcal{O}_K))
\end{aligned}$$

as Γ-modules. □

6.2 p-adic L-functions

6.2.1 Let E/F and $\mathcal{O}_K$ be as in §6.1.1. Let p be an odd prime. Let ρ be a one-dimensional $\overline{\mathbf{Q}}_p$-representation which is realised over $\mathcal{O}_K$. That is, $\mathcal{O}_K = U$ is an $\mathcal{O}_K[G(EF_\infty/F)]$-module associated to $\rho : G(E/F) \longrightarrow \overline{\mathbf{Q}}_p^*$ as in §6.1.1. The existence of the p-adic L-function, $L_p(s, \rho)$, was first shown by Deligne and Ribet [50] (see also [2] [31] [114] [120]).

Let S_p denote the set of all primes of F, $P \lhd \mathcal{O}_F$, above p and let $\sigma : \overline{\mathbf{Q}}_p \xrightarrow{\cong} \mathbf{C}$ be a fixed isomorphism. Thus $\sigma(\rho \cdot \omega^{-n})$ is a one-dimensional complex Galois representation of F, where $\omega : G(F(\mu_p)/F) \longrightarrow \mu_{p-1} \subset \mathbf{Z}_p^*$ denotes the Teichmüller character, μ_n denoting the n-th roots of unity. The function, $L_p(s, \rho)$, is characterised as the unique continuous $\overline{\mathbf{Q}}_p$-valued function of $s \in \mathbf{Z}_p - \{1\}$ (even continuous at $s = 1$ if ρ is non-trivial) such that

$$L_p(1 - n, \rho) = \sigma^{-1}(L_{F,S_p}(1 - n, \sigma(\rho \cdot \omega^{-n})))$$

for all integers $n \ge 1$. Here $L_{F,S_p}(1 - n, -)$ denotes the Artin L-function associated to F after the removal of the Euler factors attached the primes in S_p.

Let $F' = F(\mu_p)$ so that F'_∞ contains all p-primary roots of unity. The action of $\Gamma = G(F_\infty/F) \cong G(F'_\infty/F')$ on an arbitrary p-primary root of unity, ξ, is given by a homomorphism, $\kappa : \Gamma \longrightarrow \mathbf{Z}_p^*$, determined by the formula, $g(\xi) = \xi^{\kappa(g)}$ for all $g \in \Gamma$.

Set $u = \kappa(\gamma) \in \mathbf{Z}_p^*$ so that $u \equiv 1$ (modulo p^m) if $\mu_{p^m} \subset F'$. Let $\mathbf{Z}_p[\rho]$ denotes the $\mathbf{Z}_p$-algebra generated by the character-values of ρ. If ρ is one-dimensional, as above, there exists a power series

$$G_\rho(T) \in \mathbf{Z}_p[\rho][[T]]$$

such that, except for $s = 0$ when ρ is trivial,

$$L_p(1-s,\rho) = \frac{G_\rho(u^s-1)}{H_\rho(u^s-1)}$$

for $s \in \mathbf{Z}_p$. Here

$$H_\rho(T) = \begin{cases} \rho(\gamma)(1+T)-1, & \text{if } \rho \text{ is type } W, \\ 1 & \text{otherwise.} \end{cases}$$

Also, as explained in [120], if ρ' is of type W then

$$G_{\rho\rho'}(T) = G_\rho(\rho'(\gamma)(1+T)-1) \text{ and } H_{\rho\rho'}(T) = H_\rho(\rho'(\gamma)(1+T)-1).$$

If $\pi \in \mathbf{Z}_p[\rho]$ is a uniformiser then, by the Weierstrass Preparation Theorem, we may write $G_\rho(T)$ uniquely in the form

$$G_\rho(T) = \pi^{\mu(\rho)} G_\rho^*(T) U_\rho(T)$$

where $G_\rho^*(T)$ is a distinguished polynomial and $U_\rho(T)$ is a unit in the power series ring, $\mathbf{Z}_p[\rho][[T]]$. A distinguished polynomial is a monic polynomial all of whose non-leading coefficients lie in $\pi\mathbf{Z}_p[\rho]$.

If ρ is one-dimensional and of type S then it is shown in ([166] Theorem 1.3) that the polynomial, $h_\rho(T)$, of §6.1.1 satisfies

$$h_\rho(T) = G_\rho^*(T) \in \mathbf{Z}_p[\rho][T].$$

Since any one-dimensional representation may be written as the product of one of type S with one of type W, the previous discussion together with Proposition 6.1.3 shows that the above relation holds for all one-dimensional ρ (the factor $\overline{\rho}(\gamma)^{-d}$ being subsumed into the unit, $U_\rho(T)$).

6.2.2 In the following it will be wise to keep track of the identity of the base-field, F, in $G(E/F)$. Accordingly, we shall temporarily elaborate upon the notation of §6.2.1 and write $L_{p,F}(1-s,\rho)$, $\kappa_F : G(F_\infty/F) \longrightarrow \mathbf{Z}_p^*$, $\gamma_F \in G(F_\infty/F)$ and $u_F = \kappa_F(\gamma_F)$ for $L_p(1-s,\rho)$, κ, γ and u, respectively. When $F \subseteq N \subseteq E$ we choose $\gamma_N = \gamma_F^{[N\cap F_\infty:F]}$, identifying $G(N_\infty/N)$ with a subgroup of Γ as in Proposition 6.1.5.

Now let ρ be an arbitrary finite-dimensional $\overline{\mathbf{Q}}_p$-representation of $G(E/F)$. For such ρ the additivity and the invariance under inflation and induction of the Artin L-function (see §1.1.3) imply that

$$L_{p,F}(1-s,\rho_1\oplus\rho_2) = L_{p,F}(1-s,\rho_1)L_{p,F}(1-s,\rho_2),$$
$$L_{p,F}(1-s,\mathrm{Inf}_{G(N/F)}^{G(E/F)}(\rho)) = L_{p,F}(1-s,\rho),$$
$$L_{p,F}(1-s,\mathrm{Ind}_{G(E/N)}^{G(E/F)}(\rho)) = L_{p,N}(1-s,\rho)$$

for the appropriate ρ, ρ_1, ρ_2 and $F \subseteq N \subseteq E$.

By Brauer's Induction Theorem there exist integers, n_i, and one-dimensional representations of subgroups, $\rho_i : G(E/N_i) \longrightarrow \overline{\mathbf{Q}}_p^*$ such that

$$\rho = \sum_i n_i \operatorname{Ind}_{G(E/N_i)}^{G(E/F)}(\rho_i)$$

in the representation ring of $G(E/F)$. In fact, there is a canonical form for this induction formula which we may use if need be ([131] Theorem 2.3.9) in which each pair, $(G(E/N_i), \rho_i)$, is unique up to $G(E/F)$-conjugacy. Note that some of the n_i's may be negative integers. Therefore we obtain

$$\begin{aligned} L_{p,F}(1-s,\rho) &= \prod_i L_{p,N_i}(1-s,\rho_i)^{n_i} \\ &= \prod_i \left(\frac{\pi(\rho_i)^{\mu(\rho_i)} h_{\rho_i}(u_{N_i}^s - 1) U_{\rho_i}(u_{N_i}^s - 1)}{H_{\rho_i}(u_{N_i}^s - 1)}\right)^{n_i} \end{aligned}$$

where $\pi(\rho_i) \in \mathbf{Z}_p[\rho_i]$ is a uniformiser.

The following result is mentioned, without proof, in ([64] p. 82):

Proposition 6.2.3

If ρ is irreducible and $\dim_{\overline{\mathbf{Q}}_p}(\rho) \geq 2$ in §6.2.2 then

$$\prod_i H_{\rho_i}(u_{N_i}^s - 1)^{n_i} = \pm 1.$$

Proof. The appearance in the denominator of the formula for $L_{p,F}(1-s,\rho)$ of $H_{\rho_i}(u_{N_i}^s - 1)$ with $H_{\rho_i}(T)$ non-trivial occurs only when $\rho_i : G(E/N_i) \longrightarrow \overline{\mathbf{Q}}_p^*$ is of type W. This is equivalent to the restriction,

$$(\rho_i \mid G(E/E \cap F_\infty) \cap G(E/N_i) = G(E/(E \cap F_\infty)N_i)),$$

being trivial.

Since $G(E/E \cap F_\infty) \lhd G(E/F)$ we may apply the operation of taking $G(E/E \cap F_\infty)$-coinvariants (which are isomorphic to the $G(E/E \cap F_\infty)$-fixed points) to both sides of the equation for ρ (see [131] Exercises 2.5.13-2.5.15).

The $G(E/E \cap F_\infty)$-fixed points of ρ form a subrepresentation which must either be trivial or all of ρ, since ρ is irreducible. However, if ρ were trivial on $G(E/E \cap F_\infty)$ it would factorise through the cyclic quotient, $G(E \cap F_\infty/F)$, and would therefore be one-dimensional. Therefore we have the relation

$$0 = \sum_i n_i (\operatorname{Ind}_{G(E/N_i)}^{G(E/F)}(\rho_i))^{G(E/E \cap F_\infty)} = \sum_i n_i \operatorname{Ind}_{G(E/(N_i \cap F_\infty))}^{G(E/F)}(\hat{\rho}_i)$$

where $\hat{\rho}_i$ on $G(E/(N_i \cap F_\infty)) = G(E/E \cap F_\infty))G(E/N_i)$ is given by $\hat{\rho}_i(xy) = \rho_i(y)$ for $x \in G(E/E \cap F_\infty))$ and $y \in G(E/N_i)$. These representations are all inflated from the cyclic quotient group, $G(E \cap F_\infty/F) \cong \mathbf{Z}/p^m$ so that in the representation ring of $G(E \cap F_\infty/F)$ we have the relation

$$0 = \sum_i n_i \operatorname{Ind}_{G((E \cap F_\infty)/(N_i \cap F_\infty))}^{G((E \cap F_\infty)/F)}(\hat{\rho}_i).$$

Let $\mathrm{char}_i(t)$ denote the characteristic polynomial

$$\mathrm{char}_i(t) = \det(tI - \gamma \mid \mathrm{Ind}_{G((E\cap F_\infty)/(N_i\cap F_\infty))}^{G((E\cap F_\infty)/F)}(\hat{\rho}_i)) \in \overline{\mathbf{Q}}_p[t],$$

where γ is the image of $\gamma \in \Gamma$ in $G((E\cap F_\infty)/F) \cong \mathbf{Z}/p^m$. Hence we have the relation

$$1 = \prod_i \mathrm{char}_i(u_F^{-s})^{n_i}.$$

However, using the base $1\otimes 1, \gamma\otimes 1, \ldots, \gamma^{[N_i\cap F_\infty:F]-1}\otimes 1$ for the module $\mathrm{Ind}_{G((E\cap F_\infty)/(N_i\cap F_\infty))}^{G((E\cap F_\infty)/F)}(\hat{\rho}_i))$ one sees easily that

$$\begin{aligned}\mathrm{char}_i(t) &= \det\begin{pmatrix} t & 0 & 0 & \ldots & \ldots & 0 & -\rho_i(\gamma_F^{[N_i\cap F_\infty:F]}) \\ -1 & t & 0 & \ldots & \ldots & 0 & 0 \\ 0 & -1 & t & \ldots & \ldots & 0 & 0 \\ 0 & 0 & -1 & t & \ldots & 0 & 0 \\ \vdots & \vdots & \vdots & \vdots & \vdots & \vdots & \vdots \\ 0 & 0 & 0 & \ldots & \ldots & -1 & t \end{pmatrix} \\ &= (t^{[N_i\cap F_\infty:F]} - \rho_i(\gamma_F^{[N_i\cap F_\infty:F]})).\end{aligned}$$

Moreover, $\gamma_F^{[N_i\cap F_\infty:F]} = \gamma_{N_i}$ and $u_F^{[N_i\cap F_\infty:F]} = u_{N_i}$ so that the relation becomes

$$\begin{aligned} 1 &= \textstyle\prod_i (u_{N_i}^{-s} - \rho_i(\gamma_{N_i}))^{n_i} \\ &= (\textstyle\prod_i u_{N_i}^{-sn_i})(\prod_i (1-\rho_i(\gamma_{N_i})u_{N_i}^s)^{n_i}) \\ &= (u_F^{-s})^{\sum_i n_i[N_i\cap F_\infty:F]}\textstyle\prod_i (-H_{\rho_i}(u_{N_i}^s - 1))^{n_i} \end{aligned}$$

and

$$0 = \sum_i n_i \dim_{\overline{\mathbf{Q}}_p} \mathrm{Ind}_{G((E\cap F_\infty)/(N_i\cap F_\infty))}^{G((E\cap F_\infty)/F)}(\hat{\rho}_i) = \sum_i n_i[N_i\cap F_\infty : F]$$

so that

$$\prod_i H_{\rho_i}(u_{N_i}^s - 1)^{n_i} = (-1)^{\sum_i n_i} \in \{\pm 1\},$$

as required. □

6.2.4 Now let us examine the product

$$\prod_i h_{\rho_i}(u_{N_i}^s - 1)^{n_i}$$

of §6.2.2 when ρ is an arbitrary finite-dimensional $\overline{\mathbf{Q}}_p$-representation of $G(E/F)$. Recall from §6.1.1 that

$$h_{\rho_i}(T) = \det(T + 1 - ((\gamma_{N_i})\cdot -) \mid I_{E/N_i}(U(\rho_i)) \otimes_{\mathbf{Z}_p} \overline{\mathbf{Q}}_p)$$

where $U(\rho_i)$ is the underlying $G(E/N_i)$-modules of ρ_i. Hence

$$\begin{aligned} h_{\rho_i}(u^s_{N_i}-1) &= \det(u^s_{N_i}-((\gamma_{N_i})\cdot -) \mid I_{E/N_i}(U(\rho_i))\otimes_{\mathbf{Z}_p}\overline{\mathbf{Q}}_p) \\ &= \det(u_F^{s[N_i\cap F_\infty:F]}-((\gamma_F)^{[N_i\cap F_\infty:F]}\cdot -) \mid I_{E/N_i}(U(\rho_i))\otimes_{\mathbf{Z}_p}\overline{\mathbf{Q}}_p) \\ &= \det(u^s_F-((\gamma_F)\cdot -) \mid \mathrm{Ind}^{\Gamma}_{G(F_\infty/N_i\cap F_\infty)}(I_{E/N_i}(U(\rho_i))\otimes_{\mathbf{Z}_p}\overline{\mathbf{Q}}_p)), \end{aligned}$$

arguing as in the proof of Proposition 6.2.3.

Identifying $G(F_\infty/N_i\cap F_\infty)$ and $G(N_{i,\infty}/N_i)$, Proposition 6.1.5 implies that

$$h_{\rho_i}(u^s_{N_i}-1)=h_{\mathrm{Ind}^{G(E/F)}_{G(E/N_i)}(\rho_i)}(u^s_F-1).$$

Therefore we have shown that

$$\begin{aligned} \prod_i h_{\rho_i}(u^s_{N_i}-1)^{n_i} &= \prod_i h_{\mathrm{Ind}^{G(E/F)}_{G(E/N_i)}(\rho_i)}(u^s_F-1)^{n_i} \\ &= h_{\sum_i n_i \mathrm{Ind}^{G(E/F)}_{G(E/N_i)}(\rho_i)}(u^s_F-1)=h_\rho(u^s_F-1). \end{aligned}$$

Let $\Omega_{\mathbf{Q}_p}$ denote the absolute Galois group, $G(\overline{\mathbf{Q}}_p/\mathbf{Q}_p)$. Then $\Omega_{\mathbf{Q}_p}$ acts on $R(G(E/F))$, the ring of $\overline{\mathbf{Q}}_p$-representations of $G(E/F)$. If $\omega\in\Omega_{\mathbf{Q}_p}$ it is easy to see that $h_\rho(T)\in\mathcal{O}_K[T]$ satisfies $h_{\omega(\rho)}(T)=\omega(h_\rho(T))$ so that $h_\rho(T)\in\mathbf{Z}_p[\rho][T]$, where $\mathbf{Z}_p[\rho]$ is as in §6.2.1.

Also $L_p(1-n,\omega(\rho))=\omega(L_p(1-n,\rho))$ for all integers $n\geq 2$. This is seen for one-dimensional ρ by writing the Artin L-function in the form (c.f. [166] (1.2) p. 493)

$$L_F(1-n,\rho)=\sum_{\sigma\in G(E^{\mathrm{Ker}(\rho)}/F)}\rho(\sigma)\zeta_F(\sigma,1-n)$$

and observing that the partial zeta functions, $\zeta_F(\sigma,1-n)$, are rational numbers. For an arbitrary ρ one uses Brauer's Induction Theorem and the inductivity properties of the L-function to reduce to the one-dimensional case (for a similar, but more complicated, calculation see [132] pp. 28–30).

To recapitulate:

Theorem 6.2.5

Let E/F be a Galois extension of totally real number fields with Galois group, $G(E/F)$, and let p be an odd prime. Let ρ be a finite-dimensional, irreducible $\overline{\mathbf{Q}}_p$-representation of $G(E/F)$. Let $h_\rho(T)\in\mathbf{Z}_p[\rho][T]$ be the characteristic polynomial defined in §6.1.1. Then there exists a unique unit power series

$$U_\rho(T)\in\mathbf{Z}_p[\rho][[T]]^*$$

such that the p-adic L-function of ρ satisfies, for all $s\in\mathbf{Z}_p$ (except $s=0$ if ρ is trivial)

$$L_p(1-s,\rho)=\frac{\pi(\rho)^{\mu(\rho)}h_\rho(u^s-1)U_\rho(u^s-1)}{H_\rho(u^s-1)},$$

where $H_\rho(T)=\rho(\gamma)(1+T)-1$ if ρ is of type W and $H_\rho(T)=1$ otherwise. Also $\pi(\rho)\in\mathbf{Z}_p[\rho]$ is a uniformiser and $u\in\mathbf{Z}_p^*$ is as in §6.2.1.

6.3 Determinants and the Wiles unit

6.3.1 Let p be an odd prime. If G is a finite group as in §2.5.8, let $R(G)$ denote the complex representation ring of G. Let $N/\mathbf{Q}_p$ be a Galois extension containing all the $|G|$-th roots of unity. If $\chi \in R(G)$ and $g \in G$ we denote by $\chi(g)$ the value of the character function of χ on g. We say that $\chi \equiv 0$ (modulo p^t) if $\chi(g) \in p^t\mathcal{O}_N$ for all $g \in G$.

The absolute Galois group of $\mathbf{Q}_p$, $\Omega_{\mathbf{Q}_p}$, acts on $R(G)$ as in §2.5.10. The action is characterised by the formula $\omega(\chi)(g) = \omega(\chi(g))$ for all $\omega \in \Omega_{\mathbf{Q}_p}$, $\chi \in R(G)$ and $g \in G$. This makes sense because $\chi(g) \in N$. Hence we may consider the group of $\Omega_{\mathbf{Q}_p}$-equivariant homomorphisms from $R(G)$ to $\mathcal{O}_N$, $\mathrm{Hom}_{\Omega_{\mathbf{Q}_p}}(R(G), \mathcal{O}_N)$.

Define a subgroup $Cong_p(G) \subseteq \mathrm{Hom}_{\Omega_{\mathbf{Q}_p}}(R(G), \mathcal{O}_N)$, by

$$Cong_p(G) = \{f \mid f(\chi) \in p^{t+1}\mathcal{O}_N \text{ if } \chi \equiv 0 \text{ (modulo } p^t)\}.$$

Define $\mathbf{Q}_p\{G\}$ to be the $\mathbf{Q}_p$-vector space on the conjugacy classes of elements of G. There is a canonical map of vector spaces, $c : \mathbf{Q}_p[G] \longrightarrow \mathbf{Q}_p\{G\}$, sending g to its conjugacy class. Write $p^e\mathbf{Z}_p\{G\}$ for $c(p^e\mathbf{Z}_p[G])$ for $e \geq 0$. There is an isomorphism ([131] §4.5.14 p. 141; [132] §3.1.4 p. 69)

$$\psi : \mathbf{Q}_p\{G\} \xrightarrow{\cong} \mathrm{Hom}_{\Omega_{\mathbf{Q}_p}}(R(G), N)$$

given by $\psi(\sum_{g\in G} \lambda_g g)(\chi) = \sum_{g\in G} \lambda_g \chi(g)$ for $\lambda_g \in \mathbf{Q}_p$, $g \in G$ and $\chi \in R(G)$.

Consider the group, $\mathrm{Hom}_{\Omega_{\mathbf{Q}_p}}(R(G), \mathcal{O}_N^*)$, of $\Omega_{\mathbf{Q}_p}$-equivariant homomorphism from $R(G)$ to the units in the integers of N. There is a homomorphism

$$\mathrm{Det} : \mathbf{Z}_p[G]^* \longrightarrow \mathrm{Hom}_{\Omega_{\mathbf{Q}_p}}(R(G), \mathcal{O}_N^*)$$

characterised by the formula

$$\mathrm{Det}(\sum_{g\in G} \lambda_g g)(\rho) = \det(\sum_{g\in G} \lambda_g \rho(g))$$

for all representations of the form $\rho : G \longrightarrow GL_m\mathcal{O}_N$ where $u = \sum_{g\in G} \lambda_g g$ is a unit in $\mathbf{Z}_p[G]$.

The image of Det is called the subgroup of determinantal homomorphisms.

Let $\psi^p : R(G) \longrightarrow R(G)$ denote the p-th Adams operation, characterised by $\psi(\chi)(g) = \chi(g^p)$. If $u \in \mathbf{Z}_p[G]^*$ then $\mathrm{Det}(u)(\psi^p(\chi) - p\chi) \in 1 + p\mathcal{O}_N \subset \mathcal{O}_N^*$ for all $\chi \in R(G)$ ([131] §4.3.10 p. 121; [132] §3.1.1 p. 66). Hence

$$\log_p(\mathrm{Det}(u)(\psi^p(\chi) - p\chi)) \in p\mathcal{O}_N$$

and the function,

$$\chi \mapsto \log_p(\mathrm{Det}(u)(\psi^p(\chi) - p\chi))$$

lies in

$$\mathrm{Hom}_{\Omega_{\mathbf{Q}_p}}(R(G), N) \cong \mathbf{Q}_p\{G\}.$$

The resulting homomorphism

$$L_G : \mathbf{Z}_p[G]^* \longrightarrow \mathbf{Q}_p\{G\}$$

is called the *group-ring logarithm*, originally constructed in a different manner by R. Oliver and M.J. Taylor, independently.

Now let us specialise to the case when G is a p-group. Then ([113]; [155]; see also [132] p. 92) there is an exact sequence of the form

$$\mathbf{Z}_p[G]^* \xrightarrow{L_G} p\mathbf{Z}_p\{G\} \xrightarrow{\omega_G} G^{ab} \longrightarrow 0.$$

The homomorphism, ω_G, is due to R. Oliver and is described ([155] p. 62). as the composition of $1/p$ times the abelianisation map with the projection

$$\mathbf{Z}_p[G^{ab}] \longrightarrow \frac{\mathbf{Z}_p + IG^{ab}}{\mathbf{Z}_p + (IG^{ab})^2} \cong G^{ab}.$$

Here $IG^{ab} \lhd \mathbf{Z}_p[G^{ab}]$ denotes the augmentation ideal and the final isomorphism sends the coset of $\sum_i a_i(g_i - 1)$ $(a_i \in \mathbf{Z}_p, g_i \in G^{ab})$ to the element, $\prod_i g_i^{a_i} \in G^{ab}$.

In general, by ([132] §§3.3.8-3.3.9) $\psi(p\mathbf{Z}_p\{G\})$ is only a subgroup of $Cong_p(G)$ but when $G \cong \mathbf{Z}/p$ the $\psi(p\mathbf{Z}_p\{\mathbf{Z}/p\}) = Cong_p(\mathbf{Z}/p)$.

Theorem 6.3.2

Let p be an odd prime and let G be an elementary abelian p-group (i.e. of exponent p). Then there is an exact sequence of the form

$$0 \longrightarrow \mathrm{Det}(\mathbf{Z}_p[G]^*) \cap \mathrm{Hom}_{\Omega_{\mathbf{Q}_p}}(R(G), \mu(p^\infty)) \longrightarrow \mathrm{Det}(\mathbf{Z}_p[G]^*)$$
$$\xrightarrow{\psi L_G} Cong_p(G) \xrightarrow{\omega_G \psi^{-1}} G^{ab} \longrightarrow 0$$

where $\mu(p^\infty)$ denotes the group of p-primary roots of unity of N.

Proof. This is a consequence of the discussion of ([132] §3.3).

If $C \subseteq G$ is a cyclic subgroup, which is of order p by hypothesis, define $\Theta_C \in R(C)$ to be the unique element satisfying $\Theta_C(1) = 0$ and $\Theta_C(g) = p$ for non-trivial $g \in C$. If $f \in \mathrm{Hom}_{\Omega_{\mathbf{Q}_p}}(R(G), \mathcal{O}_N)$ define $f_C \in \mathrm{Hom}_{\Omega_{\mathbf{Q}_p}}(R(C), \mathcal{O}_N)$ by

$$f_C(\lambda) = [G : N_G C] f(\mathrm{Ind}_C^G(\Theta_C \lambda)).$$

Here $N_G C$ denotes the normaliser of C in G. If i_C denotes the inclusion of C into G then

$$f = \sum_{C \text{ cyclic}} |G|^{-1}(i_C)_*(f_C) \in \mathrm{Hom}_{\Omega_{\mathbf{Q}_p}}(R(G), \mathcal{O}_N)$$

where $(i_C)_*$ is the natural map induced by i_C. However, by ([132] §3.3.6 p. 84), if $f \in Cong_p(G)$ then $f_C \in |G|^{-1} f_C \in Cong_p(C)$. Since each C has order p, by ([132] §3.3.8 p. 86), $\psi(p\mathbf{Z}_p\{C\}) = Cong_p(C)$. By ([132] §3.3.1 p. 81), $(i_C)_*$ maps $\psi(p\mathbf{Z}_p\{C\})$ to $\psi(p\mathbf{Z}_p\{G\})$ so that any $f \in Cong_p(G)$ lies in $\psi(p\mathbf{Z}_p\{G\})$.

Hence we have shown that ψ induces an isomorphism

$$\psi : p\mathbf{Z}_p\{G\} \xrightarrow{\cong} Cong_p(G).$$

The result follows by substituting $Cong_p(G)$ into the exact sequence of ([132] §3.3.21 p. 93). □

6.3.3 *The Wiles Unit Conjecture*

We return now to the situation of Theorem 6.2.5. Let $n \geq 2$ be an integer and let E/F be a Galois extension of totally real number fields with Galois group $G(E/F)$. In the notation of Theorem 6.2.5 and §6.3.1, we have the function

$$(\rho \mapsto U_\rho(u^n - 1)) \in \mathrm{Hom}_{\Omega_{\mathbf{Q}_p}}(R(G(E/F)), \mathcal{O}_N^*)$$

where, if $\rho = \sum_i n_i\rho_i \in R(G(E/F))$ and the ρ_i's are $\overline{\mathbf{Q}}_p$-representations,

$$U_\rho(u^n - 1)) = \prod_i U_{\rho_i}(u^n - 1))^{n_i}.$$

The remainder of this section will be devoted to a discussion of the following:

Conjecture 6.3.4 For each $n \geq 2$, there exists a unit $\alpha_{n,E/F} \in \mathbf{Z}_p[G(E/F)]^*$ such that

$$\mathrm{Det}(\alpha_{n,E/F})(\rho) = U_\rho(u^n - 1))$$

for all $\rho \in R(G(E/F))$.

6.3.5 *Evidence for Conjecture 6.3.4*

(i) **Connections with $\Omega_n(L/K, 3)$**

If L/K is a Galois extension of number fields with group $G(L/K)$, in Definition 5.2.11, we defined a K-theory Galois module structure invariant $\Omega_m(L/K, 3) \in \mathcal{CL}(\mathbf{Z}[G(L/K)])$ for all $m \geq 1$. By analogy with the abiding conjectures concerning the classical Chinburg invariant, $\Omega_0(L/K, 3)$, Conjecture 5.2.18 predicts that

$$W_{L/K} = \Omega_m(L/K, 3) \in \mathcal{CL}(\mathbf{Z}[G(L/K)])$$

where $W_{L/K}$ is the Cassou-Noguès-Fröhlich class of Definition 5.2.16.

This conjecture is supported by the calculations of §5.5.5 (see Corollary 1.3.8). It is also supported by the calculations of [38] in the case of function fields in characteristic p.

Conjecture 6.3.4 arose out of the following approach to the proof that $W_{L/K} = \Omega_s(L/K, 3)$. Consider the Hom-description isomorphism, of Theorem 2.5.25

$$\mathrm{Det} : \mathcal{CL}(\mathbf{Z}[G]) \xrightarrow{\cong} \frac{\mathrm{Hom}_{\Omega_{\mathbf{Q}}}(R(G), J^*(M))}{\mathrm{Hom}_{\Omega_{\mathbf{Q}}}(R(G), M^*) \cdot \mathrm{Det}(U(\mathbf{Z}[G]))}.$$

in which $J^*(M)$ denotes the group of idèles of a suitably large Galois extension, $M/\mathbf{Q}$, and $U(\mathbf{Z}[G])$ denotes the unit idèles of the integral group-ring of G.

Now suppose that E/F is a Galois extension of totally real number fields, as in Conjecture 6.3.4. For the moment, suppose further that $G(E/F)$ is abelian and that the Iwasawa μ-invariant ($\mu(\rho)$ of Theorem 6.2.5) is zero for all Galois representations ρ (as is widely believed). Then, using Iwasawa theory and the results of [5], one may imitate the proof in the function field case [38] to show that the p-adic coordinate, for p odd, of a Hom-description representative of $\Omega_{n-1}(E/F,3)$ is essentially given by the function $(\rho \mapsto h_{\overline{\rho}}(u^n-1)^{-1})$ of §6.1.1, when ρ is not of type W [41].

Suppose that E/F is linearly disjoint from the cyclotomic $\mathbf{Z}_p$-extension of F, so that there are no one-dimensional representations of type W. Then, if Conjecture 6.3.4 were true and $r \geq 2$, another Hom-desription representative for $\Omega_{r-1}(E/F,3)$ would be given by $(\rho \mapsto L_p(1-r,\overline{\rho})^{-1})$ in each p-adic coordinate, for p an odd prime. Since $r \geq 2$, one can further (this is the significance of the word "essentially" in the previous paragraph) change this Hom-representative to $(\rho \mapsto L_F(1-r,\overline{\rho})^{-1})$ in each odd-primary coordinate. This function lies in $\mathrm{Hom}_{\Omega_{\mathbf{Q}}}(R(G), M^*)$, which is sufficient to imply that

$$W_{E/F} = \Omega_n(E/F,3) \in \mathcal{CL}(\mathbf{Z}[1/2][G(E/F)])$$

for all $n = r-1 \geq 1$.

(ii) **The abelian case of degree prime to p.**

When $G(E/F)$ is abelian Theorem 6.2.5 implies that $U_\rho(u^n-1)) \in \overline{\mathbf{Z}}_p^*$, where $\overline{\mathbf{Z}}_p$ is the integral closure of $\mathbf{Z}_p$ in an algebraic closure of $\mathbf{Q}_p$. which implies that there exists a unit, $\alpha_{n,E/F}$, in the maximal $\mathbf{Z}_p$-order of $\mathbf{Q}_p[G(E/F)]$ such that

$$\mathrm{Det}(\alpha_{n,E/F})(\rho) = U_\rho(u^n-1)$$

for all $\rho \in R(G(E/F))$. Further more, if p does not divide $[E:F]$ then this maximal order is equal to $\mathbf{Z}_p[G(E/F)]$, so that Conjecture 6.3.4 is true in this simple case.

We close this section with an equivalent formulation of Conjecture 6.3.4 in another simple case.

Theorem 6.3.6

Let p be an odd prime and let E/F be a Galois extension of totally real number fields whose Galois group $G(E/F)$ is an elementary abelian p-group (i.e. an abelian group of exponent p). For each $n \geq 2$, let $f(\rho) = \log_p U_{(\psi^p(\rho)-p\rho)}(u^n-1)$, when defined.

Then Conjecture 6.3.4 implies that

$$U_{(\psi^p(\rho)-p\rho)}(u^n-1) \in 1 + p^{t+1}\overline{\mathbf{Z}}_p \subset \overline{\mathbf{Z}}_p^*$$

whenever $\rho \equiv 0$ (modulo p^t), as in §6.3.1. Furthermore, when $[E:F] = p$, $f(\rho)$ is defined and satisfies

$$((p-1)/2)(f(1)/p) + \mathrm{Trace}_{\mathbf{Q}_p(\xi_p)/\mathbf{Q}_p}((\xi_p/(1-\xi_p))(f(\chi)/p)) \in p\mathbf{Z}_p$$

for all faithful $\chi : G(E/F) \cong \mathbf{Z}/p \longrightarrow \mathbf{C}^*$.

Conversely, if the above two conditions hold and $G(E/F)$ is an elementary abelian p-group, then there exists a unit, $\alpha_{n,E/F} \in \mathbf{Z}_p[G(E/F)]^*$, and a homomorphism, $h \in \mathrm{Hom}_{\Omega_{\mathbf{Q}_p}}(R(G(E/F)), \mu(p^\infty))$, such that

$$\mathrm{Det}(\alpha_{n,E/F})(\rho)h(\rho) = U_\rho(u^n - 1)$$

for all $\rho \in R(G(E/F))$.

Proof. If $(\rho \mapsto U_{(\psi^p(\rho)-p\rho)}(u^n - 1))$ is equal to $\mathrm{Det}(\alpha_{n,E/F})$ then it satisfies the determinantal congruences by ([131] §4.3.37; [132] §3.1.12 p. 73). In this case composition with the p-adic logarithm gives a function

$$(\rho \mapsto f(\rho) = \log_p(U_{(\psi^p(\rho)-p\rho)}(u^n - 1))) \in Cong_p(G(E/F)).$$

By Theorem 6.3.2, we must characterise when this function vanishes under $\omega_G\psi^{-1}$: $Cong_p(G(E/F)) \longrightarrow G(E/F)^{ab}$. However, by naturality of $U_\rho(T)$, this will happen if and only if the vanishing is true when E/F is replaced by the intermediate abelian extension, M/F corresponding to the abelian quotient $G(E/F)^{ab} \cong G(M/F)$. Since $G(E/F)$ has exponent p, $G(M/F)$ is also an elementary abelian p-group. Therefore $G(M/F)$ is faithfully detected by all the quotient maps, $G(M/F) \longrightarrow \mathbf{Z}/p$. Hence, by naturality again, $\omega_G\psi^{-1}(f)$ vanishes if and only if it does so in the case when $G(E/F) \cong \mathbf{Z}/p$.

Now we shall calculate $w_{\mathbf{Z}/p}\psi^{-1}(f)$ for $f \in Cong_p(\mathbf{Z}/p)$, assuming that $[E : F] = p$. To accomplish this we must recall the formula for $\phi = \psi^{-1}$ from ([132] p. 80). Let g be a generator of the cyclic group of order p and let χ be the faithful one-dimensional representation given by $\chi(g) = exp(2\pi i/p) = \xi_p$. Then the elements of $R(\mathbf{Z}/p) \otimes \mathbf{C}$ which are dual to the powers of g are

$$\hat{\gamma}_s = p^{-1} \sum_{i=0}^{p-1} \chi(g^{-is})\chi^i$$

for $0 \le s \le p-1$, since

$$\hat{\gamma}_s(g^j) = p^{-1} \sum_{i=0}^{p-1} \chi(g^{i(j-s)}) = \begin{cases} 1 & \text{if } j = s, \\ 0 & \text{otherwise.} \end{cases}$$

Then, by ([132] p. 80),

$$\phi(f) = \sum_{s=0}^{p-1} \sum_{i=0}^{p-1} p^{-1}\chi(g^{-is})f(\chi^i)g^s \in \mathbf{Z}_p[\mathbf{Z}/p].$$

Hence

$$\begin{aligned} \omega_{\mathbf{Z}/p}(\phi(f)) &= \prod_{s=0}^{p-1} (g^s)^{(1/p)\sum_{i=0}^{p-1} p^{-1}\chi(g^{-is})f(\chi^i)} \\ &= g^{(1/p^2)\sum_{s=0}^{p-1}\sum_{i=0}^{p-1} s\chi(g^{-is})f(\chi^i)}. \end{aligned}$$

However

$$\begin{aligned}\sum_{s=0}^{p-1}\sum_{i=0}^{p-1} s\chi(g^{-is})f(\chi^i) &= \sum_{s=1}^{p-1} sf(1) + \sum_{i=1}^{p-1}\sum_{s=1}^{p-1} s\xi_p^{-is}f(\chi^i) \\ &= (p(p-1)/2)f(1) \\ &\quad + \mathrm{Trace}_{\mathbf{Q}_p(\xi_p)/\mathbf{Q}_p}((\sum_{s=1}^{p-1} s\xi_p^{-s})f(\chi))\end{aligned}$$

and

$$\begin{aligned}\sum_{s=1}^{p-1} s\xi_p^{-s} &= (x\tfrac{d}{dx}(1+x+\ldots+x^{p-1}))|_{x=\xi_p^{-1}} \\ &= \xi_p^{-1}\prod_{j=2}^{p-1}(\xi_p^{-1}-\xi_p^{-j}) \\ &= \xi_p^{-(p-1)}\prod_{j=2}^{p-1}(1-\xi_p^{-j+1}) \\ &= \xi_p p/(1-\xi_p).\end{aligned}$$

Therefore we have

$$\omega_{\mathbf{Z}/p}(\phi(f)) = g^{((p-1)/2)(f(1)/p)+\mathrm{Trace}_{\mathbf{Q}_p(\xi_p)/\mathbf{Q}_p}((\xi_p/(1-\xi_p))(f(\chi)/p))}.$$

This completes the proof in one direction. Conversely the discussion shows that if the two conditions hold then there exists a unit, $\alpha_{n,E/F} \in \mathbf{Z}_p[G(E/F)]^*$, such that

$$\log_p \mathrm{Det}(\alpha_{n,E/F})(\psi^p\rho - p\rho) = \log_p U_{\psi^p\rho-p\rho}(u^n-1) \in p\mathcal{O}_N.$$

Therefore $\mathrm{Det}(\alpha_{n,E/F})(\psi^p\rho-p\rho)/U_{\psi^p\rho-p\rho}(u^n-1))$ is a p-primary root of unity which is congruent to 1 modulo p (cf. [131] §4.5.31). Since $G(E/F)$ has exponent p, $\psi^p(\rho) = \dim(\rho)$ and therefore $\mathrm{Det}(\alpha_{n,E/F})(\rho)/U_\rho(u^n-1))^p$ is equal to $\mathrm{Det}(\alpha_{n,E/F})(\dim(\rho))/U_{\dim(\rho)}(u^n-1))$ times a p-primary root of unity which is congruent to 1 modulo p.

On the other hand, setting $\rho = 1$ we set that $\mathrm{Det}(\alpha_{n,E/F})(1)/U_1(u^n-1))^{p-1}$ is also a p-primary root of unity which is congruent to 1 modulo p, since $\psi^p(1)-p = 1-p$. Since $\mathrm{Det}(\alpha_{n,E/F})(1)/U_1(u^n-1))$ is congruent to 1 modulo p then it must also be a p-primary root of unity, which completes the proof. □

6.4 Modular forms with coefficients in $\Lambda[G]$

In this section we consider a strategy for proving Conjecture 6.3.4. As explained in §6.3.5(ii) the technique used in [166], which uses the connection between Hida modular forms and Galois representations, one character at a time, proves that the Wiles Unit corresponds to the determinantal function made from a unit in a maximal $\mathbf{Z}_p$-order of $\mathbf{Q}_p[G(E/F)]$. Therefore one way to try to prove that this function is the determinant of a unit in $\mathbf{Z}_p[G(E/F)]^*$ would be to use as input to the arguments of [166] Hida modular forms with coefficients in $\Lambda[G(E/F)]$ rather than in Λ. Such a strategy would have three steps. The correspondence between modular forms and Galois represention, the analysis of the Galois lattices coming form the representations and the construction of the requisite modular forms.

Presumably the last step is particularly difficult for modular forms with coefficients in group-rings, since it may well involve, by analogy with [33] necessary for [166], the study of Hilbert-Blumenthal moduli spaces with G-actions.

In this section we do the first step, which is the simplest.

6.4.1 *Brauer-Nesbitt results*

Let K be a field of characteristic $p > 0$. We have in mind $K = \mathbf{F}_q((X))$, the local field given by the field of fractions of $\mathcal{O}_K = \mathbf{F}_q[[X]]$. In this section we shall be particularly interested in the case of continuous modular representations of the form

$$\Omega_{\mathbf{Q}} \longrightarrow GL_2(\mathbf{F}_q((X)))$$

which means a $GL_2(\mathbf{F}_q((X)))$-conjugacy class of homomorphisms

$$\Omega_{\mathbf{Q}} \longrightarrow GL_2(\mathbf{F}_q[[X]])$$

continuous with respect to the $X\mathbf{F}_q[[X]]$-adic topology. Here $\Omega_{\mathbf{Q}}$ is the absolute Galois group of the rationals.

Let us recall some classical results.

Theorem 6.4.2 *([20] p. 560; [48] §30.16)*

Let M, N be finitely generated $K[G]$-modules. Suppose that there exists an extension field, L/K, such that

(i) *L is a splitting field for G (i.e. for each finitely generated irreducible $K[G]$-module, Z, $\mathrm{Hom}_{K[G]}(Z, Z) \cong K$)*

and

(ii) *if V is a finitely generated, completely reducible $K[G]$-module (i.e. a sum of $K[G]$-irreducibles) then $V \otimes_K L$ is a finitely generated, completely reducible $L[G]$-module.*

Then M and N have the same decomposition factors if and only if (counting multiplicities) they have the same characteristic roots for each $g \in G$.

Proof. We replace each of M and N by the direct sum of their composition factors. This preserves the characteristic polynomial of each $g \in G$ on M and N while replacing them by completely reducible modules. Pair off as many irreducible summands of M and N as possible. The complementary summands are two completely reducible $K[G]$-modules, M' and N', with the same characteristic roots (counting multiplicities) for each $g \in G$, but they have no common irreducible summand. We must show that $M' = N' = 0$.

If not, set $M'' = M \otimes_K L$, $N'' = N \otimes_K L$, which are two completely reducible $L[G]$-modules, by the hypothesis on L.

Let $V_1, \ldots, V_s$ be a complete set of finitely generated, irreducible $L[G]$-modules. Then, by Theorem 6.4.3 below, the character functions $\{\chi_i\}$, given by $(g \mapsto trace(g|V_i))$, are linearly independent over L.

If we write $M'' = \sum_i m_i V_i$ and $N'' = \sum_j n_j V_j$ with $m_i, n_j \in \mathbf{Z}$ we have

$$0 = \sum_j (m_j - n_j)\chi_j(g)$$

for all $g \in G$. By linear independence this means that $m_j = n_j \in L$ for each j. In other words, $m_j \equiv n_j$ (modulo p) for each j. However, M'' and N'' have no irreducible summands in common because if so then

$$\dim_K \mathrm{Hom}_{K[G]}(M', N') = \dim_L \mathrm{Hom}_{L[G]}(M'', N''))$$

would be non-zero.

Hence p divides both m_j and n_j for each j, since one of m_j and n_j is zero. Now set

$$M''' = \sum_i (m_i/p)V_i \ \text{ and } N''' = \sum_j (n_j/p)V_j.$$

The characteristic polynomials of M'' and N'' are just the p-th powers of those for M''' and N'''. Since the p-th power is a field homomorphism M''' and N''' must have (counting multiplicities) the same characteristic values for each $g \in G$.

Proceeding by induction we find that p^r divides m_i for all r, which implies that each $m_i = 0$ and similarly each $n_j = 0$. □

Theorem 6.4.3 *([48] §30.12)*

Let $Z_1, \ldots, Z_t$ be a complete set of $K[G]$-irreducibles – K being a splitting field for G, with $\mathrm{char}(K) = p$. *Set* $\lambda_i = \mathrm{Trace}(Z_i)$.

The $\{\lambda_1, \ldots, \lambda_t\}$ are linearly independent over K.

Recall that being a splitting field means that

$$\mathrm{Hom}_{K[G]}(Z_i, Z_i) \cong K$$

for $i = 1, \ldots, t$. Using this we shall prove the following result, which implies Theorem 6.4.3.

Theorem 6.4.4 *([48] §27.8)*

Let K be a splitting field for G and $M_1, \ldots, M_r$ be a set of pairwise non-isomorphic $K[G]$-irreducibles of dimensions $d_1, \ldots, d_r$. Let the corresponding matrix homomorphisms be

$$\rho_i : G \longrightarrow GL_{d_i}(K).$$

Then the functions

$$\{(\rho_i)_{u,v} : G \longrightarrow K\}$$

are linearly independent over K.

Proof. We are going to assume that $\mathrm{Hom}_{K[G]}(M_i, M_i) \cong K$

Suppose that for all $g \in G$

$$\sum_{u,v} \lambda^i_{u,v}(\rho_i)_{u,v}(g) = 0$$

with $\lambda^i_{u,v} \in K$. Write $M = \oplus M_i$ then we have

$$K[G] \longrightarrow End_K(M)$$

by sending $a \in K[G]$ to $(m \mapsto a{\cdot}m)$. In fact, this map sends $K[G]$ to $\prod_{i=1}^r M_i$ since $g(M_i) \subseteq M_i$. If $\mathcal{A} = image(K[G]) \subset \prod_{i=1}^r M_i$ then each M_i must be an irreducible $\mathcal{A}$-module. By ([48] Exercise §25.4) this means that $Rad(\mathcal{A}) \cdot M = 0$. This is because $Rad(\mathcal{A}) \cdot M_i \subseteq M_i$ is a $K[G]$-submodule so that either $Rad(\mathcal{A}) \cdot M_i = 0$ or $Rad(\mathcal{A}){\cdot}M_i = M_i$. In the latter case we shall have $Rad(\mathcal{A})^m{\cdot}M_i = M_i$ for all $m \geq 0$. However, $Rad(\mathcal{A})$ acts on M_i through $M_{d_i}(K)$ so that there exists an m such that $Rad(\mathcal{A})^m$ has zero image in $M_{d_i}(K)$ which implies that $M_i = Rad(\mathcal{A}^m) \cdot M_i = 0$.

On the other hand $\mathcal{A} = image(K[G]) \subset \prod_{i=1}^r M_i$ so that $\mathcal{A}$ acts faithfully on M so $Rad A = 0$. Hence $\mathcal{A}$ is semisimple. But for a semi-simple $\mathcal{A}$ to act faithfully on a finite sum of pairwise non-isomorphic $\mathcal{A}$-modules, $M = \oplus_i M_i$ means that $M_1, \ldots, M_r$ must be a complete set of irreducible $\mathcal{A}$-modules. Then the simple components of $\mathcal{A}$ are the $End_K(M_i)$ so that

$$\mathcal{A} \cong End_K(M_i).$$

Now choose $X_s \in End_K(M)$ and $a \in \mathcal{A}$ mapping to $(0, 0, \ldots, 0, X_s, 0 \ldots, 0) \in \prod_{i=1}^r End_K(M_i)$. Then

$$0 = \sum_{u,v} \lambda^i_{u,v}(\rho_i)_{u,v}(a) = \sum_{u,v} \lambda^s_{u,v}(X_s)_{u,v}.$$

Now, if we choose X_s to have only one non-zero entry, $(X_s)_{a,b}$, then $\lambda^s_{a,b} = 0$ and varying X_s shows that $\lambda^s_{a,b} = 0$ for all a, b, s. □

6.4.5 We are still in $\mathrm{char}(K) = p > 0$ and we shall suppose that K is a local field.

If Z is a $K[G]$-module and $\dim_K(Z) = 1$ then

$$D = \mathrm{Hom}_{K[G]}(Z, Z) \subseteq \mathrm{Hom}_K(Z, Z) \cong K$$

so that $D \cong K$, the scalars.

If $\dim_K(Z) = 2$ and Z is irreducible then $\dim_K \mathrm{Hom}_{K[G]}(Z, Z) = 1$. For we have

$$D = \mathrm{Hom}_{K[G]}(Z, Z) \subseteq \mathrm{Hom}_K(Z, Z) \cong M_2(K).$$

Also D is a division algebra – for if $0 \neq f \in D$ then $0 \neq f(Z) \subseteq Z$ is a submodule so that $f(Z) = Z$ and f is an isomorphism of $K[G]$-modules, which means that $f^{-1} \in D$. Now either $D \cong K$ or $\dim_K D = 4$, by ([119] Theorem 13.7 p. 142) –

this is where K local is used. If $\dim_K D = 4$ then we would have $D = M_2(K)$, which is not a division algebra.

This discussion shows that if Z is 2-dimensional continuous $K[\Omega_{\mathbf{Q}}]$-representation and irreducible then it is determined by its trace – using (staightforward modifications of) the previous results.

6.4.6 *Some well-known pull-back diagrams*

Let p be a prime and let $n \geq 1$ be an integer. The integral group-ring of the cyclic group of order p^n is isomorphic to $\mathbf{Z}[x]/(x^{p^n} - 1)$. If $\Phi_n(x)$ denotes the n-th cyclotomic polynomial

$$\Phi_n(x) = 1 + x^{p^{n-1}} + x^{2p^{n-1}} + \ldots + x^{(p-1)p^{n-1}} \in \mathbf{Z}[x]$$

then $\mathbf{Z}[x]/(\Phi_n(x))$ is isomorphic to $\mathbf{Z}[\xi_{p^n}]$, the ring of integers in the cyclotomic field $\mathbf{Q}(\xi_{p^n})$ where ξ_m denotes a primitive m-th root of unity. We have a commutative digram of homomorphisms of commutative rings if the following form.

$$\begin{array}{ccc} \mathbf{Z}[x]/(x^{p^n} - 1) & \longrightarrow & \mathbf{Z}[x]/(x^{p^{n-1}} - 1) \\ \downarrow & & \downarrow \\ \mathbf{Z}[x]/(\Phi_n(x)) & \longrightarrow & \mathbf{Z}/p \end{array}$$

If Λ is a commutative ring which is torsion-free as an abelian group we define $\Lambda[\xi_{p^n}]$ by the tensor product

$$\Lambda[\xi_{p^n}] = \Lambda \otimes_{\mathbf{Z}} \mathbf{Z}[x]/(\Phi_n(x)).$$

The following result is well known.

Lemma 6.4.7

Let Λ be a commutative ring which is torsion-free as an abelian group. Tensoring the diagram of §6.4.6 with Λ yields a pull-back diagram of the following form.

$$\begin{array}{ccc} \Lambda[x]/(x^{p^n} - 1) & \longrightarrow & \Lambda[x]/(x^{p^{n-1}} - 1) \\ \downarrow & & \downarrow \\ \Lambda[\xi_{p^n}] & \longrightarrow & \Lambda \otimes \mathbf{Z}/p \end{array}$$

Proof. We have to show that

$$0 \longrightarrow \Lambda[x]/(x^{p^n} - 1) \longrightarrow \Lambda[x]/(x^{p^{n-1}} - 1) \oplus \Lambda[\xi_{p^n}] \longrightarrow \Lambda \otimes \mathbf{Z}/p \longrightarrow 0$$

is a short exact sequence, the homomorphisms being the sum and difference of the appropriate homomorphisms in the commutative diagram. Since tensoring with Λ is exact it suffices to prove this in the case when $\Lambda = \mathbf{Z}$, the integers.

In $\mathbf{Z}[x]$ we have an identity of the form

$$p = 1 + x + x^2 + \ldots + x^{p-1} - (x-1)((p-1) - (p-2)x - (p-3)x^2 - \ldots - z^{p-2})$$

and replacing x by $x^{p^{n-1}}$ we find that

$$p = \Phi_n(x) + (x^{p^{n-1}} - 1)f(x) \in \mathbf{Z}[x].$$

To prove exactness at the left-hand group, suppose that $\alpha = g(x) + \mathbf{Z}[x](x^{p^n} - 1)$ maps to zero in the middle direct sum. Then $g(x) \in (\Phi_n(x)) \bigcap (x^{p^{n-1}} - 1) \lhd \mathbf{Z}[x]$. However, $x^{p^n} - 1 = (x^{p^{n-1}} - 1)\Phi_n(x)$ so that $pg(x) \in \mathbf{Z}[x](x^{p^n} - 1)$. This means that $p\alpha$ is zero in a torsion-free and so $0 = \alpha \in \Lambda[x]/(x^{p^n} - 1)$, as required.

Exactness at the right-hand group is clear, from the surjectivity of the right-hand vertical homomorphism in the diagram.

Now for exactness in the centre. Suppose that $(\beta, \gamma) \in \mathbf{Z}[x]/(x^{p^{n-1}} - 1) \oplus \mathbf{Z}[\xi_{p^n}]$ maps to zero in $\mathbf{Z}/p$. By the surjectivity of the upper horizontal homomorphism in the diagram we may assume that $\beta = 0$, by subtracting the image of a suitable element of $\mathbf{Z}[x]/(x^{p^n} - 1)$. Therefore $\gamma = p\delta$ for some $\delta \in \mathbf{Z}[\xi_{p^n}] = \mathbf{Z}[x]/(\Phi_n(x))$. If $d(x) \in \mathbf{Z}[x]$ is a polynomial which maps to δ then the coset $d(x)(x^{p^{n-1}} - 1)f(x) + \mathbf{Z}[x](x^{p^n} - 1)$ maps to zero in $\mathbf{Z}[x]/(x^{p^{n-1}} - 1)$ and to $pd(x) + \mathbf{Z}[x]/(\Phi_n(x)) = \gamma \in \mathbf{Z}[\xi_{p^n}]$, as required. □

Corollary 6.4.8

Let $\mathcal{O}$ denote the ring of integers in a p-adic local field. Then there is a pull-back diagram of groups of the following form.

$$\begin{array}{ccc}
GL_2(\mathcal{O}[[X]][\mathbf{Z}/p^n]) & \longrightarrow & GL_2(\mathcal{O}[[X]][\mathbf{Z}/p^{n-1}]) \\
\downarrow & & \downarrow \\
GL_2(\mathcal{O}[\xi_{p^n}][[X]]) & \longrightarrow & GL_2(\mathcal{O}[[X]] \otimes \mathbf{Z}/p)
\end{array}$$

6.4.9 *Theorems of Carayol and Serre*

Let A be the integers of a local field with maximal ideal $M \lhd A$ and residue field $F = A/M$. For example, $A = \mathbf{Z}/p[[X]]$, $M = XA$ and $F = \mathbf{Z}/p$. Let R be an A-algebra. For example, $R = A[G]$ and $G = \Omega_{\mathbf{Q}}$.

An n-dimensional A-representation of R is the GL_nA-conjugacy class of an A-algebra homomorphism of the form

$$\rho : R \longrightarrow M_nA,$$

where M_nA denotes the $n \times n$ matrices with entries in A. When $G = \Omega_{\mathbf{Q}}$ we require that ρ be continuous with respect to the profinite topology on G and the M-adic topology on M_nA.The residual representation of such a representation is the canonical composition

$$R \stackrel{\rho}{\longrightarrow} M_nA \longrightarrow M_nF.$$

When $R = A[\Omega_{\mathbf{Q}}]$ we obtain Galois representations but only considered up to GL_nA-conjugacy rather than up to GL_nK-conjugacy, where K is the field of fractions of A.

The following result is crucial.

Theorem 6.4.10 *([29] p. 215)*
Suppose that

$$\rho_1, \rho_2 : R \longrightarrow M_nA$$

are two n-dimensional A-representations of R having the same trace function. Then, if the residual representation of ρ_1 is absolutely irreducible, there exists $X \in GL_nA$ such that

$$\rho_1 = X\rho_2X^{-1} : R \longrightarrow M_nA.$$

Proof. We begin by appealing to Brauer-Nesbitt theory to establish that the residual representations are equivalent (compare what follows with the proof of Theorem 6.4.2). For this it suffices to extend scalars to an algebraic closure of F ([48] §29.7), which preserves irreducibility, by definition of absolutely irreducibility. In characteristic zero the trace function is well known to determine the equivalence class of the representation. In characteristic p we use the linear independence of the traces of distinct irreducible representations ([48] §27.8; see Theorem 6.4.4). One considers the decomposition of the semisimplification of the residual representation of ρ_2 into the direct sum of irreducibles and shows that the multiplicities of the irreducibles appearing in the similar semisimplification for ρ_1 appear with multiplicity congruent to 1 (modulo p) and the rest appear with multiplicity congruent to 0 (modulo p). Since they have the same dimension these congruences show that they are equivalent.

Serre's Proof. If B is a commutative ring with identity then all two-sided ideals of M_nB have the form M_nI for some ideal, $I \triangleleft B$. Furthermore, if B is a local ring then all automorphisms of M_nB are of the form $(Y \mapsto XYX^{-1})$ for some $X \in GL_nA$. These results hold for Azumaya algebras in general ([88] III Corollary 5.2 and IV Corollary 1.3).

Now consider the homomorphism

$$(\rho_1, \rho_2) : R \longrightarrow M_nA \times M_nA$$

whose image is a subalgebra which will be denoted by Γ.

A result of Burnside ([48] §27.4) states that the residual representation of ρ_1 is surjective, because it is completely reducible. Since the residual representation of ρ_2 is equivalent to that of ρ_1, it is also surjective. By Nakayama's Lemma each of ρ_1 and ρ_2 are also surjective. Therefore the subalgebra, $\Gamma \subseteq M_nA \times M_nA$, maps surjectively onto each factor. Set

$$J = \Gamma \bigcap M_nA \times \{0\} \text{ and } J' = \Gamma \bigcap \{0\} \times M_nA.$$

Then both J and J' are two-sided ideals of M_nA. Also the image of Γ in $(M_nA)/J\times(M_nA)/J'$ is the graph of an isomorphism of algebras of the form

$$\iota : (M_nA)/J \xrightarrow{\cong} (M_nA)/J'$$

Also Γ is identified as the set of pairs

$$\Gamma = \{(x, x') \in M_nA) \times M_nA \mid \iota(x + J) = x' + J'\}.$$

Suppose now that $J = M_nI$ and $J' = M_nI'$ then we must have $I = I'$ since they are the annihilator ideals of the isomorphic A-modules $(M_nA)/J$ and $(M_nA)/J'$. Furthermore, if $I = A$ then $\Gamma = M_nA \times M_nA$. In $I \neq A$ we may view ι as an isomorphism of the algebra, $M_n(A/I)$. Therefore there exists $\tilde{X} \in GL_n(A/I)$ which may be lifted to $X \in GL_nA$ such that

$$\Gamma = \{(x, x') \in M_nA) \times M_nA \mid x \equiv Xx'X^{-1} \text{ (modulo } I)\}.$$

For all $(x, x') \in \Gamma$ we have $trace(x) = trace(x')$. If $y \in I$ then

$$(diag(i, 0, 0, \ldots, 0), diag(0, 0, 0, \ldots, 0)) \in \Gamma$$

so that $i = 0$. Therefore $I = I' = \{0\}$. This means that

$$\Gamma = \{(x, x') \in M_nA) \times M_nA \mid x = Xx'X^{-1}\},$$

which proves that $\rho_1 = X\rho_2X^{-1}$. □

Carayol's Proof. This uses the hypothesis that A is complete and proceeds by inductionon s for $B = A/M^s$. Burnside's result about the surjectivity of the residual representation is crucial. In addition, if A is a hensel local ring for which $Br(F) = 0$ and ρ is a representation into a semilocal extension of A, then Carayol shows that the representation actually is defined over A – providing all the components have trace functions with values in A, that the residual traces are all equal and that one residual representation is absolutely irreducible.

Definition 6.4.11 $\Lambda[G]$*-adic modular forms*

Let p be a prime, usually odd. Let $\mathcal{O}_K$ denote the ring of integers in a local field, $K/\mathbf{Q}_p$. Fix an embedding $\overline{\mathbf{Q}} \subset \overline{\mathbf{Q}}_p$ as in (Hida p. 195 et seq). Let $\omega : (\mathbf{Z}/p)^* \longrightarrow \overline{\mathbf{Q}}_p^*$ denote the Teichmüller character. Set $\Lambda = \mathcal{O}_K[[X]]$.

Following ([75] p. 195) a p-adic analytic family of character $\psi = \omega^a$ is a set of modular forms $\{f_k\}_{k=M}^{\infty}$ such that

(A1) $f_k \in \mathcal{M}_k(\Gamma_0(p), \psi\omega^{-k})$,

(A2) the q-series of f_k has the form $\sum_{n\geq 0} a(n, f_k)q^n$ with $a(n, f_k) \in \overline{\mathbf{Q}}$ for all n,

(A3) there exists a power series, $A(n, X) \in \mathcal{O}_K[[X]]$ for each $n \geq 0$, such that $a(n, k) = A(n, u^k - 1)$ for all $k \geq M$ where $u = 1 + p$.

If (A1)–(A3) hold then we say that the formal q-expansion for $\{f_k\}_{k=M}^{\infty}$

$$F(X,q) = \sum_{n\geq 0} A(n,X)q^n$$

is a Λ-adic form of character $\psi = \omega^a$ if

$$F(u^k - 1, q) = \sum_{n\geq 0} A(n, u^k - 1)q^n$$

is the q-expansion in $\mathcal{O}_K[q]$ of a modular form in $\mathcal{M}_k(\Gamma_0(p), \psi\omega^{-k}, \mathcal{O}_K)$ for all but finitely many $k \geq M$.

Now let G be a finite abelian group – we are particularly interested in the case when G is a p-group. By a $\Lambda[G]$-adic form of character $\psi = \omega^a$ we shall mean a formal q-series

$$F(X,q) = \sum_{n\geq 0} A(n,X)q^n \in \Lambda[G][q]$$

such that for every character, $\lambda : G \longrightarrow \overline{\mathbf{Q}}_p^*$,

$$\lambda(F(X,q)) = \sum_{n\geq 0} \lambda(A(n,X))q^n \in \Lambda[\lambda][q]$$

is an $\mathcal{O}_K[\lambda][[X]]$-adic modular form of character $\psi = \omega^a$. Examples of this type of modular form may be constructed by the method of ([75] p. 197).

We shall say that $F(X,q)$ is a simultaneous eigenform of all the Hecke operators if, for all p-adic characters λ, $\lambda(F(X,q))$ is an $\mathcal{O}_K[\lambda][[X]]$-adic Hida modular form which is a simultaneous eigenform of all the Hecke operators in the sense of ([75] p. 195).

From these modular forms one can construct Galois representations. The following result is not the most general, but it is typical. To state the result, recall the continuous character

$$\kappa : 1 + p\mathbf{Z}_p \longrightarrow \mathbf{Z}_p[[X]]$$

given by $\kappa((1+p)^s) = (1+X)^s$. If $x \in \mathbf{Z}_p^*$, set $\langle x\rangle = \omega(x)^{-1}x \in 1 + p\mathbf{Z}_p$ where $\omega : \mathbf{Z}_p^* \to \mathbf{F}_p^* \subset \mathbf{Z}_p^*$ is the Teichmuller character (see [75] p. 229).

Theorem 6.4.12

Let G be a finite abelian p-group. Let f be a p-adic Hida modular form of character χ with coefficients in $\mathbf{Z}_p[[X]][G]$ which is a simultaneous eigenform of all the Hecke operators. Suppose that $\epsilon'(f)$ is the p-adic modular form with coefficients in $\mathbf{Z}/p$ corresponding to the reduction (modulo the maximal ideal) homomorphism,

$$\epsilon' : \mathbf{Z}_p[[X]][G] \longrightarrow \mathbf{Z}/p.$$

Suppose that the residual representation corresponding by ([75] §7.5) to $\epsilon'(f)$

$$\pi(\epsilon'(f)) : \Omega_{\mathbf{Q}} \longrightarrow GL_2(\mathbf{Z}/p)$$

is absolutely irreducible.

Then there exists a continuous Galois representation of the form

$$\pi(f) : \Omega_{\mathbf{Q}} \longrightarrow GL_2(\mathbf{Z}_p[[X]][G])$$

which is unramified outside p. *Also, for each* p*-adic character,* $\lambda : G \to \overline{\mathbf{Z}}_p^*$, *and for each prime* q *different from* p *the image of the characteristic polynomial under the homomorphism induced by* λ *satisfies*

$$\lambda_*(\det(1 - \pi(f)(\mathrm{Frob}_q)T)) = 1 - T(q)T + \chi(q)\kappa(\langle q\rangle)q^{-1}T^2.$$

Here $T(q)$ *is the eigenvalue of the corresponding Hecke operator and* $\kappa(\langle q\rangle)$ *is as in §6.4.11.*

In addition, $\pi(f)$ *is unique up to conjugation by elements of* $GL_2(\mathbf{Z}_p[[X]][G])$.

Proof. Firstly we observe that $\mathbf{Z}_p[[X]][G]$ is a local ring with maximal ideal $\mathcal{M} = \langle p, IG, X\rangle \lhd \mathbf{Z}_p[[X]][G]$ and residue field $\mathbf{Z}/p$.

When $G = \mathbf{Z}/p^n$ we proceed by induction on n using Corollary 6.4.8. For example, when $n = 1$, we have two p-adic characters of $\mathbf{Z}/p$, the trivial one ϵ and $\phi : \mathbf{Z}/p \longrightarrow \mathbf{Z}_p[\xi_p]^*$, which sends a generator to ξ_p. Hence ([75] Theorem 1 p. 228) gives us two continuous representations

$$\pi(\epsilon(f)) : \Omega_{\mathbf{Q}} \longrightarrow GL_2(\mathbf{Z}_p[[X]])$$

and

$$\pi(\phi(f)) : \Omega_{\mathbf{Q}} \longrightarrow GL_2(\mathbf{Z}_p[\xi_p][[X]]).$$

Furthermore, by Cebotarev's Density Theorem, the residual representations of $\pi(\epsilon(f))$ and $\pi(\phi(f))$ into $GL_2(\mathbf{F}_p[[X]])$ have the same characteristic polynomial. Therefore, by Theorem 6.4.10, there exists $U \in GL_2(\mathbf{F}_p[[X]])$ which conjugates one residual representation into the other, say $U(\pi(\epsilon(f))$ modulo $p)U^{-1} = (\pi(\phi(f))$ modulo $p)$ as Galois representations into $GL_2(\mathbf{F}_p[[X]])$. We lift U to $U' \in GL_2(\mathbf{Z}_p[[X]])$ and then $U'\pi(\epsilon(f))U'^{-1}$ and $\pi(\phi(f))$ would be two continuous Galois representations into $GL_2(\mathbf{Z}_p[[X]])$ and $GL_2(\mathbf{Z}_p[\xi_p][[X]])$, reducing to the same modulo p. By Corollary 6.4.8, this gives a unique representation of the form

$$\pi(f) : \Omega_{\mathbf{Q}} \longrightarrow GL_2(\mathbf{Z}_p[[X]][\mathbf{Z}/p]),$$

as required.

The case of $G = \mathbf{Z}/p^n$ is proved in a similar manner. In general, if $G = \mathbf{Z}/p^m \times G_1$ we have a pull-back square of the form

$$\begin{array}{ccc} GL_2(\mathbf{Z}_p[[X]][G]) & \longrightarrow & GL_2(\mathbf{Z}_p[[X]][\mathbf{Z}/p^{m-1} \times G_1]) \\ \downarrow & & \downarrow \\ GL_2(\mathbf{Z}_p[\xi_{p^n}][[X]][G_1]) & \longrightarrow & GL_2(\mathbf{Z}/p[[X]][G_1]) \end{array}$$

By induction we obtain continuous Galois representations

$$\pi_1 : \Omega_F \longrightarrow GL_2(\mathbf{Z}_p[[X]][\mathbf{Z}/p^{m-1} \times G_1]$$

and

$$\pi_2 : \Omega_F \longrightarrow GL_2(\mathbf{Z}_p[\xi_{p^n}][[X]][G_1]).$$

which are equivalent when composed with the homomorphisms to

$$GL_2(\mathbf{Z}/p[[X]][G_1]).$$

Therefore there exists $X \in GL_2(\mathbf{Z}/p[[X]][G_1])$ such that $\pi_2 \equiv X\pi_1X^{-1}$ (modulo p). Lift X to $\tilde{X} \in GL_2(\mathbf{Z}_p[[X]][\mathbf{Z}/p^{m-1} \times G_1]$ and replace π_1 by $\tilde{X}\pi_1\tilde{X}^{-1}$. This $\tilde{X}\pi_1\tilde{X}^{-1}$ and π_2 fit together to give a continuous homomorphism into the pull-back, $GL_2(\mathbf{Z}_p[[X]][G])$, with the required properties. □

Remark 6.4.13 Let F be a totally real number field. Theorem 6.4.12 (and its generalization to give representations of Ω_F) is a simple application of Carayol's Theorem (Theorem 6.4.10) once one has the the existence of Galois representations into $GL_2(\mathbf{Z}_p[[X]])$ attached to Λ-adic Hida modular forms (i.e. the case when $G = \{1\}$). When $F = \mathbf{Q}$, the rationals, this is provided by ([75] Theorem 1 p. 228). The result may appear, at first sight, only to assert the existence of a representation into GL_2 of the quotient field, $\mathbf{Z}_p((X))$, but the continuity guarantees that the representation may be realised in $GL_2(\mathbf{Z}_p[[X]])$.

For a general totally real F the required input has been proved under very general conditions (see [76], [28], [156], [157], [158], [164], [165], [168]).

Chapter 7

Annihilators

In this chapter we shall relate the higher-dimensional K-theory Galois module structure invariants, $\Omega_n(N/K, 3)$, to relations between annihilator ideals of K-groups and of étale cohomology groups. The constructions are very simple and work particularly well when the K-groups are finite. In §7.1 we show how to obtain a chain of annihilator ideal relations from a class in $K_0(\mathbf{Z}[G], \mathbf{Q})$. In §§7.2.1–7.2.6 we recall Stickelberger's Theorem and the conjectures of Brumer, Coates and Sinnott. These conjectures concern annihilator relations similar to those derived in Theorem 7.1.11. Then we examine what happens when we use the lifted classes, $\hat{\Omega}_n(N/K, 3)$, which were introduced in Definition 5.6.2, as the ingredients for Theorem 7.1.11. In particular, we explain how to derive the Coates-Sinnott Conjecture from the Wiles Unit Conjecture of §6.3.4. We also offer some remarks about the Brumer Conjecture and about the generalization of the Coates-Sinnott Conjecture to the case when the K-groups are finitely generated but not finite. Finally, in Theorem 7.3.2 we prove that the radical of the annihilator of K'_{2r-2} is equal to the radical of the appropriate higher Stickelberger ideal, under mild assumptions.

7.1 $K_0(\mathbf{Z}[G], \mathbf{Q})$ and annihilator relations

7.1.1 Let G be a finite group. Suppose that

$$0 \longrightarrow F_k \xrightarrow{d_k} F_{k-1} \xrightarrow{d_{k-1}} \ldots \xrightarrow{d_2} F_1 \xrightarrow{d_1} F_0 \longrightarrow 0$$

is a perfect complex of $\mathbf{Z}[G]$-modules, as in Definition 2.2.1, having all its homology groups finite. In §§7.1.12–7.1.20 we shall briefly discuss a slightly more general situation but for the applications of this chapter the preceding case will suffice.

As usual, let $Z_t = \mathrm{Ker}(d_t : F_t \longrightarrow F_{t-1})$ and $B_t = d_{t+1}(F_{t+1}) \subseteq F_t$ denote the $\mathbf{Z}[G]$-modules of t-dimensional cycles and boundaries, respectively. We have short exact sequences of the form

$$0 \longrightarrow B_i \xrightarrow{\phi_i} Z_i \longrightarrow H_i(F_*) \longrightarrow 0$$

and

$$0 \longrightarrow Z_{i+1} \xrightarrow{\psi_{i+1}} F_{i+1} \xrightarrow{d_{i+1}} B_i \longrightarrow 0.$$

Applying $(- \otimes \mathbf{Q})$ we obtain isomorphisms

$$\phi_i \otimes 1 : B_i \otimes \mathbf{Q} \xrightarrow{\cong} Z_i \otimes \mathbf{Q}$$

and we may choose splittings of the form

$$\eta_i : B_i \otimes \mathbf{Q} \longrightarrow F_{i+1} \otimes \mathbf{Q}$$

such that

$$(d_{i+1} \otimes 1)\eta_i = 1 : B_i \otimes \mathbf{Q} \longrightarrow B_i \otimes \mathbf{Q}.$$

Then we form an isomorphism, as in Example 2.1.8(ii),

$$X : \oplus_j F_{2j} \otimes \mathbf{Q} \xrightarrow{\cong} \oplus_j F_{2j+1} \otimes \mathbf{Q}$$

given by the composition

$$\begin{aligned} \oplus_j F_{2j} \otimes \mathbf{Q} &\xrightarrow{(\sum_j \psi_{2j} + \eta_{2j-1})^{-1}} \oplus_j (Z_{2j} \otimes \mathbf{Q}) \oplus (B_{2j-1} \otimes \mathbf{Q}) \\ &\xrightarrow{(\sum_j \phi_{2j}^{-1} + \phi_{2j-1})} \oplus_j (B_{2j} \otimes \mathbf{Q}) \oplus (Z_{2j-1} \otimes \mathbf{Q}) \\ &\xrightarrow{(\sum_j \eta_{2j} + \psi_{2j-1})} \oplus_j F_{2j-1} \otimes \mathbf{Q}. \end{aligned}$$

Example 7.1.2 Suppose that $H_i(F_*) = 0$ except for $H_1(F_*)$ and $H_0(F_*)$. Then

$$\phi_i : B_i \xrightarrow{\cong} Z_i$$

for $i \geq 2$ and we have

$$0 \longrightarrow B_1 \xrightarrow{\phi_1} Z_1 \longrightarrow H_1(F_*) \longrightarrow 0$$

and

$$0 \longrightarrow d_1(F_1) = B_0 \xrightarrow{\phi_0} Z_0 = F_0 \longrightarrow H_0(F_*) \longrightarrow 0.$$

Lemma 7.1.3

Suppose, as in Example 7.1.2, that $H_i(F_*) = 0$ except for $H_1(F_*)$ and $H_0(F_*)$. Then the $\mathbf{Z}[G]$-module, B_m, is finitely generated and projective for all $m \geq 1$.

Proof. Since each F_t is torsion free and finitely generated so is the submodule, B_t. Therefore, by a result of Swan, it suffices to show that the Tate cohomology groups, $\hat{H}^i(J; B_t)$ vanish for all $t \geq 1$, $J \subseteq G$ and all i. For $m \geq 1$ we have short exact sequences of the form

$$0 \longrightarrow Z_{m+1} = B_{m+1} \xrightarrow{\psi_{m+1}} F_{m+1} \xrightarrow{d_{m+1}} B_m \longrightarrow 0$$

together with an isomorphism

$$d_k : F_k \xrightarrow{\cong} B_{k-1}.$$

By downward induction, using the long exact Tate cohomology sequences, we see that $B_{k-1}, B_{k-2}, \ldots, B_2, B_1$ are all cohomologically trivial. □

7.1.4 Now let us specialise to the case when G is abelian. In this case we are going to modify the isomorphism X to make it integral.

Returning to Example 7.1.2, Lemma 7.1.3 implies that, for $i \geq 1$, the short exact sequence

$$0 \longrightarrow B_{i+1} \stackrel{\psi_{i+1}}{\longrightarrow} F_{i+1} \stackrel{d_{i+1}}{\longrightarrow} B_i \longrightarrow 0$$

splits. Therefore we have

$$\eta_i : B_i \longrightarrow F_{i+1}$$

for $i \geq 1$ and

$$\phi_i : B_i \stackrel{=}{\longrightarrow} Z_i$$

for $i \geq 2$ without tensoring with the rationals.

Since $H_0(F_*)$ is finite the inclusion map, $\phi_0 : B_0 \longrightarrow Z_0 = F_0$ gives a rational isomorphism

$$\phi_0 \otimes 1 : B_0 \otimes \mathbf{Q} \stackrel{\cong}{\longrightarrow} Z_0 \otimes \mathbf{Q} = F_0 \otimes \mathbf{Q}$$

characterised by $(\phi_0 \otimes 1)(x \otimes a) = x \otimes a$ for all $x \in B_0, a \in \mathbf{Q}$. We also have $\eta_0 : B_0 \otimes \mathbf{Q} \longrightarrow F_1 \otimes \mathbf{Q}$ such that $(d_1 \otimes 1) \cdot \eta_0 = 1$ on $B_0 \otimes \mathbf{Q}$. Define

$$\eta = \eta_0 \cdot (\phi_0 \otimes 1)^{-1} : F_0 \otimes \mathbf{Q} \longrightarrow F_1 \otimes \mathbf{Q}.$$

Hence defining η is equivalent to defining η_0. Also

$$(d_1 \otimes 1) \cdot \eta = 1 : F_0 \otimes \mathbf{Q} \longrightarrow F_0 \otimes \mathbf{Q}$$

because, if $x \in F_0$ and $nx \in B_0$, then

$$\begin{aligned}(d_1 \otimes 1) \cdot \eta(x \otimes a) &= (d_1 \otimes 1) \cdot \eta(nx \otimes an^{-1}) \\ &= (d_1 \otimes 1) \cdot \eta \cdot (\phi_0 \otimes 1)(nx \otimes an^{-1}) \\ &= (d_1 \otimes 1) \cdot \eta_0(nx \otimes an^{-1}) \\ &= nx \otimes an^{-1} \\ &= x \otimes a.\end{aligned}$$

By definition, $X(F_0 \otimes \mathbf{Q}) \subseteq F_1 \otimes \mathbf{Q} \subseteq \oplus_j F_{2j-1} \otimes \mathbf{Q}$ and X restricted to $F_0 \otimes \mathbf{Q}$ is equal to η.

If $t \in \mathbf{Z}[G]$ and $tH_0(F_*) = 0$ we want $\eta(tF_0) \subseteq F_1$. If t is a unit in $\mathbf{Q}[G]$ this is easily accomplished. For each free generator, $z \in F_0$, we have $tz = d_1(w)$ and so we may define $\eta(z) = t^{-1}w$. Since G is abelian, multiplication by t is a $\mathbf{Z}[G]$-module homomorphism and in this case the map

$$\tilde{T} = (t \cdot -) : F_0 \longrightarrow F_0$$

would have determinant equal to

$$\det((t \cdot -) : \mathbf{Z}[G] \longrightarrow \mathbf{Z}[G])^d \quad \text{where} \quad d = \operatorname{rank}_{\mathbf{Z}[G]} F_0.$$

We may form the composition

$$X \cdot (\tilde{T} \oplus 1 \oplus 1 \oplus \ldots) : \oplus_j F_{2j} \otimes \mathbf{Q} \xrightarrow{\cong} \oplus_j F_{2j+1} \otimes \mathbf{Q}.$$

Out next result will show that $X \cdot (\tilde{T} \oplus 1 \oplus 1 \oplus \ldots)$ maps $\oplus_j F_{2j}$ to $\oplus_j F_{2j+1}$.

It will be useful to have the explicit formula for X on $(w_0, w_2, \ldots) \in \oplus_j F_{2j}$, namely

$$\begin{aligned} &X(w_0, w_2, \ldots) \\ &= (\eta(w_0) + d_2(w_2), \eta_2(w_2 - \eta_1(d_2(w_2))) + d_4(w_4), \ldots \\ &\qquad \eta_{2t}(w_{2t} - \eta_{2t-1}(d_{2t}(w_{2t}))) + d_{2t+2}(w_{2t+2}), \ldots). \end{aligned}$$

Lemma 7.1.5

Let G be any finite group. Let F_* be the perfect complex of §7.1.1 and suppose that $H_0(F_*)$ and $H_1(F_*)$ are its only non-zero homology groups. Suppose that we have a $\mathbf{Z}[G]$-module homomorphism

$$\tilde{T} : F_0 \longrightarrow F_0$$

which is a rational isomorphism and such that $\eta(\tilde{T}(F_0)) \subseteq F_1 \subseteq F_1 \otimes \mathbf{Q}$, where η is as in §7.1.4.

Then the map, $X \cdot (\tilde{T} \oplus 1 \oplus \ldots \oplus 1)$ restricts to yield an injective $\mathbf{Z}[G]$-module homomorphism of the form

$$X \cdot (\tilde{T} \oplus 1 \oplus \ldots \oplus 1) : \oplus_j F_{2j} \longrightarrow \oplus_j F_{2j-1}.$$

Proof. The injectivity follows from the fact that X and $\tilde{T}$ are rational isomorphisms. For the rest it suffices to verify that $X \cdot (\tilde{T} \oplus 1 \oplus \ldots \oplus 1)(F_{2s}) \subseteq \oplus_j F_{2j-1}$ for each $s \geq 0$. If $s = 0$ this is assured by the previous observation that X equals η on F_0. For $s \geq 1$, $\psi_{2j} + \eta_{2j-1}$ is an isomorphism between $Z_{2j} \oplus B_{2j-1}$ and F_{2j}. Then ϕ_{2j} is the identity isomorphism between B_{2j} and Z_{2j} while ϕ_{2j-1} is the inclusion of B_{2j-1} into Z_{2j-1}. Finally η_{2j} maps B_{2j} to F_{2j} while ψ_{2j-1} maps Z_{2j-1} to F_{2j-1} by the inclusion map. From these observations one readily sees that the result holds for F_{2s} when $s \geq 1$. □

Lemma 7.1.6

In the situation of Lemma 7.1.5 the kernel of the canonical homomorphism

$$\lambda : \mathrm{Ker}(d_1) \subseteq F_1 \subseteq \oplus_j F_{2j+1} \longrightarrow \frac{\oplus_j F_{2j+1}}{X \cdot (\tilde{T} \oplus 1 \oplus \ldots)(\oplus_j F_{2j})}$$

is contained in $d_2(F_2)$.

Proof. From the formula of §7.1.4 we have

$$\begin{aligned} &X \cdot (\tilde{T} \oplus 1 \oplus \ldots)(w_0, w_2, \ldots) \\ &= (\eta(\tilde{T}(w_0)) + d_2(w_2), \eta_2(w_2 - \eta_1(d_2(w_2))) + d_4(w_4), \\ &\qquad \eta_4(w_4 - \eta_3(d_4(w_4))) + d_6(w_6), \ldots). \end{aligned}$$

Suppose that $(w_1, 0, 0, \ldots) = X \cdot (\tilde{T} \oplus 1 \oplus \ldots)(w_0, w_2, \ldots)$ with $w_1 \in \mathrm{Ker}(d_1) \subseteq F_1$. This implies that $w_1 = \eta(\tilde{T}(w_0)) + d_2(w_2)$. Applying d_1 we obtain

$$\begin{aligned} 0 &= d_1(w_1) \\ &= d_1(\eta(\tilde{T}(w_0)) + d_2(w_2)) \\ &= d_1(\eta(\tilde{T}(w_0))) \\ &= \tilde{T}(w_0). \end{aligned}$$

Hence $w_1 = d_2(w_2) \in d_2(F_2)$, as required. □

7.1.7 Now suppose that G is abelian. Then we may take determinants. Let p be a prime. Although the F_i's are not free $\mathbf{Z}[G]$-modules, we may choose a $\mathbf{Z}_p[G]$-basis for the $F_i \otimes \mathbf{Z}_p$'s and write the map, $X \cdot (\tilde{T} \oplus 1 \oplus \ldots)$, as a square matrix with entries in $\mathbf{Z}_p[G]$. Similarly we may form the determinant

$$\det(X \cdot (\tilde{T} \oplus 1 \oplus \ldots)) = \det(X)\det(\tilde{T}) \in \mathbf{Z}_p[G] \cap \mathbf{Q}_p[G]^*.$$

where

$$\det(X), \det(\tilde{T}) = \det(\tilde{T} \oplus 1 \oplus \ldots) \in \mathbf{Q}_p[G]^*.$$

As explained in §2.5.34, the Hom-description representative for the element, $[\oplus_j F_{2j}, X, \oplus_j F_{2j+1}] \in K_0(\mathbf{Z}[G], \mathbf{Q})$, is the idèlic-valued function whose p-component sends a one-dimensional representation, χ, to $\chi(\det(X))$. This representing function is unique up to multiplication by functions whose p-component has the form $(\chi \mapsto \chi(\alpha_p))$ for some $\alpha_p \in \mathbf{Z}_p[G]^*$.

This proves the following result.

Proposition 7.1.8

let G be a finite abelian group and p a prime. Then, in §7.1.7, the element

$$\det(X) \in \frac{\mathbf{Q}_p[G]^*}{\mathbf{Z}_p[G]^*}$$

depends only on the class $[\oplus_j F_{2j}, X, \oplus_j F_{2j+1}] \in K_0(\mathbf{Z}[G], \mathbf{Q})$. In fact, it depends only on the p-component of

$$[\oplus_j F_{2j}, X, \oplus_j F_{2j+1}] \quad in \quad K_0(\mathbf{Z}[G], \mathbf{Q}) \cong \oplus_{p \text{ prime}} K_0(\mathbf{Z}_p[G], \mathbf{Q}_p).$$

If R is a ring and M a (left) R-module, write $\mathrm{ann}_R(M)$ for the annihilator (left) ideal $\mathrm{ann}_R(M) = \{r \in R \mid r \cdot m = 0 \ for\ all\ m \in M\}$.

Corollary 7.1.9

As in §7.1.7, let G be a finite abelian group and p a prime. Then

$$\det(X)\det(\tilde{T}) \in \mathrm{ann}_{\mathbf{Z}_p[G]}(H_1(F_*) \otimes \mathbf{Z}_p) \lhd \mathbf{Z}_p[G].$$

Proof. It is well known that

$$\det(X \cdot (\tilde{T} \oplus 1 \oplus \ldots)) \in \mathrm{ann}_{\mathbf{Z}_p[G]}(\frac{\oplus_j F_{2j+1} \otimes \mathbf{Z}_p}{X \cdot (\tilde{T} \oplus 1 \oplus \ldots)(\oplus_j F_{2j} \otimes \mathbf{Z}_p)}).$$

This is because, if we choose bases and think of $X \cdot (\tilde{T} \oplus 1 \oplus \ldots)$ in terms of multiplication by its matrix,

$$(Z \cdot -) : \mathbf{Z}[G]^t \longrightarrow \mathbf{Z}[G]^t$$

then multiplication by the scalar, $\det(X \cdot (\tilde{T} \oplus 1 \oplus \ldots))$, is equal to the composition $(Z \cdot -) \cdot (adj(Z) \cdot -)$, where $adj(Z)$ is the adjugate matrix of Z. Hence for $w \in \mathbf{Z}[G]^t$ we have $\det(X \cdot (\tilde{T} \oplus 1 \oplus \ldots))w = Z \cdot (adj(Z) \cdot w)) \in Im(Z \cdot -)$.

Therefore $\det(X)\det(\tilde{T}) = \det(X \cdot (\tilde{T} \oplus 1 \oplus \ldots))$ belongs to the annihilator

$$\mathrm{ann}_{\mathbf{Z}_p[G]}(\frac{\mathrm{Ker}(d_1) \otimes \mathbf{Z}_p}{\mathrm{Ker}(d_1) \otimes \mathbf{Z}_p \bigcap (X \cdot (\tilde{T} \oplus 1 \oplus \ldots)(\oplus_j F_{2j} \otimes \mathbf{Z}_p)}) = \frac{\mathrm{Ker}(d_1)}{\mathrm{Ker}(\lambda)} \otimes \mathbf{Z}_p.$$

Since tensoring with $\mathbf{Z}_p$ is exact, Lemma 7.1.6 implies that this last $\mathbf{Z}_p[G]$-module has $H^1(F_*) \otimes \mathbf{Z}_p = (\mathrm{Ker}(d_1) \otimes \mathbf{Z}_p)/d_2(F_2 \otimes \mathbf{Z}_p)$ as a quotient, which implies the result. □

7.1.10 *Dualisation*

Recall from Proposition 2.5.35 that the class,

$$[\oplus_j F_{2j}, X, \oplus_j F_{2j+1}] \in K_0(\mathbf{Z}[G], \mathbf{Q}),$$

is independent of the choices of rational spittings used to construct X and depends only on the quasi-isomorphism class of the complex, F_*. Hence we may replace F_* by the quasi-isomorphic truncated complex

$$0 \longrightarrow B_1 \longrightarrow F_1 \longrightarrow F_0 \longrightarrow 0$$

which is perfect by Lemma 7.1.3.

This may also be seen directly. If we choose the same rational splittings as in §7.1.4, we obtain two isomorphisms

$$Y : F_0 \otimes \mathbf{Q} \oplus B_1 \otimes \mathbf{Q} \xrightarrow{\cong} F_0 \otimes \mathbf{Q}$$

and

$$Z : B_1 \otimes \mathbf{Q} \oplus F_3 \otimes \mathbf{Q} \oplus F_5 \otimes \mathbf{Q} \oplus \ldots \xrightarrow{\cong} F_2 \otimes \mathbf{Q} \oplus F_4 \otimes \mathbf{Q} \oplus \ldots$$

and $X = (Y \oplus 1 \oplus 1 \oplus \ldots)(1 \oplus Z^{-1})$.

Hence for each prime p, $\det(X) = \det(Y)\det(Z)^{-1} \in \mathbf{Q}_p[G]^*$. However, the splittings used to define Z are all integral and so $\det(Z) \in \mathbf{Z}_p[G]^*$. Therefore

$$\det(X) = \det(Y) \in \frac{\mathbf{Q}_p[G]^*}{\mathbf{Z}_p[G]^*}$$

and so we may replace F_* by the three-term complex, as claimed.

By the Hom-description of §2.5.34, $[\oplus_j F_{2j}, X, \oplus_j F_{2j+1}] = [F_0 \oplus B_1, Y, F_1]$, providing that Y is defined using the same splitting, $\eta : F_0 \longrightarrow F_1$, as X.

Now set $F^0 = \mathrm{Hom}(F_0, \mathbf{Z})$, $F^1 = \mathrm{Hom}(F_1, \mathbf{Z})$ and $F^2 = \mathrm{Hom}(B_1, \mathbf{Z})$ and consider the cochain complex

$$0 \longrightarrow F^0 \xrightarrow{d_1^*} F^1 \longrightarrow F^2 \longrightarrow 0.$$

From the universal coefficient formula the cohomology groups of this cochain complex vanish except for

$$H^1(F^*) \cong \mathrm{Ext}^1(H_0(F_*), \mathbf{Z}) \cong \mathrm{Hom}(H_0(F_*), \mathbf{Q}/\mathbf{Z})$$

and

$$H^2(F^*) \cong \mathrm{Ext}^1(H_1(F_*), \mathbf{Z}) \cong \mathrm{Hom}(H_1(F_*), \mathbf{Q}/\mathbf{Z}).$$

Theorem 7.1.11 [1]

Let G be a finite abelian group and p a prime. Suppose that

$$0 \longrightarrow F_k \xrightarrow{d_k} F_{k-1} \xrightarrow{d_{k-1}} \ldots \xrightarrow{d_2} F_1 \xrightarrow{d_1} F_0 \longrightarrow 0$$

is a perfect complex $\mathbf{Z}[G]$-modules, as in Definition 2.2.1, having $H_i(F_*)$ finite for $i = 0, 1$ and zero otherwise. Let $[\oplus_j F_{2j}, X, \oplus_j F_{2j+1}] \in K_0(\mathbf{Z}[G], \mathbf{Q})$. be as in Proposition 7.1.8.

Then, if $t_i \in \mathbf{Q}_p[G]^* \bigcap \mathrm{ann}_{\mathbf{Z}_p[G]}(H_i(F_*) \otimes \mathbf{Z}_p)$,

$$\det(X)^{(-1)^i} t_i^{m_{p,i}} \in \mathrm{ann}_{\mathbf{Z}_p[G]}(H_{j(i)}(F_*) \otimes \mathbf{Z}_p) \lhd \mathbf{Z}_p[G]$$

for $i = 0, 1$. Here $m_{p,i}$ is the minimal number of generators required for the $\mathbf{Z}_p[G]$-module $H_i(F_*) \otimes \mathbf{Z}_p$ and $j(1) = 0, j(0) = 1$.

In fact, this annihilator relation remains true for all $t_i \in \mathrm{ann}_{\mathbf{Z}_p[G]}(H_i(F_*) \otimes \mathbf{Z}_p)$.

Proof. If $t_i \in \mathrm{ann}_{\mathbf{Z}_p[G]}(H_i(F_*) \otimes \mathbf{Z}_p)$ but is not a unit in $\mathbf{Q}_p[G]^*$ we may find a large multiple, n, of $|H_0(F_*)| \cdot |H_1(F_*)|$ so that $n + t_i \in \mathbf{Q}_p[G]^*$. If n is larger enough to ensure integrality the result for $t_i^{m_{p,i}}$ follows from the that for $(n + t_i)^{m_{p,i}}$, by the binomial theorem. Henceforth we assume that $t_i \in \mathbf{Q}_p[G]^*$.

The part of the statement concerning $t_0 \in \mathbf{Q}_p[G]^* \bigcap \mathrm{ann}_{\mathbf{Z}_p[G]}(H_i(F_*) \otimes \mathbf{Z}_p)$, but with rank F_0 instead of $m_{p,0}$, would follow from Proposition 7.1.9 when we set $\tilde{T}$ equal to multiplication by t_0 on $F_0 \otimes \mathbf{Z}_p$, as in §7.1.4 (after tensoring with $\mathbf{Z}_p$). To obtain the better estimate, by the p-adic version of Proposition 2.5.35, we may replace $F_* \otimes \mathbf{Z}_p$ by a quasi-isomorphic bounded complex of $\mathbf{Z}_p[G]$-modules, P_* with $P_j = 0$ for $j < 0$ and $\mathrm{rank}_{\mathbf{Z}_p[G]}(P_0) = m_{p,0}$.

We shall obtain the statement concerning t_1 by dualisation.

Firstly, by the discussion of §7.1.10, we may replace F_* by the perfect complex

$$0 \longrightarrow B_1 \longrightarrow F_1 \longrightarrow F_0 \longrightarrow 0.$$

which is perfect by Lemma 7.1.3.

[1] It can be shown, in fact, that $\det(X)^{(1)^i} t_i^{m_{p,i}}$ lies in the fitting ideal of $H_{j(i)}(F_*) \otimes \mathbf{Z}_p$.

As in §7.1.10, set $F^0 = \mathrm{Hom}(F_0, \mathbf{Z})$, $F^1 = \mathrm{Hom}(F_1, \mathbf{Z})$ and $F^2 = \mathrm{Hom}(B_1, \mathbf{Z})$ and consider the cochain complex

$$0 \longrightarrow F^0 \xrightarrow{d_1^*} F^1 \longrightarrow F^2 \longrightarrow 0.$$

From the universal coefficient formula the cohomology groups of this cochain complex vanish except for

$$H^1(F^*) \cong \mathrm{Hom}(H_0(F_*), \mathbf{Q}/\mathbf{Z}) \qquad \text{and} \qquad H^2(F^*) \cong \mathrm{Hom}(H_1(F_*), \mathbf{Q}/\mathbf{Z}).$$

In order to apply the first part to this perfect complex we first observe that, after tensoring with $\mathbf{Q}_p$, the matrix of the dual of Y, in §7.1.10, with respect to the dual $\mathbf{Z}_p[G]$-basis is equal to $\tau(Y^{tr})$ where τ is the automorphism of the group-ring induced by sending each element of G to its inverse, $\tau(\sum_{g\in G} n_g \cdot g) = \sum_{g\in G} n_g \cdot g^{-1}$, and Y^{tr} is the transpose of Y.

Hence, if $t \in \mathbf{Q}_p[G]^* \bigcap \mathrm{ann}_{\mathbf{Z}_p[G]}(H^2(F_*) \otimes \mathbf{Z}_p)$,

$$\det(\tau(Y^{tr}))^{-1} t^{m_{p,1}} \in \mathrm{ann}_{\mathbf{Z}_p[G]}(H^1(F_*) \otimes \mathbf{Z}_p) \lhd \mathbf{Z}_p[G].$$

Of course, using F^*, we would only get this relation with $m_{p,1}$ replaced by $\mathrm{rank}_{\mathbf{Z}_p[G]}(F^2)$ but this can be improved to $m_{p,1}$ as we did for $m_{p,0}$.

However, the action of G on $\mathrm{Hom}(H_i(F_*), \mathbf{Q}/\mathbf{Z})$ is given by $g(f) = f(g^{-1} \cdot -)$ which may identified with G acting on $H_i(F_*)$ with the new action, $g(z) = \tau(g) \cdot z$. This means that

$$\tau(\mathrm{ann}_{\mathbf{Z}_p[G]}(H^i(F_*) \otimes \mathbf{Z}_p)) = \mathrm{ann}_{\mathbf{Z}_p[G]}(H_i(F_*) \otimes \mathbf{Z}_p).$$

Since $\det(\tau(Y^{tr})) = \det(\tau(Y))$ differs from $\tau(\det(X))$ by a unit in the p-adic group ring, we may apply τ to the previous relation to obtain

$$\det(X)^{-1} t_1^{m_{p,1}} \in \mathrm{ann}_{\mathbf{Z}_p[G]}(H_1(F_*) \otimes \mathbf{Z}_p) \lhd \mathbf{Z}_p[G]$$

where $t_1 = \tau(t)$. □

7.1.12 *Incorporating a regulator*

Suppose now that we have a chain complex of finitely generated, projective $\mathbf{Z}[G]$-modules

$$0 \longrightarrow F_k \xrightarrow{d_k} F_{k-1} \xrightarrow{d_{k-1}} \ldots \xrightarrow{d_2} F_1 \xrightarrow{d_1} F_0 \longrightarrow 0$$

for which all homology groups are trivial except for $H_0(F_*)$ and $H_1(F_*)$. Of course, these are finitely generated $\mathbf{Z}[G]$-modules and now we shall study the case where they are *not finite*. In fact, we shall suppose that we have a $\mathbf{Z}[G]$-module homomorphism

$$R : H_0(F_*) \longrightarrow H_1(F_*)$$

such that

$$R \otimes 1 : H_0(F_*) \otimes \mathbf{Q} \longrightarrow H_1(F_*) \otimes \mathbf{Q}$$

is an isomorphism. This isomorphism will be referred to as a *regulator*.

Recall that we have a family of $\mathbf{Z}[G]$-module homomorphisms

$$\phi_i : B_i \xrightarrow{\cong} Z_i$$

when $i \geq 2$ and

$$0 \longrightarrow Z_{i+1} \xrightarrow{\psi_{i+1}} F_{i+1} \xrightarrow{d_{i+1}} B_i \longrightarrow 0$$

which are split short exact for $i \geq 1$ with splittings

$$\eta_i : B_i \longrightarrow F_{i+1}$$

so that $d_{i+1} \cdot \eta_i = 1 : B_i \longrightarrow B_i$ for $i \geq 1$. For the remaining two dimensions we have short exact sequences of the form

$$0 \longrightarrow B_1 \otimes \mathbf{Q} \xrightarrow{\phi_1} Z_1 \otimes \mathbf{Q} \xrightarrow{\pi_1} H_1(F_*) \otimes \mathbf{Q} \longrightarrow 0$$

and

$$0 \longrightarrow B_0 \otimes \mathbf{Q} \xrightarrow{\phi_0} Z_0 \otimes \mathbf{Q} \xrightarrow{\pi_0} H_0(F_*) \otimes \mathbf{Q} \longrightarrow 0$$

together with rational splittings

$$\rho_1 : H_1(F_*) \otimes \mathbf{Q} \longrightarrow Z_1 \otimes \mathbf{Q}$$

and

$$\rho_0 : H_0(F_*) \otimes \mathbf{Q} \longrightarrow Z_0 \otimes \mathbf{Q}$$

such that $\rho_1 \cdot \pi_1 = 1$, $\rho_0 \cdot \pi_0 = 1$.

Then, as in Example 2.1.8(ii), we form

$$X : \oplus_j F_{2j} \otimes \mathbf{Q} \xrightarrow{\cong} \oplus_j F_{2j+1} \otimes \mathbf{Q}$$

given by the composition

$$\begin{array}{l} F_0 \otimes \mathbf{Q} \oplus_{j\geq 1} F_{2j} \otimes \mathbf{Q} \\ \xrightarrow{(\phi_0+\rho_0)^{-1}\oplus(\sum_{j\geq 1} \psi_{2j}+\eta_{2j-1})^{-1}} B_0 \otimes \mathbf{Q} \oplus H_0(F_*) \otimes \mathbf{Q} \\ \qquad \oplus_{j\geq 1}(Z_{2j} \otimes \mathbf{Q}) \oplus (B_{2j-1} \otimes \mathbf{Q}) \\ \xrightarrow{1\oplus R\oplus 1} B_0 \otimes \mathbf{Q} \oplus H_1(F_*) \otimes \mathbf{Q} \oplus_{j\geq 1} (Z_{2j} \otimes \mathbf{Q}) \oplus (B_{2j-1} \otimes \mathbf{Q}) \\ \xrightarrow{1\oplus 1\oplus 1\oplus(\sum_{j\geq 1} \phi_{2j}^{-1}+\phi_{2j+1})} B_0 \otimes \mathbf{Q} \oplus H_1(F_*) \otimes \mathbf{Q} \oplus B_1 \otimes \mathbf{Q} \\ \qquad \oplus_{j\geq 1}(B_{2j} \otimes \mathbf{Q}) \oplus (Z_{2j+1} \otimes \mathbf{Q}) \\ \xrightarrow{1\oplus(\rho_1+\phi_1)\oplus(\sum_{j\geq 1} \psi_{2j+1}+\eta_{2j})} B_0 \otimes \mathbf{Q} \oplus Z_1 \otimes \mathbf{Q} \oplus_{j\geq 1} F_{2j+1} \otimes \mathbf{Q} \\ \xrightarrow{(\psi_1+\eta_0)\oplus 1} F_1 \oplus \mathbf{Q} \oplus_{j\geq 1} F_{2j+1} \otimes \mathbf{Q} = \oplus_j F_{2j+1} \otimes \mathbf{Q}. \end{array}$$

Therefore we obtain a K-theory class

$$[\oplus_j F_{2j}, X, \oplus_j F_{2j+1}] \in K_0(\mathbf{Z}[G], \mathbf{Q}).$$

Now let us calculate X. If $w_0 \in F_0 = Z_0 \subset F_0 \otimes \mathbf{Q}$ is mapped via X to

$$(w_0 - \rho_0(\pi_0(w_0)), \pi_0(w_0)) \in (B_0 \oplus H_0(F_*)) \otimes \mathbf{Q}$$

then to

$$(w_0 - \rho_0(\pi_0(w_0)), R(\pi_0(w_0))) \in (B_0 \oplus H_1(F_*)) \otimes \mathbf{Q}$$

then to

$$(w_0 - \rho_0(\pi_0(w_0)), \rho_1(R(\pi_0(w_0)))) \in (B_0 \oplus Z_1) \otimes \mathbf{Q}$$

and finally to

$$(\eta_0(w_0 - \rho_0(\pi_0(w_0))) + \psi_1(\rho_1(R(\pi_0(w_0)))) \in F_1 \otimes \mathbf{Q}.$$

For $j \geq 1$ and $w_{2j} \in F_{2j}$ – no tensoring with the rationals – we have w_{2j} sent to

$$(w_{2j} - \eta_{2j-1}(d_{2j}(w_{2j})), d_{2j}(w_{2j})) \in Z_{2j} \oplus B_{2j-1}$$

then to

$$(\phi_{2j}^{-1}(w_{2j} - \eta_{2j-1}(d_{2j}(w_{2j}))), \phi_{2j-1}(d_{2j}(w_{2j}))) \in B_{2j} \oplus Z_{2j-1}$$

then to

$$(d_{2j}(w_{2j}), \eta_{2j}(\phi_{2j}^{-1}(w_{2j} - \eta_{2j-1}(d_{2j}(w_{2j}))))) \in F_{2j-1} \oplus F_{2j+1}.$$

Next, since F_0 is projective, there exists a $\mathbf{Z}[G]$-module homomorphism

$$\tilde{R} : F_0 \longrightarrow Z_1$$

such that $\pi_1 \tilde{R} = R\pi_0$.

Notice that, if $w_0 \in F_0 \otimes \mathbf{Q}$, then

$$\tilde{R}(w_0) - \psi_1(\rho_1(R(\pi_0(w_0)))) \in B_1 \otimes \mathbf{Q}$$

because this element belongs to $Z_1 \otimes \mathbf{Q}$, by definition, but in $H_1(F_*) \otimes \mathbf{Q}$

$$\begin{aligned} &\pi_1(\tilde{R}(w_0)) - \pi_1(\psi_1(\rho_1(R(\pi_0(w_0))))) \\ &= R(\pi_0(w_0)) - R(\pi_0(w_0)) \\ &= 0. \end{aligned}$$

Now consider the analogue of Lemma 7.1.6 in the case with non-trivial regulator. Suppose that we have a $\mathbf{Z}[G]$-module homomorphism

$$\tilde{T} : F_0 \longrightarrow F_0$$

such that the image of

$$X \cdot (\tilde{T} \oplus 1 \oplus \ldots \oplus 1) : \oplus_j F_{2j} \longrightarrow \oplus_j F_{2j-1} \otimes \mathbf{Q}$$

lands in $\oplus_j F_{2j-1}$. Hence we have a $\mathbf{Z}[G]$-module homomorphism of the form

$$X \cdot (\tilde{T} \oplus 1 \oplus \ldots \oplus 1) : \oplus_j F_{2j} \longrightarrow \oplus_j F_{2j-1}.$$

Suppose that $w_1 \in \operatorname{Ker}(d_1)$ and that

$$(w_1, 0, 0, \ldots, 0) = X \cdot (\tilde{T} \oplus 1 \oplus \ldots \oplus 1)(w_0, w_2, w_4, \ldots).$$

Then

$$w_1 = \eta_0(\tilde{T}(w_0) - \rho_0(\pi_0(\tilde{T}(w_0)))) + \psi_1(\rho_1(R(\pi_0(\tilde{T}(w_0))))) + d_2(w_2)$$

and applying d_1 yields

$$0 = \tilde{T}(w_0) - \rho_0(\pi_0(\tilde{T}(w_0)))$$

since $Im(\psi_1) \subseteq \operatorname{Ker}(d_1)$. Therefore

$$w_1 = \psi_1(\rho_1(R(\pi_0(\tilde{T}(w_0))))) + d_2(w_2) \in F_1 \otimes \mathbf{Q}$$

or, equivalently,

$$w_1 = \rho_1(R(\pi_0(\tilde{T}(w_0)))) + d_2(w_2) \in Z_1 \otimes \mathbf{Q}.$$

If the homology class $[w_1] = [w_1 + d_2(w_2)] \in H_1(F_*)$ has finite order then

$$0 = \pi_1(w_1) = \pi_1(w_1 + d_2(w_2)) = \pi_1(\rho_1(R(\pi_0(\tilde{T}(w_0))))) \in H_1(F_*) \otimes \mathbf{Q}$$

which implies that

$$0 = R(\pi_0(\tilde{T}(w_0))) \in H_1(F_*) \otimes \mathbf{Q}$$

so that

$$w_1 = d_2(w_2) \in Z_1 \otimes \mathbf{Q}.$$

Hence we have proved the following analogue of Lemma 7.1.6:

Lemma 7.1.13

In the circumstances of §7.1.12, let $T\operatorname{Ker}(d_1) \subseteq \operatorname{Ker}(d_1)$ denote the $\mathbf{Z}[G]$-submodule consisting of those elements whose homology classes have finite order in $H_1(F_*)$. Then the kernel of the canonical homomorphism

$$\lambda : T\operatorname{Ker}(d_1) \subseteq \operatorname{Ker}(d_1) \subseteq F_1 \subseteq \oplus_j F_{2j+1} \longrightarrow \frac{\oplus_j F_{2j+1}}{X \cdot \tilde{T}(\oplus_j F_{2j})}$$

is contained in $d_2(F_2)$.

7.1.14 *Finding $\tilde{T}$ in the regulator case*

If $w_{2i} \in F_{2i}$ then we have

$$\begin{aligned}
&X(w_0, w_2, \ldots) \\
&= (\eta_0(w_0 - \rho_0(\pi_0(w_0))) + \psi_1(\rho_1(R(\pi_0(w_0)))) + d_2(w_2), \ldots, \\
&\qquad , \ldots, \eta_{2j}(\phi_{2j}^{-1}(w_{2j} - \eta_{2j-1}(d_{2j}(w_{2j})))) + d_{2j+2}(w_{2j+2}), \ldots)
\end{aligned}$$

and our first step will be to replace X by $\hat{X}$ given by

$$\begin{aligned}&\hat{X}(w_0, w_2, \ldots)\\&= (\eta_0(w_0 - \rho_0(\pi_0(w_0))) + \tilde{R}(w_0) + d_2(w_2), \ldots,\\&\qquad ,\ldots, \eta_{2j}(\phi_{2j}^{-1}(w_{2j} - \eta_{2j-1}(d_{2j}(w_{2j})))) + d_{2j+2}(w_{2j+2}), \ldots).\end{aligned}$$

Lemma 7.1.15

In §7.1.14

$$\hat{X} : \oplus_j\, F_{2j} \otimes \mathbf{Q} \longrightarrow \oplus_j\, F_{2j+1} \otimes \mathbf{Q}$$

is an isomorphism.

Proof. We have only to show that $\hat{X}$ is one-one. If $(w_0, w_2, \ldots) \in \mathrm{Ker}(\hat{X})$ then the F_1-coordinate gives

$$0 = \eta_0(w_0 - \rho_0(\pi_0(w_0))) + \tilde{R}(w_0) + d_2(w_2) \in F_1 \otimes \mathbf{Q}.$$

Applying d_1 we obtain

$$0 = w_0 - \rho_0(\pi_0(w_0)) + d_1(\tilde{R}(w_0)) = w_0 - \rho_0(\pi_0(w_0)).$$

Therefore

$$0 = \tilde{R}(w_0) + d_2(w_2) \in F_1 \otimes \mathbf{Q}.$$

Applying π_1 we obtain

$$0 = \pi_1(\tilde{R}(w_0)) = R(\pi_0(w_0)) \in H_1(F_*) \otimes \mathbf{Q}.$$

Therefore

$$\pi_0(w_0) = 0 \in H_0(F_*) \otimes \mathbf{Q}$$

because R is an isomorphism. Since $w_0 - \rho_0(\pi_0(w_0)) = 0$ we find that $w_0 = 0$ and

$$0 = \hat{X}(0, w_2, \ldots) = X(0, w_2, \ldots)$$

which implies that $0 = w_2 = \ldots$, since X is an isomorphism. □

Proposition 7.1.16

Let $t \in \mathrm{ann}(\mathrm{TorsH}_0(F_*))$ and let $\tilde{T} : F_0 \longrightarrow F_0$ be multiplication by t. Then

$$\hat{X}(\tilde{T} \oplus 1 \oplus 1 \oplus \ldots) : \oplus_j\, F_{2j} \longrightarrow \oplus_j\, F_{2j+1} \otimes \mathbf{Q}$$

satisfies

$$\hat{X}(\tilde{T} \oplus 1 \oplus 1 \oplus \ldots)(\oplus_j\, F_{2j}) \subseteq \oplus_j\, F_{2j+1}.$$

Proof. We may arrange that η_0 maps tx into F_1 for all $x \in F_0$ whose class in $H_0(F_*)$ is torsion.

If $(w_0, w_2, \ldots) \in \oplus_j\, F_{2j}$ then all coordinates in $\hat{X}(tw_0, w_2, \ldots) \in \oplus_j\, F_{2j+1} \otimes \mathbf{Q}$ are integral, except possibly the F_1-coordinate which is given by

$$\eta_0(tw_0 - \rho_0(\pi_0(tw_0))) + \tilde{R}(tw_0) + d_2(w_2).$$

This element lies in F_1 if and only if $\eta_0(tw_0 - \rho_0(\pi_0(tw_0)))$ does. However, $w_0 - \rho_0(\pi_0(w_0)) \in F_0$ maps to a torsion class in $H_0(F_*)$ so that $\eta_0(tw_0 - \rho_0(\pi_0(tw_0))) \in F_1$, as required. □

Corollary 7.1.17
In §7.1.14

$$\mathrm{ann}(\mathrm{TorsH}_0(F_*))^{\mathrm{rank}(F_0)}\det(\hat{X}) \subseteq \mathrm{ann}(\mathrm{TorsH}_1(F_*)).$$

7.1.18 *The dualisation of the regulator case*

We now attempt the dualisation (in the sense of §7.1.10) of the non-trivial regulator case of §7.1.12.

Suppose, as in §7.1.12, that we have a chain complex of finitely generated, projective $\mathbf{Z}[G]$-modules

$$0 \longrightarrow F_k \xrightarrow{d_k} F_{k-1} \xrightarrow{d_{k-1}} \dots \xrightarrow{d_2} F_1 \xrightarrow{d_1} F_0 \longrightarrow 0$$

for which all homology groups are trivial except for $H_0(F_*)$ and $H_1(F_*)$. Of course, these are finitely generated $\mathbf{Z}[G]$-modules. Suppose that we have a non-trivial $\mathbf{Z}[G]$-module homomorphism

$$R : H_0(F_*) \longrightarrow H_1(F_*)$$

such that

$$R \otimes 1 : H_0(F_*) \otimes \mathbf{Q} \longrightarrow H_1(F_*) \otimes \mathbf{Q}$$

is an isomorphism – the *regulator*.

Write $F^i = \mathrm{Hom}(F_i, \mathbf{Z})$ with the left $\mathbf{Z}[G]$-module structure given by

$$g(f)(z) = f(g^{-1}(z)).$$

Then we may form the dual cochain complex

$$0 \longrightarrow F^0 \xrightarrow{d_1^*} F^1 \xrightarrow{d_2^*} \dots \xrightarrow{d_{k-1}^*} F^{k-1} \xrightarrow{d_k^*} F^k \longrightarrow 0.$$

From the universal coefficient formula ([145] p. 243) the cohomology groups of this cochain fit into short exact sequences of $\mathbf{Z}[G]$-modules of the form

$$0 \longrightarrow \mathrm{Ext}^1(H_{q-1}(F_*), \mathbf{Z}) \longrightarrow H^q(F^*) \longrightarrow \mathrm{Hom}(H_q(F_*), \mathbf{Z}) \longrightarrow 0.$$

Since $H_i(F_*) = 0$ for $i \neq 0, 1$ we have

$$H^0(F^*) \cong \mathrm{Hom}(H_0(F_*), \mathbf{Z}),$$

$$0 \longrightarrow \mathrm{Ext}^1(\mathrm{TorsH}_0(F_*), \mathbf{Z}) \longrightarrow H^1(F^*) \longrightarrow \mathrm{Hom}(H_1(F_*), \mathbf{Z}) \longrightarrow 0,$$

$$\mathrm{Ext}^1(\mathrm{TorsH}_1(F_*), \mathbf{Z}) \cong H^2(F^*)$$

$0 = H^q(F^*)$ for $q \geq 3$. Also there is an isomorphism of $\mathbf{Z}[G]$-modules of the form

$$\mathrm{Ext}^1(\mathrm{TorsH}_0(F_*), \mathbf{Z}) \cong \mathrm{Hom}(\mathrm{TorsH}_0(F_*), \mathbf{Q}/\mathbf{Z})$$

as mentioned in §7.1.10. In addition, the regulator induces

$$R^* : H^1(F^*) \longrightarrow \mathrm{Hom}(H_1(F_*), \mathbf{Z}) \xrightarrow{R^*} \mathrm{Hom}(H_0(F_*), \mathbf{Z}) \cong H^0(F^*)$$

such that

$$R^* \otimes 1 : H^1(F^*) \otimes \mathbf{Q} \longrightarrow H^0(F^*) \otimes \mathbf{Q}$$

is an isomorphism.

Write $B^i \subseteq Z^i \subseteq F^i$ for the coboundaries and cocycles in F^*. For $i \geq 2$ the short exact sequence

$$0 \longrightarrow Z^i \xrightarrow{\psi_i} F^i \xrightarrow{d^*_{i+1}} B^{i+1} \longrightarrow 0$$

is split by $\eta_{i+1} : B^{i+1} \longrightarrow F^i$ so that $d^*_{i+1} \cdot \eta_{i+1} = 1$. Also, after tensoring with the rationals, we have

$$0 \longrightarrow Z^1 \otimes \mathbf{Q} \xrightarrow{\psi_1} F^1 \otimes \mathbf{Q} \xrightarrow{d^*_2} B^2 \otimes \mathbf{Q} \longrightarrow 0$$

is split by $\eta_2 : B^2 \otimes \mathbf{Q} \longrightarrow F^1 \otimes \mathbf{Q}$ with $d^*_2 \cdot \eta_2 = 1$ and

$$0 \longrightarrow Z^0 \otimes \mathbf{Q} \xrightarrow{\psi_0} F^0 \otimes \mathbf{Q} \xrightarrow{d^*_1} B^1 \otimes \mathbf{Q} \longrightarrow 0$$

is split by $\eta_1 : B^1 \otimes \mathbf{Q} \longrightarrow F^0 \otimes \mathbf{Q}$ with $d^*_1 \cdot \eta_1 = 1$. In addition the short exact sequences

$$0 \longrightarrow B^2 \xrightarrow{\phi_2} Z^2 \xrightarrow{\pi_2} H^2(F^*) \longrightarrow 0$$

and

$$0 \longrightarrow B^1 \xrightarrow{\phi_1} Z^1 \xrightarrow{\pi_1} H^1(F^*) \longrightarrow 0$$

become, upon rationalisation, an isomorphism

$$\phi_2 \otimes 1 : B^2 \otimes \mathbf{Q} \xrightarrow{\cong} Z^2 \otimes \mathbf{Q}$$

and a short exact sequence

$$0 \longrightarrow B^1 \otimes \mathbf{Q} \xrightarrow{\phi_1} Z^1 \otimes \mathbf{Q} \xrightarrow{\pi_1} H^1(F^*) \otimes \mathbf{Q} \longrightarrow 0$$

split by $\rho_1 : H^1(F^*) \otimes \mathbf{Q} \longrightarrow Z^1 \otimes \mathbf{Q}$ so that $\pi_1 \cdot \rho_1 = 1$.

Now let us calculate a formula for the isomorphism

$$X : \oplus_j F^{2j} \otimes \mathbf{Q} \longrightarrow \oplus_j F^{2j+1} \otimes \mathbf{Q}$$

associated to these splittings. We wish to calculate $X(w_0, w_2, w_4, \ldots)$ where $w_{2j} \in F^{2j} \otimes \mathbf{Q}$.

An element $w_0 \in F^0 \otimes \mathbf{Q}$ is mapped first to

$$(w_0 - \eta_1(d^*_1(w_0)), d^*_1(w_0)) \in Z^0 \otimes \mathbf{Q} \oplus B^1 \otimes \mathbf{Q} = H^0(F^*) \otimes \mathbf{Q} \oplus B^1 \otimes \mathbf{Q}$$

and then to

$$((R^*)^{-1}(w_0 - \eta_1(d_1^*(w_0))), d_1^*(w_0)) \in H^1(F^*) \otimes \mathbf{Q} \oplus B^1 \otimes \mathbf{Q}$$

and then to

$$\rho_1((R^*)^{-1}(w_0 - \eta_1(d_1^*(w_0)))) + \phi_1(d_1^*(w_0)) \in Z^1 \otimes \mathbf{Q} \subseteq F^1 \otimes \mathbf{Q}.$$

An element $w_2 \in F^2 \otimes \mathbf{Q}$ is mapped first to

$$(w_2 - \eta_3(d_3^*(w_2)), d_3^*(w_2)) \in Z^2 \otimes \mathbf{Q} \oplus B^3 \otimes \mathbf{Q}$$

then to

$$(w_2 - \eta_3(d_3^*(w_2)), \phi_3(d_3^*(w_2))) \in Z^2 \otimes \mathbf{Q} \oplus Z^3 \otimes \mathbf{Q}$$

and then

$$(\phi_2^{-1}(w_2 - \eta_3(d_3^*(w_2))), \phi_3(d_3^*(w_2))) \in B^2 \otimes \mathbf{Q} \oplus Z^3 \otimes \mathbf{Q}$$

and

$$(\eta_2(\phi_2^{-1}(w_2 - \eta_3(d_3^*(w_2)))), \psi_3(\phi_3(d_3^*(w_2)))) \in F^1 \otimes \mathbf{Q} \oplus F^3 \otimes \mathbf{Q}.$$

If $j \geq 2$ an element $w_{2j} \in F^{2j} \otimes \mathbf{Q}$ is mapped first to

$$(w_{2j} - \eta_{2j+1}(d_{2j+1}^*(w_{2j})), d_{2j+1}^*(w_{2j})) \in Z^{2j} \otimes \mathbf{Q} \oplus B^{2j+1} \otimes \mathbf{Q}$$

then to

$$(w_{2j} - \eta_{2j+1}(d_{2j+1}^*(w_{2j})), \phi_{2j+1}(d_{2j+1}^*(w_{2j}))) \in Z^{2j} \otimes \mathbf{Q} \oplus Z^{2j+1} \otimes \mathbf{Q}$$

and then

$$(\phi_{2j}^{-1}(w_{2j} - \eta_{2j+1}(d_{2j+1}^*(w_{2j}))), \phi_{2j+1}(d_{2j+1}^*(w_{2j}))) \in B^{2j} \otimes \mathbf{Q} \oplus Z^{2j+1} \otimes \mathbf{Q}$$

and finally to $(\eta_{2j}(\phi_{2j}^{-1}(w_{2j} - \eta_{2j+1}(d_{2j+1}^*(w_{2j})))), \psi_{2j+1}(\phi_{2j+1}(d_{2j+1}^*(w_{2j}))))$ in $F^{2j-1} \otimes \mathbf{Q} \oplus F^{2j+1} \otimes \mathbf{Q}$.

Notice that, if $j \geq 2$ and $w_{2j} \in F^{2j}$ then $d_{2j+1}^*(w_{2j}) \in B^{2j+1}$, $w_{2j} - \eta_{2j+1}(d_{2j+1}^*(w_{2j})) \in Z^{2j} = B^{2j}$ and so $\eta_{2j}(\phi_{2j}^{-1}(w_{2j} - \eta_{2j+1}(d_{2j+1}^*(w_{2j})))) \in F^{2j-1}$. Also, if $j \geq 2$, $\psi_{2j+1}(\phi_{2j+1}(d_{2j+1}^*(w_{2j}))) = d_{2j+1}^*(w_{2j}) \in F^{2j+1}$.

If $t \in \mathbf{Q}[G]^* \bigcap \mathrm{ann}(H^2(F^*))$ we may, as in §7.1.4, choose η_2 so that

$$t\eta_2(\phi_2^{-1}(w_2 - \eta_3(d_3^*(w_2)))) \in F^1 \subset F^1 \otimes \mathbf{Q}.$$

Then define $T : F^2 \longrightarrow F^2$ to be multiplication by t on the summand Z^2 and the identity on the summand B^3. This ensures that $(X \cdot T)(F^2) \subset F^1 \subset F^1 \otimes \mathbf{Q}$. The map

$$X : \oplus_j F^{2j} \otimes \mathbf{Q} \longrightarrow \oplus_j F^{2j+1} \otimes \mathbf{Q}$$

is given explicitly by

$$\begin{aligned}
&X(w_0, w_2, \ldots) \\
&= (\rho_1((R^*)^{-1}(w_0 - \eta_1(d_1^*(w_0)))) + d_1^*(w_0) + \eta_2(\phi_2^{-1}(w_2 - \eta_3(d_3^*(w_2)))), \\
&d_3^*(w_2) + \eta_4(\phi_4^{-1}(w_4 - \eta_5(d_5^*(w_4)))), d_5^*(w_4) + \eta_6(\phi_6^{-1}(w_6 - \eta_7(d_7^*(w_7)))), \ldots).
\end{aligned}$$

This formula has all the coordinates integral when $(w_0, w_2, \ldots) \in \oplus_j F^{2j}$ with the possible exception of $\rho_1((R^*)^{-1}(w_0 - \eta_1(d_1^*(w_0)))) \in F^1 \otimes \mathbf{Q}$.

Lemma 7.1.19 *Any* $\mathbf{Z}[G]$*-module homomorphism*

$$X : \oplus_j F^{2j} \longrightarrow F^1 \otimes \mathbf{Q} \oplus F^3 \oplus F^5 \oplus \ldots$$

given by the formula

$$\begin{aligned}
&X(w_0, w_2, \ldots) \\
&= (\Psi(w_0) + d_1^*(w_0) + \eta_2(\phi_2^{-1}(w_2 - \eta_3(d_3^*(w_2)))), \\
&d_3^*(w_2) + \eta_4(\phi_4^{-1}(w_4 - \eta_5(d_5^*(w_4)))), d_5^*(w_4) + \eta_6(\phi_6^{-1}(w_6 - \eta_7(d_7^*(w_7)))), \ldots),
\end{aligned}$$

with $\Psi : F^0 \longrightarrow Z^1$ *and* $\Psi + d_1^*$ *injective, induces a rational isomorphism*

$$X : \oplus_j F^{2j} \otimes \mathbf{Q} \longrightarrow \oplus_j F^{2j+1} \otimes \mathbf{Q}$$

and a $\mathbf{Z}[G]$*-module homomorphism*

$$X \cdot (1 \oplus T \oplus 1 \oplus \ldots) : \oplus_j F^{2j} \longrightarrow \oplus_j F^{2j+1}.$$

Proof. We have discussed the integrality in §7.1.18 so we have only to prove that X is injective. However, if $X(w_0, w_2, \ldots)$ is trivial then $\Psi(w_0) + d_1^*(w_0) + \eta_2(\phi_2^{-1}(w_2 - \eta_3(d_3^*(w_2)))) = 0 \in F^1$ and applying d_2^* yields $w_2 = \eta_3(d_3^*(w_2))$ and $\Psi(w_0) + d_1^*(w_0) = 0 \in F^1$ and the result follows easily. □

7.1.20 *Construction of a* Ψ *for Lemma 7.1.19*

The following construction is rather unsatisfactory, but it will have to serve pro tem.

We have an integral $\mathbf{Z}[G]$-module homomorphism

$$R^* : \frac{H^1(F^*)}{\mathrm{TorsH}^1(F^*)} \longrightarrow H^0(F^*)$$

between finitely generated, torsion free abelian groups. This map is a rational isomorphism and

$$\det(R^*)(R^*)^{-1} : H^0(F^*) \otimes \mathbf{Q} \longrightarrow \frac{H^1(F^*)}{\mathrm{TorsH}^1(F^*)} \otimes \mathbf{Q}$$

induces a $\mathbf{Z}[G]$-module homomorphism of the form

$$\det(R^*)(R^*)^{-1} : H^0(F^*) \longrightarrow \frac{H^1(F^*)}{\mathrm{TorsH}^1(F^*)}$$

and hence a $\mathbf{Z}[G]$-module homomorphism

$$\det(R^*)(R^*)^{-1} \cdot \pi : Z^0 \longrightarrow H^0(F^*) \longrightarrow \frac{H^1(F^*)}{\mathrm{TorsH}^1(F^*)}$$

in which π sends a cocycle to its cohomology class. Also, since

$$0 \longrightarrow Z^0 \xrightarrow{\psi_0} F^0 \xrightarrow{d_1^*} B^1 \longrightarrow 0$$

is split as a sequence of abelian groups, we may choose η_1 so $|G|(w_0 - \eta_1(d_1^*(w_0)) \in Z^0$ for all $w_0 \in F^0$. Therefore we have a $\mathbf{Z}[G]$-module homomorphism

$$\Psi' : F^0 \xrightarrow{|G|(1-\eta_1 d_1^*)} Z^0 \longrightarrow H^0(F^*) \longrightarrow \frac{H^1(F^*)}{\mathrm{TorsH}^1(F^*)}.$$

Finally, since F^0 is projective and Z^1 surjects canonically onto $\frac{H^1(F^*)}{\mathrm{TorsH}^1(F^*)}$, we may lift this composition to obtain a suitable Ψ – *but only if we can ensure that no non-zero* $y \in B^1$ satisfies $\Psi(\eta_1(y)) = -y$. This is because, to show that $\Psi + d_1^*$ is injective, we may work rationally and in this case any $z \in F^0 \otimes \mathbf{Q}$ may be written as $z = w + \eta_1(y)$ for $y \in B^1 \otimes \mathbf{Q}$ and $w \in Z^0 \otimes \mathbf{Q}$. If $0 = \Psi(z) + d_1^*(z)$ then $\Psi'(w) = \Psi'(z) = 0$ in $\frac{H^1(F^*)}{\mathrm{TorsH}^1(F^*)} \otimes \mathbf{Q}$, which implies that $w = 0$. Therefore we are reduced to the condition $0 = \Psi(\eta_1(y))) + d_1^*(\eta_1(y)) = \Psi(\eta_1(y))) + y$.

We can overcome this problem by multiplying by $|G|$ a second time and then lifting $|G|$ times

$$Z^0 \longrightarrow H^0(F^*) \longrightarrow \frac{H^1(F^*)}{\mathrm{TorsH}^1(F^*)}$$

to a $\mathbf{Z}[G]$-module homomorphism of the form $Z^0 \longrightarrow Z^1$.

In particular cases we might conceivably do better!

7.2 Conjectures of Brumer, Coates and Sinnott

7.2.1 *Stickelberger elements*

In this section we are going to recall some conjectures concerning annihilator ideals which appear in [21], [44] and [167]. Suppose that L/K is a Galois extension of number fields with $G(L/K)$ abelian. Suppose also that K is totally real and that L is totally real or is a CM field; that is, L is a totally imaginary quadratic extension of a totally real field (see [163] p. 38). Let S be a finite set of primes of

$\mathcal{O}_K$ including those which ramify in L/K. The reciprocity map of class field theory sends the class of a proper ideal, $\mathcal{A} \triangleleft \mathcal{O}_K$, to its Artin symbol $(\mathcal{A}, L/K) \in G(L/K)$. To this data is associated the partial zeta function which is defined for complex numbers, s, having $\mathcal{R}e(s) > 1$ by

$$\zeta_{K,S}(g,s) = \sum_{(\mathcal{A},L/K)=g,\ \mathcal{A}\ \text{prime}\ \ to\ S} N\mathcal{A}^{-s}.$$

Here $g \in G(L/K)$ and the sum is over all ideals coprime to all primes in S. These functions have a meromorphic continuation to the whole complex plane and at $s = 0$ the corresponding Stickelberger element is defined to be

$$\Theta_{L/K,S}(1) = \sum_{g \in G(L/K)} \zeta_{K,S}(g,0) \cdot g^{-1} \in \mathbf{Q}[G(L/K)].$$

More generally, set

$$\Theta_{L/K,S}(n) = \sum_{g \in G(L/K)} \zeta_{K,S}(g,1-n) \cdot g^{-1} \in \mathbf{Q}[G(L/K)].$$

If χ is a one-dimensional representation of $G(L/K)$ (for example, a p-adic character $\chi : G(L/K) \to \overline{\mathbf{Q}}_p^*$) then (cf. §6.2.1)

$$\chi(\Theta_{L/K,S}(n)) = L_{K,S}(1-n, \chi^{-1}).$$

The following integrality result gives the first connection between Stickelberger elements and annihilators. When $K = \mathbf{Q}$ this was proved in [93], for K real quadratic in [45] and in general in [31], [50].

Theorem 7.2.2 *(See Proposition 7.3.11)*
Let L/K be an abelian Galois extension of number fields, as in $7.2.1, then

$$I_1(L/K,S) = \text{ann}_{\mathbf{Z}[G(L/K)]}(\mu(L)) \cdot \Theta_{L/K,S}(1) \subseteq \mathbf{Z}[G(L/K)].$$

Here $\mu(L)$ denotes the $\mathbf{Z}[G(L/K)]$-module given by the roots of unity in L. The ideal, $I_1(L/K,S)$, is called the (first) Stickelberger ideal.

Conjecture 7.2.3 The Brumer Conjecture goes further than the mere integrality statement of Theorem 7.2.2 to predict more about the Stickelberger ideal. Namely it states that

$$I_1(L/K,S) = \text{ann}_{\mathbf{Z}[G(L/K)]}(\mu(L)) \cdot \Theta_{L/K,S}(1) \subseteq \text{ann}_{\mathbf{Z}[G(L/K)]}(\mathcal{CL}(\mathcal{O}_L)).$$

When $K = \mathbf{Q}$ this is Stickelberger's Theorem ([43] p. 298; [163] p. 94). Since the class-group is finite one may study this conjecture prime by prime. In [167], using the results of [166], Brumer's Conjecture concerning the annihilator of $\mathcal{CL}(\mathcal{O}_L) \otimes \mathbf{Z}_p$ is proved (under mild assumptions) when p does not divide $[L:K]$. In fact, more is claimed in [167] but not proved. As we shall explain in §7.3.1, the results of [167] do imply the equality of the radicals of these ideals.

7.2.4 Theorem 7.2.2 and Conjecture 7.2.3 are part of a larger picture, which involves a generalization of the roots of unity module, $\mu(L)$.

Let L be a number field and let $\mu(E)$ denote the group of roots of unity in a Galois extension, E/L. Then

$$H^0(G(E/L); \mu(E)^{\otimes n}) = (\mu(E)^{\otimes n})^{G(E/L)}$$

is a cyclic group. Each element of $G(E/L)$ acts on $\mu(E)^{\otimes n}$ through the canonical projection to the abelian quotient, $G(L(\mu(E))/L)$. Furthermore we can identify $\mu(E)^{\otimes n}$ with $\mu(E)$ as a group in such a way that the action of $g \in G(L(\mu(E))/L)$ translates to the canonical action of g^n on $\mu(E)$. Hence $g \in G(L(\mu(E))/L)$ acts trivially on $\mu(E)^{\otimes n}$ if and only if the order of g in $G(L(\mu(E))/L)$ divides n. In other words the cyclic group

$$(\mu(E)^{\otimes n})^{G(E/L)} = (\mu(E)^{\otimes n})^{G(L(\mu(E))/L)}$$

has order $w_n(L)$ of ([43] p. 293), which is the largest integer m such that $G(L(\mu_m)/L)$ has exponent dividing n. Here μ_m denotes the group of m-th roots of unity.

Clearly the group $(\mu(E)^{\otimes n})^{G(E/L)}$ is independent of E providing, for example, that E contains a primitive $w_n(L)$ -th root of unity.

Proposition 7.2.5

Let L/K be an abelian extension of number fields with Galois group, $G(L/K)$. Let $w_n(L)$ be as in §7.2.4 and let S_K be a finite set of primes of K containing all Archimedean primes, all primes which ramify in L/K and all primes which divide $w_n(L)$. Then:

(i) *For all Galois extensions, E/L, containing a primitive $w_n(L)$-th root of unity $(\mu(E)^{\otimes n})^{G(E/L)}$ is a cyclic group of order $w_n(L)$.*

(ii) *If $(P, L/K) \in G(L/K)$ denotes the Artin symbol of $P \notin S_K$ then*

$$HCF_{P \notin S_K,\ (P,L/K)=1}(1 - NP^n) = w_n(L).$$

(iii) *The annihilator ideal, $\mathrm{ann}_{\mathbf{Z}[G(L/K)]}((\mu(E)^{\otimes n})^{G(E/L)}) \lhd \mathbf{Z}[G(L/K)]$, is given by*

$$\mathbf{Z}[G(L/K)]\langle (P, L/K) - NP^n \mid P \lhd \mathcal{O}_K,\ P \notin S_K \rangle \lhd \mathbf{Z}[G(L/K)].$$

Proof. Part (i) was established in §7.2.4. Part (ii) is proved in ([43] Lemma 2.3). Part (iii) is proved from (ii) by the argument of ([151] p. 82). □

Conjecture 7.2.6 The following conjecture was introduced in [44] (see also [1], [47]). Let $L/\mathbf{Q}$ be a totally real, abelian Galois extension with conductor f. Then, if S_∞ is just the set of infinite primes, we have

$$\Theta_{\mathbf{Q}(\mu_f)/\mathbf{Q},S_\infty}(n) = \sum_{g \in G(\mathbf{Q}(\mu_f)/\mathbf{Q})} \zeta_{\mathbf{Q},S_\infty}(g, 1-n) \cdot g^{-1} \in \mathbf{Q}[G(\mathbf{Q}(\mu_f)/\mathbf{Q})]$$

and for each integer, b coprime to f, by ([44] Theorem 1.2)

$$S_n(b) = w_{n+1}(\mathbf{Q}) \cdot (b^{n+1} - (b, \mathbf{Q}(\mu_f)/\mathbf{Q}))\Theta_{\mathbf{Q}(\mu_f)/\mathbf{Q},S_\infty}(n+1) \in \mathbf{Z}[G(\mathbf{Q}(\mu_f)/\mathbf{Q})].$$

The conjecture of [44] predicts that the image of $S_n(b)$ in $\mathbf{Z}[G(L/\mathbf{Q})]$ lies in $\mathrm{ann}_{\mathbf{Z}[G(L/\mathbf{Q})]}(K_{2n}(\mathcal{O}_L))$ for each $n \geq 1$.

It should be sufficient to treat the case when L is the totally real subfield of $\mathbf{Q}(\mu_f)$, since the K-groups, $K_{2n}(\mathcal{O}_L)$, should satisfy Galois codescent in this situation. More precisely, if S is a finite, Galois invariant set of primes of the totally real subfield, $\mathbf{Q}(\mu_f)^+$, then one expects an isomorphism of the form

$$K_{2n}(\mathcal{O}_{L,S'}) \cong K_{2n}(\mathcal{O}_{\mathbf{Q}(\mu_f)^+,S})_{G(\mathbf{Q}(\mu_f)^+/L)}$$

where S' is the set of primes below those of S and $A_G = A \otimes_{\mathbf{Z}[G]} \mathbf{Z}$ denotes the G-coinvariants of A.

In [44] this conjecture is proved when $n = 1$ when b is coprime to the order of $K_2(\mathcal{O}_L)$. In this case, by [96] and [106], $K_3^{\mathrm{ind}}(L) \cong (\mu(E)^{\otimes 2})^{G(E/L)}$ of §7.2.4 and, by Proposition 7.2.5, the image of $(b^2 - (b, \mathbf{Q}(\mu_f)/\mathbf{Q}))$ lies in $\mathrm{ann}_{\mathbf{Z}[G(L/\mathbf{Q})]}(K_3^{\mathrm{ind}}(L))$.

Therefore, Conjecture 7.2.3 and Conjecture 7.2.6 could both be subsumed into a conjecture that for each integer, $n \geq 0$, and each abelian extension of number fields, L/K, with K totally real

$$\mathrm{ann}_{\mathbf{Z}[G(L/K)]}(\mathrm{Tors}K_{2n+1}^{\mathrm{ind}}(L)) \cdot \Theta_{L/K,S}(n+1) \subseteq \mathrm{ann}_{\mathbf{Z}[G(L/K)]}(\mathrm{Tors}K_{2n}(\mathcal{O}_L)).$$

There are two important things to note. Firstly, $K_{2n}(\mathcal{O}_L)$ is finite for $n > 0$ and it is expected that $\mathrm{Tors}K_{2n+1}^{\mathrm{ind}}(L)$ is isomorphic to the module $(\mu(E)^{\otimes n+1})^{G(E/L)}$ of §7.2.4. Secondly, I have not included the integer, $w_{n+1}(K)$, in the conjecture, even though it appears in the conjecture of [44], which only refers to the case $K = \mathbf{Q}$.

In fact, suppose that S is as in §7.2.1 and S_1 is the set of primes of L above those in S. Then, if $n > 0$, one expects (for example, see Theorem 7.2.9) that

$$\mathrm{ann}_{\mathbf{Z}[G(L/K)]}(\mathrm{Tors}K_{2n+1}^{\mathrm{ind}}(L)) \cdot \Theta_{L/K,S}(n+1) \subseteq \mathrm{ann}_{\mathbf{Z}[G(L/K)]}(K_{2n}(\mathcal{O}_{L,S_1})).$$

This is consistent with the changing of S. If we add one more prime, $P \lhd \mathcal{O}_K$ to S to give S' then, because P is unramified in L/K,

$$\Theta_{L/K,S'}(n+1) = \Theta_{L/K,S}(n+1)(1 - NP^n\,\mathrm{Frob}_P^{-1}).$$

Let S_1' be the set of primes of L above those of S'. There is a short exact localisation sequence of the form

$$0 \to K_{2n}(\mathcal{O}_{L,S_1}) \to K_{2n}(\mathcal{O}_{L,S_1'}) \to \oplus_{Q \lhd \mathcal{O}_L,\ \text{prime}\ ,\ Q|P} K_{2n-1}(\mathcal{O}_L/Q) \to 0.$$

The Frobenius element, Frob_P, acts on $\oplus_{Q \lhd \mathcal{O}_L,\ \text{prime}\ ,\ Q|P} K_{2n-1}(\mathcal{O}_L/Q)$ preserving each factor and $1 - NP^n\,\mathrm{Frob}_P^{-1}$ annihilates each factor, by [115]. Hence the conjecture for S would imply it for S'.

Example 7.2.7 *(Stickelberger Examples)*

Consider the Galois extensions, $F = \mathbf{Q}(\sqrt{2}) \subset E = \mathbf{Q}(\sqrt{2}, \sqrt{p}) \subset N = E(\beta)$ of Example 1.3.2, where $p \equiv 3$ (modulo 8) is a prime and $\beta^2 = ((\sqrt{p})/(u + v\sqrt{p})) \cdot (\sqrt{2}(1+\sqrt{2}))$. We are going to consider the cases when $p = 3, 11, 19$. The Galois group of N/F is cyclic of order four, generated by y such that $y(\beta) = \beta((u + v\sqrt{p})/\sqrt{2})$. Let $S = \{P_2, P_p\}$ where $P_q \lhd \mathcal{O}_F$ denotes the prime above q. Hence we have, in $\mathbf{Q}[\langle y\rangle]$,

$$\Theta_{N/F,S}(2) = \zeta_{F,S}(1,-1) + \zeta_{F,S}(y,-1)y^3 + \zeta_{F,S}(y^3,-1)y + \zeta_{F,S}(y^2,-1)y^2.$$

Since $\zeta_{F,S}(y,-1) = \zeta_{F,S}(y^3,-1)$, by Galois invariance, we may write $\theta_{N/F,S}(1) = a + b(y+y^3) + cy^2$ with $a, b, c \in \mathbf{Q}$. Let $\chi : \langle y\rangle \longrightarrow \mathbf{C}^*$ be the character given by $\chi(y) = \sqrt{-1}$. Then $\chi^\alpha(\Theta_{N/F,S}(2)) = L_{F,S}(-1, \chi^{-\alpha})$ for $\alpha = 0, 1, 2, 3$. Therefore

$$\begin{aligned} a + 2b + c &= \chi^0(\Theta_{N/F,S}(2)) &&= L_{F,S}(-1,1), \\ a - c &= \chi(\Theta_{N/F,S}(2)) &&= L_{F,S}(-1,\chi), \\ a - 2b + c &= \chi^2(\Theta_{N/F,S}(2)) &&= L_{F,S}(-1,\chi^2). \end{aligned}$$

Hence

$$\begin{aligned} 4a &= L_{F,S}(-1,1) + L_{F,S}(-1,\chi^2) + 2L_{F,S}(-1,\chi), \\ 4b &= L_{F,S}(-1,1) - L_{F,S}(-1,\chi^2), \\ 4c &= L_{F,S}(-1,1) + L_{F,S}(-1,\chi^2) - 2L_{F,S}(-1,\chi). \end{aligned}$$

Now $L_{F,S}(-1,1) = (p^2-1)\zeta_F(-1) = (p^2-1)|K_2(\mathcal{O}_F)|/|K_3^{\text{ind}}(F)|$, since the Birch-Tate conjecture holds for these fields [37]. Also $K_2(\mathcal{O}_F)$ is an elementary abelian 2-group of rank two while $|K_3^{\text{ind}}(F)| = 48$ so that $L_{F,S}(-1,1) = 4(p^2-1)/48 = (p^2-1)/12$. Similarly,

$$\begin{aligned} L_{F,S}(-1,1)L_{F,S}(-1,\chi^2) &= (p^2-1)\zeta_E(-1) \\ &= \begin{cases} 8 & \text{if } p = 3, \\ 120 \cdot 7 \cdot 23 \cdot 3^{-1} & \text{if } p = 11, \\ 360 \cdot 19 \cdot 41 \cdot 3^{-1} & \text{if } p = 19 \end{cases} \end{aligned}$$

from the table in §1.3.7.

Case (i): $p = 3$:
In this case we have $L_{F,S}(-1,1) = 2/3$, $L_{F,S}(-1,\chi^2) = 8/L_{F,S}(-1,1) = 12$ and $L_{F,S}(-1,\chi) = 284$ by §1.3.7 so that $a = 1/4(2/3 + 12 + 568) = 1/4(580 + 2/3) = 1742/12 = 871/6$, $b = 1/4(2/3 - 12) = -34/12 = -17/6$ and $c = 1/4(2/3 + 12 - 568) = 1/4(2/3 - 556) = -1666/12 = -833/6$. Therefore, when $p = 3$, $\Theta_{N/F,S}(2) = 871/6 - 17/6(y + y^3) - 833/6y^2$.

On the other hand $K_3^{\text{ind}}(N) \cong (\mu(E)^{\otimes 2})^{G(E/N)}$ of 7.2.4 is isomorphic to $\mathbf{Z}/48$ with trivial y-action. Since $\zeta_N(-1) = 2^4 \cdot 71^2$, $|K_2(\mathcal{O}_N)| = 48 \cdot 2^4 \cdot 71^2 = 2^8 \cdot 3 \cdot 71^2$. Since the 3-torsion in $K_2(\mathcal{O}_E)$ has order three with y acting like -1 the same is true for $K_2(\mathcal{O}_N)$. The 2-primary torsion is an elementary abelian 2-group [37].

Now let us examine what the conjecture predicts. Since $y-1$ annihilates $K_3^{\mathrm{ind}}(N)$ we calculate

$$\begin{aligned}(y-1)\theta_{N/F,S}(1) &= 1/6(871y - 17(y^2+1) - 833/6y^3 \\ &\quad -871 + 17(y+y^3) + 833y^2) \\ &= 1/6(888(y-1) + 816y^2(1-y)) \\ &= (y-1)(148 - 136y^2).\end{aligned}$$

This element of $\mathbf{Z}[\langle y\rangle]$ certainly annihilates the 2- and 3-torsion of $K_2(\mathcal{O}_N)$, since the coefficients are even and $-24 \equiv 0$ (module 3). If the 71-primary torsion were cyclic then y (which has to act non-trivially to ensure that the coinvariants, $K_2(\mathcal{O}_E)$, has no 71-torsion) must act on $\mathbf{Z}/71^2$ like -1. This refutes the conjecture, since $(-2)(12)$ is not trivial modulo 71^2. On the other hand, if the 71-primary torsion is elementary abelian we can have y act on $\mathbf{Z}/71 \oplus \mathbf{Z}/71$ by switching factors and changing one sign. This satisfies $y^2=-1$, which is consistent with the conjecture since $148-136(-1)) \equiv 0$ (modulo 71)! Therefore, when $p=3$, the conjecture predicts that

$$K_2(\mathcal{O}_N) \cong (\oplus_{i=1}^{8}\mathbf{Z}/2) \oplus \mathbf{Z}/3 \oplus \mathbf{Z}/71 \oplus \mathbf{Z}/71$$

with the Galois action described above.

Case (ii): $p=11$:
In this case we have $L_{F,S}(-1,1) = 10$, $L_{F,S}(-1,\chi^2) = 120\cdot 7\cdot 23\cdot 3^{-1}/10 = 4\cdot 7\cdot 23 = 644$ and $L_{F,S}(-1,\chi) = 4\cdot 3\cdot 11\cdot 109 = 14388$ by §1.3.7 so that $a = 1/4(10+644+28776) = 14715/2$, $b = 1/4(10-644) = -317/2$ and $c = 1/4(10+644-28776) = -14061/2$. Therefore, when $p=11$, $\Theta_{N/F,S}(2) = 14715/2 - 317/2(y+y^3) - 14061/2y^2$.

As before $K_3^{\mathrm{ind}}(N)$ is isomorphic to $\mathbf{Z}/48$ with trivial y-action. Since $\zeta_N(-1) = 2^4\cdot 3\cdot 11^2\cdot 109^2\cdot 7\cdot 23$, $|K_2(\mathcal{O}_N)| = 2^8\cdot 3^2\cdot 7\cdot 11^2\cdot 23\cdot 109^2$. Since the 3-torsion in $K_2(\mathcal{O}_E)$ has order three with y acting like -1. The 2-primary torsion is an elementary abelian 2-group [37]. The action must be non-trivial on 11- and 109-primary torsion.

Now to what the conjecture predicts. Since $y-1$ annihilates $K_3^{\mathrm{ind}}(N)$ we calculate

$$\begin{aligned}(y-1)\Theta_{N/F,S}(2) &= 1/2(14715y - 317(y^2+1) - 14061y^3 \\ &\quad -14715 + 317(y+y^3) + 14061y^2) \\ &= 1/2(15032(y-1) + 13744y^2(1-y)) \\ &= (y-1)(7516 - 6872y^2).\end{aligned}$$

This element of $\mathbf{Z}[\langle y\rangle]$ certainly annihilates the 2-torsion of $K_2(\mathcal{O}_N)$, since the coefficients are even. If y^2 acts trivially on the 7- and 23-torsion then they are annihilated because $7516-6872 = 644 = 4\cdot 7\cdot 23$ and if y acts on $\mathbf{Z}/3\cdot 11\cdot$

$109 \oplus \mathbf{Z}/3 \cdot 11 \cdot 109$ by switching the factors and changing one sign then this is annihilated because $7516+6872 = 14388 = 2^2 \cdot 3 \cdot 11 \cdot 109$. Therefore, when $p = 11$, the conjecture predicts that

$$K_2(\mathcal{O}_N) \cong (\oplus_{i=1}^{8} \mathbf{Z}/2) \oplus \mathbf{Z}/7 \cdot 23 \oplus \mathbf{Z}/3 \cdot 11 \cdot 109 \oplus \mathbf{Z}/3 \cdot 11 \cdot 109$$

with the Galois action described above.

Case (iii): $p = 19$:
In this case we have $L_{F,S}(-1,1) = 30$, $L_{F,S}(-1,\chi^2) = 360 \cdot 19 \cdot 41 \cdot 3^{-1} \cdot 30^{-1} = 2^2 \cdot 19 \cdot 41 = 3116$ and $L_{F,S}(-1,\chi) = 2^2 \cdot 3^2 \cdot 1993 = 71748$ by §1.3.7 so that $a = 1/4(30 + 3116 + 143496) = 73321/2$, $b = 1/4(30 - 3116) = -1543/2$ and $c = 1/4(30 + 3116 - 143496) = -70175/2$. Therefore, when $p = 19$, $\Theta_{N/F,S}(2) = 73321/2 - 1543/2(y + y^3) - 70175/2y^2$.

As before $K_3^{\mathrm{ind}}(N)$ is isomorphic to $\mathbf{Z}/48$ with trivial y-action. Since $\zeta_N(-1) = 2^4 \cdot 3^4 \cdot 1993^2 \cdot 3^{-1} \cdot 19 \cdot 41$, $|K_2(\mathcal{O}_N)| = 2^8 \cdot 3^4 \cdot 19 \cdot 41 \cdot 1993^2$. The 2-primary torsion again has index 2 and the action must be non-trivial on 109-primary torsion.

Now to what the conjecture predicts. Since $y - 1$ annihilates $K_3^{\mathrm{ind}}(N)$ we calculate

$$\begin{aligned}(y-1)\Theta_{N/F,S}(2) &= 1/2(73321y - 1543(y^2+1) - 70175y^3 \\ &\qquad -73321 + 1543(y+y^3) + 70175y^2) \\ &= 1/2(74864(y-1) + 68632y^2(1-y)) \\ &= (y-1)(37432 - 34316y^2).\end{aligned}$$

This element of $\mathbf{Z}[\langle y\rangle]$ certainly annihilates the 2-torsion of $K_2(\mathcal{O}_N)$, since the coefficients are even. If the 3-primary and 1993-primary torsion is $(H_9 \oplus \mathbf{Z}/1993) \oplus (H_9 \oplus \mathbf{Z}/1993)$, with H_9 of order nine, on which y acts by switching the factors and changing one sign then this is annihilated because $37432 + 34316 = 2^2 \cdot 3^2 \cdot 1993$. If y^2 acts trivially on the 19- and 41-torsion then they are annihilated because $37432 - 34316 = 2^2 \cdot 19 \cdot 41$. Therefore, when $p = 19$, the conjecture predicts that

$$K_2(\mathcal{O}_N) \cong (\oplus_{i=1}^{8} \mathbf{Z}/2) \oplus \mathbf{Z}/19 \cdot 41 \oplus (H_9 \oplus \mathbf{Z}/1993) \oplus (H_9 \oplus \mathbf{Z}/1993)$$

with the Galois action described above.

The predicted Galois actions in Cases (i)–(iii) seem to be consistent with the cardinalities of the odd-primary eigen-groups which are given by ([166] Thereom 1.5) and, interestingly, they are also consistent with leaving out the factor, $w_{n+1}(K)$, in the conjecture, even though it appears in the conjecture of [44].

In each of the Cases (i)–(iii) $K_2'(\mathcal{O}_N) \cong K_2(\mathcal{O}_N)[1/2]$ and, if S_1 consists of the two primes of N above those of S, we have an isomorphism of the form $K_2'(\mathcal{O}_{N,S_1}) \cong K_2(\mathcal{O}_N)[1/2] \oplus \mathbf{Z}/8$ where y acts trivially on $\mathbf{Z}/8$. Therefore, when $p = 3, 11$ or 19, $(y-1)\Theta_{N/F,S}(2) \in \mathrm{ann}_{\mathbf{Z}[\langle y\rangle]}(K_2'(\mathcal{O}_{N,S_1}))$. This means that the final version of Conjecture 7.2.6 is true for these examples.

7.2.8 $\hat{\Omega}_1(L/K,3)$ *and annihilator relations*

Now we shall apply Theorem 7.1.11 to the element

$$\hat{\Omega}_1(L/K,3) = [\oplus_j F_{2j}, X, \oplus_j F_{2j+1}] \in K_0(\mathbf{Z}[G(L/K)], \mathbf{Q})$$

of Definition 5.6.2 in the case when L/K is a totally real extension of number fields with $G(L/K)$ abelian. In this case the non-zero homology groups of F_* are both finite, being given by

$$H_0(F_*) \cong K_2'(\mathcal{O}_{L,S}),\ H_1(F_*) \cong K_3^{\text{ind}}(L) \cong (\mu(E)^{\otimes 2})^{G(E/L)}$$

where the last module is as in Proposition 7.2.5. Here S is a finite Galois invariant set of primes of L containing all those above primes which ramify in L/K. Applying Theorem 7.1.11 to the complex F_* and the rational isomorphism X we obtain a chain of annihilator relations of the form

$$\begin{aligned}&\operatorname{ann}_{\mathbf{Z}_p[G(L/K)]}(K_2'(\mathcal{O}_{L,S}) \otimes \mathbf{Z}_p)^{m_{p,0}} \\ &\subseteq \det(X)^{-1} \operatorname{ann}_{\mathbf{Z}_p[G(L/K)]}((\mu(E)^{\otimes 2})^{G(E/L)} \otimes \mathbf{Z}_p) \\ &\subseteq \operatorname{ann}_{\mathbf{Z}_p[G(L/K)]}(K_2'(\mathcal{O}_{L,S}) \otimes \mathbf{Z}_p).\end{aligned}$$

Here $m_{p,0}$ is the minimal number of generators of $K_2'(\mathcal{O}_{L,S}) \otimes \mathbf{Z}_p$.

Note that $m_{p,1} = 1$ because $K_3^{\text{ind}}(L)$ is a cyclic group.

Now suppose that $r \geq 2$ and consider the 2-extension of finitely generated $\mathbf{Z}[G(L/K)]$-modules Proposition 5.2.10(16)

$$K_{2r-1}' \longrightarrow A_r \longrightarrow B_r \longrightarrow K_{2r-2}'.$$

The group, K_{2r-2}', is finite and is expected to be a good approximation to $K_{2r-2}'(\mathcal{O}_{L,S})$. Also there is a short exact sequence of the form

$$0 \longrightarrow Y_{2r+1} \longrightarrow K_{2r-1}' \longrightarrow (\mu(E)^{\otimes r})^{G(E/L)} \longrightarrow 0$$

by Proposition 5.2.9. However, if r is even then $Y_{2r+1} = \mathbf{Z}[S_\infty]$ is a free module and if r is odd $Y_{2r+1} = 0$ ([38] Lemma 6.19) so that in all cases we have a canonical isomorphism of the form

$$\operatorname{Ext}^2_{\mathbf{Z}[G(L/K)]}(K_{2r-2}', K_{2r-1}') \cong \operatorname{Ext}^2_{\mathbf{Z}[G(L/K)]}(K_{2r-2}', (\mu(E)^{\otimes r})^{G(E/L)}).$$

In fact, the corresponding 2-extension is represented by

$$\frac{K_{2r-1}'}{Y_{2r+1}} \longrightarrow \frac{A_r}{Y_{2r+1}} \longrightarrow B_r \longrightarrow K_{2r-2}'.$$

Now choose a bounded perfect complex, F_*, which is quasi-isomorphic to the complex

$$0 \longrightarrow \frac{A_r}{Y_{2r+1}} \longrightarrow B_r \longrightarrow 0$$

with B_r in dimension zero. Choosing a rational isomorphism, X_r, we may construct an element

$$\hat{\Omega}_{r-1}(L/K,3) = [\oplus_j F_{2j}, X_r, \oplus_j F_{2j+1}] \in K_0(\mathbf{Z}[G(L/K)], \mathbf{Q})$$

such that $\pi(\hat{\Omega}_{r-1}(L/K,3)) = \Omega_{r-1}(L/K,3) \in \mathcal{CL}(\mathbf{Z}[G(L/K)])$ (cf. §5.6.1).

Applying Theorem 7.1.11 to $\hat{\Omega}_{r-1}(L/K,3)$ for $r = 2, 3, 4, \ldots$, as we did in the case $r = 2$, we obtain the following result.

Theorem 7.2.9

Let L/K is a totally real extension of number fields with $G(L/K)$ abelian and suppose that $r \geq 2$. In the notation of §7.2.8, if

$$\hat{\Omega}_{r-1}(L/K,3) = [\oplus_j F_{2j}, X_r, \oplus_j F_{2j+1}]$$

then, for each prime p,

$$\begin{aligned} &\mathrm{ann}_{\mathbf{Z}_p[G(L/K)]}(K'_{2r-2} \otimes \mathbf{Z}_p)^{m_{p,0}} \\ &\subseteq \det(X_r)^{-1}\, \mathrm{ann}_{\mathbf{Z}_p[G(L/K)]}((\mu(E)^{\otimes r})^{G(E/L)} \otimes \mathbf{Z}_p) \\ &\subseteq \mathrm{ann}_{\mathbf{Z}_p[G(L/K)]}(K'_{2r-2} \otimes \mathbf{Z}_p). \end{aligned}$$

Here $m_{p,0}$ is the minimal number of generators of the $\mathbf{Z}[G(L/K)]$-module $K'_{2r-2} \otimes \mathbf{Z}_p$ and $K'_2 = K'_2(\mathcal{O}_{L,S})$.

7.2.10 *Annihilators and the Wiles Unit – explanation and speculation*

Let L/K be a totally real extension of number fields with $G(L/K)$ abelian. I shall conclude this chapter by giving a speculative explanation to account for the truth of Conjecture 7.2.6. In fact, we shall study a close approximation to Conjecture 7.2.6. Firstly, from the localisation sequence of §2.4.9 $K_{2r-2}(\mathcal{O}_L)$ is a $\mathbf{Z}[G(L/K)]$-submodule of $K_{2r-2}(\mathcal{O}_{L,S})$ for any finite, Galois-invariant set of primes of L, S. Hence the annihilator of $K_{2r-2}(\mathcal{O}_L)$ is contained in that of $K_{2r-2}(\mathcal{O}_{L,S})$. Secondly, the image of $K_{2r-2}(\mathcal{O}_{L,S})$ in $\oplus_{v \in S_\infty} K_{2r-2}(L_v)$ is equal to the $\mathbf{F}_2$-vector space on S_∞, the infinite places of L. Hence we may as well study the annihilator of $K'_{2r-2}(\mathcal{O}_{L,S}) = \mathrm{Ker}(K_{2r-2}(\mathcal{O}_{L,S}) \rightarrow \oplus_{v \in S_\infty} K_{2r-2}(L_v))$. Furthermore the Lichtenbaum-Quillen conjecture predicts that $K'_{2r-2}(\mathcal{O}_{L,S})$ is very closely approximated by K'_{2r-2} of Proposition 5.2.10 and Definition 5.2.11 ([160]; see also [123]). Note that K'_{2r-2} depends on a finite, Galois invariant set of primes, S, containing all finite primes lying over those which ramify in L/K.

Now we turn our attention to the connection between the Wiles Unit of Conjecture 6.3.4 and Theorem 7.2.9.

Let p be an odd prime. Suppose, as is widely believed, that the Iwasawa μ-invariant ($\mu(\rho)$ of Theorem 6.2.5) is zero for all Galois representations ρ (cf. [53]). Suppose also the L/K is linearly disjoint from the cyclotomic $\mathbf{Z}_p$-extension of K, so that there are no characters of Type W on $G(L/K)$. Then, as mentioned in §6.3.5, using Iwasawa theory and the results of [5], one may imitate the proof in the function field case [38] to show that the p-adic coordinate of a Hom-description representative of $\Omega_{r-1}(L/K,3)$ in Theorem 2.5.25 is essentially given by the function $(\rho \mapsto h_{\overline{\rho}}(u^r-1))^{-1}$ of §6.1.1 [41]. The proof of [41] seems to generalize to show that this same function is a Hom-description representative of $\hat{\Omega}_n(L/K,3) \in K_0(\mathbf{Z}[G(L/K)],\mathbf{Q})$ in §2.5.34. Let us assume this to be true – this is where the speculation comes in!

To continue, making these assumptions, if we define $H_r \in \mathbf{Q}_p[G(L/K)]^*$ by

$$\chi(H_r) = h_{\chi^{-1}}(u^r-1)$$

for all characters, $\chi : G(L/K) \longrightarrow \overline{\mathbf{Q}}_p^*$ then

$$\det(X_r) = H_r^{-1} \in \frac{\mathbf{Q}_p[G(L/K)]^*}{\mathbf{Z}_p[G(L/K)]^*}.$$

If, in addition, the Wiles Unit Conjecture (Conjecture 6.3.4) is true then

$$H_r = \Theta_{L/K,S\bigcup S_p}(r) \in \frac{\mathbf{Q}_p[G(L/K)]^*}{\mathbf{Z}_p[G(L/K)]^*},$$

since $\chi(\Theta_{L/K,S\bigcup S_p}(r)) = L_{K,S\bigcup S_p}(1-r,\chi^{-1})$. As a consequence of Theorem 7.2.9 we then obtain annihilator relations of the form

$$\begin{aligned}
&\mathrm{ann}_{\mathbf{Z}_p[G(L/K)]}(K'_{2r-2}\otimes\mathbf{Z}_p)^{m_{p,0}}\\
&\subseteq \Theta_{L/K,S\bigcup S_p}(r)\,\mathrm{ann}_{\mathbf{Z}_p[G(L/K)]}((\mu(E)^{\otimes r})^{G(E/L)}\otimes\mathbf{Z}_p)\\
&\subseteq \mathrm{ann}_{\mathbf{Z}_p[G(L/K)]}(K'_{2r-2}\otimes\mathbf{Z}_p).
\end{aligned}$$

Ths is reassuringly similar to the Coates-Sinnott Conjecture and Conjecture 7.2.6.

Remark 7.2.11 (i) *Euler factors*: As explained in the discussion of Conjecture 7.2.6, the addition of the primes S_p changes the L-function, $L_{K,S}(1-n,\chi)$ to the p-adic L-function by multiplying by the Euler factors corresponding to the primes above p and changes $\theta_{L/K,S}(n)$. For example, if a prime P of K above p has trivial inertia group then this prime contributes a factor equal to $(1-NP^{n-1}\cdot\mathrm{Frob}_P^{-1}) \in \mathbf{Q}_p[G(L/K)]^*$ and we have

$$\Theta_{L/K,S\bigcup\{P\}}(n) = \Theta_{L/K,S}(n)(1-NP^{n-1}\,\mathrm{Frob}_P^{-1}).$$

On the other hand the extra group which this prime introduces to K'_{2n} is the sum of $K_{2n-1}(\mathcal{O}_L/Q)\otimes\mathbf{Z}_p$ over primes Q dividing P (see Conjecture 7.2.6). This additional module is annihilated by the $(1-NP^{n-1}\cdot\mathrm{Frob}_P^{-1})$ so that the truth of our annihilator conjectures for S imply them for $S\bigcup\{P\}$.

(ii) *The Brumer Conjecture and the presence of non-trivial regulators*: In §§7.1.12–7.1.19 (particularly Corollary 7.1.17) we examined the type of annihilator relations which we would get if we applied the discussion of §7.2.10 to a lifting of $\Omega_n(L/K,3)$ to an element – as yet we have no canonical definition to off (see §5.6.12) – $\hat{\Omega}_n(L/K,3) \in K_0(\mathbf{Z}[G(L/K)],\mathbf{Q})$ in the case when $K_{2n+1}(\mathcal{O}_L)$ is not finite. In this case there would be non-trivial regulators and the annihilator relations, which are not as good in the presence of regulators, would involve the annihilators of the torsion subgroups.

In [122] a 2-extension of finitely generated $\mathbf{Z}[G(L/K)]$-modules is constructed which has cohomologically trivial modules in the middle and for which the torsion subgroups of the ends are $\mathcal{CL}(\mathcal{O}_L)$ and $\mu(L)$. Applying the discussion analogous to that of §7.2.10 – with the estimates from the non-trivial regulator case (§§7.1.12–7.1.19) – would give annihilator relations of the type predicted by the Brumer conjecture (Conjecture 7.2.3).

(iii) *Galois descent for the Stickelberger ideals*: In their study of the Brumer conjecture and the Stickelberger ideals

$$I_1(L/K,S) = \mathrm{ann}_{\mathbf{Z}[G(L/K)]}(\mu(L)) \cdot \Theta_{L/K,S}(1) \lhd \mathbf{Z}[G(L/K)]$$

Hayes, Popescu and Sands ([72], [125], [126]) have shown that $I_1(L/K,S)$ is natural with respect to change of fields. The discussion of §7.2.10 would predict this, in the following sense.

Define the higher Stickelberger ideals, $I_n(L/K,S) \lhd \mathbf{Z}[G(L/K)]$, by the formulae

$$I_n(L/K,S) = \mathbf{Z}[G(L/K)] \bigcap (\Theta_{L/K,S}(n)\, \mathrm{ann}_{\mathbf{Z}_p[G(L/K)]}((\mu(E)^{\otimes n})^{G(E/L)}).$$

If L/K is totally real and if we are in the dimensions of Theorem 7.2.9 then the Wiles Unit Conjecture of §6.3.4 together with the discussion of §7.2.10 relates $I_r(L/K,S)$ to the 2-extension of Proposition 5.2.10(16). This 2-extension is natural in the sense of §3.1.1 (cf. Theorem 3.1.20; [34], [131] Chapter VII).

In particular, the 2-extension satisfies Galois descent and so Conjecture 6.3.4 would imply the same for $I_r(L/K,S)$ when L/K is an abelian Galois extension of totally number fields and $r \geq 2$.

(iv) *Abelian fields*: There is one case in which we can show that, for r even, $r \geq 2$ and p an odd prime,

$$\det(X_r)^{-1} = \Theta_{L/K,S\bigcup S_p}(r) \in \frac{\mathbf{Q}_p[G(L/K)]^*}{\mathbf{Z}_p[G(L/K)]^*}$$

without assumng the truth of the Wiles Unit Conjecture (Conjecture 6.3.4). This is the case when $K = \mathbf{Q}$ so that $L/\mathbf{Q}$ is a totally real, abelian Galois extension of number fields. In fact, it suffices to take $L = \mathbf{Q}(\xi_m)^+$, the real subfield of a cyclotomic field. In this case each of K'_{2r-1}/Y_{2r+1} and K'_{2r-2} are finite and Y_{2r+1} is free (cf. Proposition 5.2.9 and §5.2.13) so that we have a lifted invariant

$$\hat{\Omega}_{r-1}(L/\mathbf{Q},3) \in K_0(\mathbf{Z}[G(L/\mathbf{Q})],\mathbf{Q})$$

whose Hom-description representative is $\det(X_r)$.

On the other hand, the "equivariant Tamagawa number" invariant of $L/\mathbf{Q}$, constructed in ([22], [23]), also lies in this group. As mentioned in §5.6.12, these invariants are very closely related and the calculations of [27], which describe the invariant of [22] in terms of equivariant Stickelberger elements, can be used to give a similar description for $\hat{\Omega}_{r-1}(L/\mathbf{Q},3)$.

(v) *Function fields*: We may also apply these results in the case of function fields.

Suppose that N/K is a finite Galois extension of global function fields with abelian Galois group $G(N/K)$ and let $p = \mathrm{char}(N)$. Let S be a finite, non-empty set of primes of K, including all those primes which ramify in N/K. Let S' denote the set of primes of N lying over those of S. Let $O_{N,S'}$ be the ring of S'-integers of N; that is, the elements of N which are regular away from S'. Then $W = \mathrm{Spec}(O_{N,S'})$ is a smooth affine curve over $\mathbf{F}_p$.

Let l be a prime different from p. If $\overline{\mathbf{Q}_l}$ denotes an algebraic closure of $\mathbf{Q}_l$ and $\chi : G(L/K) \longrightarrow \overline{\mathbf{Q}_l}^*$ is a one-dimensional l-adic representation we denote by $L_{K,S}(t,\chi)$ denote the Artin L-function ([107] p. 293; see also [101]) with the Euler factors associated to primes of S removed. In the notation of [107] $L_K(t,\chi)$ would be denoted by $L(W,\chi,t)$.

Under these circumstances, for each integer $n \geq 2$, there exists a (unique) Stickelberger element

$$\Theta_{N/K,S}(n) \in \mathbf{Q}_l[G(N/K)]^*$$

such that

$$\chi(\Theta_{N/K,S}(n)) = L_{K,S}(1-n,\chi^{-1})$$

for all one-dimensional l-adic representations, χ (see [38] Proposition 7.13).

For $i = 0,1$ and $n \geq 1$ we have étale cohomology groups ([38] §7)

$$H^i_{\text{ét}}(W;\mathbf{Q}_l/\mathbf{Z}_l(n+1)) \cong H^{i+1}_{\text{ét}}(W;\mathbf{Z}_l(n+1))$$

where $W = \mathrm{Spec}(O_{N,S'})$. Each of these finite groups is a $\mathbf{Z}_l[G(N/K)]$-module.

Suppose that E/K is a Galois extension containing N and having $\mu_{l^\infty}(E)$, the group of l-primary roots of unity in E, very large. Then

$$H^0_{\text{ét}}(W;\mathbf{Q}_l/\mathbf{Z}_l(n+1)) \cong (\mu_{l^\infty}(E)^{\otimes n+1})^{G(E/N)},$$

which is a cyclic group.

As in the number field case the annihilator, $\mathrm{ann}_{\mathbf{Z}_l[G(N/K)]}(H^0_{\text{ét}}(W;\mathbf{Q}_l/\mathbf{Z}_l(n+1)))$ is well known, being computed as in ([43] Lemma 2.3; [152] p. 82).

Theorem 7.2.12

Suppose that N/K is a finite Galois extension of global function fields with abelian Galois group, $G(N/K)$. If $n \geq 1$ with l and $W = \mathrm{Spec}(O_{N,S'})$ as in §7.2.11(v) then

$$\begin{aligned}&\{t^m \mid t \in \mathrm{ann}_{\mathbf{Z}_l[G(N/K)]}(H^1_{\text{ét}}(W;\mathbf{Q}_l/\mathbf{Z}_l(n+1)))\}\\ &\subseteq \Theta_{N/K,S}(n+1)\cdot \mathrm{ann}_{\mathbf{Z}_l[G(N/K)]}(H^0_{\text{ét}}(W;\mathbf{Q}_l/\mathbf{Z}_l(n+1)))\\ &\subseteq \mathrm{ann}_{\mathbf{Z}_l[G(N/K)]}(H^1_{\text{ét}}(W;\mathbf{Q}_l/\mathbf{Z}_l(n+1)))\end{aligned}$$

where m is the minimum number of generators for the $\mathbf{Z}_l[G(N/K)]$-module,

$$H^1_{\text{ét}}(W; \mathbf{Q}_l/\mathbf{Z}_l(n+1)).$$

Proof. In ([38] Proposition 7.7) it is shown that there exists a bounded, perfect cochain complex of $\mathbf{Z}_l[G(N/K)]$-modules of the form

$$M^* : 0 \to M^s \to M^{s+1} \to \ldots \to M^t \to 0$$

together with a $\mathbf{Z}_l[G(N/K)]$-module automorphism, $\alpha : M^* \to M^*$, such that for each j the homomorphism $1-\alpha : M^j \to M^j$ is injective with finite cokernel. In addition, the mapping cone, $C^* = Cone(1-\alpha)$ has trivial cohomology with the exception of $H^i(C^*) \cong H^i_{\text{ét}}(X; \mathbf{Q}_l/\mathbf{Z}_l(n+1))$ for $i = 0, 1$. The complex C^* gives rise to an element of

$$K_0(\mathbf{Z}_l[G], \mathbf{Q}_l) \cong \frac{\mathbf{Q}_l[G]^*}{\mathbf{Z}_l[G]^*}.$$

In fact the truncated complex, D^*, given by $D^j = C^j$ for $j \leq 0$, $D^1 = \text{Ker}(d : C^1 \to C^2)$ and $D^j = 0$ for $j \geq 2$ defines the same element. Now setting $F_{1-j} = D^j$ and applying the l-adic version of Theorem 7.1.11 to F_* yields

$$\begin{aligned}
&\{z^t \mid z \in \text{ann}_{\mathbf{Z}_l[G(N/K)]}(H^1_{\text{ét}}(X; \mathbf{Q}_l/\mathbf{Z}_l(n+1)))\} \\
&\subseteq \det(X)^{-1} \cdot \text{ann}_{\mathbf{Z}_l[G(N/K)]}(H^0_{\text{ét}}(X; \mathbf{Q}_l/\mathbf{Z}_l(n+1))) \\
&\subseteq \text{ann}_{\mathbf{Z}_l[G(N/K)]}(H^1_{\text{ét}}(X; \mathbf{Q}_l/\mathbf{Z}_l(n+1)))
\end{aligned}$$

where $[\oplus_j F_{2j}, X, \oplus_j F_{2j+1}]$ is the element of $K_0(\mathbf{Z}_l[G], \mathbf{Q}_l)$ given by the construction of §7.1.1. However, for the cochain complex C^* it is shown in ([38] Proposition 7.17) that this element satisfies $\chi(\det(X)) = L_{K,S}(-n, \chi^{-1})^{-1}$ for all one-dimensional l-adic representations, χ. Hence

$$\det(X) = \Theta_{N/K,S}(n+1)^{-1},$$

as required. □

7.3 The radical of the Stickelberger ideal

7.3.1 As in Conjecture 7.2.6, let L/K be a Galois extension of totally real number fields with abelian Galois group, $G(L/K)$ and let $r \geq 2$ be an even integer. As in §7.2.4, let $w_r(L)$ be the largest integer m such that, if μ_m is the group of mth roots of unity in L^{sep}, then $\text{Gal}(L(\mu_m)/L)$ has exponenent dividing r. As in Proposition 7.2.5, let S_K be a finite set of places of K containing all archimedean places, all primes ramifying in L/K and all primes dividing $w_r(L)$. Let S be the set of places of L over S_K. Let p be an odd rational prime and set $X_p = \text{Spec}(O_{L,S}[1/p])$, as in §5.1.1, where $O_{L,S}$ is the ring of S-integers of L.

Since L is totally real, $p > 2$ and $r \geq 2$ Lemma 5.2.3, Corollary 5.2.4 and ([38] Lemma 6.19) imply that $H^{i+1}(X_p; \mathbf{Z}_p(r)) = H^i(X_p; \mathbf{Z}_p(r)_+)$ for $i = 0, 1$, where the étale pro-sheaf $\mathbf{Z}_p(r)_+$ is defined in §5.2.1. These same references also show that the above cohomology groups are finite G-modules and that $H^1(X_p; \mathbf{Z}_p(r))$ is naturally isomorphic to $(\mu(E)^{\otimes r})^{\mathrm{Gal}(E/L)}$ of §7.2.4.

By a result of Siegel, as in Conjecture 7.2.6, there exists a unique element $\Theta_{L/K,S}(r) \in \mathbf{Q}[G(L/K)]$ such that for all p-adic characters, $\chi : G(L/K) \longrightarrow \overline{\mathbf{Q}}_p^*$,

$$\chi(\Theta_{L/K,S}(r)) = L_{K,S}(1-r, \chi^{-1})$$

where, as in §7.2.10, the right-hand side is the Artin L-function of χ^{-1} with the Euler factors associated to the primes in S removed. Note that since χ is associated to an abelian extension L/K of totally real fields, $L_S(-n, \chi) = 0$ if $n > 0$ is even.

For $i = 1, 2$, consider the annihilators, $\mathrm{ann}_{\mathbf{Z}_p[G(L/K)]}(H^i(X_p; \mathbf{Z}_p(r)))$.

Let K_∞ be the cyclotomic $\mathbf{Z}_p$ extension of K, as in §6.1.1. It is well known (see Theorem 7.2.2 and Proposition 7.3.11) that for p odd, the results of [50] imply that

$$\Theta_{L/K,S}(r) \cdot \mathrm{ann}_{\mathbf{Z}_p[G(L/K)]}(H^1(X_p; \mathbf{Z}_p(r))) \subset \mathbf{Z}_p[G(L/K)]$$

provided that $L \cap K_\infty = K$. Incidentally, it is expected that this is also true without the assumption that $L \cap K_\infty = K$.

The Lichtenbaum-Quillen conjecture predicts an isomorphism between

$$H^i(X_p; \mathbf{Z}_p(r)) \quad \text{and} \quad K_{2r}(\mathcal{O}_{L,S}) \otimes \mathbf{Z}_p \quad \text{for} \quad i = 1, 2$$

so that, combining this with the Conjecture 7.2.6 would imply that

$$\begin{aligned}\Theta_{L/K,S}(r) &\cdot \mathrm{ann}_{\mathbf{Z}_p[G(L/K)]}(H^1(X_p; \mathbf{Z}_p(r))) \\ &\subseteq \mathrm{ann}_{\mathbf{Z}_p[G(L/K)]}(H^2(X_p; \mathbf{Z}_p(r))) \lhd \mathbf{Z}_p[G(L/K)].\end{aligned}$$

The following is the main result of this section. If $I \lhd \mathbf{Z}_p[G(L/K)]$ recall that the radical of I is defined to be equal to $\sqrt{I} = \{z \in \mathbf{Z}_p[G(L/K)] \mid z^m \in I \textit{ for some } m \geq 1\}$.

Theorem 7.3.2

Let p be an odd prime and $L \cap K_\infty = K$, as in §7.3.1, then

$$\begin{aligned}&\sqrt{\Theta_{L/K,S}(r) \cdot \mathrm{ann}_{\mathbf{Z}_p[G(L/K)]}(H^1(X_p; \mathbf{Z}_p(r)))} \\ &= \sqrt{\mathrm{ann}_{\mathbf{Z}_p[G(L/K)]}(H^2(X_p; \mathbf{Z}_p(r)))}.\end{aligned}$$

7.3.3 *Identifying radicals*

The proof of Theorem 7.3.2 will take the form of a discussion occupying the remainder of this section. We begin with some results which are useful in identifying radicals. Suppose that G is a finite abelian group and that M is a finite (left) $\mathbf{Z}[G]$-module. Let $I \subset \mathbf{Z}[G]$ be the annihilator ideal of M, $I = \mathrm{ann}_{\mathbf{Z}[G]}(M)$.

Lemma 7.3.4

Let $\mathcal{S}$ be the set of all prime ideals of $\mathbf{Z}[G]$ which contain I in §7.3.3. Then each prime in $\mathcal{S}$ is maximal and

$$\sqrt{I} = \bigcap_{\mathcal{P}\in\mathcal{S}} \mathcal{P}.$$

Proof. Sending $\alpha \in \mathbf{Z}[G]$ to the product over elements of M of copies of M embeds $\mathbf{Z}|[G]/I$ into a finite product of copies of M, since M is finite. Therefore $\mathbf{Z}|[G]/I$ is finite and so is each $\mathbf{Z}[G]/\mathcal{P}$ for $\mathcal{P} \in \mathcal{S}$. Therefore any such $\mathcal{P}$ contains a prime number, p. As $\mathbf{Z}[G]$ is a finitely generated abelian group, this forces $\mathbf{Z}[G]/\mathcal{P}$ to be a finite integral domain and hence a field. Therefore $\mathcal{P}$ is maximal.

Since the commutative ring, $\mathbf{Z}[G]/\sqrt{I}$, has no nilpotent elements the intersection of all the prime ideals of $\mathbf{Z}[G]/\sqrt{I}$ is $\{0\}$, which completes the proof. □

Lemma 7.3.5

Suppose that $\mathcal{P}$ is a maximal ideal of $\mathbf{Z}[G]$ in Lemma 7.3.4. Then there is a rational prime p and a p-adic character, $\chi : G \longrightarrow \overline{\mathbf{Q}}_p^*$, of order prime to p such that

$$\mathcal{P} = \{\alpha \in \mathbf{Z}[G] \mid \tilde{\chi}(\alpha) \in \overline{\mathbf{Z}}_p \text{ is not a unit}\}.$$

Here $\overline{\mathbf{Z}}_p$ is the integral closure of $\mathbf{Z}_p$ in $\overline{\mathbf{Q}}_p$ and $\tilde{\chi}(\sum_{g\in G} n_g \cdot g) = \sum_{g\in G} n_g\chi(g)$.

Proof. Let $\hat{G}$ denote the Pontryagin dual of G so that G is the Pontryagin dual of $\hat{G}$. The absolute representation ring $R(\hat{G}) = R_{\overline{\mathbf{Q}}}(\hat{G}) = K_0(\overline{\mathbf{Q}}[G])$ is the $\mathbf{Z}$-algebra generated by G and is thus isomorphic to $\mathbf{Z}[G]$. Let ζ be a root of unity of order equal to $|G|$, the order of G, and set $A = \mathbf{Z}[\zeta]$. The ring $A \otimes_{\mathbf{Z}} R(\hat{G})$ is finite over $R(\hat{G})$. Therefore, by ([94] Proposition I.9) every prime ideal of $R_{\hat{G}}$ is the intersection of $R_{\hat{G}}$ with a prime ideal of $A \otimes_{\mathbf{Z}} R_{\hat{G}}$. The result now follows from the description of the prime ideals of $A \otimes_{\mathbf{Z}} R_{\hat{G}}$ given in ([129] Proposition 30). □

Definition 7.3.6 Let $\mathcal{P}$, p and χ be as in Lemma 7.3.5 and let $\mathbf{Z}_p[\chi]$ be the ring generated over $\mathbf{Z}_p$ by the values of χ (cf. §6.2.1). Let T_χ be a rank one $\mathbf{Z}_p[\chi]$-module having an action of $g \in G$ given by $\chi(g)$. Suppose L is a $\mathbf{Z}[G]$-module. Then $\mathbf{Z}_p[\chi] \otimes_{\mathbf{Z}} L$ becomes a $\mathbf{Z}[G]$-module via the action, $g(\alpha \otimes z) = \alpha \otimes g(z)$. Define $L(\chi) = \mathrm{Hom}_{\mathbf{Z}_p[\chi]}(T_\chi, \mathbf{Z}_p[\chi] \otimes_{\mathbf{Z}} L)$. Then $L(\chi)$ is a $\mathbf{Z}_p[G]$-module via the action of G on T_χ and on $\mathbf{Z}_p(\chi) \otimes_{\mathbf{Z}} L$, by $g(f) = g(f(g^{-1} \cdot -)$. As usual, L^G and L_G are, respectively, the invariants and co-invariants of the G-action on L.

Lemma 7.3.7

Suppose that M is a finite simple $\mathbf{Z}[G]$-module and that $\mathcal{P}$, p, χ are as in Lemma 7.3.5. If $M(\chi)^G = 0$ or $M(\chi)_G = 0$ then $H^i(G; M(\chi)) = H_i(G; M(\chi)) = 0$ for all integers i.

Proof. If $pM = 0$ then $M(\chi) = 0$ so we may suppose $pM \neq 0$. Then M is inflated from a simple $\mathbf{Z}/p[G/G_p]$-module, where G_p is the p-Sylow subgroup of G. Since

$\mathbf{Z}/p[G/G_p]$ is semi-simple, we have $M(\chi)^G = 0$ if and only if $M(\chi)_G = 0$ and so we may assume that $M(\chi)^G = M(\chi)_G = 0$.

Let L^* be a resolution by projective $\mathbf{Z}/p[G_p]$-module of the one-dimensional module, $\mathbf{Z}/p$, with trivial G-action. Since G_p is also quotient of G, we can view L^* as a projective resolution of $\mathbf{Z}/p$ as a $\mathbf{Z}/p[G]$-module. Since $M(\chi)^G = M(\chi)_G = 0$, we find that the G-invariants of the terms of the complex $\mathrm{Hom}_{\mathbf{Z}/p}(L^*, M(\chi))$ and the G-coinvariants of the terms of the complex $L^* \otimes_{\mathbf{Z}/p} M(\chi)$ are trivial. Thus $H^i(G; M(\chi)) = H_i(G; M(\chi)) = 0$ for all integers i, as required. □

Lemma 7.3.8

Let M be a finite $\mathbf{Z}[G]$-module. Suppose $\mathcal{P}$ is a maximal ideal of $\mathbf{Z}[G]$ and that p and χ are as in Lemma 7.3.5 and Definition 7.3.6. Then the localisation, $M_{\mathcal{P}}$, of M at $\mathcal{P}$ is a $\mathbf{Z}_p[\chi]$-module, as are $M(\chi)^G$ and $M(\chi)_G$. If L is a finite $\mathbf{Z}_p[\chi]$-module, let $l_\chi(L)$ be the length of a composition series for L as a $\mathbf{Z}_p[\chi]$-module. Also

$$l_\chi(M_{\mathcal{P}}) \geq l_\chi(M(\chi)^G) \quad \text{and } l_\chi(M_{\mathcal{P}}) \geq l_\chi(M(\chi)_G)$$

with equalities if $p \not| |G|$, $M(\chi)^G = 0$ or $M(\chi)_G = 0$.

Proof. By Lemma 7.3.5, the algebra map $\mathbf{Z}[G] \to \mathbf{Z}_p[\chi]$ associated to χ induces a ring homomorphism, $\mathbf{Z}[G]_{\mathcal{P}} \to \mathbf{Z}_p[\chi]$ having dense image in the p-adic topology of $\mathbf{Z}_p[\chi]$. Since $M_{\mathcal{P}}$ is a finite $\mathbf{Z}[G]_{\mathcal{P}}$- module, we may view $M_{\mathcal{P}}$ as a $\mathbf{Z}_p[\chi]$-module. Since the action of G on $\mathbf{Z}_p[\chi]$ was defined to be trivial, Definition 7.3.6 shows that $M(\chi)^G$ and $M(\chi)_G$ are also $\mathbf{Z}_p[\chi]$-modules.

For the length relations, we suppose first that M is a simple $\mathbf{Z}[G]$-module. Then M is isomorphic to $\mathbf{Z}[G]/\mathcal{P}'$ for some maximal ideal $\mathcal{P}' \lhd \mathbf{Z}[G]$. Then we have stated equalities between lengths, by the description of such $\mathcal{P}'$ in Lemma 7.3.5.

Now suppose that M has a composition series of length $n > 1$ and that the result holds for modules of length strictly less than n. There is an exact sequence of the form

$$0 \longrightarrow M' \longrightarrow M \longrightarrow M/M' \longrightarrow 0$$

in which M' is a simple $\mathbf{Z}[G]$-module and M/M' has length $n-1$. We have exact sequences

$$0 \longrightarrow (M')_{\mathcal{P}} \longrightarrow M_{\mathcal{P}} \longrightarrow (M/M')_{\mathcal{P}} \longrightarrow 0$$

and

$$0 \to M'(\chi)^G \to M(\chi)^G \to (M/M')(\chi)^G \to H^1(G, M'(\chi)).$$

These sequences and the induction hypothesis immediately give the bound $l_\chi(M_{\mathcal{P}}) \geq l_\chi(M(\chi)^G)$, with equality, by induction, if p does not divide $|G|$, since then $H^1(G; M'(\chi)) = 0$. Suppose now that $M(\chi)^G = 0$. Then it must be the case that $M'(\chi)^G = 0$, so by Lemma 7.3.7 $H^1(G; M'(\chi)) = 0$. Thus the exact sequences show that $(M/M')(\chi)^G = 0$ together with the induction hypothesis imply that $M_{\mathcal{P}} = 0$. Note that the bound $l_\chi(M_{\mathcal{P}}) \geq l_\chi((M(\chi))_G)$ also shows in this case that $M(\chi)_G = 0$.

The proof that $M_{\mathcal{P}} = 0$ when $M(\chi)_G = 0$ is similar. □

Corollary 7.3.9

With the notation of Lemma 7.3.8, let G_p be the p-Sylow subgroup of G. Let $M' = M^{G_p}$ and $M'' = M_{G_p}$. Then M' and M'' are $\mathbf{Z}[G/G_p]$-modules. The image, $\mathcal{P}'$, of $\mathcal{P}$ in $\mathbf{Z}[G/G_p]$ is a maximal ideal. Also the vanishing of any one of $M_{\mathcal{P}}$, $M'_{\mathcal{P}'}$ or $M''_{\mathcal{P}'}$ implies the vanishing of the other two.

Proof. Since the character χ associated to $\mathcal{P}$ in Lemma 7.3.5 has order prime to p, it is trivial on G_p. Thus $\mathcal{P}'$ is the prime ideal of $\mathbf{Z}[G/G_p]$ associated to the character χ' of G/G_p induced by χ. Since $M(\chi)^G = M'(\chi)^{G/G_p}$, Lemma 7.3.8 shows that $M_{\mathcal{P}} = 0$ if and only if $M'_{\mathcal{P}'} = 0$. The equivalence of $M_{\mathcal{P}} = 0$ and $M''_{\mathcal{P}'} = 0$ is proved similarly. □

7.3.10 *A result of Deligne and Ribet [50]*

As in §7.3.1, assume that p is an odd prime and that $L \cap K_\infty = K$ where K_∞ is the cyclotomic $\mathbf{Z}_p$-extension of K. Our next objective is to prove the following result, mentioned in §7.3.1.

Proposition 7.3.11

In §7.3.10

$$\Theta_{L/K,S}(r) \cdot \mathrm{ann}_{\mathbf{Z}_p[G(L/K)]}(H^1(X_p; \mathbf{Z}_p(r))) \subseteq \mathbf{Z}_p[G(L/K)].$$

Proof. Let L' be the totally real subfield $L(\mu_p)^+$ of the CM field obtained by adjoining the pth roots of unity to K. Then $L' \cap K_\infty = K$ since $[L' : L]$ divides $p-1$. As in ([166] §2), the Galois group $G(L'_\infty/K)$ is a product

$$G(L'_\infty/K) = G(L'/K) \times G(K_\infty/K).$$

The associated completed group ring is $\mathbf{Z}_p[[G(L'_\infty/K)]] \cong R[[T]]$ where $R = \mathbf{Z}_p[G(L'/K)]$ and a (fixed choice of a) topological generator, $\gamma \in G(K_\infty/K) \cong \mathbf{Z}_p$, is sent to $1+T$ (see §6.1.1).

We have an isomorphism of the form $R = \prod_{m\in\Pi} R_m$ where Π is the set of maximal ideals of R and R_m is the completion of R at m. The characters ψ of $G(L'/K)$ associated to $m \in \Pi$ are the restrictions to $G(L'/K)$ of the compositions $R \to R_m \to \overline{\mathbf{Q}}_p$ for some $\mathbf{Z}_p$-algebra homomorphism from R_m to $\overline{\mathbf{Q}}_p$. Among these ψ there is a unique $G(\overline{\mathbf{Q}}_p/\mathbf{Q}_p)$-conjugacy class of characters of order prime to p. These are called the basic characters of m in ([166] §2). Let $e_\psi : R_m \to \mathbf{Z}_p[\psi]$ be the natural algebra map associated to a p-adic character ψ of $G(L'/K)$. Setting $e_\psi(T) = T$ extends e_ψ to an algebra map $e_\psi : R_m[[T]] \to \mathbf{Z}_p[[T]]$ between formal power series rings. The p-adic L-function $L_{p,S}(1-n,\psi)$ has the property (cf. §6.2.1) that

$$L_{p,S}(1-n,\psi) = L_{K,S\bigcup S_p}(1-n,\psi\omega^{-n})$$

for $n \geq 1$. Here $\omega : G(K(\mu_p)/K) \to \mu_{p-1} \subset \mathbf{Z}_p^*$ is the Teichmüller character. As in ([166] p. 494), define $u \in \mathbf{Z}_p^*$ by requiring $\gamma(\zeta) = \zeta^u$ for all p-power roots of unity ζ, where γ is the chosen topological generator of $G(K_\infty/K)$. Here one lifts

γ to $G(K(\mu_{p^\infty}/K) = G(K(\mu_p)/K) \times G(K_\infty/K)$ by requiring the projection of γ to $G(K(\mu_p)/K)$ to be trivial.

The results of [50] have the following consequences (see [166] §2; see also §6.2.1). If there is non-trivial basic character associated to m then there is a power series $G_{m,S}(T) \in R_m[[T]]$ such that

$$e_\psi(G_{m,S})(u^r - 1) = L_{p,S}(1-r,\psi) = L_{K,S\bigcup S_p}(1-r,\psi\omega^{-r})$$

for all ψ belonging to m. Suppose now that the only basic character associated to m is trivial. Pick an ideal c of K prime to S and p, and let σ_c be the image in $G(K_\infty/K)$ of c under the Artin homomorphism. Let $H_{m,c}(T)$ be the power series corresponding to $1-\sigma_c \in \mathbf{Z}_p[[G(L'_\infty/K)]]$. Then there is a power series $G_{m,c,S}(T) \in R_m[[T]]$ such that

$$\frac{e_\psi(G_{m,S})(u^r-1))}{e_\psi(H_{m,c})(u^r-1)} = L_{p,S}(1-r,\psi) = L_{K,S\bigcup S_p}(1-r,\psi\omega^{-r})$$

for all ψ belonging to m.

Recall that $\Theta(L/K,S,r) \in \mathbf{Q}[G(L/K)]$ was defined by the requirement that $\chi(\Theta_{L/K,S}(r)) = L_{K,S}(1-r,\chi^{-1})$ for all characters χ of $G = G(L/K)$. Define $\Theta' = \Theta(L/K, S\bigcup S_p, r)$. Since S contained all the primes of K which ramify in L, each prime $\mathcal{P}$ of K over p is unramified in L. For all characters χ of $G(L/K)$ one has

$$L_{K,S\bigcup S_p}(1-r,\chi^{-1}) = L_{K,S}(1-r,\chi^{-1}) \cdot \prod_{\mathcal{P}\in S_p} (1-\chi^{-1}((\mathcal{P},L/K)N\mathcal{P}^{r-1})$$

It follows that

$$\Theta' = \Theta(r) \cdot \prod_{\mathcal{P}\in Sp} (1-(\mathcal{P},L/K)^{-1}N\mathcal{P}^{r-1}).$$

Since $r > 1$, the product over $\mathcal{P}$ in the formula for Θ' is a unit in $\mathbf{Z}_p[G]$. Hence to prove Proposition 7.3.11, Proposition 7.2.5 shows it will suffice to prove that

$$\Theta' \cdot (1-(C,L/K)^{-1} \cdot NC^r) \subset \mathbf{Z}_p[G(L/K)]$$

for all primes C of K not in S_p.

If m has a non-trivial basic character, then $H_{m,C}(u^s - 1) = 1$ by definition. If m has trivial basic character, then $H_{m,C}(T)$ equals the power series, introduced earlier, corresponding to $1 - \sigma_C = 1 - (C, L/K)$ in $R_m[[T]]$. Now

$$\sigma_C = ((C,L'/K),\tau(C)) \in G(L'_\infty/K)G(L'/K) \times G(K_\infty/K) \cong G(L'_\infty/K).$$

where $\tau(C) \in G(K_\infty/K)$ is defined in the following manner. Restriction of Galois automorphisms induces an injection $G(K_\infty/K) \subset G(\mathbf{Q}_\infty/\mathbf{Q})$ and we have a natural isomorphism between $G(\mathbf{Q}_\infty/\mathbf{Q})$ and the multiplicative subgroup $1+p\mathbf{Z}_p \subset \mathbf{Z}_p^*$.

The image of $\tau(C)$ in $(1+p\mathbf{Z}_p)^*$ is the projection of $NC \in \mathbf{Z}_p^*$ to the second factor under the canonical isomorphism $\mathbf{Z}_p^* \equiv \mu(\mathbf{Z}_p^*) \times (1+p\mathbf{Z}_p)^*$. Using this we find

$$H_{m,C}(u^r - 1) = 1 - (C, L'/K) \cdot \tau(C)^r.$$

Observe now that if $\psi = \chi \cdot \omega^r$ then

$$\begin{aligned} e_\psi(H_{m,C}(u^r-1)) &= 1 - \chi((C,L'/K) \cdot \omega^r((C,L'/K)) \cdot \tau(C)^r \\ &= 1 - \chi((C,L'/K)) \cdot NC^r. \end{aligned}$$

Since $K \subset L \subset L'$, we have a natural surjection $G(L'/K) \to G(L/K)$ inducing a surjection of algebras, $\pi : R = \mathbf{Z}_p[G(L'/K)] \to \mathbf{Z}_p[G(L/K)]$. Define a twist $\pi_r : R \to \mathbf{Z}_p[G]$ by $\pi_r(h) = \omega^r(h) \cdot \pi(h)$ for $h \in G(L'/K)$ and extending to R by $\mathbf{Z}_p$-linearity. Then π_r is a homomorphism of algebras. If χ is a character of $G(L/K)$, we can regard χ also as a character of $G(L'/K)$ by inflation. The previous relation implies, for all $\alpha \in R$,

$$\psi(\alpha) = \chi(\pi_r(\alpha)).$$

We are now in a position to complete the proof of Proposition 7.3.11. Define $\alpha \in R$ by specifying its R_m-component, α_m, in the following manner. Define

$$\alpha_m = \begin{cases} G_{m,S}(u^r-1) \cdot (1 - (C,L'/K) \cdot \omega((C,L'/K))^{-r} \cdot NC^r) \\ \qquad\qquad \text{if } m \text{ has trivial basic character} \\ G_{m,S}(u^r-1) \\ \qquad\qquad \text{otherwise.} \end{cases}$$

Suppose χ and ψ are characters of $G(L/K)$ and $G(L'/K)$, respectively, satisfying $\psi = \chi \cdot \omega^r$, as before. If m has a non-trivial basic character and ψ belongs to m, then

$$\begin{aligned} \psi(\alpha_m) &= e_\psi(G_{m,S}(u^r-1) \cdot (1 - (C,L'/K) \cdot \omega((C,L'/K))^{-r} \cdot NC^r)) \\ &= L_{K,S\cup S_p}(1-r,\chi) \cdot \chi(1 - (C,L'/K) \cdot NC^r)) \end{aligned}$$

If m has trivial basic character and ψ belongs to m, then

$$\begin{aligned} \psi(\alpha_m) &= e_\psi(G_{m,S}(u^r-1)/H_{m,C}(u^r-1)) \cdot e_\psi(H_{m,C}(u^r-1)) \\ &= L_{K,S\cup S_p}(1-r,\chi) \cdot \chi(1 - \sigma(C,L'/K) \cdot NC^r). \end{aligned}$$

Define $\beta \in \mathbf{Z}_p[G(L/K)]$ by $\beta = \tau(\pi_r(\alpha))$ where, as in the proof of Theorem 7.1.11, τ is the automorphism of $\mathbf{Z}_p[G(L/K)]$ induced by sending each element of $G(L/K)$ to its inverse. Then we find, for all characters χ of $G(L/K)$, that

$$\begin{aligned} \chi(\beta) &= \chi^{-1}(\pi_r(\alpha)) \\ &= L_{K,S\cup S_p}(1-r,\chi^{-1}) \cdot (1 - \chi((C,L'/K)^{-1}) \cdot NC^r) \\ &= \chi(\Theta_{L/K,S}(r) \cdot (1 - (C,L'/K)^{-1} \cdot NC^r)) \end{aligned}$$

It follows that

$$\Theta_{L/K,S}(r) \cdot (1 - (C,L'/K)^{-1} \cdot NC^r) = \beta \in \mathbf{Z}_p[G(L/K)],$$

completes the proof of Proposition 7.3.11. □

7.3.12 *Proof of Theorem 7.3.2*

In view of Lemma 7.3.5, to prove Theorem 7.3.2 it will suffice to show, if p is an odd prime and $\mathcal{P}$ is a prime ideal of $\mathbf{Z}[G(L/K)]$ containing p, then $\mathcal{P}$ contains $\Theta_{L/K,S}(r) \cdot \mathrm{ann}_{\mathbf{Z}_p[G(L/K)]}(H^1(X_p; \mathbf{Z}_p(r)))$ if and only if it contains $\mathrm{ann}_{\mathbf{Z}_p[G(L/K)]}(H^2(X_p; \mathbf{Z}_p(r)))$.

As before, let G_p be the p-Sylow subgroup of $G(L/K)$. For $i = 1, 2$, let $J'_{i,r,p}$ for be the image of $\mathrm{ann}_{\mathbf{Z}_p[G(L/K)]}(H^i(X_p; \mathbf{Z}_p(r)))$ in $\mathbf{Z}[G(L/K)/G_p]$ and let $\mathcal{P}'$ be the image of $\mathcal{P}$. Let $J''_{i,r,p}$ be the ideal of $\mathbf{Z}[G(L/K)/G_p]$ which results on replacing L/K in the definition of $J'_{i,r,p}$ by L'/K, where $L' = L^{G_p}$.

By the definition of $\Theta_{L/K,S}(r)$ and the invariance of Artin L-functions under inflation of characters (see Proposition 1.1.8), the image $\Theta'(r)$ of $\Theta_{L/K,S}(r)$ in $\mathbf{Q}[G/G_p]$ is the element which results from replacing L/K in the definition of $\Theta_{L/K,S}(r)$ by L'/K, where $L' = L^{G_p}$.

Proposition 7.2.5 shows that $J'_{1,r,p} = J''_{1,r,p}$. Lemma 7.3.5 shows that $\mathcal{P}$ is the unique prime of $\mathbf{Z}[G(L/K)]$ lying over the image, $\mathcal{P}'$, of $\mathcal{P}$ in $\mathbf{Z}[G(L/K)/G_p]$ so that $\mathcal{P}$ contains $\Theta_{L/K,S}(r)\cdot\mathrm{ann}_{\mathbf{Z}_p[G(L/K)]}(H^1(X_p; \mathbf{Z}_p(r)))$ if and only if $\mathcal{P}'$ contains $\Theta'(r)J''_{1,r,p}$.

It follows from [38] Lemma 6.20) that the covariants $H^2(X_p; \mathbf{Z}_p(r))_{G_p}$ are isomorphic as a $\mathbf{Z}[G(L/K)/G_p]$-module to the corresponding cohomology group when L/K is replaced by L'/K. Thus Lemma 7.3.8 and Corollary 7.3.9 imply that $\mathcal{P}$ contains $\mathrm{ann}_{\mathbf{Z}_p[G(L/K)]}(H^2(X_p; \mathbf{Z}_p(r)))$ if and only if $\mathcal{P}'$ contains $J''_{2,r,p}$. Combining this with the above remarks allows us to replace L/K by L'/K and thus reduce to the case where p is prime to the order of $G(L/K)$. In this case, Theorem 7.3.2 follows from Lemma 7.3.8 and the expression for the ideal of $\mathbf{Z}[\chi]$ generated by $L_S(1-r,\chi)$ given in ([38] Proposition 6.15, eq. 6.32) when χ is a character of $G(L'/K)$. □

Bibliography

[1] G. Banaszak: Algebraic K-theory of number fields and rings of integers and the Stickelberger ideal; Annals of Math. 135 (1992) 325–360.

[2] D. Barsky: Fonctions Zeta p-adiques d'une Classe de Rayon des Corps de Nombres Totalement-réels; Groupe d'étude d'analyse ultrametrique (1977–78).

[3] H. Bass: *Algebraic K-theory*; (1968) W. A. Benjamin.

[4] H. Bass, J. Milnor and J-P. Serre: The solution of the congruence subgroup problem for SL_n ($n \geq 3$) and Sp_n ($n \geq 2$); Pub. Math. I. H. E. S. (Paris) 33 (1967).

[5] P. Bayer and J. Neukirch: On values of zeta functions and l-adic Euler characteristics; Inventiones Math. 50 (1978) 35–64.

[6] A. A. Beilinson: Height pairings between algebraic cycles; Contemp. Math. 67 (1987) 1–24.

[7] A. A. Beilinson: Higher regulators and values of L-functions of curves; Funct. Anal. Appl. 14 (1980) 116–118.

[8] A. A. Beilinson: Higher regulators and values of L-functions; Sovremennye Problemy Matematiki 24 Moscow, VINITI (1984) Trans. in J. Soviet Math. 30 (1985) 2036–2070.

[9] A. A. Beilinson: Polylogarithms and cyclotomic elements; preprint (1990).

[10] A. A. Beilinson, R. MacPherson and V. Schechtman: Notes on motivic cohomology; Duke Math. J. 54 (1987) 679–710.

[11] S. Bentzen and I. Madsen: Trace maps in algebraic K-theory and the Coates-Wiles homomorphism; J. reine angew. Math. 411 (1990) 171–195.

[12] W. Bley and D. Burns: Equivariant epsilon constants, discriminants and étale cohomology; King's College preprint (10-2000).

[13] S. Bloch: Higher regulators, algebraic K-theory and zeta functions of elliptic curves; Irvine University Lecture Notes.

[14] S. Bloch and K. Kato: L-functions and Tamagawa numbers; Grothendieck Festschrift vol 1, Progress in Mathematics 86, Birkhäuser (1990) 333–400.

[15] R. A. Bogomolov: Two theorems on the divisibility and torsion in the Milnor K-groups; Mat. Sb. 130 (1986) 3, 404–412; English transl. Math. USSR-Sb. 58 (1987) 2, 407–416.

[16] M. Böksted and I. Madsen: Algebraic K-theory of local fields: the unramified case; Aarhus preprint #20 (1994).

[17] A. Borel: Cohomologie de SL_n et valeurs de fonctions zêta; Ann. Scuola Normale Sup. Ser. 4 7 (1974) 613–636 and

[18] A. Borel: Cohomology of arithmetic groups, Proc. 1974 ICM vol. I; Can. Math. Soc. (1975) 435–442.

[19] A. Borel: Stable real cohomology of arithmetic groups; Ann. Sci. Éc. Norm. Sup. (Paris) 7 (1974) 235–272.

[20] R. Brauer and C. Nesbitt: On the modular characters of groups; Annals of Math. 42 (1941) 556–590.

[21] A. Brumer: On the units of algebraic number fields; Mathematika 14 (1967) 121–124.

[22] D. Burns: Iwasawa theory and p-adic Hodge theory over non-commutative algebras I; King's College preprint (30-4-97).

[23] D. Burns: Equivariant Tamagawa numbers and Galois module theory II; King's College preprint (16-10-1998) – to appear in Compositio Math.

[24] D. Burns and M. Flach: Motivic L-functions and Galois module structure invariants; Math. Annalen 305 (1996) 65–102.

[25] D. Burns and M. Flach: On Galois structure invariants associated to Tate motives; Amer. J. Math. 120 (1998) 1343–1397.

[26] D. Burns and M. Flach: Tamagawa numbers for motives with non-commutative coefficients; King's College preprint (2000).

[27] D. Burns and C. Greither: On the equivariant Tamagawa number conjecture for Tate motives; King's College preprint (10-2000).

[28] K. Buzzard and R. Taylor: Companion forms and weight 1 forms; Annals of Math. 149 (1999) 905–919.

[29] H. Carayol: Formes modulaires er représentations Galoisiennes à valeurs dans un anneau local complet; Contemp. Math. 165 (1994) 213–237.

[30] Ph. Cassou-Noguès: Quelques théorèmes de base normal d'entiers; Ann. Inst. Fourier 28 (1978) 1–33.

[31] P. Cassou-Nogués: Valeurs aux entiéres negatif des fonctions zeta et fonctions zeta p-adique; Inventiones Math. 51 (1979) 29–59.

[32] Ph. Cassou-Nogués, T. Chinburg, A. Fröhlich and M. J. Taylor: L-functions and Galois modules; London Math. Soc. 1989 Durham Symposium, *L-functions and Arithmetic* Cambridge University Press (1991) 75–139.

[33] C.-L. Chai: Arithmetic minimal compactification of the Hilbert-Blumenthal spaces; Annals of Math. 131 (1990) 541–554.

[34] T. Chinburg: Exact sequences and Galois module structure; Annals Math. 121 (1985) 351–376.

[35] T. Chinburg: *The Analytic Theory of Multiplicative Galois Structure*; Mem. A. M. Soc. 395 (1989).

[36] T. Chinburg, M. Kolster, G. Pappas and V. P. Snaith: Galois structure of K-groups of rings of integers; C. R. Acad. Sci. Paris t. 320, Séries I (1995) 1435–1440.

[37] T. Chinburg, G. Pappas, M. Kolster and V. Snaith: Quaternionic exercises in K-theory Galois module structure; Proc. Great Lakes K-theory Conf., Fields Institute Communications Series #16 (A. M. Soc. Publications) 1–29 (1997).

[38] T. Chinburg, M. Kolster, G. Pappas and V. P. Snaith: Galois module structure of K-groups of rings of integers; K-Theory 14 (1998) (4), 319–369.

[39] T. Chinburg, M. Kolster and V. Snaith: Quaternionic exercises in K-theory Galois module structure II; Algebraic K-theory and its Applications (eds. H. Bass, A. O. Kuku and C. Pedrini); Proc. Symp. I. C. T. P. Trieste World Scientific Press (1999) 337–369.

[40] T. Chinburg, M. Kolster and V. Snaith: Comparison of K-theory Galois module structure invariants; Can. J. Math. (1) 52 (2000) 47–91.

[41] T. Chinburg and V. Snaith: On the Wiles Unit; talk at Oberwolfach (1998), preprint (1999).

[42] L. G. Chouinard: Projectivity and relative projectivity over group rings; J. Pure Appl. Alg. 7 (1976) 287–302.

[43] J. H. Coates: p-adic L-functions and Iwasawa theory; *Algebraic Number Fields* (ed. A. Fröhlich Academic Press (1977).

[44] J. H. Coates and W. Sinnott: An analogue of Stickelberger's theorem for the higher K-groups; Inventiones Math. 24 (1974) 149–161.

[45] J. H. Coates and W. Sinnott: On p-adic L-functions over real quadratic fields; Inventiones Math. 25 (1974) 253–279.

[46] R. F. Coleman: Dilogarithms, regulators and p-adic L-functions; Inventiones Math. 69 (1982) 171–208.

[47] P. Cornacchia and P. A. Ostvaer: On the Coates-Sinnott conjecture; K-theory (2) 19 (2000) 195–209.

[48] C. W. Curtis and I. Reiner: *Representation Theory of Finite Groups and Associative Algebras*; Interscience (1962).

[49] C. W. Curtis and I. Reiner: *Methods of Representation Theory* vols. I & II, Wiley (1981, 1987).

[50] P. Deligne and K. Ribet: Values of abelian L-functions at negative integers over totally real fields; Invent. Math. 59 (1980) 227–286.

[51] W. G. Dwyer and E. M. Friedlander: Algebraic and étale K-theory; Trans. A. M. Soc. 292 (1) 247–280 (1985).

[52] W. Dwyer, E. M. Friedlander, V. P. Snaith and R. W. Thomason: Algebraic K-theory eventually surjects onto topological K-theory; Inventiones Math. 66 (1982) 481–491.

[53] B. Ferrero and L. Washington: The Iwasawa invariant μ_p vanishes for abelian fields; Annals of Math. 109 (1979) 377–395.

[54] I. B. Fesenko and S. V. Vostokov: *Local fields and their extensions – a constructive approach*; Trans. of Math. Monographs vol. 121, Amer. Math. Soc. (1991).

[55] J-M. Fontaine and W. Messing: p-adic periods and p-adic étale cohomology; Current Trends in Arithmetical Algebraic Geometry (ed. K. A. Ribet) Contemp. Math. A. M. Soc. 67 (1987) 179–207.

[56] E. M. Friedlander: Motivic complexes of Suslin and Voevodsky; Séminaire Bourbaki no. 833 (1996–97) 355–378.

[57] A. Fröhlich: *Galois Module Structure of Algebraic Integers*; Ergeb. Math. 3 Folge-Band 1 Springer-Verlag (1983).

[58] A. Fröhlich and J. Queyrut: On the functional equation for Artin L-functions for characters of real representations; Inventiones Math. 20 (1973) 125–138.

[59] A. Fröhlich and M. J. Taylor: *Algebraic Number Theory*; Cambridge Studies in Advanced Math. #27.

[60] T. Geisser and M. Levine: The K-theory of fields in characteristic p; Inventiones Math. (3) 139 (2000) 459–494.

[61] P. G. Goerss and J. F. Jardine: *Simplicial Homomotopy Theory*; Progress in Math. #174(1999) Birkhäuser.

[62] D. Grayson: Higher algebraic K-theory II (after D. G. Quillen); Algebraic K-theory (1976) 217–240, Lecture Notes in Math. #551, Springer Verlag.

[63] D. R. Grayson: On the K-theory of fields; Proc. Conf. algebraic K-theory and algebraic number theory (Honolulu 1987) A. M. Soc. Contemp. Math. 83 (1989) 31–55.

[64] R. Greenberg: On p-adic Artin L-functions; Nagoya J. Math. 89 (1983) 77–87.

[65] C. Greither: Some cases of Brumer's conjecture for abelian CM extensions of totally real fields; Math. Zeit. 233 (2000) 515–534.

[66] M. Gros: Régulateurs syntomiques et valeurs de fonctions L p-adiques I (with an appendix by M Kurihara); Inventiones Math. 99 (1990) 293–320.

[67] M. Gros: Régulateurs syntomiques et valeurs de fonctions L p-adiques II; Inventiones Math. 115 (1994) 61–79.

[68] K. W. Gruenberg, J. Ritter and A. Weiss: A local approach to Chinburg's root number conjecture; Proc. L. M. Soc. 79 (1999) 47–80.

[69] D. Haran: Cohomology theory of Artin-Schreier structures; J. Pure Appl. Algebra 69 (1990) 141–160.

[70] R. Hartshorne: *Algebraic Geometry*; 3rd edition (1983) Springer Verlag Grad. Texts in Maths #52.

[71] R. Hartshorne: *Residues and duality*; Lecture Notes in Math. 20 (1966) Springer Verlag.

[72] D. R. Hayes: Base change for the conjecture of Brumer-Stark; J. Reine. Angew. Math. 497 (1998) 83–89.

[73] L. Hesselholt and I. Madsen: On the K-theory of finite algebras over Witt vectors of perfect fields; Topology 36 (1997) 29–102.

[74] L. Hesselholt and I. Madsen: K theory of local fields; Arhus preprint (1999).

[75] H. Hida: *Elementary theory of L-functions and Eisenstein series*; L. M. Soc. Student Texts 26 (Cambridge University Press) 1993.

[76] H. Hida: *Modular forms and Galois cohomology*; Cambridge Studies in Advanced Math. #69 (2000) Cambridge University Press.

[77] J. Hooper, V. Snaith and M. van Tran: The second Chinburg conjecture for quaternion fields; Mem. A. M. Soc. vol. 148 #704 (2000).

[78] A. Huber and J. Wildeshaus: Classical polylogarithms according to Beilinson and Deligne; Doc. Math. J. DMV 3 (1998) 27–133.

[79] K. Iwasawa: *Local Class Field Theory*; Oxford Math. Monographs, Oxford University Press (1986).

[80] B. Kahn: Descente Galoisienne et K_2 des corps de nombres; K-theory 7 (1) 55–100 (1993).

[81] B. Kahn: The decomposable part of motivic cohomology and bijectivity of the norm residue homomorphism; Contemp. Math. 126 (1992) 79–87.

[82] B. Kahn: On the Lichtenbaum-Quillen Conjecture; Proc. Lake Louis K-theory Conf. *Algebraic K-Theory and Algebraic Topology* (eds. J. F. Jardine and P. Goerss) Kluwer NATO series (1993) 147–166.

[83] I. Kaplansky: *Fields and Rings*; University of Chicago Press 2nd. edition (1972).

[84] M. Karoubi: *K-Theory, an introduction*; Grund. Math. Wiss. #226 (1978) Springer Verlag.

[85] K. Kato: Iwasawa theory and p-adic Hodge theory; Kodai Math. J. 16 (1993) 1–31.

[86] K. Kato: Lectures on the approach to Iwasawa theory of Hasse-Weil L-functions via B_{dR} Part I; Arithmetical Algebraic Geometry (ed. E. Ballico), Springer-Verlag Lecture Notes in Math. 1553 (1993) 50–163.

[87] K. Kato: Lectures on the approach to Iwasawa theory of Hasse-Weil L-functions via B_{dR} Part II; preprint (1993).

[88] M.-A. Knus and M. Ojanguren: *Théorie de la descente et algèbres d'Azumaya*; Lecture Notes in Math. 389 (Springer-Verlag) 1974.

[89] M. Kolster: A relation between the 2-primary parts of the Main Conjecture and the Birch-Tate Conjecture; Can. Math. Bull. 32 (12) 248–251 (1989).

[90] M. Kolster, T. Nguyen Quang Do and V. Fleckinger: Twisted S-units, p-adic class number formulas and the Lichtenbaum conjectures; Duke J. Math. 84 (1996) 679–717 (erratum Duke J. Math. 90 (1997) 641–643 plus further errata).

[91] M. Kolster and T. Nguyen Quang Do: Syntomic regulators and special values of p-adic L-functions; Inventiones Math. (2) 133 (1998) 417–448.

[92] Y. Koya: A generalization of class formations using hypercohomology; Inventiones Math. 101 (1990) 705–715.

[93] T. Kubota and H. Leopoldt: Eine p-adische Theorie der Zetawerte; J. Reine. Angew. Math. 213 (1964) 328–339.

[94] S. Lang: *Algebraic Number Theory*; Addison-Wesley (1970).

[95] S. Lang: *Algebra* 2nd ed. ; Addison-Wesley (1984).

[96] M. Levine: The indecomposable K_3 of a field; Ann. Sci. École Norm. Sup. 22 (1989) 255–344.

[97] S. Lichtenbaum: The construction of weight-two arithmetic cohomology; Inventiones Math. 88 (1987) 183–215.

[98] S. Lichtenbaum: New results on weight-two motivic cohomology; The Grothendieck Festschrift vol. III (Birkhäuser 1990) pp. 35–55.

[99] J-L. Loday and D. G. Quillen: Cyclic homology and the Lie algebra homology of matrices; Comm. Math. Helv. 59 (1984) 565–591.

[100] J-L. Loday and D. G. Quillen: Homologie cyclique et homologie de l'algèbre de Lie des matrices; C. R. Acad. Sci. Paris t. 296 (1983) 295–297.

[101] J. Martinet: Character theory and Artin L-functions; London Math. Soc. 1975 Durham Symposium, *Algebraic Number Fields* Academic Press (1977) 1–87.

[102] B. Mazur: Deforming Galois representations pp. 385–437 in *Galois groups over* **Q** (eds. Y. Ihara, K. Ribet and J-P. Serre) MSRI Pub. no. 16 (Springer-Verlag) 1989.

[103] A. S. Merkurjev: On the torsion of K_2 of local fields; Annals of Maths. (2) 118 (1983) 375–381.

[104] A. S. Merkurjev: K_2 of fields and the Brauer group; Applications of algebraic K-theory to geometry and number theory (Part II) Contemp. Math. #55 (1986) 529–546.

[105] A. S. Merkurjev and A. A. Suslin: K-cohomology of Severi-Brauer varieties and the norm residue homomorphism; Izv. Nauk. SSSR 46 (1982) 1011–1046 (Eng. trans. Math. USSR 21 (1983) 307–340).

[106] A. S. Merkurjev and A. A. Suslin: The K_3 group of a field; Izv. Nauk. SSSR 54 (1990) 339–356 (Eng. trans. Math. USSR 36 (1990) 541–565).

[107] J. Milne: *Étale cohomology*; Princeton University Press (1980).

[108] J. Milne: *Arithmetic Duality Theorems*, Perspectives in Mathematics #1 (1986) Academic Press, New York.

[109] J. W. Milnor: *Introduction to Algebraic K-theory*; Ann. Math. Studies #72 (1971) Princeton University Press.

[110] J. W. Milnor: Algebraic K-theory and quadratic forms; Inventiones Math. 9 (1970) 318–344.

[111] J. W. Milnor: The realisation of a semi-simplicial complex; Annals of Math. 65 (1957) 357–362.

[112] J. Neukirch: The Beilinson conjecture for algebraic number fields; Perspectives in Math. #4 (eds. M. Rapoport, N. Schappacher, P. Schneider) Academic Press (1988) 193–247.

[113] R. Oliver: SK_1 for finite group rings II; Math. Scand. 47 (1980) 195–231.

[114] C. Queen: The existence of p-adic L-functions; *Number Theory and Algebra*, Collected papers dedicated to Henry B. Mann (Academic Press, New York and London) 263–288 (1977).

[115] D. G. Quillen: On the cohomology and K-theory of the general linear groups over a finite field; Annals of Math. 96 (1972) 552–586.

[116] D. G. Quillen: On the finite generation of the groups K_i of rings of algebraic integers; Algebraic K-theory I (1973) 179–198, Lecture Notes in Math. # 341, Springer-Verlag.

[117] D. G. Quillen: High Algebraic K-theory I; Algebraic K-theory I (1973) 85–147, Lecture Notes in Math. # 341, Springer-Verlag.

[118] M. Rapoport: Comparison of the regulators of Beilinson and Borel; Perspectives in Math. #4 (eds. M. Rapoport, N. Schappacher, P. Schneider) Academic Press (1988) 169–192

[119] I. Reiner: *Maximal Orders*; L. M. Soc. Monographs #5 (1975) Academic Press.

[120] K. Ribet: Report on p-adic L-functions over totally real fields; Astérisque 61 (1979) 177–192.

[121] J. Ritter and A. Weiss: The lifted root number conjecture and Iwasawa theory; preprint University of Augsburg (2000).

[122] J. Ritter and A. Weiss: A Tate sequence for global units; Compositio Math. 102 (1996) 147–178.

[123] J. Rognes and C. A. Weibel (with an appendix by M. Kolster): Journal A. M. Soc. (1) 13 (2000) 1–54.

[124] J. Rosenberg: *Algebraic K-theory and its applications*; Grad. Texts in Math. #147 (1994) Springer Verlag.

[125] J. Sands: Abelian fields and the Brumer-Stark conjecture; Comp. Math. 53 (1984) 337–346.

[126] J. Sands: Base change for higher Stickelbeger ideals; J. Number. Th. 73 (1998) 518–526.

[127] G. B. Segal: Classifying spaces and spectral sequences; Publ. Math. I. H. E. S. 34 (1968) 105–112.

[128] J-P. Serre: *Local Fields*; Grad. Texts in Math. # 67 (1979) Springer-Verlag.

[129] J-P. Serre: *Linear Representations of Finite Groups*; Grad. Texts in Math. # 42 (1977) Springer-Verlag.

[130] V. P. Snaith: *Topological Methods in Galois Representation Theory*; C. M. Soc. Monographs (1989) Wiley.

[131] V. P. Snaith: *Explicit Brauer Induction (with applications to algebra and number theory)*; Cambridge Studies in Advanced Math. #40 (1994) Cambridge University Press.

[132] V. P. Snaith: *Galois Module Structure*; Fields Institute Monographs # 2 A. M. Soc. (1994).

[133] V. P. Snaith: On the localisation sequence in K-theory; Proc. A. M. Soc. (3) 79 (1980) 359–364.

[134] V. P. Snaith: Local fundamental classes derived from higher dimensional K-groups; Proc. Great Lakes K-theory Conf., Fields Institute Communications Series #16 (A. M. Soc. Publications) 285–324 (1997).

[135] V. P. Snaith: Local fundamental classes derived from higher dimensional K-groups II; Proc. Great Lakes K-theory Conf., Fields Institute Communications Series #16 (A. M. Soc. Publications) 325–344 (1997).

[136] V. P. Snaith: Local fundamental classes derived from higher K-groups in characteristic p; McMaster preprint #2 (1996/7).

[137] V. P. Snaith: Local fundamental classes derived from higher dimensional K-groups III; Proc. Dublin Conf. on Quadratic Forms (1999); Contemp. Math. series A. M. Soc. 272 (2000).

[138] V. P. Snaith: Cyclotomic Galois Module Structure and the Second Chinburg invariant; Math. Proc. Camb. Phil. Soc. 117 (1995) 57–82.

[139] V. P. Snaith: The second Chinburg invariant for cyclotomic fields via the Hom-description; Comptes Rendus Royal Soc. Can. (1) XVII (1995) 25–30.

[140] V. P. Snaith: Burns' equivariant Tamagawa number conjecture for quaternion fields; University of Southampton preprint (1998).

[141] M. Somekawa: Log-syntomic regulators and p-adic polylogarithms; preprint Tokyo Inst. Tech. (1995).

[142] C. Soulé: K-théorie des anneaux d'entiers de corps de nombres et cohomologie étale; Inventiones Math. 55 (1979) 251–295.

[143] C. Soulé: On higher p-adic regulators; Springer-Verlag Lecture Notes in Mathematics #854 (1981) 372–401.

[144] C. Soulé: Éléments cyclotomiques en K-théorie; Soc. Math. Fr. Astérisque 147–148 (1987) 225–257.

[145] E. H. Spanier: *Algebraic Topology*; (1966) McGraw-Hill.

[146] M. Spiess: Artin-Verdier duality for arithmetic surfaces; Math. Annalen 305 (1996) 705–792.

[147] M. Spiess: Private communication (13-11-2000).

[148] A. A. Suslin: On the K-theory of local fields; J. P. A. Alg. (2–3) 34 (1984) 301–318.

[149] R. G. Swan: Induced representations and projective modules; Annals of Math. 71 (1960) 552–578.

[150] R. G. Swan: *Algebraic K-theory*; Lecture Notes in Math. #76 Springer Verlag (1968).

[151] J. Tate: On the torsion in K_2 of fields; *Algebraic Number Theory* Proc. International Symposium, Kyoto (1976) 243–261, published by the Japanese Society for the Promotion of Science.

[152] J. T. Tate: *Les Conjectures de Stark sur les Fonctions L d'Artin en $s = 0$*; Progress in Mathematics # 47 (1984) Birkhauser.

[153] J. T. Tate: Local Constants; London Math. Soc. 1975 Durham Symposium, *Algebraic Number Fields* (ed. A. Fröhlich) 89–131 (1977) Academic Press.

[154] J. T. Tate: Relations between K_2 and Galois cohomology; Inventiones Math. 36 (1976) 257–274.

[155] Taylor, M. J. (1984), *Class Groups of Group Rings*, L. M. Soc. Lecture Notes Series #91, Cambridge University Press.

[156] R. Taylor: On Galois representations associated to Hilbert modular forms; Inventiones Math. 98 (1989) 265–280.

[157] R. Taylor: Galois representations associated to Siegel modular forms of low weight; Duke Math. J. 63 (1991) 281–332.

[158] R. Taylor: l-adic representations associated to modular forms over imaginary quadratic fields II; Inventiones Math. 116 (1994) 619–643.

[159] T. Tsuji: Semi-local units modulo cyclotomic units; J. Number Th. 78 (1999) 1–26.

[160] V. Voevodsky: The Milnor conjecture; Max-Planck Inst. preprint series or http: //www. math. uiuc. edu/K-theory (170) 1996.

[161] V. Voevodsky: Triangulated categories and motives over a field; Annals of Math. Study 143 Princeton University Press (2000) 188–239.

[162] J. B. Wagoner: Continuous cohomology and p-adic K-theory; Springer-Verlag Lecture Notes in Mathematics #551 (1976) 241–248.

[163] L. Washington: *Introduction to Cyclotomic Fields*; Springer-Verlag Grad. Texts in Math. # 83.

[164] A. Wiles: On p-adic representations for totally real fields; Annals of Math. 123 (1986) 407–456.

[165] A. Wiles: On ordinary λ-adic representations associated to modular forms; Inventiones Math. 94 (1988) 529–573.

[166] A. Wiles: The Iwasawa conjecture for totally real fields; Annals of Math. 131 (1990) 493–540.

[167] A. Wiles: On a conjecture of Brumer; Annals of Math. 131 (1990) 555–565.

[168] A. Wiles: Modular elliptic curves and Fermat's last theorem; Annals of Math. 141 (1995) 443–551.

[169] T. Zink: Étale cohomology and duality in algebraic number fields; Appendix 2 to *Galois cohomology of algebraic number fields*, by K. Haberland, Deutsche Verlag der Wissenschaften (1978).

Index